# Four-place Common Logarithms of Numbers

| N | 0 | 1 | 2 | 3 | 4 | 5 | 6 | 7 | 8 | 9 |
|---|---|---|---|---|---|---|---|---|---|---|
| 55 | 7404 | 7412 | 7419 | 7427 | 7435 | 7443 | 7451 | 7459 | 7466 | 7474 |
| 56 | 7482 | 7490 | 7497 | 7505 | 7513 | 7520 | 7528 | 7536 | 7543 | 7551 |
| 57 | 7559 | 7566 | 7574 | 7582 | 7589 | 7597 | 7604 | 7612 | 7619 | 7627 |
| 58 | 7634 | 7642 | 7649 | 7657 | 7664 | 7672 | 7679 | 7686 | 7694 | 7701 |
| 59 | 7709 | 7716 | 7723 | 7731 | 7738 | 7745 | 7752 | 7760 | 7767 | 7774 |
| 60 | 7782 | 7789 | 7796 | 7803 | 7810 | 7818 | 7825 | 7832 | 7839 | 7846 |
| 61 | 7853 | 7860 | 7868 | 7875 | 7882 | 7889 | 7896 | 7903 | 7910 | 7917 |
| 62 | 7924 | 7931 | 7938 | 7945 | 7952 | 7959 | 7966 | 7973 | 7980 | 7987 |
| 63 | 7993 | 8000 | 8007 | 8014 | 8021 | 8028 | 8035 | 8041 | 8048 | 8055 |
| 64 | 8062 | 8069 | 8075 | 8082 | 8089 | 8096 | 8102 | 8109 | 8116 | 8122 |
| 65 | 8129 | 8136 | 8142 | 8149 | 8156 | 8162 | 8169 | 8176 | 8182 | 8189 |
| 66 | 8195 | 8202 | 8209 | 8215 | 8222 | 8228 | 8235 | 8241 | 8248 | 8254 |
| 67 | 8261 | 8267 | 8274 | 8280 | 8287 | 8293 | 8299 | 8306 | 8312 | 8319 |
| 68 | 8325 | 8331 | 8338 | 8344 | 8351 | 8357 | 8363 | 8370 | 8376 | 8382 |
| 69 | 8388 | 8395 | 8401 | 8407 | 8414 | 8420 | 8426 | 8432 | 8439 | 8445 |
| 70 | 8451 | 8457 | 8463 | 8470 | 8476 | 8482 | 8488 | 8494 | 8500 | 8506 |
| 71 | 8513 | 8519 | 8525 | 8531 | 8537 | 8543 | 8549 | 8555 | 8561 | 8567 |
| 72 | 8573 | 8579 | 8585 | 8591 | 8597 | 8603 | 8609 | 8615 | 8621 | 8627 |
| 73 | 8633 | 8639 | 8645 | 8651 | 8657 | 8663 | 8669 | 8675 | 8681 | 8686 |
| 74 | 8692 | 8698 | 8704 | 8710 | 8716 | 8722 | 8727 | 8733 | 8739 | 8745 |
| 75 | 8751 | 8756 | 8762 | 8768 | 8774 | 8779 | 8785 | 8791 | 8797 | 8802 |
| 76 | 8808 | 8814 | 8820 | 8825 | 8831 | 8837 | 8842 | 8848 | 8854 | 8859 |
| 77 | 8865 | 8871 | 8876 | 8882 | 8887 | 8893 | 8899 | 8904 | 8910 | 8915 |
| 78 | 8921 | 8927 | 8932 | 8938 | 8943 | 8949 | 8954 | 8960 | 8965 | 8971 |
| 79 | 8976 | 8982 | 8987 | 8993 | 8998 | 9004 | 9009 | 9015 | 9020 | 9025 |
| 80 | 9031 | 9036 | 9042 | 9047 | 9053 | 9058 | 9063 | 9069 | 9074 | 9079 |
| 81 | 9085 | 9090 | 9096 | 9101 | 9106 | 9112 | 9117 | 9122 | 9128 | 9133 |
| 82 | 9138 | 9143 | 9149 | 9154 | 9159 | 9165 | 9170 | 9175 | 9180 | 9186 |
| 83 | 9191 | 9196 | 9201 | 9206 | 9212 | 9217 | 9222 | 9227 | 9232 | 9238 |
| 84 | 9243 | 9248 | 9253 | 9258 | 9263 | 9269 | 9274 | 9279 | 9284 | 9289 |
| 85 | 9294 | 9299 | 9304 | 9309 | 9315 | 9320 | 9325 | 9330 | 9335 | 9340 |
| 86 | 9345 | 9350 | 9355 | 9360 | 9365 | 9370 | 9375 | 9380 | 9385 | 9390 |
| 87 | 9395 | 9400 | 9405 | 9410 | 9415 | 9420 | 9425 | 9430 | 9435 | 9440 |
| 88 | 9445 | 9450 | 9455 | 9460 | 9465 | 9469 | 9474 | 9479 | 9484 | 9489 |
| 89 | 9494 | 9499 | 9504 | 9509 | 9513 | 9518 | 9523 | 9528 | 9533 | 9538 |
| 90 | 9542 | 9547 | 9552 | 9557 | 9562 | 9566 | 9571 | 9576 | 9581 | 9586 |
| 91 | 9590 | 9595 | 9600 | 9605 | 9609 | 9614 | 9619 | 9624 | 9628 | 9633 |
| 92 | 9638 | 9643 | 9647 | 9652 | 9657 | 9661 | 9666 | 9671 | 9675 | 9680 |
| 93 | 9685 | 9689 | 9694 | 9699 | 9703 | 9708 | 9713 | 9717 | 9722 | 9727 |
| 94 | 9731 | 9736 | 9741 | 9745 | 9750 | 9754 | 9759 | 9763 | 9768 | 9773 |
| 95 | 9777 | 9782 | 9786 | 9791 | 9795 | 9800 | 9805 | 9809 | 9814 | 9818 |
| 96 | 9823 | 9827 | 9832 | 9836 | 9841 | 9845 | 9850 | 9854 | 9859 | 9863 |
| 97 | 9868 | 9872 | 9877 | 9881 | 9886 | 9890 | 9894 | 9899 | 9903 | 9908 |
| 98 | 9912 | 9917 | 9921 | 9926 | 9930 | 9934 | 9939 | 9943 | 9948 | 9952 |
| 99 | 9956 | 9961 | 9965 | 9969 | 9974 | 9978 | 9983 | 9987 | 9991 | 9996 |
| N | 0 | 1 | 2 | 3 | 4 | 5 | 6 | 7 | 8 | 9 |

Basic Physical Chemistry
For The
Life Sciences

# Basic Physical Chemistry

SECOND EDITION

# For The

# Life Sciences

**Virginia R. Williams**

Louisiana State University

**Hulen B. Williams**

Louisiana State University

Mathematical Appendix by Bill B. Townsend

**W. H. Freeman and Company**   San Francisco

© Copyright 1967, 1973 by W. H. Freeman
and Company

Printed in the United States of America


2   3   4   5   6   7   8   9

Library of Congress Cataloging in Publication Data


Williams, Virginia R.
  Basic physical chemistry for the life sciences.

  Includes bibliographical references.
  1. Chemistry, Physical and theoretical.
I. Williams, Hulen B , 1920–     joint author.
II. Title.   [DNLM:  1. Chemistry, Physical.
QD453 W727b   1973]
QD453.2.W54   1973     541'.3'02457     73–7514
ISBN 0-7167-0171–5

This revised volume is dedicated to the memory of


*Virginia Rice Williams*
1919-1970


Professor of Biochemistry
Louisiana State University, Baton Rouge
—a 'woman for all seasons'—
unforgettable teacher, gifted scientist and musician,
cherished wife and mother.


Hulen B. Williams

# Preface
# to the Second Edition

The second edition of this text should be of vastly greater service to its users. Much material from the previous edition has been extensively re-written and many changes, corrections, and additions have been made. Perhaps the most significant of these is the new chapter on spectroscopy entitled "Electromagnetic Radiation and Matter." Many who used the first edition had asked for a chapter on this topic.

More and better problems have been included, and answers are pro-vided for most of them. Where appropriate, references have been brought up to date, and many new ones added. The material dealing with attrac-tive forces between ions, molecules, and atoms has been greatly ex-panded. A section on temperature and the zeroth law of thermodynamics has been included in Chapter 1. In Chapter 2, the student is introduced to a little fellow known as "Maxwell's Demon," provided with a more rigorous development of the kinetic equation for gases, and is given a glimpse of statistical thermodynamics. In Chapter 3, the section on free energy and chemical equilibrium has been expanded, and the discussion of the effect of temperature on chemical equilibrium has been greatly enlarged. Material on the hydrolysis of ATP and on cellulose ion ex-changers for protein purification has been added to Chapter 4, and Chap-ter 5 now contains a discussion of the oxygen electrode and a new section on electrical potentials and ion movement through membranes. Discus-sions of the general velocity equation, enzyme inhibition, and the analy-sis of enzyme mechanisms have been added to Chapter 6.

Before her untimely death, Dr. Virginia R. Williams did much of the work of revising the first edition. Since then, I have been ably assisted by several persons. Without the help of Dr. Wayne Mattice, Assistant Professor of Biochemistry at Louisiana State University, the job would not have been completed. He worked on many segments of the book, submitted additional problems and, in particular, wrote much of the material in Chapter 9. My special thanks go to him. The comments of Dr. Ralph L. Seifert, Indiana University, Bloomington, with respect to the first edition, and those of Dr. Park S. Nobel, University of California, Los Angeles, with regard to the revision, were especially helpful and are greatly appreciated. Many other users of the text have generously contributed constructive comments on the first edition. To them also, I express my thanks. To Mrs. Margaret Ann Boudreaux, who spent many hours in typing and helping to organize the revised manuscript, I extend my thanks and affection.

I wish to use this opportunity to express my deep sense of loss at the death of my friend and colleague, Dr. Bill Townsend, who wrote the Mathematical Appendix to the first edition. Its supporting value to the text has been widely recognized, and it continues to be an important part of this edition.

Hulen B. Williams

June 1972

# Preface
# to the First Edition

*Basic Physical Chemistry for the Life Sciences* offers a core of concepts for students whose major interests and needs lie in the fields of medicine, biology, agriculture, veterinary science, microbiology, and the like. It is not a text for the biochemistry major, but may be of value to him as a source of thermodynamic data on biochemical compounds. With the ascendancy of molecular biology there has come the necessity of providing all students in the life sciences with a foundation in physical chemistry so that the biochemistry of oxidative phosphorylation, high energy compounds, enzyme kinetics, homeostasis, multiple equilibria — to cite a few examples — can be presented in a meaningful and quantitative way.

Many topics covered in the usual year-course in physical chemistry have been omitted either because they are not sufficiently germane to the life sciences or because they could not be included in a book designed for a one-semester course. In our presentation we assume that the student has the usual foundation in organic chemistry, elementary physics, and calculus, and some background in biochemistry. Neither the calculus nor the biochemistry is indispensable. The necessary mathematical foundation, by Dr. B. B. Townsend, is found in the appendix. The student who is willing to master this material may successfully use the remainder of the text without formal preparation in calculus. We would be dishonest not to admit, however, that the self-taught student will find himself at some disadvantage compared with the student who has had one or two semesters of calculus. We subscribe to the philosophy that physical chemistry cannot be taught shorn of its mathematical foundations. A sincere effort has been made, however, to make the mathematics simple, understandable, and interesting. Those who prefer a fuller, more rigorous treatment should turn to one of the standard texts for a year-course in physical chemistry.

*Basic Physical Chemistry for the Life Sciences* is the outgrowth of approximately fifteen years of teaching such a course to students of these allied disciplines. Both authors have taught the course and feel that the chapters which follow represent a selection of the most indispensable and pertinent topics. Suggestions and criticisms of the presentation will be welcome and will serve as a guide to the preparation of an eventual revision. In addition, we would especially appreciate having our readers point out any errors which have escaped us in proofreading.

We wish to express our gratitude to the following colleagues who have offered helpful criticisms on various sections of the material: Dr. Jordan G. Lee, Dr. Joel Selbin, Dr. Eugene Berg, Dr. Edgar Steele, Dr. Robert C. McIlhenny, and Dr. Sue Hanlon (University of Illinois). The following persons have generously granted permission to include original problems: Mrs. Maud B. Purdy and Dr. J. B. Neilands (University of California). Special material was contributed by Dr. Edgar Steele and Dr. A. T. Phillips. Finally, we wish especially to acknowledge the invaluable assistance of Dr. W. C. Deal, Michigan State University.

Virginia R. Williams
Hulen B. Williams

June 1967

# Contents

# The Nature of Physical Laws

Although the methods of mathematics have been generally acknowl-edged to be indispensable in an exact science such as physics, the tech-niques of mathematics and the laws of physical science have not been similarly regarded in the life sciences. Some of this past indifference must be laid at the door of the traditionally qualitative academic prepara-tion in these disciplines, and some to the slow development of our un-derstanding of the complexities of living systems. The day appears to be rapidly approaching, however, when the standard freshman mathematics course will be the calculus, unless the student has already mastered it in high school. One foreseeable dividend of a general upgrading in mathe-matical preparation will be greater ease of inter-disciplinary communi-cation. When this time arrives, a knowledge of physical chemistry may well be a prerequisite for the study of elementary biology everywhere, not just in a few universities and colleges. That day, however, has not yet dawned.

The late G. N. Lewis is reported to have defined physical chemistry as the study of everything in science that is interesting. (He was a phys-ical chemist.) Hoping for an appreciation of this point of view we will begin by developing a concept of the various kinds of physical laws. Since physical chemistry, and by corollary biochemistry, is the applica-tion of physical laws to chemical and biochemical systems, such a be-ginning is an eminently logical one.

A physical law familiar to almost everyone is the Law of Conservation of Energy. It is frequently referred to as a "Law of Nature" because it arises out of human experience as well as from scientific experimenta-tion. Its metaphysical interpretation attracted philosophers, who found it, on the whole, rather reassuring. The illustrious French mathemati-cian, Henri Poincaré, summarized the matter succinctly (attributing the words to a renowned physicist), " 'Tout le monde y croit fermement parce que les mathématiciens s'imaginent que c'est un fait d'observation et les observateurs que c'est un théorème de mathématiques!' "* For nonnuclear reactions, the Law of Conservation of Energy holds within the limits of experimental error as far as we know; that is, no exceptions have ever been discovered. It holds independently of the number of molecules, or particles, or components of the particular system being studied. On the other hand, certain physical laws are called "statistical laws" and can be relied upon only when the particular event being observed occurs a sufficient number of times. The familiar "laws of chance" are of this type. When tossing a penny, we can be assured of a 50–50 distribution of heads and tails only if we have the patience to toss it enough times. If we toss ten pennies 500 times, the laws of chance predict that the coins will come up 9 tails and 1 head in about ten percent of the tosses, but the distribution of 5 heads and 5 tails is predicted to occur 123 times out of 500. Such a law of chance has its parallel in physical science. A physical law of great significance (though its implications were disturbing metaphysically) states that all natural systems tend to become more random or mixed-up in the course of time—to run "downhill." Such a law, like that for the tossing of a penny, may prove invalid if there are not enough events.

You will discover that most physical laws have appended to them rather precise statements about the conditions under which the laws hold. These conditions are often logical and readily understandable, like the temperature dependence of the laws of chemical equilibrium. In addition the physical chemist and the biochemist find it advantageous to simplify the mathematical expression of certain laws; their application is thus restricted, but at the same time they become enormously more useful. For example, the most useful equations dealing with solvent-solute relationships have been so simplified that they are applicable only to dilute solutions. Thus in dilute aqueous systems we can use molarities and molalities interchangeably, convert mole fraction solute to a molarity, and take other welcome short cuts. Such simplified laws are often called "limiting laws." In the purest sense, as implied above, most physical laws are "limiting laws" in that they are restricted in some way. In the following chapters, however, the term "limiting law" will be reserved for those laws that have undergone drastic mathematical surgery.

Physical laws are derived in two ways: through experimental observation and through mathematical development of theoretical principles. The theoretical derivation must eventually accompany the generalizations based on observation if a law is to have fundamental significance.

---

*Preface to *Thermodynamique*. G. Carré, Paris, 1892. Translated: "Everyone believes firmly in it because the mathematicians think it is an experimental fact and the experimentalists think it is a theorem of mathematics!"

The general pattern for this procedure is the scientific method, that aspect of science which redeemed it from the doldrums of the Middle Ages and made of it a discipline.

The scientific method is the essence of modern science; without it the science and technology of today could not exist. Let us examine this all-important method, its implications, its demands, and its rigors. Those who qualify as scientists must have certain attitudes and gifts that enable them to use the scientific method, that is, the ability to observe, curiosity about the phenomena they encounter, and intellectual honesty. On these foundations they build from the blueprint of the scientific method.

The scientist first observes that under certain conditions a particular system behaves in reproducible and predictable ways. He can usually describe this behavior by formulating a mathematical equation, which we term *empirical*.

His curiosity next leads him to ask the question, "Is there anything about the fundamental nature of the matter in the system that can be deduced from the observed behavior?" As he pursues the answer, he finds it necessary to reword the question — a basic technique in application of the method. "Can one speculate about the intrinsic nature of matter in a manner that is consistent with the way we observed it to behave?" The speculation is an intelligent guess called a hypothesis.

The hypothesis is examined in the light of reason and of appropriate mathematical principles and techniques, and an equation is derived (behavior is predicted), which now may be compared with the empirical equation that initially described the observed behavior of the system. Now the scientist must determine how well the deduced and the observed agree, for according to the scientific method, insight into the fundamental nature of matter is obtained only to the extent that the derived predictions agree with experimental observations.

In practically all cases derived results suggest new experiments to check further the validity of the original assumptions. Usually the scientist finds that the derived predictions about the behavior of matter are sufficiently at variance with experimental results that he must either refine the old hypothesis by making it more sophisticated or develop completely new ideas. The process continues with new experiments and refinement of basic concepts until, within experimental limitations, little value is gained from additional sophistication.

We should indicate that it seems possible that several completely different sets of postulates *might* be equally useful in predicting the behavior of a system. Indeed, in some cases this *appears* to be true, *e.g.*, the two theories that have led us to think of the "dual nature" of radiant energy. Scientists do not always require that a set of postulates be unique to be useful, and when more than one set exist that seem to have acceptable features, the serviceable aspects of all are utilized.

But, although the scientist is constantly challenged to apply his method to as many reasonable sets of ideas as his mind can conceive for a single phenomenon, there is probably only one completely correct explanation. Instances where two or more sets of postulates seem to be required indicate that all contain some measure of truth. As progress is made there will emerge a unified assembly of precepts that have passed the tests of mathematical interpretation, dialectic argument, and experimental proof.

It is the authors' belief that students of science should always be alert for possible application of the ideas and concepts of science in other areas of human endeavor, *e.g.*, economics, political theory, and sociology. It is not our purpose to state that scientific method is applicable in human relations, but we believe one cannot avoid consideration of its possible applicability in arriving at whatever decision seems correct to him.

Basic Physical Chemistry
For The
Life Sciences

# 1 | THE GAS LAWS

Science has its cathedrals, built by the efforts of a few architects
and of many workers.

G. N. Lewis

The study of physical chemistry may follow different routes, just as the
traveler from New York to San Francisco may choose any one of several
itineraries. We have elected to begin with thermodynamics and to assume that the student has mastered the fundamentals of atomic and
molecular structure in his general chemistry course. Preliminary to a
consideration of the laws of thermodynamics, however, we must study
the laws governing the behavior of gases, because our simple thermodynamic models will be systems of gases.

## 1.1 Attractive Forces Between Ions, Molecules, and Atoms

The physical world which surrounds us displays a staggering degree of
complexity in its composition. Nevertheless, matter occurs in only three
states: gaseous, liquid, and solid. All atoms and molecules include
protons and electrons in their structure. If the numbers of protons and
electrons are not equal, the atom or molecule will bear a net charge and
is called an ion. As a consequence of their net charge, ions generate
electrical fields. Atoms or molecules which possess zero net charge may
nevertheless generate electrical fields if their centers of positive and
negative charge do not coincide. Molecules characterized by this kind
of charge separation are referred to as being polar; this means that
because of their structure they are small dipoles, as opposed to molecules with nonpolar structures having coincident centers of positive
and negative charge. These categories are not mutually exclusive. For
example, several atoms chemically bound can constitute an ion and a
dipole if it bears a net charge and has a separation of the centers of
positive and negative charge.

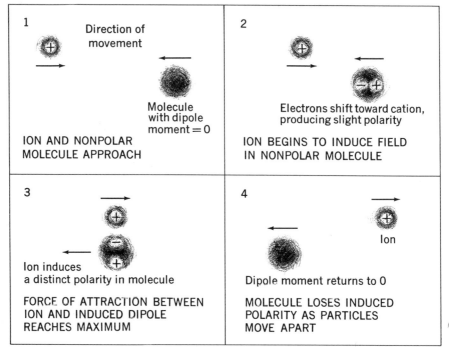

**Figure 1.1** Ion-induced dipole attraction. Induced polarization in the ion has been neglected in the interest of simplicity.

Since the energy of a charged species is affected by an electrical field, the energy of an ion or dipole will be affected by the presence of other ions or dipoles. The quantitative role of dipoles in these interactions depends upon whether they are permanent, as a result of fixed molecular configuration, or temporary, as a consequence of electrical induction. The temporary ones are called induced dipoles and are formed when a nonpolar atom or molecule wanders close to the electrical field of an ion or a dipole, as illustrated in Fig. 1.1. Ions, permanent dipoles, and induced dipoles all have electrical effects on one another, making the following six categories of electrostatic interactions possible: ion-ion, ion-permanent dipole, ion-induced dipole, permanent dipole-permanent dipole, permanent dipole-induced dipole, and induced dipole-induced dipole. The interaction forces are not equal.

The fundamental point is that the potential energy of attraction (or repulsion) varies inversely with the distance of separation (or with some power of it), and the resultant electrical force is that rate at which that potential energy is changing with distance: force $= -dE/dr$, where $E$ is energy and $r$ is the distance of separation between particles. If the force is one of attraction, the potential energy decreases as the particles approach, causing $dE/dr$ to be positive and the resulting force negative.

If the particles repel each other, $dE/dr$ is negative and the force is positive. The force of attraction for an ion-ion pair can be calculated from Coulomb's Law:

$$\text{force} = \frac{Q_1 Q_2}{Dr^2},$$

where $Q_1$ and $Q_2$ are the net charges on the particles and $D$ is the dielectric constant of the medium in which the charges exist. The dielectric constant is a dimensionless quantity which relates to the insulating properties of the medium. If the medium is placed between the plates of a condenser, the dielectric constant is the ratio of the capacity of the condenser in this state to its capacity when there is a vacuum between the same plates. The dielectric constant of water is high, about 80, whereas the dielectric constant of air is essentially that of a vacuum, or unity.

If a sodium ion of radius 0.95 Å approaches to within 10 Å of a chloride ion of radius 1.81 Å, the distance separating their centers will be 12.76 Å as shown in Fig. 1.2. The force of attraction in air will be,

$$\text{force} = \frac{(4.80 \times 10^{-10} \text{ esu})(-4.80 \times 10^{-10} \text{ esu})}{1.0 \times (12.76 \times 10^{-8} \text{ cm})^2} = -1.52 \times 10^{-5} \text{ dyne},$$

where the charges of the ions are given in esu (electrostatic units) and 1.0 is the dielectric constant for air. The negative sign for the force means that the force acts to decrease the distance of separation between the ions, that is charges of opposite sign attract each other.

The potential energy of attraction may be thought of as the work required to bring the ions from infinite distance to the actual internuclear distance of separation. From the original definition of the force, we see that the energy — or work — is equal to the integral of the force over the distance:

$$E = - \int \text{force } dr,$$

or more specifically for the sodium and chloride ions separated in the vapor phase by a distance of 12.76 Å,

$$E = - \int_{\infty}^{r} \frac{Q_1 Q_2}{Dr^2}\, dr = \frac{Q_1 Q_2}{Dr} - \frac{Q_1 Q_2}{D\infty}$$

$$= \frac{(4.80 \times 10^{-10} \text{ esu})(-4.80 \times 10^{-10} \text{ esu})}{1.0 \times (12.76 \times 10^{-8} \text{ cm})} = -1.81 \times 10^{-12} \text{ erg},$$

or

$$\frac{-1.81 \times 10^{-12} \text{ erg}}{1.602 \times 10^{-12} \text{ erg/electron volt}} = -1.12 \text{ electron volt}.$$

The negative sign for potential energy indicates that the potential energy of attraction between a sodium ion and a chloride ion is 1.12 electron volts lower when they are separated by 12.76 Å than when they are separated by an infinite distance.

The previous equation and the illustrative calculation take account of the primary interaction that dominates the potential energy at distances of separation which exceed the sum of the ionic radii. If this interaction were the only contribution to the potential energy at all possible distances of separation, the potential energy should become increasingly more negative as the distance of separation decreases. We do not, however, find this to be true at very small distances of separation because additional terms make significant contributions to the potential energy when the distance of separation is only a few Angstroms. Inclusion of these terms in the expression for the potential energy has two important effects: (1) the potential energy will attain a minimum value at a distance of separation which is greater than zero, and (2) the potential energy will increase rapidly as the distance of separation becomes less than the value corresponding to the potential energy minimum. These features are presented schematically in Fig. 1.2.

The minimum in the potential energy curve occurs approximately where the ions touch, or, more elegantly stated, when the internuclear distance of separation is equal to the sum of the ionic radii. We recognize that because of multiple ion interactions in the crystal there is a difference between the ionic distance for $Na^+Cl^-$ in the crystal (2.81 Å) and in the vapor phase (2.51 Å).[*] The sum of the radii calculation gives 2.76 Å. The ions are said to be separated by a distance equal to the sum of their van der Waals radii, a term that will be explained more fully below. The ionic bond is most stable at this separation distance because here the potential energy is the lowest. If the two ions are now forced closer together so that their electron clouds overlap, the two positive nuclei will be close enough to interact and repulsion results. The potential energy rises sharply and becomes more positive in sign, as repulsion increases. The ions will accordingly move away from each other.

The exact mathematical relationship between the potential energy of attraction and the distance of separation depends on the type of interaction, that is, whether it is ion-ion, ion-dipole, or one of the other types. Table 1.1 lists the six types of interaction and gives for each the dependence of $E$ on $r$, the distance of separation. The entries in Table 1.1 do not include the repulsive contributions to the potential energy that become important at distances of separation comparable to the sum of the van der Waals radii. Values for $E$ in terms of the bond energies or interaction energies range from some 0.1 to 0.3 kilocalorie/mole for the

---

[*]Moore, Walter J., *Physical Chemistry*, 4th Edition, Prentice-Hall Inc., p. 718, 1972.

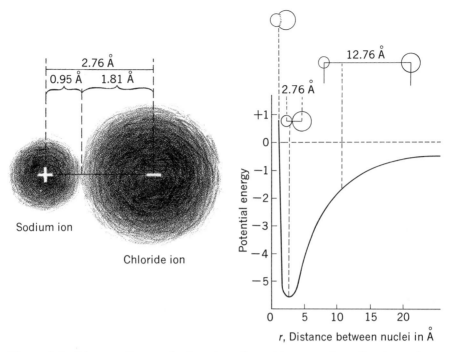

**Figure 1.2**   Distance of approach of two ions. Potential energy of a sodium ion and chloride ion as they collide in the vapor phase.

induced dipole-induced dipole "bond" to some 40 to 50 kilocalories/mole for the ion-ion bond.

   The potential energy of attraction between an ion and a permanent dipole is proportional to the inverse square of the distance of separation, whereas that for an ion and an induced dipole is proportional to the inverse fourth power and therefore decreases much more rapidly with increasing distance of separation. Ion-dipole interactions are important in many biological systems. For example, amino acids exist in solution at pH 7 in the form of dipolar ions that participate in interactions with ions from buffers and inorganic salts. The electrostatics of these effects have been discussed in depth in the reference by Edsall and Wyman cited at the end of this chapter. Also, ion-dipole interactions lead to the "salting in" and "salting out" of proteins and will be discussed in Chapter 4.

   The last three categories shown in Table 1.1 are collectively known as van der Waals interactions, because the van der Waals Equation for gases (Eq. 1.26, p. 26) contains a pressure correction term related to these attractive forces. Van der Waals interactions are largely responsible for the condensation of gases, the freezing of liquids, and the failure of

**TABLE 1.1** THE RELATIONSHIP OF DISTANCE OF
SEPARATION ($r$) TO POTENTIAL ENERGY OF ATTRACTION ($E$)

| Type of interaction | $E \propto$ |
|---|---|
| Ion-ion | $1/r$ |
| Ion-permanent dipole | $1/r^2$ |
| Ion-induced dipole | $1/r^4$ |
| Permanent dipole-permanent dipole° | $1/r^3 - 1/r^6$ |
| Permanent dipole-induced dipole | $1/r^6$ |
| Induced dipole-induced dipole | $1/r^6$ |

°The interaction of two permanent dipoles with fixed orientation and definite angular relationship is inversely cubed. When the interaction is averaged over all orientations and relative angular positions, it becomes inverse 6th power. It is the latter that corresponds to van der Waals interaction.

gases and solutions to obey certain limiting laws perfectly. They also contribute significantly to forces responsible for the three-dimensional folding of biological macromolecules. Attractive forces between induced dipoles are sometimes referred to as London forces or dispersion forces and are of particular interest in such phenomena as the liquefaction of the monatomic gases. A further source of molecular interaction is the extremely weak force resulting when a positive nucleus and an electron cloud of one molecule overlap with those of another. At very short distances this overlap produces a repulsion with the repulsion energy approximately proportional to a term between $1/r^9 - 1/r^{12}$. These effects are too small to consider in the elementary discussions that follow, but have been included in this summary for the sake of completeness.

Although it does not fall precisely in any of the categories listed in Table 1.1, the *hydrogen bond* is of special historical interest. This bond results when a hydrogen atom bridges two electronegative atoms. It is principally of the permanent dipole-permanent dipole type, but it is often strengthened also by contributions from resonating ionic structures. In biological systems, the electronegative atoms are usually nitrogen or oxygen, although one of the electronegative pair may be carbon, sulfur, or another element. Hydrogen bonds are also called *hydrogen bridges* or *proton bonds*. The simplest illustration is the hydrogen-bonded structure of water, as shown in Fig. 1.3. Hydrogen bonding causes some low molecular weight substances — water and methanol for example — to be liquids rather than gases at room temperature and to have higher boiling points than other compounds of similar molecular weight. A discussion of current water structure concepts is presented in Chapter 2, Section 2.6, following a presentation of some of the basic principles of thermodynamics.

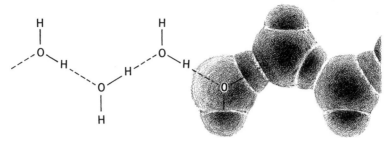

**Figure 1.3** A short chain of hydrogen-bonded water molecules.

Because a number of structural types are possible, hydrogen bonds differ in length, in energy, and in symmetry—and therefore in polarity. In organic compounds hydrogen bonds are generally unsymmetrical: the hydrogen will lie closer to one of the bridgehead atoms than to the other and will be more tightly bonded to the closer electronegative atom. The more unsymmetrical the bond, the more easily it is broken and therefore, the lower the bond energy. The energy of rupture of hydrogen bonds is about 2 to 7 kilocalories per mole versus 30 to 150 kilocalories per mole for ordinary covalent bonds. Here we are considering only the energy of rupture and assuming that no new bonds are formed. The distance between bridgehead atoms of the hydrogen bonds found in biological molecules is usually 2.5 to 3.2 Å. The hydrogen— or proton—is ordinarily about 0.95 to 1.1 Å from one electronegative atom and 1.4 to 2.2 Å from the other one. Some examples of hydrogen bonds commonly found in biological systems are shown in Fig. 1.4.

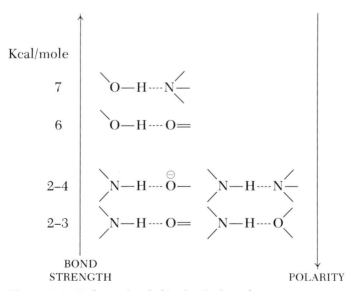

**Figure 1.4** Hydrogen bonds found in biological systems.

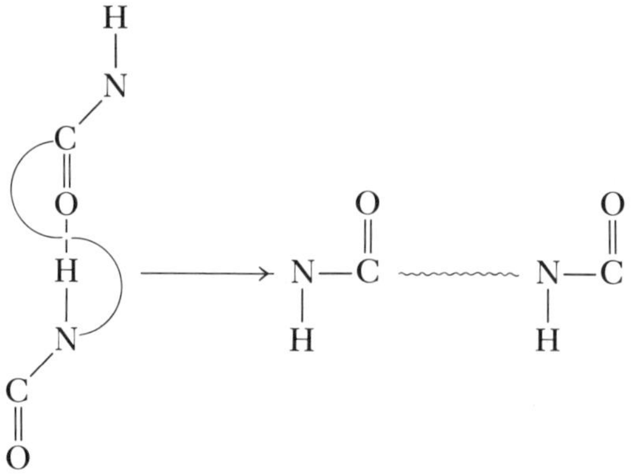

$(\alpha)$

More recent work suggests that the contribution of the H-bond to the determination of the preferred conformations of biological molecules may have been overrated. The solvent here is water and is therefore capable of forming hydrogen bonds (see Fig. 1.3), so it may not be appropriate to view the disruption of a hydrogen bond in, for example, a protein as shown in Eq. ($\alpha$).

When the solvent is considered, the actual process might be better approximated by equation ($\beta$), where we recognize the possibility that the units which had participated in an amide-amide hydrogen bond are freed to form hydrogen bonds with the solvent.

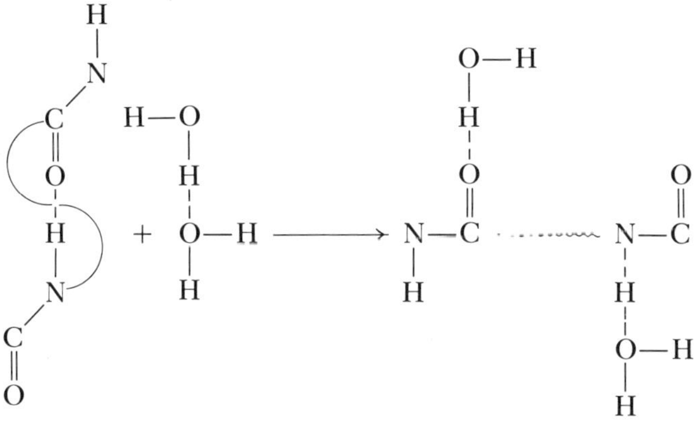

$(\beta)$

Equation ($\alpha$) implies the net destruction of a hydrogen bond, whereas equation ($\beta$) shows no net destruction of hydrogen-bonds. Instead there has simply been an alteration in the types of hydrogen bonds. Attempts have been made to measure the equilibrium state for interamide hydrogen bonding in an aqueous medium using urea, N-methylacetamide,

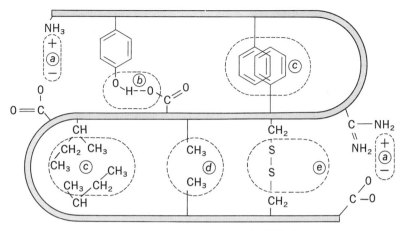

**Figure 1.5** Some types of bonds that stabilize the folded structure of proteins: (a) ionic bond, (b) hydrogen bond (c) and (d) several types of van der Waals interactions (e) covalent disulfide bridge. [Slightly modified from the original, reprinted by permission from Anfinsen, C., *The Molecular Basis of Evolution.* John Wiley and Sons, New York, 1959, p. 102.]

and δ-valerolactam as model amides.\* The results are somewhat dependent on the amide used, and suggest that the preference for interamide hydrogen bonds in water is small and may even be zero. The best assessment of the role of the hydrogen bond in biological systems is that a stable system will maximize the total number of hydrogen bonds but may be relatively insensitive to the types of hydrogen bonds contained therein.

Although the presence of water in living organisms raises questions about the importance of the H-bond in determining the structure of biological molecules, its presence is responsible for another factor, the hydrophobic bond† — an extremely important factor in determining what shapes biological macromolecules will assume. It is common knowledge that hydrocarbons (such as oil) and water do not mix. Some of the side chains of proteins, such as those indicated by (c) in Fig. 1.5, are hydrocarbons. If the shape of the molecule were such that these hydrocarbon side chains were forced to be in contact with water, the interaction of the water and hydrocarbon would cause a stable configuration that was not of lowest energy. Ordinarily thermodynamic (see Chapter 2) considerations would lead us to prefer a "lowest energy"

\*Schellman, J. A., *Compt. Rend. Trav. Lab. Carlburg, Ser. Chim.,* **29**, 223 (1955); I. Klotz and J. Franzen, *J. Amer. Chem. Soc.,* **84**, 3461 (1962); H. Susi, S. Timasheff, and J. Ard, *J. Biol. Chem.,* **239**, 3051 (1964).

†Kauzmann, W., *Adv. Protein Chem.,* **14**, 1 (1959).

configuration. This energy problem could be avoided if it were possible for the protein to arrange itself so that the hydrocarbon side chains were shielded from contact with water. For the globular proteins particularly, increasing experimental evidence favors an elliptical model with ionic and polar groups on its exterior and alkyl side chains folded toward the interior, away from the surrounding water. All of the major types of bonding contributing to the structure of a protein are shown in Fig. 1.5. The hydrophobic bond is illustrated by (*c*) in Fig. 1.5. It must be emphasized, however, that the important feature of the hydrophobic bond is not the interaction of one alkyl side chain with another, but rather the avoidance of the interaction of an alkyl side chain with water.

Both the hydrogen bond and the hydrophobic (apolar) bond are low in bond energy, but a sufficient number of them can bind a large molecule into a stable architecture just as the many feeble bonds of the Lilliputians held Gulliver fast to the ground. Despite the energy level of these bonds, they possess sufficient strength to resist disruption by molecular collision at room temperature. As temperature rises, however, the increased force of molecular collisions and the increased vibrational energy of the molecules cause both hydrogen and apolar bonds to be broken easily and contribute to such phenomena as the thermal denaturation of proteins and the "melting" of DNA molecules.

## 1.2   Temperature and the Zeroth Law of Thermodynamics

The study of physical chemistry usually begins with the laws governing the behavior of gases because these laws are fundamental to the treatment of more-involved systems. Such "systems" can range in complexity from a raindrop to the entire universe. They are whatever parts of the physical world we dissect out for study, provided we adequately define them. Systems of gases are simple because (1) mixtures of gases always form true solutions, in which the composition is uniform throughout, that is, consists of one "phase," and (2) at the same temperature and pressure, equal volumes of gases contain equal numbers of moles (Avogadro's Law) thereby permitting us to handle volumes of gases in the same way as moles of gases.

Systems of gases have certain common characteristics, called *properties*, including mass, volume, temperature, pressure, density, and the like. Obviously some of these properties are interdependent. If we know the mass and volume of a gas, then we do not need independent information about its density, because density = mass/volume. Other interrelationships exist also, and the influence of external fields can be neglected in dealing with the mechanical properties of the gas.

The chemist strives to economize by discovering the minimum number of properties needed to describe the system completely. Once described or defined as being identical in terms of this minimum number of properties, the system of a gas will be in the same state whether it exists in Canberra or in Cincinnati.

Of the properties listed above, temperature is one of the most difficult conceptually. The manifestation that we call temperature is, in fact, a measure of the average kinetic energy of the molecules of a body. A flow of heat from one body to another simply refers to the transfer[*] of these motions on the molecular level to the molecules of the second body. These clear statistical concepts of temperature and heat were not always recognized. Earlier understandings of temperature were not based on a concept of molecules but concerned certain properties of bodies at equilibrium. Heat was thought to be an invisible fluid called "caloric" that flowed from the hotter body to the cooler one when in contact, thus bringing about temperature changes.

As a preview of what is to follow in Section 1.3, let us try to grasp the essential aspects of the interrelationships between the properties and state of a gas. If external field effects are negligible, the state of a given mass of gas is determined when we specify the pressure and the volume. Furthermore, these two characteristics are related to the temperature in such a way that the state of the gas can be determined by specifying any two of the three properties, $P$, $V$, and $T$. That is to say, of these three variables, only two are independent. Mathematically we write

$$T = f \ (P,V). \tag{1.1}$$

The function $f \ (P,V)$ defined in this way is called the empirical temperature and does not refer to any specific temperature scale. The definition of temperature given in Eq. 1.1 is sometimes called the *Zeroth Law of Thermodynamics*.[†]

Now, if two different bodies possessing what we might describe as different degrees of "hotness" are brought into contact and allowed to come to equilibrium, it is found that under the equilibrium conditions,

$$f \ (P_1, V_1) = f \ (P_2, V_2)$$

or

$$T_1 = T_2.$$

If one of the bodies, *e.g.*, the second one, is a properly contained gas or liquid at constant pressure, then the change in volume produced in the second body measures the temperature of the first, and it may act as a

---

[*] Maxwell's word was "communication." Klein, Martin J., *American Scientist,* **58**: No. 1, 84, 1970.

[†] Often the zeroth law of thermodynamics is stated in terms of temperature and thermal equilibrium. ("Two systems in thermal equilibrium with a third are in thermal equilibrium with each other," M. W. Zemansky, *Heat and Thermodynamics,* 5th ed., McGraw-Hill, 1968.)

thermometer. To "read" the temperature a body or "fluid," preferably one that expands linearly with temperature, and an acceptably defined and calibrated scale are needed. The scale commonly used in routine scientific temperature measurements is the centigrade* or Celsius scale which defines the melting point of ice as 0° and the boiling point of water at 100°, both at one atmosphere pressure.

If the molecular motion referred to earlier is communicated in such a manner that the resulting "motions and displacements are those of visible bodies consisting of great numbers of molecules moving together, the communication of energy is called work."† From this point of view, a thermometer as we commonly know it is one simple device that converts heat energy into work with the amount of work done being a function of the difference of temperature between the object body and the measuring device, the thermometer. Heat flowing from a warmer body to the thermometer causing an expansion of the thermometric fluid which at constant pressure does work in pushing back the atmosphere. For object bodies at lower temperatures compression of the thermometric fluid will result and the atmosphere will do work on the fluid.

### 1.3 The Perfect Gas Law or Ideal Gas Equation

Two important relationships known as the Law of Boyle and the Law of Charles were discovered experimentally rather than being deduced from scientific precepts, and therefore are known as empirical laws. In 1662 Sir Robert Boyle obtained the data given in Table 1.2 by measuring the compression of air in a U-tube, the short end of which was sealed and calibrated and the long open end of which served for the addition of mercury. The length of the air column $A$ was proportional to volume, as the tube was of uniform cross section. Since the $(P \times A)$ terms were essentially constant,

$$P_1 V_1 = P_2 V_2 \qquad \text{(at constant temperature).} \qquad (1.2)$$

This is, of course, Boyle's Law.

The Law of Charles, also discovered independently by Gay-Lussac, is similarly empirical and was proposed more than a hundred years later. The modern statement of Charles' Law is: for each degree Celsius that a gas is cooled its volume shrinks 1/273.15 of the volume at 0°. Expressed as a mathematical statement,

$$V_T = V_0 + \frac{t}{273.15} \cdot V_0 \qquad \text{(at constant pressure).} \qquad (1.3)$$

---

*By international agreement, the term "centigrade" for a temperature scale is to be discontinued, and the Celsius scale is to be universally adopted.

†Maxwell, J. C., "Tait's Thermodynamics," *Nature*, **17**: 257, 1878.

**TABLE 1.2** SIR ROBERT BOYLE'S TABLE OF THE CONDENSATION OF THE AIR

| A<br>Length (inches) of a column of air in short arm (proportional to V) | Length (inches) of a column of mercury in the long arm | P<br>Total pressure (inches of mercury) | P × A |
|---|---|---|---|
| 60 | 0 | $29\frac{1}{8}$ | 1747 |
| 50 | $6\frac{3}{16}$ | $35\frac{5}{16}$ | 1766 |
| 40 | $15\frac{1}{16}$ | $44\frac{3}{16}$ | 1762 |
| 30 | $29\frac{11}{16}$ | $58\frac{13}{16}$ | 1746 |
| 20 | $58\frac{2}{16}$ | $87\frac{14}{16}$ | 1758 |
| 15 | $88\frac{7}{16}$ | $117\frac{9}{16}$ | 1759 |

*Source:* Quoted from "A Defence of the Physico-Mechanical Experiments," *Works of Robert Boyle,* Vol. II, by Peter Shaw. London, 1725, p. 671.

This equation predicts that reducing the temperature to −273.15°C would cause the gas to disappear, or to have no volume. The unlikelihood of this event made it apparent that the law could not hold exactly. The bottom of the temperature scale, −273.15°C became the 0° of a new scale, the absolute-temperature scale. Later this became known as the Kelvin scale, after the scientist who established its theoretical importance in thermodynamics. The degree intervals are the same as those of the Celsius scale, but the zero reference point is 273.15° lower.

Rearranging the above equation gives

$$V_T = V_0 \cdot \frac{T}{273.15} \qquad \text{(where } T = t + 273.15\text{).}\qquad (1.4)$$

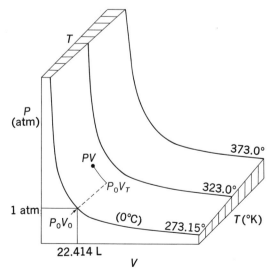

**Figure 1.6** *PV* isotherms for a perfect gas.

The Laws of Boyle and Charles can now be combined by examining the behavior of a gas along the three coordinates of pressure, volume, and temperature (Fig. 1.6). The point $P_0V_0$ on the 0°C constant temperature line (isotherm) lies at one atmosphere pressure and 22.414 liters volume — in other words, a mole of gas at standard conditions. If we move along the path $P_0V_0 \ldots P_0V_T$ (a constant pressure path known as an isobar), the path is defined by the equation

$$V_T = V_0 \cdot \frac{T}{273.15} \qquad \text{(Charles' Law, Eq. 1.4).}$$

Multiplying each side by $P_0$ gives

$$P_0V_T = P_0V_0 \cdot \frac{T}{273.15}. \tag{1.5}$$

Since by Boyle's Law, along the isotherm containing the points $PV$ and $P_0V_T$, $PV$ must equal $P_0V_T$,

$$PV = P_0V_0 \cdot \frac{T}{273.15}, \tag{1.6}$$

and substituting for 273.15 the symbol $T_0$, *i.e.*, standard temperature, we obtain

$$PV = \frac{P_0V_0}{T_0} \cdot T, \tag{1.7}$$

the familiar combination form of Boyle's and Charles' Laws. Since the factor $P_0V_0/T_0$ represents nothing more than a proportionality constant for converting degrees absolute into an energy term (it will be shown that $PV$ has the dimensions of energy), it is replaced by the general symbol $R$ and thus we have the *Perfect Gas Law*, or *Ideal Gas Equation* as it is also called:

$$P\bar{V} = RT \qquad \text{(for 1 mole of gas, where } \bar{V} \text{ is the molar volume) (1.8)}$$

and

$$PV = nRT \qquad \text{(for } n \text{ moles of gas).} \tag{1.9}$$

The name Perfect Gas Law implies, not that the law is perfect, but that it is a law for a hypothetical perfect gas, or the synonymous term, ideal gas. Conversely, the ideal gas is defined as one which obeys this equation exactly. As implied earlier, the equation has its imperfections when applied to real, or nonideal, gases. Not only does the volume of the gas molecules themselves become a problem in systems of high density, but the electrostatic forces described earlier produce additional deviations.

**TABLE 1.3** TYPES OF ENERGY

| Energy | Intensity factor | Capacity factor |
|---|---|---|
| Kinetic | $\frac{1}{2}$ velocity$^2$ | mass |
| Surface | surface tension | surface area |
| Heat | temperature | heat capacity |
| Electrical | potential (volts) | charge (coulombs) |
| Chemical | chemical potential | moles of component |

These imperfections occur over a wide temperature range. An ideal gas may therefore be redefined as one in which (1) no attractive forces exist between molecules, and (2) the volume of the molecules is negligible compared with the total volume.

Let us proceed to examine the left-hand term of the Ideal Gas Equation, $PV$. It has the dimensions of energy with pressure and volume expressed in any convenient units. One example is to express pressure in dynes/cm$^2$ and volume in cm$^3$, resulting in dynes-cm, which are ergs, the unit of energy in the CGS (centimeter-gram-second) system. Another useful formulation has pressure in atmospheres and volume in liters. The product here is also energy since an atmosphere is the force per unit area exerted by a column of mercury 76 cm high, or equivalent to dynes/cm$^2$ and liters is cm$^3$/1000. All forms of energy can be factored into a *capacity* term and an *intensity* term, as we see exemplified by $PV$. In this case, $P$, is the intensity factor and $V$ is the capacity factor. For kinetic energy, mass is the capacity term, and one-half the square of velocity is the intensity term; for electrical energy, charge is the capacity term and potential the intensity term. Additional examples are shown in Table 1.3.

By using appropriate conversion factors we may equate the various types of energy units. However, intensity factors alone (or capacity factors alone) cannot be equated or exchanged in an energy expression. The student may find it helpful in distinguishing intensity factors from capacity factors to remember that the intensity factor is independent of the size or extensiveness of the system and is a property that is uniform throughout the system at equilibrium.

The proportionality factor of the Perfect Gas Law, $R$, is very useful in converting one type of energy unit into another. To illustrate, let us calculate $R$ from the Laws of Boyle and Charles:

$$R = P_0V_0/T_0 = (1 \text{ atm})(22.414 \text{ liters})/273.15° \text{ per mole, or}$$
$$R = 0.08206 \text{ liter-atm/mole-degree.} \tag{1.10}$$

As will be shown, expressed in calories, $R = 1.987$ cal/mole-degree. Therefore, we can conveniently change energy expressed in liter-atmospheres to calories by multiplying by the fraction 1.987/0.08206. Other values of

$R$ are calculated in the following ways. Since $R$ is the gas constant per mole, "mole" always appears in the denominator of the expression.

$$R = \frac{(1 \text{ atmosphere})(22,414 \text{ cc})}{273.15°} = 82.06 \text{ cc-atm/mole-degree}. \quad (1.11)$$

When pressure is expressed in dynes per cm², we multiply the barometer height, 76 cm of mercury, by the density of mercury, 13.596 at 0° C, and by the acceleration of gravity, 980.6 cm per sec². The pressure in dynes per cm² multiplied by volume in cm³ gives the energy in ergs. Ergs may be converted into joules by dividing by $10^7$, and joules may be converted into calories by dividing by the mechanical equivalent of heat, 4.184 joules per calorie:

$$R = \frac{(76.00)(13.596)(980.6)(22,414)}{273.15°}$$

$$= 8.314 \times 10^7 \text{ ergs/mole-degree}, \quad (1.12)$$

$$R = \frac{8.314 \times 10^7}{10^7} = 8.314 \text{ joules/mole-degree}, \quad (1.13)$$

$$R = \frac{8.314}{4.184} = 1.987 \text{ calories/mole-degree}. \quad (1.14)$$

The term $R$ appears in a number of useful relationships, usually as a consequence of our assuming that a gas behaves ideally and that we may therefore substitute $RT/P$ for $V$ or $RT/V$ for $P$. Such a substitution is made with the full knowledge that by so doing we automatically impose restrictions, thus making it a limiting law.

The gas constant has an important relationship to the heat capacities of gases. The heat capacity of a substance is the quantity of heat (usually expressed in calories) required to raise the temperature of that substance one degree centigrade. Heat capacities are usually expressed per mole; the heat capacity per gram is given a special designation, the *specific heat*. The heat capacity of a given material will depend on whether the substance is maintained at constant volume or at constant pressure during the heating process. If constant volume is maintained, the symbol for the molar heat capacity is $\bar{C}_V$; if constant pressure, the symbol is $\bar{C}_P$. For liquids and solids, $\bar{C}_V$ and $\bar{C}_P$ are very nearly the same because the volume change with temperature is quite small. With gases, on the other hand, $\bar{C}_P$ is larger than $\bar{C}_V$ by the amount of heat necessary to expand the gas. For all ideal gases, this quantity of heat is equal to $R$, or approximately two calories. That is, $\bar{C}_P - \bar{C}_V = R$. The values of $\bar{C}_P$ and $\bar{C}_V$ for ideal monatomic gases are about 5 cal/degree and about 3 cal/degree respectively. For polyatomic gases the heat capacities are temperature dependent and are not easily calculated. Table 1.4 contains data for

**TABLE 1.4** MOLAR HEAT CAPACITIES OF SELECTED GASES

| Formula | Class | $\bar{C}_p$ | $\bar{C}_v$ | $\dfrac{\bar{C}_p}{\bar{C}_v} = \gamma$ | Type of motion |
|---|---|---|---|---|---|
| | | (cal. deg.$^{-1}$ mole$^{-1}$ at 25°C) | | | |
| He | | 4.97 | 2.98 | 1.67 | Translational (only) |
| Ne | Monatomic | 4.97 | 2.98 | 1.67 | |
| Ar | | 4.97 | 2.98 | 1.67 | |
| CO | | 6.97 | 4.97 | 1.40 | Translational |
| N$_2$ | Diatomic | 6.94 | 4.95 | 1.40 | |
| H$_2$ | | 6.90 | 4.91 | 1.41 | |
| N$_2$O | Triatomic | 9.33 | 7.29 | 1.28 | Vibrational |
| CO$_2$ | | 8.96 | 6.92 | 1.29 | |
| NH$_3$ | Tetratomic | 8.63 | 6.57 | 1.31 | Rotational |
| C$_2$H$_6$ | Polyatomic | 12.71 | 10.65 | 1.19 | |

several molecules of different complexity. Notice how the molar heat capacity values increase as the number of atoms in the gas molecule increases and hence the number of modes of rotation and vibration of the molecule. This increase is because rotational and vibrational motions are also ways by which a molecule can utilize heat. The data refer to temperatures near 25° and a pressure of 1 atm. The ratio of the two heat capacities, $\bar{C}_P/\bar{C}_V$, is given the special designation $\gamma$. We will return to a discussion of heat capacities in Chapter 2.

Several illustrations of the application of the gas laws follow.

*Example 1.1* In the Van Slyke method for determining the CO$_2$ capacity of blood, the sample is placed over mercury in a closed flask and CO$_2$ is released by the addition of acid. The released gas is compressed to an accurately known volume (*e.g.*, 0.5 cc) by means of a leveling bulb containing mercury, and its pressure determined by the difference between the initial and final manometer readings (see Fig. 1.7).

If this pressure difference corresponds to 162 mm at a temperature of 27°C when the gas is compressed to 0.5 cc, what is the corresponding volume of CO$_2$ at standard temperature and pressure? If the volume of the blood sample is 0.2 ml, what is its CO$_2$ capacity expressed as volumes percent? (Note: ignore two factors in the calculation: the solubility of CO$_2$ in the sample, and the reabsorption of CO$_2$ by the sample as a result of the increase in gas pressure.) Since $n$, the number of moles of CO$_2$, is unknown, the Perfect Gas Law cannot be used here. Instead, solve the problem by applying the Laws of Boyle and Charles as given in Eq. 1.7:

$$V_0 = \frac{PVT_0}{P_0 T} = \frac{(162 \text{ mm})(0.5 \text{ cc})(273°)}{(760 \text{ mm})(273° + 27°)} = 0.097 \text{ cc.}$$

This is the volume of CO$_2$ released from 0.2 ml of blood. The CO$_2$ capacity expressed as volumes percent will be

Volumes percent CO$_2$ = $(100)(5)V_0 = (100)(5)(0.097) = 48.5.$

Sample is
introduced here

Gas

Solution

To shaking
machine

0.5 cc

2 cc

Calibrations

Mercury

50 cc

To leveling bulb

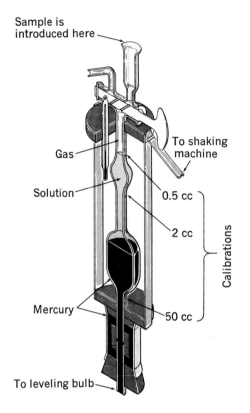

**Figure 1.7** Extraction chamber for Van Slyke
and Neill Manometric Gas Analysis Apparatus.
Blood sample is introduced into chamber and
acid is added. Pressure is determined before
mixing. Following release of $CO_2$, gas is
compressed to original volume inside chamber
by means of a mercury leveling bulb which
connects with lower tube of the chamber.
Pressure of system is read on manometer (not
shown). Chamber is calibrated at 0.5 cc, 2.0 cc,
and 50 cc. Gas volume is maintained constant
for initial and final pressure readings.

*Example 1.2*   Using the Perfect Gas Law, $PV = nRT$, calculate the volume
occupied by 65.0 g of methane at 110°C and 720 mm pressure.

$$V = nRT/P = \frac{(65.0 \text{ g})}{16.0 \text{ g/mole}} \frac{(0.08206 \text{ liter-atm/mole-degree})(110° + 273°)}{(720 \text{ mm})(1 \text{ atm/760 mm})}$$

$$= 135 \text{ liters.}$$

*Example 1.3*   A known weight of a volatile liquid can be vaporized and the
volume of the gas measured in a Victor Meyer apparatus, which is shown in
Fig. 1.8. From the volume of the gas, its molecular weight can be calculated.
A small sample (0.1437 g) of an organic liquid (boiling point 72°C) is released
from its glass ampule over a boiling solvent bath (see Fig. 1.8). The vapor displaces 22.9 cc of air, which is measured over mercury at a temperature of
25.0°C and a pressure of 744 mm. What is the molecular weight of the liquid?
    First, substitute g/M for $n$ in the Perfect Gas Equation and solve the
equation for M:

$$M = gRT/PV.$$

Now substitute the given data into the right-hand term:

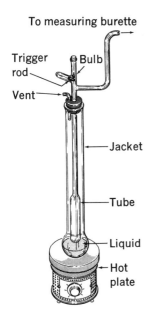

To measuring burette

Trigger rod

Bulb

Vent

Jacket

Tube

Liquid

Hot plate

**Figure 1.8**  Victor Meyer apparatus for determination of molecular weights. A weighed sample is introduced into the apparatus in a small bulb, which is supported on an iron trigger rod. The rod can be pulled to the left with a magnet, thereby causing the bulb to fall to the bottom of the inner tube. The bulb breaks and the contents volatilize because of heat supplied by the boiling solvent in the jacket. The vapor from the compound displaces a corresponding volume of air out of the tube, through the connecting tube, and into the measuring burette which is filled with mercury (not shown). The volume of displaced air is adjusted to atmospheric pressure by means of a mercury leveling bulb, and the pressure, volume, and temperature are recorded. The liquid in the jacket must be boiling strongly to keep the vapor at a uniform temperature. Vapors from the boiling solvent bath may escape through the vent in the stopper.

$$M = \frac{(0.1437 \text{ g})(0.08206 \text{ liter-atm/mole-degree})(25.0° + 273.1°)}{(744 \text{ mm})(1 \text{ atm/760 mm})(0.0229 \text{ liters})}$$

$$= 157 \text{ grams/mole.}$$

Another application of the gas laws is encountered in the Barcroft-Warburg Manometric Method for studying gas exchange in metabolic systems. This subject is treated in Section 1.8, p. 32.

### 1.4  Another Route to the Perfect Gas Law: The Kinetic Equation for Gases

Physical laws may be derived from theoretical tenets, as has been discussed in considerable detail in the Introduction, pp. xv to xviii. As an example of the direct application of scientific method we turn our attention to the experimental Laws of Boyle and Charles and the equation that empirically combines them in the form $PV = nRT$. As early as 1738, Daniel Bernoulli made a theoretical derivation of Boyle's Law by considering the collisions of gas molecules with the wall of a container. The collection of ideas that conceive of temperature and pressure as manifestations of molecular motion is called the kinetic theory of gases. Pressure is viewed as the result of molecular bombardment against the walls of containers, and temperature is regarded as being proportional to the average translational energy of the molecules. We begin the derivation at this point and make certain simplifying assumptions:

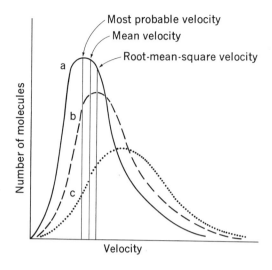

**Figure 1.9** Distribution of velocities among molecules of a gas. Temperature increases as we move from curve a to curve c.

1. Molecules are negligibly small compared with the distance separating them.
2. There are no forces of attraction between molecules.
3. Molecules move in straight lines and their collisions with one another and with the sides of a container are perfectly elastic. This simply means that the kinetic energy of the molecules is conserved when they collide, although it may be transferred from one molecule to another, causing one to speed up and the other to slow down. The average kinetic energy for all of the molecules will remain constant as long as the temperature is constant.

Under conditions of low density, these assumptions will not cause any difficulties.

Although the average kinetic energy of an assembly of molecules is constant at a specified temperature (as is their average velocity if they are of equal mass), there is a distribution of velocities which results from intermolecular collisions. This is shown schematically in Fig. 1.9 at three different temperatures, "a" being the lowest and "c" the highest. Notice that there are several types of average speed, the root-mean-square velocity, the arithmetic mean velocity and the most probable velocity. These differ somewhat for a particular gas at a certain temperature; for example, for hydrogen molecules at 0°C, the most probable velocity is $1.50 \times 10^5$ cm/sec, the arithmetic mean velocity is $1.69 \times 10^5$ cm/sec, and the root-mean-square velocity is $1.84 \times 10^5$ cm/sec.

Consider the simple system of $N$ molecules of a gas, each of mass $m$, at temperature, $T$, in a cubical box whose edge is $a$ cm., Figure 1.10.

The molecules are moving with velocities of $v_1, v_2, v_3 \cdots v_N$ respectively. These velocities all can be expressed in terms of their three

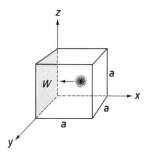

**Figure 1.10**  A molecule in a cubic container of edge *a*, moving toward the wall, W.

respective components normal to the faces of the cube as placed in the framework of the *x*, *y*, and *z* axes in Fig. 1.10. For all the molecules the components are given by $v_{x_1}, v_{x_2}, v_{x_3} \cdots v_{x_N}; v_{y_1}, v_{y_2}, v_{y_3} \cdots v_{y_N}; v_{z_1}, v_{z_2}, v_{z_3} \cdots v_{z_N}$ where $v_{x_1}$ means the *x* component of the velocity of molecule number 1, etc.

$$v_1^2 = v_{x_1}^2 + v_{y_1}^2 + v_{z_1}^2$$

$$v_2^2 = v_{x_2}^2 + v_{y_2}^2 + v_{z_2}^2$$

$$v_3^2 = v_{x_3}^2 + v_{y_3}^2 + v_{z_3}^2 \tag{1.15}$$

$$v_N^2 = v_{x_N}^2 + v_{y_N}^2 + v_{z_N}^2.$$

There is a significant relationship between $v_1^2, v_2^2, v_3^2 \cdots v_N^2$ called the root-mean-square velocity given by

$$u = \sqrt{\frac{v_1^2 + v_2^2 + v_3^2 + \cdots v_N^2}{N}} \tag{1.16}$$

We will see how this relationship is involved in the kinetic equation and ultimately in the perfect gas law.

As a molecule moves back and forth across the box, it will strike face W every 2*a* cm, and if its velocity in the direction of W, the *x* direction, is $v_{x_1}$ cm/sec, it will strike W, $v_{x_1}/2a$ times per second (distance = rate × time, and the frequency of collision will be the reciprocal of the time per collision). After each perfectly elastic collision with the wall the molecule will rebound with a velocity $-v_{x_1}$, suffering no loss in kinetic energy. Momentum is the product of mass and velocity and the time rate of change of momentum is force,[*]

$$f_{x_1} = \frac{d(mv_{x_1})}{dt}.$$

---

[*]The fundamental equation of mechanics is $f = ma$. Since $a$, the acceleration, is $dv/dt$, the time rate of change of velocity, $f = mdv/dt$ or $f = d(mv)/dt$.

The momentum of the molecule before collision is $mv_{x_1}$, and after collision is $-mv_{x_1}$ with a resultant change in momentum per collision (for molecule 1) of $2mv_{x_1}$. If this molecule collides $v_{x_1}/2a$ times per second with face $W$, its total change in momentum per unit time will be

$$\frac{d(mv_{x_1})}{dt} = f_{x_1} = \frac{v_{x_1}}{2a} \cdot 2mv_{x_1} = \frac{mv_{x_1}^2}{a}.$$

The total force exerted on $W$ by $N$ molecules will be the sum of the forces exerted by the individual molecules or

$$f = \frac{m(v_{x_1}^2 + v_{x_2}^2 + v_{x_3}^2 + \cdots + v_{x_N}^2)}{a},$$

and the total pressure will be the force per unit area,

$$P_W = m \frac{(v_{x_1}^2 + v_{x_2}^2 + v_{x_3}^2 + \cdots + v_{x_N}^2)}{a^3}$$

$$P_W = m \frac{(v_{x_1}^2 + v_{x_2}^2 + v_{x_3}^2 + \cdots + v_{x_N}^2)}{V}$$

as $a^3$ is the volume of the cube. But we still do not have the pressure expressed in terms of the "velocity" of the molecules in the box. To accomplish this it is necessary to consider any other two faces perpendicular to face $W$ and to each other, and since the pressure must be the same on all faces, state

$$3P = m[(v_{x_1}^2 + v_{x_2}^2 + v_{x_3}^2 + \cdots + v_{x_N}^2) + (v_{y_1}^2 + v_{y_2}^2 + v_{y_3}^2 + \cdots + v_{y_N}^2)$$
$$+ (v_{z_1}^2 + v_{z_2}^2 + v_{z_3}^2 + \cdots + v_{z_N}^2)]/V$$

$$= m[(v_{x_1}^2 + v_{y_1}^2 + v_{z_1}^2) + (v_{x_2}^2 + v_{y_2}^2 + v_{z_2}^2)$$
$$+ (v_{x_3}^2 + v_{y_3}^2 + v_{z_3}^2) + \cdots + (v_{x_N}^2 + v_{y_N}^2 + v_{z_N}^2)]/V$$

which means that from Eq. 1.15

$$3P = m(v_1^2 + v_2^2 + v_3^2 + \cdots + v_N^2)/V$$

and that from Eq. 1.16

$$P = mu^2 N/3V$$

or

$$PV = \tfrac{1}{3}Nmu^2 \qquad\qquad (1.17)$$

where $N$ is the number of molecules in the container and $u$ is the root-mean-square velocity of the molecules.

Since the kinetic energy of a molecule is defined as

$$K.E. = \tfrac{1}{2}mu^2,$$

we may write

$$PV = \tfrac{1}{3}Nmu^2 = \tfrac{2}{3}K.E. \tag{1.18}$$

Equation 1.18 is called the Kinetic Equation for Gases.

Further, since the kinetic energy of a system is a function of its temperature only, we may proceed to say

$$PV = \tfrac{1}{3}Nmu^2 = \tfrac{2}{3}K.E. = nAT, \tag{1.19}$$

where $n$ is the number of moles present and $A$ is a proportionality constant. We recognize this equation as being identical in form with the Perfect Gas Law and furthermore know that the constant $A$ is none other than the gas constant $R$. When $n$ is unity, $N$ must be equal to Avogadro's number, since one mole of any substance is Avogadro's number of molecules.

## 1.5 Other Relationships Derivable from the Kinetic Equation for Gases

Relationships other than the Laws of Boyle and Charles can be derived from the Kinetic Equation.

In 1831 Sir Thomas Graham studied the rate of escape of gases through various orifices and concluded that the rate was inversely proportional to the square root of the density of the gas—a principle now known as Graham's Law. The Kinetic Equation for Gases predicts the same result. Since the rate at which the gas molecules leak out through a pinhole depends only on their velocity (at constant temperature), we will solve the kinetic equation for $u$, the velocity term:

$$u = \sqrt{\frac{3PV}{Nm}}. \tag{1.20}$$

A convenient way to compare different gases is to select a mole of each. If $N$ is allowed to equal Avogadro's number, then $Nm$ will be equal to M, the molecular weight of the gas. Making this substitution, we write

$$u = \sqrt{\frac{3PV}{M}}. \tag{1.21}$$

Immediately we see that this expression contains an inverted density term: $V/M$. Replacing $V/M$ with $1/d$ gives

$$u = \sqrt{\frac{3P}{d}}. \tag{1.22}$$

If we compare two gases, such as hydrogen and nitrogen, at the same temperature and pressure, we find that

$$\frac{u_{H2}}{u_{N2}} = \sqrt{\frac{d_{N2}}{d_{H2}}},$$   (1.23)

in other words, Graham's Law of Effusion.

Another useful application of the kinetic equation lies in using our knowledge of $PV$ and M to calculate the speed at which gas molecules travel at various temperatures. In calculating the velocity (root-mean-square velocity) of the molecules of a gas, we must exercise care in inserting data into the equation to insure that it will be dimensionally correct: velocity is usually expressed in centimeters per second, and therefore all the other terms must be expressed in the units of the CGS system. Pressure must be given in dynes per cm² and the gas constant $R$ must be expressed in ergs per mole-degree. We can make the calculation in two ways,

$$u = \sqrt{\frac{3PV}{Nm}} = \sqrt{\frac{3PV}{M}},$$

or

$$u = \sqrt{\frac{3RT}{Nm}} = \sqrt{\frac{3RT}{M}}.$$   (1.24)

*Example 1.4*   Calculate the root-mean-square velocity of hydrogen molecules at standard conditions from Eq. 1.21.

$P = (76)(13.6)(980.6)$ dynes per cm²
$V = 22,414$ cc
$M = 2.016$

$$u = \sqrt{\frac{3PV}{M}} = \left[\frac{(3)\,(76)\,(13.6)\,(980.6)\,(22,414)}{2.016}\right]^{1/2} = 183,900 \text{ cm per sec. or } 4080 \text{ miles per hour}$$

If Eq. 1.24 is used, $R$ must be entered as $8.314 \times 10^7$ ergs per mole-degree.

The kinetic equation for gases permits us to calculate the translational energy of the molecules from a knowledge of $PV$ or $RT$. Since the translational energy is a type of kinetic energy, for a mole of gas of molecular (or atomic for the rare gases) weight M,

$$K.E. = \tfrac{1}{2}Mu^2 = \tfrac{3}{2}PV = \tfrac{3}{2}RT.$$

The translational energy can be thought of as the total energy of motion about the three coordinates in space, each coordinate being assigned an energy component equal to $\tfrac{1}{2}RT$. For a monatomic gas such as helium, no other type of motion is possible. For diatomic and polyatomic mole-

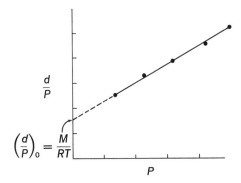

**Figure 1.11** A plot of the density: pressure ratio versus pressure for a gas. At $P = 0$, significant deviations from the Perfect Gas Law disappear and the intercept $(d/P)_0$ is equal to M/RT.

cules, however, one must also consider the energy of rotation about various axes and the energy of vibration. These other types of energy are responsible for the differences between the heat capacities of monatomic gases and of more complex molecules.

> *Example 1.5* Calculate the translational energy of argon at 100°C.
>
> $$K.E. = \tfrac{3}{2}RT = (\tfrac{3}{2})(1.987)(373.2) = 1{,}112 \text{ cal per mole.}$$

In calculations involving gas densities it is often convenient to rearrange the Perfect Gas Equation so as to introduce a density term. If we write

$$PV = nRT = \left(\frac{g}{M}\right)RT,$$

we can substitute $d$, the density, for g/V, and solve for M:

$$M = \frac{dRT}{P}. \tag{1.25}$$

For an ideal gas, $d/P$ is constant. For real gases, ideal behavior is approached as the pressure is lowered. Hence, common practice is to plot density-pressure ratios obtained at different pressures in the manner illustrated in Fig. 1.11. The molecular weight is calculated from the limiting ratio, which is the intercept obtained when $d/P$ is extrapolated to $P = 0$.

> *Example 1.6* By extrapolating $d/P$ values to zero pressure, an intercept value equal to 2.861 was obtained for a gas at 0°C. Calculate the molecular weight of the gas.
>
> $$M = \frac{dRT}{P} = (2.861)(0.08206)(273.15) = 64.13 \text{ grams/mole.}$$

Unlike the densities of liquids and solids, the densities of gases are always expressed in *grams per liter*.

### 1.6 An Equation of State for Real (Nonideal) Gases: The van der Waals Equation

We have already had occasion to mention that real gases — unlike the hypothetical ideal gas — rarely obey the Perfect Gas Law. Maximum deviations from ideal behavior occur at high pressures and/or low temperatures. Under these conditions the volume of the system is relatively small, and the net volume of the molecules is a significant part of the total volume. Moreover, the experimental pressure is noticeably less than ideal when molecules are close together because of a decrease in pressure resulting from increased intermolecular attraction. Various gases do not deviate in the same way or to the same extent, as may be seen from the accompanying diagram (Fig. 1.12). For a perfect gas, the product of pressure and volume, *i.e.*, *PV*, will be a constant as long as the temperature is unchanged. Therefore, if we plot *PV* versus *P* at a constant temperature, we will obtain a straight line running parallel to the abscissa (*P*). Inspection of Fig. 1.12 shows that hydrogen, oxygen, and carbon dioxide all depart significantly from the ideal, and all three do so in different ways. As we have already predicted, high pressures produce the greatest deviations. A plot identical in appearance with Fig. 1.12 will be obtained if we express the ordinate values in terms of *PV/nRT*, the *compressibility factor*, rather than simply *PV*. Whereas the various lines converge at 22.4 liter-atm in Fig. 1.12, in a compressibility factor plot, the point of convergence would be unity, since $PV/nRT = 1.0$ for an ideal gas.

To correct for the inadequacies of the Perfect Gas Equation, van der Waals proposed in 1879 the incorporation of two additional terms: a constant *a* to be added to *P*, to correct for the pressure decrease resulting from intermolecular attraction, and a constant *b* to represent the "effective" volume of the gas molecules and to be subtracted from V. Both *a* and *b* are empirical constants determined from Perfect Gas Law deviations. The equation for one mole of gas follows:

$$\left(P + \frac{a}{\bar{V}^2}\right)(\bar{V} - b) = RT. \tag{1.26}$$

As we can see, something else has been added, besides *a* and *b*. The pressure correction term is $a/V^2$ for the following reason. The molecules on the surface of the gas are being pulled in by the inner molecules. The effectiveness of the pull is proportional to both the surface density and the inner density of the gas. The higher these densities, the greater will be the forces of attraction. Since density is mass/volume, the effect of density can be included by multiplying *a* by $1/V^2$. The term *b* is actually about four times the volume of a mole of molecules and is sometimes called the *excluded volume* because it is the space actually swept out

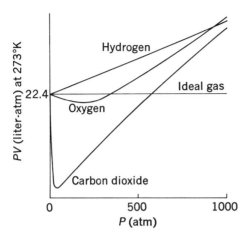

**Figure 1.12**   Deviations of real gases from ideality under conditions of high pressure.

by a mole of closely packed molecules. For any particular gas, *a* and *b* are tabulated for one mole. If more or less than one mole is involved, we must use the form

$$(P + n^2a/V^2)(V - nb) = nRT. \qquad (1.27)$$

Van der Waals constants for several gases are listed in Table 1.5. Note that the constants for carbon dioxide are among the largest of those reported, as might be expected from the deviations from ideality shown in Fig. 1.12. A three-dimensional phase diagram for carbon dioxide is shown in Fig. 1.13. The upper part of the figure was calculated from the $PV$ isotherms for carbon dioxide, just as Fig. 1.6 was drawn from the isotherms for an ideal gas: note the similarities. In Fig. 1.13 we are particularly interested in the intersection of the liquid-vapor region with the gas phase isotherm at the critical point. This occurs at the maximum of the two-phase region; at this point, carbon dioxide is at its critical temperature (31.01°C), critical volume, and critical pressure. The student will recall that above the critical temperature a substance can exist in only one fluid state and under these circumstances a gas cannot be distinguished from a liquid. The critical temperature for a pure liquid can be determined by observing under pressure the temperature at which the meniscus between the gas and liquid phases disappears when the system is warmed and the temperature at which it reappears when it is cooled. The critical volume – the volume of a mole of substance at the critical temperature and pressure – may be obtained indirectly from measurements of the density of the liquid and its saturated vapor at temperatures just below critical, although with difficulty and with relatively low accuracy.

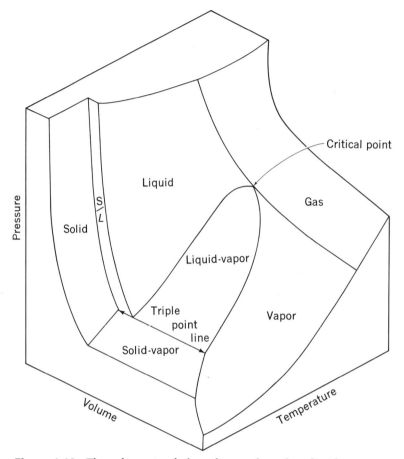

**Figure 1.13** Three dimensional phase diagram for carbon dioxide.

One of the reasons that Fig. 1.13 is important is that it leads us to a method for evaluating the van der Waals constants. If we take Eq. 1.26, expand it and arrange it in descending powers of $\bar{V}$, we obtain

$$\bar{V}^3 + (b + \frac{RT}{P})\bar{V}^2 + \frac{a}{P}\bar{V} - \frac{ab}{P} = 0. \tag{1.28}$$

At the critical point, we introduce the critical temperature, $T$, and the critical pressure, $P$:

$$\bar{V}^3 + \left(b + \frac{RT_c}{P_c}\right)\bar{V}^2 + \frac{a}{P_c}\bar{V} - \frac{ab}{P_c} = 0. \tag{1.29}$$

This equation has three roots, but at the critical point, the roots will all be equal, since there is only one volume, $\bar{V}_c$, corresponding to $P_c$.

**TABLE 1.5** VAN DER WAALS CONSTANTS FOR
CERTAIN GASES

| Gas | a in $L^2$-atm/mole$^2$ | b in $L$/mole |
|---|---|---|
| Helium | 0.0341 | 0.0237 |
| Hydrogen | 0.224 | 0.0266 |
| Nitrogen | 1.39 | 0.0391 |
| Oxygen | 1.36 | 0.0318 |
| Carbon monoxide | 1.49 | 0.0399 |
| Carbon dioxide | 3.59 | 0.0427 |
| Ammonia | 4.17 | 0.0371 |

Therefore, if $\bar{V}_c$ denotes the critical volume, we may say, $\bar{V} = \bar{V}_c$, and

$$(\bar{V} - \bar{V}_c)^3 = 0. \tag{1.30}$$

By expanding Eq. 1.30 it yields,

$$\bar{V}^3 - 3\bar{V}_c\bar{V}^2 + 3\bar{V}_c^2\bar{V} - \bar{V}_c^3 = 0. \tag{1.31}$$

Eq. 1.29, like Eq. 1.31, is also a series of descending powers of $\bar{V}$ and we can therefore equate the coefficients of like powers of the variable, $\bar{V}$, and also equate the constant terms, resulting in:

$$\frac{1}{3}\left(b + \frac{RT_c}{P_c}\right) = \bar{V}_c \qquad \frac{a}{3P_c} = \bar{V}_c^2 \qquad \frac{ab}{P_c} = \bar{V}_c^3. \tag{1.32}$$

From the known values of $\bar{V}_c$, $P_c$, and $T_c$ for carbon dioxide, we can evaluate $a$ and $b$:

$$b = \frac{\bar{V}_c}{3}; \qquad a = 3P_c\bar{V}_c^2. \tag{1.33}$$

Since a number of manometric procedures of biological interest involve the measurement of $CO_2$, let us compare the results obtained by using the Ideal Gas Equation and by using the van der Waals Equation in a calculation.

*Example 1.7* In a metabolic study we are ordinarily interested in determining the volume of $CO_2$ produced or the number of moles of organic acid to which it corresponds. However, using the van der Waals Equation to find $V$ or $n$ requires solving a cubic equation; we will therefore, in the interest of simplicity, calculate the corresponding pressure.

If 1.50 millimoles of lactic acid react with excess $Na_2CO_2$ to produce 20.4 milliliters of dry $CO_2$ at a temperature of 25°C, what would be the pressure of the $CO_2$ as calculated by the Perfect Gas Law, and by the van der Waals Equation? What is the percentage error in using the former?

According to the Perfect Gas Equation,

$$P = nRT/V = \frac{(1.50 \times 10^{-3})\,(0.082)\,(298)}{20.4 \times 10^{-3}}$$

$$= 1.80 \text{ atm.}$$

According to the van der Waals Equation,

$$P = \frac{nRT}{V - nb} - \frac{n^2a}{V^2}$$

$$= \frac{(1.50 \times 10^{-3})\,(0.082)\,(298)}{[20.4 \times 10^{-3} - (1.50 \times 10^{-3})\,(0.0427)]} - \frac{(3.59)\,(1.50 \times 10^{-3})^2}{(20.4 \times 10^{-3})^2}$$

$$= 1.78 \text{ atm.}$$

$$\text{Percent error} = \frac{1.80 - 1.78}{1.78} \times 100 = 1\%.$$

An error of this magnitude can be tolerated in the manometric procedures commonly used in biological studies. For this reason, the calculations of corrected volumes or pressure are based on the simple Laws of Boyle and Charles.

Through his success in writing a more workable gas equation, van der Waals immortalized himself in two ways: the equation now bears his name, and the intermolecular forces that give rise to the $a/V^2$ term are called van der Waals forces, as we have already indicated. These are the last three types listed in Table 1.1.

Equations such as the Perfect Gas Law and the van der Waals Equation for gases are called "equations of state." Others, notably the Equation of Berthelot and the Equation of Dieterici, have been devised.*

## 1.7 Dalton's Law of Partial Pressures

When a container is filled with a mixture of gases, each gas exerts the same pressure as it would if it were the only gas in the vessel. The pressure of each of the components of the gas mixture will depend solely on the number of moles of it present, since the temperature and volume will be the same for all. These individual pressures are called *partial pressures,* their sum being equal to the total pressure. This is Dalton's Law of Partial Pressures.

$$P_{\text{total}} = P_1 + P_2 + P_3 + \cdots P_N. \tag{1.34}$$

---

*For further pursuit of this matter the student is advised to consult the general references at the end of the chapter.

If $V$ is the total volume and $n_1$, $n_2$, $n_3$, etc. represent the number of moles of each component,

$$PV = n_1RT + n_2RT + n_3RT + \cdots n_NRT \tag{1.35}$$

$$= RT(n_1 + n_2 + n_3 + \cdots n_N).$$

The partial pressure of an individual gas – for example, gas 1 – will be

$$P_1 = \frac{n_1RT}{V}. \tag{1.36}$$

We now see that the ratio of the partial pressure of a gas to the total pressure is equal to the ratio of the number of moles of that component to the total number of moles:

$$\frac{P_1}{P_{total}} = \frac{n_1}{n_1 + n_2 + n_3 + \cdots n_N}. \tag{1.37}$$

This ratio is called the *mole fraction* and is designated by the symbol X. For gas 1, the mole fraction is $X_1$, etc. Considerations of partial pressures are involved in many calculations. Volumetric procedures involving the collection of gases over water must take into account the partial pressure of water vapor in the gas mixture. The low vapor pressure of mercury as compared with that of water gives it a distinct advantage as a displacing fluid, since the partial pressure of mercury vapor may be neglected in such calculations.

The air we inhale is a mixture of gases, and the composition of expired air reflects the gas-exchange process occurring in the lungs. Alveolar air is saturated with water vapor at body temperature; the partial pressure of this water vapor is about 48 mm. If the total pressure of alveolar air should happen to be 760 mm, the pressure due to $O_2 + CO_2 + N_2$ would be $760 - 48 = 712$ mm. The partial pressure of each gas may be calculated if the percentage composition is known, as the following example illustrates.

*Example 1.8* The percentage composition of dry alveolar air is: $O_2$, 14%; $CO_2$, 5.6%; and $N_2$, 80%. If the total pressure is 712 mm, what are the respective partial pressures?

$$P_{O_2} = (0.14)(712) = 100 \text{ mm};$$
$$P_{CO_2} = (0.056)(712) = 40 \text{ mm};$$
$$P_{N_2} = (0.80)(712) = 570 \text{ mm}.$$

In Table 1.6 are given values for the composition of dry inspired, expired, and alveolar air, representative of man.

**TABLE 1.6** COMPOSITION OF
INSPIRED, EXPIRED, AND ALVEOLAR
AIR IN MAN AT REST IN VOLUMES
PERCENT

| Gas | Inspired air | Expired air | Alveolar air |
|-----|--------------|-------------|--------------|
| $O_2$ | 20.95 | 16.1 | 14.0 |
| $CO_2$ | 0.04 | 4.5 | 5.6 |
| $N_2$ | 79.0 | 79.2 | 80.0 |

*Source:* Reproduced from *Textbook of Biochemistry*, 3rd Ed., by West, E. S., and W. R. Todd, Macmillan, New York, 1961, p. 562.

### 1.8 The Barcroft-Warburg Manometric Method as an Application of the Gas Laws

The most familiar application of the gas laws in biological science appears in the calculation of the "flask constants" used in the Barcroft-Warburg manometric method for measuring gas exchange in metabolic processes. The most common determination is that of oxygen consumption by a tissue or by an enzyme preparation in the presence of substrate, buffer, etc. A manometer and flask suitable for the measurement of oxygen uptake are shown in Fig. 1.14. One must know accurately the volume of the flask and attached manometer capillary down to the reference level of the manometer fluid (shown at 150 mm in Fig. 1.14). Precalibrated flasks and manometers are now available commercially, thus relieving the research worker of the burden of hours of calibration time.

The components of the experimental system are divided between the center compartment of the assay flask (A) and the side arm (B) so that no reaction can proceed until the contents of the side arm are tipped into the center compartment. Potassium hydroxide solution is placed in the center well (C), and a small piece of fluted filter paper is inserted there to increase the surface of the caustic solution. The potassium hydroxide serves as a $CO_2$ trapping agent and thus permits the measurement of $O_2$ disappearance uncomplicated by the appearance of a second gas. The manometer and flask are attached to an oscillating support which is so aligned as to immerse the flask in a constant temperature bath. The bath is vigorously stirred, and the manometer and flask are shaken to permit equilibration at the bath temperature. During this time the stopcock (D) is open to the air. Once equilibration is reached, the shaking is stopped, the flask and manometer momentarily removed from the bath, the manometer fluid in the right arm rapidly adjusted to a reference level (*e.g.,* 150 mm) by means of the screw (E), the stopcock (D)

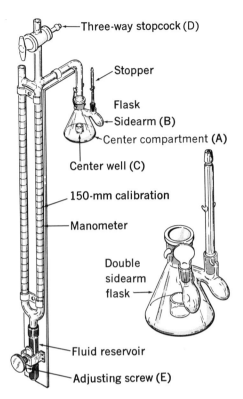

Three-way stopcock (D)

Stopper

Flask

Sidearm (B)

Center compartment (A)

Center well (C)

150-mm calibration

Manometer

Double sidearm flask

Fluid reservoir

Adjusting screw (E)

**Figure 1.14** The Warburg constant volume manometer and flasks.

is closed, and the contents of the side arm (B) tipped into the center compartment of the flask (A). The manometer and flask are returned to the bath, shaking is resumed, and the reaction is under way. From time to time (or, only once—at the end of the experiment), the manometer fluid is readjusted to the 150-mm mark in the closed right arm and the fluid level in the open left arm read and recorded. As the system consumes oxygen, the readings on the open arm scale will drop since the system is being adjusted to constant volume each time a reading is taken. The difference in the fluid level in the two arms corresponds to the pressure drop in the system.

Any pressure fluctuations from causes other than oxygen consumption are detected by simultaneously reading the pressures on a control flask containing the same volume of reactants as the experimental flask but undergoing no metabolic reaction (usually, one essential component is omitted). This vessel is known as the thermobarometer flask.

The flask constant is the number by which $h$, the observed pressure change in the open arm of the manometer, is multiplied to obtain the volume of $O_2$ consumed, corrected to standard conditions. Inspection of the expression containing the flask constant reveals that two principles are involved: (1) the application of the Laws of Boyle and Charles in converting an observed pressure to a standard gas volume and (2) a correction for the volume of $O_2$ dissolved in the vessel fluid. The equation is:

$$V_0 = h \left[ \frac{V_g \frac{273}{T} + V_f \alpha}{P_0} \right] = h \cdot k,$$

flask constant

where

$V_0 =$ volume of gas utilized at 0°C and 760 mm;

$h =$ pressure difference measured in mm of manometer fluid;

$V_g =$ volume of gas phase in flask and manometer arm down to fluid level (*e.g.*, 150 mm). This must already be known or obtained by previous calibration;

$V_f =$ volume of fluid in flask;

$\alpha =$ solubility of gas in fluid in flask;

$P_0 =$ standard pressure in mm of manometer fluid (usually an aqueous solution of sodium chloride and sodium choleate of density 1.033).

$$P_0 = (760)\frac{13.6}{1.03} = 10,000 \text{ mm.}$$

(N.B.: Volumes are customarily given in microliters.)

Ignoring for the moment the second numerator term, $V_f \alpha$, we have

$$V_0 = h \left[ \frac{V_g \frac{273}{T}}{P_0} \right],$$

which can be rearranged to

$$\frac{V_0 P_0}{273} = \frac{h V_g}{T}.$$

Since $273 = T_0$, this is the combined form of the Laws of Boyle and Charles (Eq. 1.7).

*Example 1.9*  Let us calculate $V_0$ for a particular experiment. A Warburg flask has been previously calibrated and found to possess a total volume of 12.930 ml down to the 150-mm mark on the manometer. The total volume of buffer, substrate, and enzyme is 3.00 ml and 0.20 ml of 20% KOH is placed in the center well. Oxygen uptake is measured at 25°C, at which temperature $\alpha = 0.0285$. What will the flask constant be?

$V_g$ = (total volume − fluid volume) = (12.930 − 3.20)
   = 9.730 ml = 9730 $\mu$l;
$V_f$ = 3.20 ml = 3200 $\mu$l;
$\alpha$ = 0.0285 $\mu$l gas per $\mu$l fluid at 25°C;
$T$ = 273° + 25° = 298°;
$P_0$ = 10,000 mm manometer fluid;

$$k = \frac{(9730)\left(\dfrac{273}{298}\right) + (3200)(0.0285)}{10,000} = \frac{8910 + 91.2}{10,000} = 0.900.$$

## Suggested Additional Reading

### General: The Gas Laws

Daniels, F., and R. A. Alberty, *Physical Chemistry*, 3rd Ed. John Wiley and Sons, New York, 1966.
Moore, W. J., *Physical Chemistry*, 4th Ed. Prentice-Hall, Englewood Cliffs, N.J., 1972.

### Electrostatics

Edsall, J. T., and Wyman, J., *Biophysical Chemistry*. Academic Press, N.Y., 1958, Chap. V.
Jencks, W. P., *Catalysis in Chemistry and Enzymology*, McGraw-Hill, N.Y., 1969, Chapt. 7.

### Medical Applications

Clark, W. M., *Topics in Physical Chemistry*. Williams and Wilkins, Baltimore, 1952.

### Manometry

Umbreit, W. W., R. Burris, and J. F. Stauffer, *Manometric Techniques*. Burgess, Minneapolis, 1957.

### The Hydrogen Bond

Pimentel, G. C., and A. L. McClellan, *The Hydrogen Bond*. W. H. Freeman and Company, San Francisco, 1960.
Jencks, W. P., *Catalysis in Chemistry and Enzymology*. McGraw-Hill, N.Y., 1969, Chapt. 6.

### The Hydrophobic Bond

Jencks, W. P., *Catalysis in Chemistry and Enzymology*. McGraw-Hill, N.Y., 1969, Chapt. 8.

## Problems

**1.1** Calculate the volume occupied by 2 moles of an ideal gas at 2 atm pressure and 25°C.

**1.2** At 27°C, 250 cc of $H_2$, measured under a pressure of 400 mm of Hg, and 1200 cc of $N_2$, measured under a pressure of 500 mm of Hg, are introduced into an evacuated 3-liter flask. Calculate the resulting pressure.

**1.3** A flask contains 5 g of a mixture of $O_2$ and $N_2$ at S.T.P. Magnesium ribbon is introduced into the flask and ignited. All of the oxygen is removed, leaving a residue which weighs 5 g. What will be the final pressure in the container at 0°C? Assume that there is no reaction between $N_2$ and Mg.

**1.4** Derive a relationship for the percentage change in volume per degree rise in temperature at constant pressure for an ideal gas.

*Answer.* $\dfrac{100}{T}$.

**1.5** A flask contains 5 moles of an ideal gas at 25°C and 2 atm pressure. The temperature of the gas is raised to 50°C. What fraction of the gas must be removed to keep the pressure in the flask constant?

**1.6** Find an expression for the increase in pressure when the volume of a gas is halved and the Kelvin temperature doubled.

*Answer.* $\Delta P = \dfrac{3nRT_1}{V_1} = 3P_1$.

**1.7** Twenty grams of an unknown solid is placed in a 2 L flask. The flask is evacuated and then heated to 150°C at which temperature all of the unknown solid is converted to a gas. The pressure is 5 atm. What is the molecular weight of the substance?

*Answer.* 70.

**1.8** What is the molarity of oxygen in the inspired air in Table 1.6 at 25°C and 1 atm?

*Answer.* 0.00857 molar.

**1.9** What is the molarity of an ideal gas at 25°C and 1 atm?

*Answer.* 0.0409 molar.

**1.10** The density of ammonia was determined at various pressures by weighing the gas in large glass bulbs. The values of the density in grams per liter at 0°C were as follows: 0.77169 at 1 atm; 0.51515 at $\frac{2}{3}$ atm; 0.38293 at $\frac{1}{2}$ atm; 0.25461 at $\frac{1}{3}$ atm. What is the molecular weight of ammonia?

**1.11** Calculate the ratio of the velocities of He and Ne at 25°C and 1 atm pressure. Also calculate the ratio of their kinetic energies under these conditions.

*Answer.* 2.23:1.00; 1:1.

**1.12** Example 1.4 shows that gas molecules move with high velocities at normal environmental temperatures. Explain how these high velocities can be consistent with the much lower wind velocities experienced on earth.

**1.13** The gypsy moth produces a natural attractant, $C_{18}H_{34}O_3$. If a female moth is trapped behind a cellophane screen containing a pinhole and the carbon dioxide she produces diffuses through the pinhole at the rate of 1 milli micromole per 90 seconds, how long will it require for the same quantity of attractant to diffuse through the orifice?

**1.14** What is the ratio of the diffusion rates of $^2H_2O$ vapor and $^3H_2O$ vapor?

**1.15** Calculate the root-mean-square velocity of the molecules in cm per sec in a 1-liter container filled with neon to a pressure of 5 atm at 25°C.

*Answer.* $6.09 \times 10^4$ cm/sec.

**1.16** Calculate the pressure of 2 moles of an ideal gas at 27°C in a 24 L container. Next, assume that the gas is oxygen and calculate the pressure using the van der Waals Equation.

**1.17** The van der Waals constants for water vapor are $a = 5.464$ atm $L^2$/mole$^2$ and $b = 0.0305$ L/mole. Compare these to the van der Waals constants for He, $N_2$, and $CO_2$ in Table 1.5.
(a) Explain the relative sizes of the constant $b$ for these four gases.
(b) Explain the relative sizes of the constant $a$ in terms of the molecular properties of He, $N_2$, $H_2O$, and $CO_2$.

**1.18** A snowball at $-15$°C is thrown at a large tree trunk. If we disregard all frictional heat loss, at what speed must the snowball travel so as to melt on impact? The specific heat of ice is 0.5 cal per gram; the heat of fusion of ice is 79.7 cal per gram.

**1.19** A mixture consisting of 0.200 g of $H_2$, 0.210 g of $N_2$, and 0.510 g of $NH_3$ has a total pressure of 1 atm at 27°C. Calculate (a) the mole fraction, (b) the partial pressure of each gas, and (c) the total volume.

*Answer.*
(a) $X_{H_2} = 0.727$;          $X_{N_2} = 0.055$;          $X_{NH_3} = 0.218$
(b) $P_{H_2} = 0.727$ atm;     $P_{N_2} = 0.055$ atm;     $P_{NH_3} = 0.218$ atm;
(c) $V = 3.38$ liters.

**1.20** The maximum acceptable concentration (MAC) of harmful vapors in the atmosphere is frequently given in parts per million (ppm). The MAC of carbon tetrachloride vapor in air is given as 25 ppm by the Merck Index. How does this number compare with the equilibrium concentration of carbon tetrachloride vapor in a laboratory at 23°C and 1 atm? The vapor pressure of carbon tetrachloride is 100 mm Hg at 23°C. Assume air is 80 vol % $N_2$ and 20 vol % $O_2$.

*Answer.* equilibrium conc = 446,000 ppm

**1.21** Use the van der Waals Equation to calculate the temperature of a 50-liter container that is filled with 3 moles of methane at a pressure of 7.5 atm. The van der Waals constants for methane are: $a = 2.253$ $L^2$ atm mole$^{-2}$; $b = 0.04278$ $L$ mole$^{-1}$.

**1.22** Show that $R = (8/3)(P_cV_c/T_c)$.

**1.23** At the critical point, $(\partial P/\partial V) = 0$ and $(\partial^2 P/\partial V^2) = 0$. Applying this information to the van der Waals Equation, show that $T_c = 8a/27bR$, $V_c = 3b$, and $P_c = a/27b^2$.

**1.24** How many 1 gram balls moving with a velocity of 1000 cm per sec would have to strike a 0.5 by 0.5 meter 1000 gram plate per second to keep it suspended in air?

**1.25** Calculate the difference in gas pressures above and below the plate (Problem 1.24) needed to hold it up.

*Answer.* 3.92 dynes/cm$^2$.

**1.26** Calculations of gas volumes by the Van Slyke manometric method are usually based on previously tabulated constants published by Van Slyke and co-workers. Show that the pressure of $O_2$ in mm Hg should be multiplied by a factor equal to 0.300 to give the corrected gas volume per 100 ml of sample at standard conditions for 0.5 ml of gas obtained from a 0.2 ml sample at 26°C.

**1.27** A Warburg flask is calibrated by measuring the pressure change when the following reaction occurs:

$$N_2H_4 + 4Fe(CN)_6^{3-} \longrightarrow N_2 + 4Fe(CN)_6^{4-} + 4H^+.$$

In the main compartment of the flask are placed 0.4 ml of 0.1 $M$ potassium ferricyanide, 0.4 ml of 4 $N$ sodium hydroxide, and 1.2 ml of water. In the side arm are placed 0.5 ml of 0.1 $M$ hydrazine sulfate (excess) and 0.5 ml of 4 N NaOH. Nothing is placed in the center well. After equilibration, the contents of the side arm are tipped in. At equilibrium the change in the height of the manometer fluid in the outer arm is 160 mm. The temperature of the bath is 37°C.

(a) What is the flask constant, $k$?

(b) If the solubility of nitrogen is 0.0123 ml per ml of water at 37°, what is the volume of gas ($V_g$) in the flask and manometer arm?

*Answer.*

(a) $k = 1.4$;

(b) $V_g = 15.9$ ml.

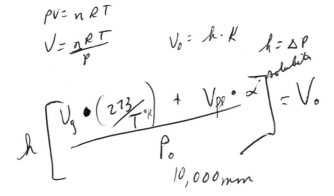

$$pV = n RT$$

$$V = \frac{nRT}{p} \qquad V_o = h \cdot k \qquad h = \Delta p$$

$$h \left[ \frac{V_g \bullet \left( 273 \big/ T °K \right) + V_{\mathscr{P}} \bullet \alpha \cdot \frac{\text{solubility}}{P_o} }{10,000 mm} \right] = V_o$$

# 2 | THE LAWS OF THERMODYNAMICS AND AN INTRODUCTION TO THE FREE ENERGY CONCEPT

Die Energie der Welt ist konstant. Die Entropie der Welt strebt einem Maximum zu.
*Clausius.*

*Thermodynamics* is the study of heat and its transformations. The science of thermodynamics arose in the late eighteenth century as a consequence of growing interest in everyday processes involving work, heat, and material transformations. The Industrial Revolution and the growth of technology provided the initial impetus. Just as the frictional heat generated in the boring of cannons attracted the attention of Benjamin Thompson (Count Rumford) and prompted him to design the first quantitative thermodynamic experiments, so the energetic balance in other processes occupied the thoughts of a celebrated group including Black, Lavoisier, Hess, Carnot, and Mayer. These innovative theoreticians were followed by Joule, Maxwell, Clausius, Kelvin (William Thomson), and Helmholtz, all of whom were, in their later years, contemporaries of Boltzmann, van't Hoff, and Gibbs. The last of these, J. Willard Gibbs, possessed the most productive mind in the history of thermodynamic theory and is the only American in the group.[*]

Whereas the word thermodynamics was an appropriate descriptive term for early studies involving the quantitative relationships between heat and other forms of energy, modern emphasis has shifted from the analysis of energy changes accompanying material transformations to the employment of thermodynamics as a tool for understanding and predicting the behavior of systems in terms of their *energetics*. It is used to estimate the direction in which forces will be acting in chemical systems and the magnitude of these driving forces. Thermodynamics has

---

[*]An engrossing account of the development of these concepts is given in Chapter 23 of the book by Wightman included in the *Suggested Additional Readings*.

become less *descriptive* and more *predictive* in emphasis, and consequently more useful.

An additional change is seen in the rise of *statistical* thermodynamics (as opposed to *classical* thermodynamics), which deals with the behavior of molecules themselves rather than such macroscopic systems as steam engines, distillations and crystallizations, electrolyses, and the like. Statistical thermodynamics begins with individual atoms and molecules and sums up their collective behavior and properties, whereas classical thermodynamics deals with bulk properties only and does not consider individual particles. Both approaches have much to say to the chemist and biochemist, but we shall confine ourselves to an exploration of the classical approach.

Although thermodynamics, or energetics, as it is also called, is a powerful tool, it does not possess omnipotence. It can predict the maximum work obtainable in a defined process; it can predict stabilities, yields, optimum temperatures and pressures, the best choices of solvent, etc. Thermodynamics will reveal whether a certain chemical reaction will proceed favorably in the desired direction, but it will not tell the time required or the path (mechanism) by which such a reaction proceeds. It is common knowledge that the cellulose in a wooden table will, at the kindling temperature, react spontaneously with the oxygen of the air to give carbon dioxide, water, and heat, and that this is the *favored* direction of the reaction. We can even calculate how much heat will be evolved. Thermodynamics does not tell us, however, the magnitude and properties of the thermal energy barrier which must be hurdled before the reaction proceeds spontaneously. Reaction rates and mechanisms are treated in that branch of physical chemistry called *kinetics*, which will be considered in a later chapter. In the main, thermodynamics focuses its attention on equilibrium states; kinetics is more concerned with transient states.

## 2.1 The First Law of Thermodynamics

**Definitions**  *Energy* is familiarly defined as "the capacity to do work." In Chapter 1, Table 1.3, a number of different types of energy were listed and it was shown that each could be factored into a capacity term and an intensity term. Although various capacity terms or intensity terms cannot be directly equated, any energy unit may be equated with another if a proportionality factor is known. Thus, calories can be converted to ergs or to liter-atmospheres, just as was done with the units of $R$, the gas constant. The concept of energy gives us relatively little trouble in modern times because we have learned to associate it with work done by a moving piston, a falling weight, or an electric cell, and we can detect heat energy through our sensory receptors. Mayer, Carnot, and their associates, on the other hand, did not clearly differentiate energy from force and momentum.[*]

In solving a problem by the methods of thermodynamics, one must define the *system,* that is, whatever part of the physical world is severed from its *surroundings* to be examined thermodynamically. If a system is homogeneous, all properties are the same in all parts and it is continuous from point to point. The properties of a system are its characteristics: size, temperature, density, pressure, color, the presence of magnetic or electrical or gravitational fields, etc. If a system is heterogeneous, two or more distinct regions called *phases* are present and are separated by surfaces called interfaces. The thermodynamics of heterogeneous systems is presented in Chapter 3; in this chapter we shall confine ourselves to extremely simple homogeneous systems. Systems are of two types, *open systems* and *closed systems.* In a closed system, no mass may leave or enter during a particular change, known as a *process,* although energy may leave or enter. In an open system, both mass and energy may enter or leave.

We shall have many occasions to refer to systems at *equilibrium,* and we shall try to discover appropriate criteria for equilibrium. When the composition and properties of a system undergo no detectable change after the lapse of a sufficiently long period of time, the system is considered to be in a state of equilibrium. However, a chemical equilibrium is by no means a state of inactivity. Rather, it is that balance point at which the reaction proceeds from left to right with a velocity identical to its velocity from right to left.

As was said in Chapter 1, the properties of a system determine its *state.* If a change occurs in one of the properties, for example, the temperature, the state of the system changes. We usually restrict ourselves to those properties which can be measured with relative ease: temperature, pressure, volume, and composition. The interdependency of some of these eliminates the need for measuring them all.

Since our knowledge of the nature of matter does not yet permit us to calculate the absolute energy (grand total) present in a system, we must content ourselves with measuring energy changes between two states of a system or between one state and an arbitrary standard. For example, if we are dealing with a gas, we could compare it with the ideal gas at one atmosphere pressure at the temperature concerned. A liquid solvent could be compared with the pure liquid, and a solid substance with its most stable crystalline allotropic form, all of these being at a specified temperature and one atmosphere pressure. These would be called *standard states* and afford us reference points from which changes

---

[*]Our sense of security is on the whole a false one, arising from an oversimplification of the nature of energy. Considering that the Einstein Equation, $E = mc^2$, leads us to regard mass and energy as two aspects of the same phenomenon, we may have to admit that we do not really understand energy after all.

could be measured. Without the concept of standard states, thermodynamics would be seriously handicapped. The idea of standard states
is illustrated by the following analogy: We are watching climbers strug-
gling up the north wall of the Eiger, the bottom of which is obscured
from our view. We can perceive how much the climbers move up or slip
back with respect to one another or how far they move from a clearly
visible ledge extending across the face of the wall. The position of a
climber with respect to the ledge is analogous to the energy of a chemical
species relative to its energy in the standard state. Just as we do not know
the absolute height of the climbers above the valley floor because we
cannot measure that distance from our vantage point, so we do not know
the absolute energies of atoms and molecules. Standard states are chosen
arbitrarily, and complete agreement does not prevail regarding the best
choice for certain substances.

**The statement of the First Law**   The First Law of Thermodynamics is
the familiar Law of Conservation of Energy. Carnot and Mayer had
grasped the significance of the law, but it was first clearly stated by
Helmholtz in 1847.[*] It cannot be proved in an exact way and is a product of human experience. Such laws are often called "laws of nature."
Henri Poincaré, the noted mathematician, made a wry observation in
1892, saying in effect that everyone believes firmly in the Law of Conservation of Energy because the mathematicians think it is an experimental fact and the experimentalists think it is a theorem of mathematics.[†]
Our present knowledge leads us to modernize the First Law by including
mass as a form of energy or to state the law as the conservation of energy
and mass, consistent with the Einstein relationship: $E = mc^2$. However,
in applying it to ordinary (nonnuclear) processes we will use the First
Law in the form of the Law of Conservation of Energy.

Philosophically, the First Law was well received, because we find
a certain reassurance in believing in the permanence of things. However, the First Law says nothing about the ultimate usefulness of the
energy of the universe. This statement is left for the Second Law, as will
be shown. The First Law, however, does forbid the existence of a perpetual motion machine that could constantly produce more energy
(work) than was utilized to operate it, *i.e.*, that would "create energy."
The earliest recorded drawings for a perpetual motion machine of the
"first type," or those in violation of the First Law, belong to the 13th
Century. Scientific theories of the Middle Ages did not forbid the existence of philosopher's stones, universal solvents, or perpetual motion
machines. Figure 2.1 illustrates a sophisticated "perpetuum mobile"

---

[*] H. von Helmholtz, *Über die Erhaltung der Kraft,* G. Reimer, Berlin, 1847.
[†] H. Poincaré, Preface to *Thermodynamique,* Paris, G. Carré, 1892.

**Figure 2.1**   Proposed perpetual motion machine utilizing continuous water power for the grinding of knives (1580). Water flowing from the upper reservoir turned the water wheel, which turned the gear mechanism operating the Archimedes screw, thereby lifting water from the lower to the upper reservoir. The rotating shaft of the water wheel was also supposed to turn the grinding wheels for sharpening knives. The device did not work because the screw is incapable of raising as much water as the machine requires for its operation. (After Illustration 508 in F. M. Feldhaus, *Die Technik der Vorzeit, der Geschichtlichenzeit und der Naturvölker,* Verlag von Wilhelm Engelmann, Berlin, 1914. Many other designs and their history may be found in H. Dircks, *Perpetuum Mobile,* I and II, reprinted from the original (1861 and 1870) by B. M. Israel, Amsterdam, 1968.)

designed in the 16th Century for the purpose of grinding knives. Needless to say, it did not work.

Thermodynamics makes use of a notation which arises out of the necessity for writing equations involving heat and other forms of energy. The symbols represent the *thermodynamic functions* and are

$E$, the internal energy;

$H$, the enthalpy or heat content;

$S$, the entropy;

$G$, the Gibbs free energy;[*]

$A$, the Helmholtz free energy;[†]

$w$, work;

$q$, heat.

These symbols have definitions which we shall not attempt to defend or trace to their origins. Our purpose here is to acquire the essential vocabulary so that we may converse in the language of thermodynamics.

---

[*]$G$ has been recently chosen by the International Union of Pure and Applied Chemistry as the most desirable of several symbols for the Gibbs free energy currently in use, another of these being $F$.

[†]$A$, the work function, or Helmholtz free energy, will not be discussed in this presentation for reasons which are given in the next paragraph.

The thermodynamic functions are related through the following fundamental equations, presented in their simplest form:

$$H = E + PV$$
$$A = E - TS$$
$$G = H - TS$$
$$dE = T\ dS - P\ dV$$
$$dH = T\ dS + V\ dP$$
$$dA = -S\ dT - P\ dV$$
$$dG = -S\ dT + V\ dP$$

Most of these equations will be introduced in the ensuing development of the concepts of thermodynamics. The second and sixth equations will be omitted from the discussion, however, because $A$, the work function, is similar to, but not as useful as, $G$, the free energy. The modern thermodynamic literature of the life sciences is devoid of references to $A$. Since our purpose is to present those fundamental concepts most necessary and most valuable to the student rather than to treat the subject exhaustively, no further mention of $A$ will be made.

The functions $H$, $E$, $G$, $q$ and $w$ all have the units of energy and are most commonly expressed in calories. The function $S$ has the units of calories divided by temperature; the energy term is $TS$.

The four functions $H$, $E$, $G$, and $S$ differ from $q$ and $w$ in a highly important respect. The former are *state functions*, which is to say that their values depend only on the state of the system. Changes in their values depend only on the initial and final states of the system and not on the path lying between. On the other hand, $q$ and $w$ are dependent upon the path and consequently, the path along which a change occurs must be known in order to calculate them. However, if we specify the path by holding the system at constant volume or at constant pressure, and no other work is done by the system, such as by way of electrical energy or mechanical energy (rotating a paddle wheel by the gas moving from one part of the system to another, for example), then $q$ becomes a state function. Examples will be cited following the mathematical presentation of the First Law.

Since the value of a state function such as $E$ depends only on the state of the system, we can always perform the integration $\int_{E_1}^{E_2} dE$ to obtain $E_2 - E_1$. On the other hand, $dq$ and $dw$ cannot be integrated to yield a "$q$" or a "$w$" because heat and work are dependent upon the path and $dq$ and $dw$ are *inexact differentials*. Their sum, $dq + dw$, however, is an *exact differential*, equal to $dE$. These ideas may be stated in the following equivalent ways:

1. $E$ is a function of the state of the system.
2. $dE$ is an exact differential.

3. The integral of $dE$ about a closed path (the line integral) is equal to zero, *i.e.*, $\oint dE = 0$. If a system goes through a succession of changes in state and finally returns to its original state, the net amount of energy transferred to or from the system is zero.

A further important and useful expression that applies to exact differentials is the Euler reciprocity relationship, which states that if an exact differential is expressed as $dE = y\ dx + z\ dw$, then

$$\left(\frac{\partial y}{\partial w}\right)_x = \left(\frac{\partial z}{\partial x}\right)_w$$

The relationship is a consequence of the unimportance of the *order* of differentiation in the calculus.

We shall use the symbol $d$ to represent an infinitesimal change or increment and the symbol $\Delta$ for a macroscopic change. In the mathematical statement of the First Law the energy of the system is symbolized by $E$. $E$ includes the potential energy, kinetic energy, heat energy, and any other form of energy possessed by the system. Since the absolute energy (total energy) of the system is unknown, we must restrict ourselves to energy changes as the system goes from one state (a set of conditions) to another (a second set of conditions). Thus the energy change between the system and its surroundings will be $E_2 - E_1$, or $\Delta E$. The subscripts 2 and 1 will be used throughout the discussion to refer to the values of the function in the final and initial states respectively. In any alteration in the state of the system all possible energy changes must be accounted for: (a) the system can absorb or lose heat and (b) the system can do work or have work done upon it. As a mathematical statement of First Law, we write *for the system and surroundings considered together*

$$\int dE = 0.$$

*For the system alone*, we describe any alteration in its internal energy by writing

$$\Delta E = q + w, \tag{2.1}$$

which may be read, "The increase in internal energy of a system is equal to the heat supplied to the system plus the work done by the surroundings on the system." If the system loses internal energy to the surroundings, the sign of $\Delta E$ will be negative. According to the above mathematical convention of stating the First Law, if $w$ is positive, the surroundings do work on the system; if $w$ is negative, the system does work on the surroundings. A positive value of $q$ means that heat is supplied to the system; a negative value indicates heat transfer to the surroundings by the system.*

Another way of thinking of the mathematical statement of the First Law is to view it as an accounting of all forms of energy in transit through a system. Any change, gain or loss, can be described by the simple equation given above. As stated before, the energy change will depend only on the initial and final states of the system, but $q$ and $w$ will depend on the path. An illustration of this fundamental difference between $\Delta E$ and $q$ (or $w$) is given by Klotz.†

. . . let us consider a heavy boulder resting on the edge of a steep cliff. This boulder could be brought to the floor of the valley in many ways. One method would be to attach it to one end of a rope, attach another boulder of almost equal weight, resting at the bottom of the valley, to the other end of the rope and by stringing the rope over a pulley wheel suitably mounted at the brink of the cliff, to allow the high boulder to be lowered to the ground while the second boulder is raised to the top of the cliff. In this procedure the first boulder does a positive amount of work, but no heat is absorbed or evolved in the process. Another method for getting the boulder down might be to let it slide down some pathway to the valley. Less work could be obtained by this method, but a definite quantity of heat would be evolved. Many other paths could be chosen differing in the frictional resistance they would offer to the motion of the boulder. For each path the work ($w$) done may differ from that for any other path, and the heat ($q$) evolved will be different. Nevertheless, despite great variations in comparative values of $q$ and $w$, the difference $q - w$ ($q + w$ in the convention used in this text), has been found in all experience to be the same, so long as the respective starting points and final points of the boulder's travel are the same for each path. (Klotz uses the $\Delta E = q - w$ convention.)

**Illustrations of the First Law**   For the present we will exclude all forms of work other than pressure-volume work, principally for the sake of simplicity. It may be readily seen that the pressure-volume product has the units of mechanical work.

$$PV = \frac{\text{force}}{(\text{distance})^2} \cdot (\text{distance})^3$$

$$= \text{force} \cdot \text{distance, which is mechanical work.}$$

---

*These sign conventions are those recently recommended by the International Union of Pure and Applied Chemistry. Some readers will be more familiar with the convention in which the First Law is written

$$\Delta E = q - w.$$

†Klotz, I. M., *Energetics in Biochemical Reactions*, pp. 5–7. Academic Press, New York, 1957. Reprinted by permission.

The work done in the expansion of a gas is always given by the product of the pressure *against which* the gas is expanding and the volume change which the gas undergoes. $P_{ext}$ is the external pressure.

$$w = -\int_{V_1}^{V_2} P_{ext}\, dV. \qquad (2.2)$$

Solution of the general equation in this instance yields the *work done by the system on the surroundings* since in the expansion of a gas $V_2 > V_1$.

We will now proceed with some sample calculations of thermodynamic quantities for the expansion of a gas under different conditions (along different paths).

*First case: isothermal expansion of a perfect gas:* $\Delta E = 0$. From Eq. 1.18 we know that $PV = \frac{2}{3}K.E.$; furthermore, kinetic theory states that the kinetic energy is a function of temperature alone. Therefore, in an isothermal system (temperature is constant), there will be no change in the energy of the gas. If $\Delta E = 0$, then $q + w = 0$ and $q = -w$. The expansion can take place in several ways:

1. Into a vacuum. Because no work can be done against a vacuum, if the gas expands isothermally into an evacuated vessel, $w = 0$. Since $q = -w$, it follows that $q$ also is equal to 0.
2. Against constant pressure. If the gas expands isothermally against a constant external pressure, $w = -P \int_{V_1}^{V_2} dV$ or $-P(V_2 - V_1)$. The heat supplied will be equal to the work of expansion since $\Delta E = 0$. $q = -w = P\Delta V$.
3. Against changing external pressure. If we allow the pressure to drop as the volume increases, the calculation of $w$ is more complicated. The accompanying diagram (Fig. 2.2) illustrates the problem. In a system expanding against constant pressure, the work is represented by the rectangular area of height $P_2$ and width $(V_2 - V_1)$. However, if the external pressure is dropping along the curve $P_1 P_2$, more work is done and is represented by the area lying under the curve. It is this area we wish to calculate. We must know the value of $P$ at every point during the expansion or be able to express the pressure as an integrable function of the volume so that it may be substituted in Eq. 2.2 for calculation of the work.

In the special case in which during the expansion, the external pressure never differs from the internal pressure by more than the infinitesimal increment $dP$, $P_{internal}$ may be substituted for $P_{external}$ in the calculation of work, as will now be shown. At every stage in this kind of expansion, $P_{ext} = (P_{int} - dP)$. It follows that $P_{ext}\, dV = (P_{int} - dP)\, dV = P_{int}\, dV - dP\, dV$.

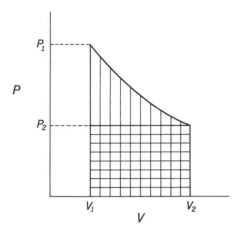

**Figure 2.2** $P_{ext}V$ relationships (work) for a gas expanding against a constant external pressure ($P_2$) and against a changing pressure ($P$).

The term $dP\, dV$ is insignificant since it is the product of two infinitesimals and, therefore, it may be neglected. If we sum all of the $P_{int}\, dV$ terms ($P_{int}$ is a variable), we calculate the work:

$$w = -\int_{V_1}^{V_2} P_{int}\, dV \qquad \text{(This is Eq. 2.2)}.$$

We cannot handle the integration as it is stated and must first eliminate one of the variables. If the gas is ideal and temperature is constant, the Perfect Gas Law permits an appropriate substitution for $P$ in terms of the second variable $V$. For one mole of gas, $P = RT/V$. If the change is isothermal, $RT$ will be a constant and

$$w = -RT \int_{V_1}^{V_2} \frac{dV}{V}. \qquad (2.3)$$

Integration of Eq. 2.3 gives

$$w = -RT \ln\frac{V_2}{V_1}. \qquad (2.4)$$

From the Perfect Gas Law we know that at constant temperature an equivalent statement is

$$w = -RT \ln\frac{P_1}{P_2}. \qquad (2.5)$$

Work calculated in this way is the *maximum* work which the system can do in going from the initial to the final state. Maximum work is obtained by carrying out the expansion or contraction in increments sufficiently small that the direction of the process can be reversed at any

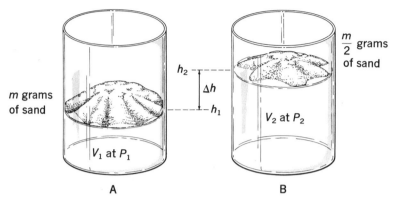

**Figure 2.3** An imaginary system in which a weightless, frictionless membrane serves as a support for a certain weight of sand that is exactly balanced by the pressure of the gas contained within the volume $V_1$ or $V_2$.

instant by a minute pressure change. The process is visualized as being a series of gradual changes, each one almost an equilibrium in itself. For each stage in the process the pressure of the expanding gas will not differ from external pressure by more than $dP$. This type of operation is called a *reversible* one and has a specific meaning in thermodynamics. The implication is not that the process is to be reversed in a gross sense, but that it *could* be reversed at any instant by applying minute forces opposite in sign to those acting to produce the initial process. Only under these circumstances will maximum work be obtained. If the process is not carried out reversibly, $w$ will be smaller than $w_{max}$, but we cannot say exactly how much.

The example shown in Fig. 2.3 should help the student understand the difference between reversible and irreversible processes and clarify the concept of maximum work. We see in A a volume containing one mole of gas, $V_1$, confined at a total pressure $P_1$, which is atmospheric pressure plus the pressure due to $m$ grams of sand resting on the weightless and frictionless membrane confining the gas. In B, $V_2$ is the volume which results from the mass of sand being reduced to $\frac{1}{2}m$ at the final total pressure $P_2$. Two ways (paths) of going from state A to state B may be visualized. In both, we will supply whatever heat is necessary to restore the system to its original temperature. The gas is the system, and everything else constitutes the surroundings.

By one path, an irreversible one, we suddenly remove one half of the sand from the membrane and the gas expands against a constant pressure $P_2$ to a volume $V_2$. The work done in the process is given in numbered line 2, page 48.

$$w = -P \, \Delta V. \tag{2.6}$$

In this specific case,

$$w = -P_2(V_2 - V_1).$$

The work obviously *includes* $-\frac{1}{2}(mg\,\Delta h)$, where $g$ is the acceleration of gravity and $\Delta h$ is the distance moved by the membrane, as shown in Fig. 2.3.

By a second path, a reversible one, we remove one grain of sand of mass $dm$ at a time, permitting the gas to expand a corresponding but changing amount, $dV$, against the decreasing pressures, $(P_1 - dP)$, $(P_1 - 2dP)$, etc. The process approaches true reversibility and, indeed, would be reversible if the grains were sufficiently small. When we have removed one half of the sand by the second procedure, the work done by the gas, $w$, will have been (assuming the gas is ideal)

$$w = -\int_{V_1}^{V_2} P\,dV = nRT \int_{V_1}^{V_2} \frac{dV}{V},$$

as already given in Eq. 2.2 and Eq. 2.3, which leads to

$$w = nRT \ln \frac{V_2}{V_1} \quad \text{(Eq. 2.4).}$$

$$w_{net} = -nRT\left(\ln V_2 - \ln V_1\right)$$
$$= -nRT \ln \frac{P_1}{P_2}$$

The work *includes* $-\int$ (weight of sand)$g\,dh$, but the weight of sand is an unknown function of $h$ and the term as expressed cannot be integrated. However, it should be clear that by the second path more mass, on the average, is raised a greater height in the earth's gravitational field than is raised by the irreversible one, and consequently more work is done by the system in the reversible isothermal expansion of the gas.

---

*Example 2.1* It may readily be calculated that $P_2(V_2 - V_1) < RT \ln(V_2/V_1)$ if a mole of an ideal gas is taken through two expansions involving the same pressure changes, the first being an irreversible expansion and the second a reversible one. To make the calculation simple, consider the example in Fig. 2.3 and imagine that all the sand is removed instead of only half. Assume the external pressure ($P_2$) is one atmosphere and that the weight of the sand adds a pressure of one atmosphere so that $P_1$ is two atmospheres. If the temperature is constant at 0°C, the final volume of the mole of perfect gas will be 22.4 liters ($V_2$). Under two atmospheres pressure the volume will be 11.2 liters ($V_1$). All that remains is to calculate the work. When all the sand is removed at one time in the irreversible process the gas expands against the atmospheric pressure to its final volume, and the work done is

$$w = -P_2(V_2 - V_1) = -(1)(22.4 - 11.2) = -11.2 \text{ liter-atm.}$$

When the sand is removed one grain at a time,

$$w_{max} = -2.303 \, RT \log\frac{V_2}{V_1} = -(2.303)(0.082)(273)\left(\log\frac{22.4}{11.2}\right)$$

$$= -15.5 \text{ liter-atm.}$$

**Second case: the isothermal vaporization of a liquid at constant pressure:** $\Delta E \neq 0$. This change is both isothermal and isobaric (pressure is constant). Two types of energy change are occurring: heat is flowing into the system to vaporize the liquid, and the vapor which is formed expands against the atmosphere and does work. If we choose a mole of liquid as our system, $q =$ the molar heat of vaporization. The work done will be $-P(V_v - V_l)$ where the volume terms denote molar volumes of vapor and liquid respectively. Since $V_v >>> V_l$, the work will very nearly equal $-PV_v$, and, if the vapor obeys the Perfect Gas Law, will equal $-RT$. In general, work in such cases will be $-\Delta n \, RT$ where $\Delta n$ is the number of gaseous moles in the final state minus the number of gaseous moles in the initial state. Since we are vaporizing only one mole of liquid and no gas was present initially, the change in the number of gaseous moles, $\Delta n$, is equal to unity $(1 - 0 = 1)$.

*Example 2.2* The following illustration shows the calculation of $w$ and $\Delta E$ for a mole of water at 100°C and 1 atmosphere pressure. The heat of vaporization, $q$, will be 539 calories per gram (the specific heat of vaporization of water at 100°C) multiplied by the gram molecular weight of water, 18.

$$q = (18)(539) = 9702 \text{ cal;}$$
$$w = -P \, \Delta V = -(1)(22.4)(373/273) = -30.6 \text{ liter-atm.}$$

Also,

$$w = -\Delta n \, RT = -(1)(1.987)(373) = -741 \text{ cal,}$$

and

$$\Delta E = q + w = 9702 - 741 = 8961 \text{ cal.}$$

If $-30.6$ liter-atmospheres is converted into calories, the two methods for calculating $w$ give identical results:

$$(-30.6)\left(\frac{1.987}{0.082}\right) = -741 \text{ cal.}$$

This use of the various forms of the gas constant, $R$, to make conversions from one type of energy unit to another was discussed in Chapter 1. In the preceding examples most of the terms were rounded off to three significant figures, as they would be for a slide rule calculation.

**Third case: the adiabatic system:** $q = 0$. An adiabatic change occurs in a system that is so insulated from its environment that no heat may enter or leave the system. Any work done by the system must be

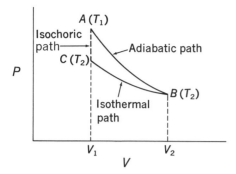

**Figure 2.4**  A comparison of two reversible paths for going from $A$ to $B$: an adiabatic path, $AB$, and a path composed of isochoric step, $AC$, and an isothermal step, $CB$.

performed at the expense of its internal energy. Since, by definition, $q = 0$, it follows that $\Delta E = w$. Equation 2.2, $w = \int P\, dV$, holds for the adiabatic system, but Eq. 2.3 is not applicable, since temperature is not constant. As work is done by the system and its internal energy decreases, the temperature falls. Since $P$, $V$, and $T$ are all changing, the work expression cannot be integrated as it now stands. An equation is needed from which we can calculate the properties of the system in its initial and final states. The problem can be solved if we can find an expression for $\Delta E$ in terms of the three variables in the adiabatic change, $P$, $V$, and $T$. Let us assume that we have $n$ moles of ideal gas at initial pressure and volume, $P_1$ and $V_1$, which expand reversibly and adiabatically to a final pressure and volume, $P_2$ and $V_2$. In the process the temperature falls from $T_1$ to $T_2$. Since $E$ is a state function, $\Delta E$ will depend only upon the initial and final states of the system. We can therefore evaluate it, if we can invent a pathway composed of individual steps along which $\Delta E$ is readily calculated. Consider the diagram shown in Fig. 2.4. The path $AB$ represents the adiabatic change from the initial to the final state. Instead of going by this route, let us take an alternate path composed of two separate steps. First we will cool the gas at constant volume from $T_1$ to $T_2$. On the diagram, this step is the line $AC$. The change in internal energy for the constant volume step is given by the relationship

$$\Delta E = n \int_{T_1}^{T_2} \bar{C}_V\, dT = n\bar{C}_V(T_2 - T_1) = n\bar{C}_V\, \Delta T \text{ (if } \bar{C}_V \text{ is constant)}, \quad (2.7)$$

where $\bar{C}_V$ is the heat capacity of one mole of gas at constant volume. If $\bar{C}_V$ is not constant over the temperature range of the system, an expression for $\bar{C}_V$ as a function of temperature must be used in the integration. We will not introduce this complication into the consideration of the adiabatic system, but the student should bear in mind that the final expressions in Eq. 2.7 are for a limiting case.

Now all that remains is to go from $C$ to $B$ along the path $CB$, which represents the isothermal expansion of the gas, already at $T_2$, to its final volume $V_2$. What will be the value of $\Delta E$ for this step? We have already shown that $\Delta E = 0$ for a perfect gas undergoing a volume change at constant temperature (see page 48). Therefore, for the two steps, the value of $\Delta E$ is given by Eq. 2.7.

Certain useful relationships between any two of the state functions, $P$, $V$ and $T$ can be obtained for adiabatic changes if $\Delta E = w$ is written in the differential form, $dE = dw$ and treated in the following way,

$$dE = dw$$

$$n\bar{C}_V \, dT = -P \, dV.$$

From the Perfect Gas Law we may substitute $nRT/V$ for $P$:

$$\bar{C}_V \, dT = -RT\frac{dV}{V}.$$

Collecting variables, indicating that the operation of integration is to be performed between limits and taking the constant terms outside the integral signs, we get

$$\int_{T_1}^{T_2} \frac{dT}{T} = -\frac{R}{\bar{C}_V} \int_{V_1}^{V_2} \frac{dV}{V}.$$

The integration is now a simple matter and gives

$$\ln\frac{T_2}{T_1} = -\frac{R}{\bar{C}_V} \ln\frac{V_2}{V_1}. \tag{2.8}$$

We recall from Chapter 1 that $R = \bar{C}_P - \bar{C}_V$ and $\bar{C}_P/\bar{C}_V = \gamma$. With these substitutions Eq. 2.8 becomes

$$\ln\frac{T_2}{T_1} = -\left(\frac{\bar{C}_P - \bar{C}_V}{\bar{C}_V}\right) \ln\frac{V_2}{V_1} = (1 - \gamma) \ln\frac{V_2}{V_1},$$

or

$$\frac{T_2}{T_1} = \left(\frac{V_2}{V_1}\right)^{(1-\gamma)} \tag{2.9}$$

If the relationship desired is one between initial and final values of $P$ and $T$ instead of $T$ and $V$, $P_1 T_2/P_2 T_1$ should be substituted for $V_2/V_1$ in Eq. 2.8. Be careful not to substitute $P_1/P_2$, as this would be correct only for the isothermal case. The derivation is

$$\ln\frac{T_2}{T_1} = -\frac{R}{\bar{C}_V}\ln\frac{P_1 T_2}{P_2 T_1} = -\frac{R}{\bar{C}_V}\ln\frac{P_1}{P_2} - \frac{R}{\bar{C}_V}\ln\frac{T_2}{T_1},$$

$$\ln\frac{T_2}{T_1} + \frac{R}{\bar{C}_V}\ln\frac{T_2}{T_1} = -\frac{R}{\bar{C}_V}\ln\frac{P_1}{P_2},$$

$$\left(1 + \frac{R}{\bar{C}_V}\right)\ln\frac{T_2}{T_1} = -\frac{R}{\bar{C}_V}\ln\frac{P_1}{P_2},$$

$$\frac{\bar{C}_P}{\bar{C}_V}\ln\frac{T_2}{T_1} = -\frac{R}{\bar{C}_V}\ln\frac{P_1}{P_2},$$

$$\left(\frac{T_2}{T_1}\right)^\gamma = \left(\frac{P_1}{P_2}\right)^{(1-\gamma)} \tag{2.10}$$

Similar treatment of Eq. 2.8, but with the substitution of $\dfrac{P_2 V_2}{P_1 V_1}$ for $\dfrac{T_2}{T_1}$ will give the relationship in terms of $P$ and $V$, and as the final result,

$$P_1 V_1^\gamma = P_2 V_2^\gamma,$$

or,

$$PV^\gamma = \text{constant.} \tag{2.11}$$

The comparable expression for the isothermal case is: $PV = \text{constant}$.

An adiabatic system may also expand along an irreversible path. Suppose that the pressure is suddenly released and the gas expands adiabatically against a constant external pressure. Since the expansion is not a reversible one, Eqs. 2.8–2.11 cannot be used. However, our original statements concerning the adiabatic system still hold: $q = 0$, and $\Delta E = w$. The value of $\Delta E$ still depends only on the initial and final states and Eq. 2.7 is still applicable:

$$\Delta E = w = n\bar{C}_V(T_2 - T_1).$$

*Example 2.3* Calculate the work done and the final volume in the following irreversible adiabatic expansion: Three moles of an ideal monatomic gas ($\bar{C}_V = 3$ cal/mole-degree) expand irreversibly from an initial pressure of 5 atm, against a constant external pressure of one atm, until the temperature drops from the initial value of 350°K to a final value of 275°K. How much work is done and what is the final volume?

$$\Delta E = n\bar{C}_V(T_2 - T_1) = (3)(3)(275 - 350) = -675 \text{ cal.}$$

The loss in internal energy must equal the work done in the adiabatic expansion. To calculate the final volume, it is necessary to convert calories to energy units in liter-atm: $-675$ calories $= -27.8$ liter-atm. Since work is done against constant pressure,

$$w = -P(V_2 - V_1) = -(1)(V_2 - V_1) = -27.8 \text{ liter-atm.}$$

In order to solve for $V_2$ first calculate $V_1$ from the Perfect Gas Law.

$$V_1 = \frac{nRT}{P} = \frac{(3)(0.082)(350)}{5} = 17.2 \text{ liters},$$

and therefore,

$$V_2 = 27.8 + 17.2 = 45.0 \text{ liters}.$$

*Example 2.4* Calculate the work done and the final volume of the gas in a reversible adiabatic expansion. Consider that the system holds exactly 0.5 mole of a perfect monatomic gas at an initial pressure of 10 atm and an initial volume of one liter. The gas is expanded reversibly and adiabatically until a final pressure of one atm is reached. What will be $V_2$, $T_1$, $T_2$, and $w$? The exponent $\gamma$ in Eq. 2.11 is $\bar{C}_P/\bar{C}_V$; for an ideal monatomic gas, $\gamma = 5/3$ (see Table 1.4). The final volume in a reversible adiabatic expansion is given by Eq. 2.11:

$$P_1(V_1)^{5/3} = P_2(V_2)^{5/3}.$$

From the data given,

$$(10)(1)^{5/3} = (1)(V_2)^{5/3}.$$

Cubing both sides of the equation gives

$$(1000)(1)^5 = (V_2)^5.$$

Taking the fifth root of both sides yields

$$V_2 = 4 \text{ liters}.$$

$T_2$ is calculated from the Perfect Gas Law:

$$T_2 = \frac{P_2 V_2}{nR} = \frac{(1)(4)}{(0.5)(0.082)} = 97.5°\text{K}.$$

In the same way, $T_1$ is obtained:

$$T_1 = \frac{P_1 V_1}{nR} = \frac{(10)(1)}{(0.5)(0.082)} = 243.9°\text{K}.$$

$$\Delta E = w = n\bar{C}_V(T_2 - T_1) = (0.5)(3)(97.5 - 243.9) = -220 \text{ cal}.$$

## 2.2 The Enthalpy, H

The enthalpy or heat content has been defined as

$$H = E + PV. \tag{2.12}$$

No attempt should be made to equate this expression with the First Law. It is simply a useful definition, developed early in the history of thermodynamics, and describes $H$ as a state function. Since we do not know the absolute values of $H$ or $E$, we use the definition in terms of changes

$$\Delta H = \Delta E + \Delta(PV). \tag{2.13}$$

Like $\Delta E$, $\Delta H$ depends only on the initial and final states of a system and is independent of the path along which a change may occur. The equation reveals several useful things on inspection. For the case of the perfect gas changing volume at constant temperature, pressure and volume change reciprocally: All the $PV$ terms are equal and $\Delta(PV) = 0$. For this case, then, the enthalpy and the internal energy changes are equal. If the gas is expanding or contracting against a constant pressure,

$$\Delta H = \Delta E + P \, \Delta V. \tag{2.14}$$

From the examples considered in the previous section, we know that $-P \, \Delta V$ is the work term for a system restricted to expansion work at constant pressure. Therefore, if $-P \, \Delta V = w$, then $\Delta H = q$, since $\Delta E = q + w$. We can write, for this situation,

$$\Delta H = q_P, \tag{2.15}$$

where $q_P$ specifically denotes the heat absorbed at constant pressure. The corresponding term for a constant volume process is $q_V$. Note that since the changes in state are reversible, the terms $q_P$ and $q_V$ are state functions equal to $\Delta H$ and $\Delta E$ respectively.

If for a process, $q$ is positive, the system absorbs heat and the change is *endothermic*; when $q$ is negative, heat is evolved and the change is *exothermic*.

If the process under consideration is a phase transition (*e.g.*, solid to liquid or vapor, liquid to vapor) occurring at constant temperature and pressure, or a transformation within the solid phase, $\Delta H$ denotes the appropriate molar heat of fusion, sublimation, etc. For those transitions in which no significant volume change occurs at constant temperature and pressure $\Delta H$ is essentially equal to $\Delta E$.

When a substance merely undergoes a temperature change at constant pressure, the heat supplied is $q_P$ and equals $n \int \bar{C}_P \, dT$ where $\bar{C}_P$ is the heat capacity at constant pressure. We recall the definitions of $E$ and $H$ and differentiate these functions:

$$dE = dq - P \, dV,$$

$$dH = dE + P \, dV + V \, dP = dq - P \, dV + P \, dV + V \, dP,$$

and on imposing constant pressure we obtain

$$dH = dq_P.$$

Therefore, $dH = n\bar{C}_P \, dT$, and if the molar heat capacity of the substance is constant over the temperature range in question,

$$\Delta H = n\bar{C}_P(T_2 - T_1) = n\bar{C}_P \, \Delta T. \tag{2.16}$$

The student should compare this equation with Eq. 2.7.

$$H \Rightarrow C_p$$
$$E \Rightarrow C_v$$

$$\Delta H = n\bar{C}_p \Delta T$$
$$\Delta E = n\bar{C}_v \Delta T$$

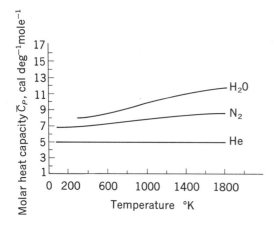

**Figure 2.5**  Effect of temperature on the heat capacities of helium, nitrogen, and water vapor at constant pressure.

In many cases, and in all cases where broad temperature ranges are involved, it cannot be safely assumed that $\bar{C}_P$ and $\bar{C}_V$ are independent of temperature. They will usually be expressed by empirical equations of the form $\bar{C}_P = a + bT + cT^2 + \cdots$. The equation we integrated to yield 2.16 should be written for the general case, $dH = n(a + bT + cT^2 + \cdots)dT$ which integrates to give

$$[\Delta H]_{T_1}^{T_2} = n[aT + \tfrac{1}{2}bT^2 + \tfrac{1}{3}cT^3 + \cdots]_{T_1}^{T_2}$$

between the limits of the higher and lower temperatures. The influence of temperature on the molar heat capacities of helium, nitrogen, and water vapor at constant pressure is shown in Fig. 2.5.

The heat absorbed or evolved in any chemical reaction at constant pressure is always represented by the symbol $\Delta H$. As pointed out above, it can also be used to symbolize the molar heat exchange in a physical process such as melting or vaporization. Some examples of the different uses of $\Delta H$ are given below. The subscripts (g), (l), and (s) refer to the physical state of the substance; (aq) means that the substance is in aqueous solution. A further stipulation remains: $\Delta H$ (written to the right of the equations) equals

$$\Sigma H_{\text{products}} - \Sigma H_{\text{reactants}} \cdot \ \gtrsim \Delta H$$

When products and reactants are in their standard states, the enthalpy change is called a standard enthalpy change and is designated as $\Delta H°$. For gases, the standard state is the ideal gas at one atm pressure and at a designated temperature; for liquids, it is the pure liquid under the same conditions, and for solids, it is an agreed upon crystalline state under the same conditions. Properly, the temperature should always be specified as a subscript on $\Delta H°$, but unless otherwise specified, the standard state refers to a temperature of 25°C.

Some of the different types of enthalpy changes are as follows:

*Heat of reaction*

$$MgCl_{2(s)} + 2Na_{(s)} \rightarrow$$
$$2NaCl_{(s)} + Mg_{(s)}, \qquad \Delta H° = -43.06 \text{ kcal./mole } MgCl_{2(s)}$$

*(handwritten: STD. STATE, 1 atm, 25°C)*

*Heat of combustion* — *(handwritten: using $O_2$)*

*(handwritten: EXOTHERMIC LOSES HEAT TO SURROUNDINGS)*

$$C_2H_{6(g)} + \tfrac{7}{2}O_{2(g)} \rightarrow$$
$$2CO_{2(g)} + 3H_2O_{(l)}, \qquad \Delta H° = -372.8 \text{ kcal./mole } C_6H_{5(g)}$$

*Heat of formation* — *(handwritten: make comp out of its elements)*

$$Pb_{(s)} + S_{(s)} \rightarrow PbS_{(s)}, \qquad \Delta H° = -22.5 \text{ kcal./mole}$$

*Heat of ionization*

$$H_2O_{(l)} \rightarrow H^+_{(aq)} + OH^-_{(aq)}, \qquad \Delta H° = 13.7 \text{ kcal./mole}$$

*Heat of solution**

$$K^+Cl^-_{(s)} + aq \rightarrow K^+_{(aq)} + Cl^-_{(aq)}, \qquad \Delta H° = 4.44 \text{ kcal./mole } K^+Cl^-_{(s)}$$
(200 moles)

*Heat of neutralization*

$$H^+Cl^-_{(aq)} + Na^+OH^-_{(aq)} \rightarrow$$
$$Na^+Cl^-_{(aq)} + H_2O_{(l)}, \qquad \Delta H° = -13.7 \text{ kcal./mole}$$

*Heat of vaporization* (100°C)

$$H_2O_{(l)} \rightarrow H_2O_{(g)}, \qquad \Delta H°_{373} = 9.72 \text{ kcal./mole}$$

*Heat of fusion* (0°C)

$$H_2O_{(s)} \rightarrow H_2O_{(l)}, \qquad \Delta H°_{273} = 1.44 \text{ kcal./mole}$$

The first equation in the preceding list illustrates a reaction for which $\Delta E = \Delta H$ since it occurs at constant pressure and there is no significant volume change. Most biochemical reactions are of this type, since only a few physiological processes involve the uptake or release of gas. Measurements of $\Delta H$ are frequently and conveniently made in a sealed calorimeter, an apparatus of constant volume. If a $P\,\Delta V$ term is involved in the reaction, proper correction must be made for $P\,\Delta V$ to evaluate $\Delta H$ at one atm pressure.†

---

*The heat of solution given for the example is an *integral heat of solution,* that is, the heat absorbed when one mole of the solute is dissolved in a sufficient quantity of solvent to give a specified concentration. In the example, 200 moles of solvent is specified.

†For a general discussion of the subject of calorimetry, the student is referred to an article by Sturtevant, J. M., in Weissberger, A., *Physical Methods of Organic Chemistry,* 3rd Ed., Vol. 1, Part 1, 523–654, Interscience, New York, 1959.

**Thermochemistry**   That branch of thermodynamics which deals specifically with enthalpy changes accompanying chemical reactions is called *thermochemistry.* Two important statements, first made many years ago, are called the Laws of Thermochemistry. These are

*The Law of Lavoisier and Laplace (1780).* The heat required to decompose a compound is equal to the heat evolved in its formation.

*Hess' Law, or the Law of Constant Heat Summation (1840).* The heat evolved in a chemical process at constant pressure is the same whether the process takes place in one or several steps.

The Law of Lavoisier and Laplace enables us to write chemical reactions in either direction if we merely reverse the sign of the $\Delta H$ term as products and reactants exchange places.

Hess' Law is a formal statement of the principle that has already been emphasized several times, *i.e.*, the enthalpy change in a process depends only on the initial and final states of the system and does not depend on the route between them. A fortunate consequence of this law is that enthalpy changes can be calculated for reactions that cannot be studied experimentally, for one reason or another, but for which the appropriate equations can be obtained by the expedient of adding chemical equations as one does algebraic equations.

The concept of the *heat of formation* is so useful that it should be singled out for special emphasis. It is, by definition, the heat absorbed or evolved in the synthesis of one mole of a compound from its elements, all components being in their standard states. The heat of formation of an *element* in its standard state is arbitrarily taken as zero. The standard state for an element is its most stable form at 25°C and at a pressure of one atm. For $Br_{2(l)}$, $\Delta H_f^\circ = 0$; but for $Br_{2(g)}$, $\Delta H_f^\circ = +7.4$ kcal. per mole. Little ambiguity exists in the choice of the standard reference state for most of the elements, one exception being carbon. Carbon exists in two solid crystalline forms at 25°C and one atmosphere pressure: diamond and graphite. For most calculations, graphite is used as the standard state, but we must make certain that the same reference state for carbon has been used before combining heats of formation of its compounds.

The utility of heat of formation data is summarized in the following equation:

$$\Delta H_{reaction} = \Sigma \Delta H_{formation\ of\ products} - \Sigma \Delta H_{formation\ of\ reactants}. \quad (2.17)$$

Ideally, this relationship would permit us to calculate $\Delta H$ for any chemical reaction. The limiting factor, obviously, is our incomplete knowledge of heats of formation of compounds. Some values can be obtained by direct calorimetry, for example,

$$H_{2(g)} + \tfrac{1}{2}O_{2(g)} \rightarrow H_2O_{(l)}, \qquad \Delta H^\circ = -68.32 \text{ kcal./mole } H_2O_{(l)}$$

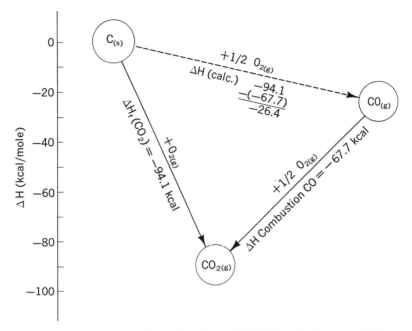

**Figure 2.6** Calculation of $\Delta H_f^\circ(CO)$ from $\Delta H_f^\circ(CO_2)$ and $\Delta H_{combustion}^\circ$ of CO.

Other values for heats of formation are acquired by applying Hess' Law:

$$C_{(s)} + O_{2(g)} \rightarrow CO_{2(g)}, \qquad \Delta H^\circ = -94.1 \text{ kcal./mole } CO_{2(g)}$$
$$CO_{(g)} + \tfrac{1}{2}O_{2(g)} \rightarrow CO_{2(g)}, \qquad \Delta H^\circ = -67.7 \text{ kcal./mole } CO_{2(g)}$$

Subtracting the two equations, as well as the two enthalpy terms, gives

$$C_{(s)} + \tfrac{1}{2}O_{2(g)} \rightarrow CO_{(g)}, \qquad \Delta H^\circ = -26.4 \text{ kcal./mole } CO_{(g)}$$

We thereby obtain the heat of formation of carbon monoxide. The final equation represents a chemical reaction that cannot be studied experimentally because some of the carbon will inevitably be oxidized to carbon dioxide. Figure 2.6 shows diagrammatically how we have used an alternate route to reach our destination in this calculation.

Heats of formation have been determined by direct or indirect methods for a number of compounds, and a selection of these is presented in Table 2.1. Note that the heat of formation depends on the physical state of the compound, e.g., $H_2O_{(l)}$ and $H_2O_{(g)}$. Less heat is evolved in the formation of a mole of water vapor than a mole of liquid water because, in the former process, part of the heat is utilized as heat of vaporization of the water. All the heats of formation are for the compounds at 25°C and one atm pressure, which permits them to be combined in various applications of Eq. 2.17. They are, therefore, standard heats of formation and so designated by the notation $\Delta H_f^\circ$. If we wish to combine a particular value of a heat of formation that is not corrected to 25°C with other standard heats of formation, we must obtain the appropriate heat capacity data and correct the nonstandard heat of formation to 25°C.

> *Example 2.5*  Calculate $\Delta H^\circ$ for the hydrolysis of urea to $CO_2$ and $NH_3$. Although the reaction does not proceed at a detectable rate in the absence of the enzyme, urease, $\Delta H_{reaction}^\circ$ may be calculated from thermodynamic data if the appropriate heats of formation are available. If an enzyme behaves as a true catalyst, it will not alter the position of equilibrium or the thermodynamic characteristics of the reaction. The equation for the hydrolysis is
>
> $$H_2O_{(l)} + H_2N—CO—NH_{2(aq)} \rightarrow CO_{2(aq)} + 2NH_{3(aq)}.$$
>
> Applying Eq. 2.17 to the data in Table 2.1 gives
>
> $$\Delta H_{reaction}^\circ = (-98.69 + 2 \times -19.32) - (-68.32 + -76.30)$$
>
> $$= 7.29 \text{ kcal./mole } CO_{2(aq)}.$$

Although the aqueous solutions given in Table 2.1 are the hypothetical one-molal solutions of urea, $CO_2$, and $NH_3$, the heats of formation in these specific cases are essentially the same for infinitely dilute solutions of the compounds. On the other hand, if data for urea$_{(s)}$, $CO_{2(g)}$, and $NH_{3(g)}$ are used, a substantially different heat of reaction is obtained$-31.82$ kcal. The student should therefore be warned that thermochemical calculations are of little value unless the data are appropriate to the system.

A calculation of $\Delta H_{reaction}^\circ$ from thermochemical data can be of great value as an independent check on the standard enthalpy change determined by another method. From the standpoint of theory, direct calorimetry is certainly the method of choice, provided the reaction will follow the path indicated by the written equation. Until recently, the calorimetric determination of the $\Delta H^\circ$ of enzymatic reactions was not feasible because suitable instrumentation had not been developed. A new technique known as *heatburst calorimetry*, however, permits measurements of very minute quantities of heat. With the heat-pulse microcalorimeter, $\Delta H^\circ$ has been measured and $\Delta G^\circ$ calculated for the hydrolysis of adenosine triphosphate (ATP), an achievement of exciting proportions.[*] Figure 2.7 shows the experimental curves from which $\Delta H_{hydrolysis}^\circ$ of ATP was calculated. The heatburst microcalorimeter as originally designed is a batch calorimeter, but it can be modified for flow operation. The flow calorimeter is simple in principle and suitable for the small samples typical of biochemical experiments.[†]

---

[*]Kitzinger, V. C., and T. H. Benzinger, Z. *Naturforsch.*, **10B**:375, 1955. Benzinger, T. H., and R. Hems, *Proc. Nat. Acad. Sci. (U.S.)*, **42**:896, 1956.

[†]Sturtevant, J. M., "Flow Calorimetry," *Fractions*, No. 1, 1, 1969, Beckman Instruments, Inc., Palo Alto, Calif.

**TABLE 2.1** STANDARD HEATS OF FORMATION OF SELECTED COMPOUNDS AT 25°C

| Compound | $\Delta H_f^0$ in kilocalories per mole | Compound | $\Delta H_f^0$ in kilocalories per mole |
|---|---|---|---|
| Acetaldehyde$_{(g)}$ | −39.76 | Glycerol$_{(l)}$ | −159.80 |
| Acetic acid$_{(l)}$ | −116.4 | n-Hexane$_{(l)}$ | −47.52 |
| Acetone$_{(l)}$ | −59.34 | Methane$_{(g)}$ | −17.89 |
| Acetylene$_{(g)}$ | 54.19 | Methanol$_{(l)}$ | −57.02 |
| L-Alanine$_{(s)}$ | −134.17 | Propane$_{(g)}$ | −24.82 |
| L-Aspartic acid$_{(s)}$ | −233.75 | Propylene$_{(g)}$ | 4.88 |
| Benzene$_{(l)}$ | 12.3 | Succinic acid$_{(s)}$ | −224.92 |
| Benzoic acid$_{(s)}$ | −91.7 | Urea$_{(s)}$ | −79.63 |
| n-Butane$_{(g)}$ | −30.15 | Urea $_{(aq,\ hypothetical\ 1\ m)}$ | −76.30 |
| n-Butanol$_{(l)}$ | −79.59 | $CO_{(g)}$ | −26.42 |
| 1-Butene$_{(g)}$ | −0.03 | $CO_{2(g)}$ | −94.05 |
| n-Butyric acid$_{(l)}$ | −128.09 | $CO_{2(aq,\ hypothetical\ 1\ m)}$ | −98.69 |
| Cyclohexane$_{(l)}$ | −37.34 | $H_2O_{(l)}$ | −68.32 |
| Ethane$_{(g)}$ | −20.24 | $H_2O_{(g)}$ | −57.80 |
| Ethanol$_{(l)}$ | −66.36 | $NH_{3(g)}$ | −11.04 |
| Ethylene$_{(g)}$ | 12.50 | $NH_{3(aq,\ hypothetical\ 1\ m)}$ | −19.32 |
| Fumaric acid$_{(s)}$ | −194.13 | $SO_{2(g)}$ | −70.96 |
| α-D-glucose$_{(s)}$ | −304.64 | $NaCl_{(s)}$ | −98.23 |
| L-Glutamic acid$_{(s)}$ | −240.38 | $HCl_{(g)}$ | −22.02 |

*Sources:* Data for L-alanine, L-aspartic acid, fumaric acid, D-glucose, L-glutamic acid, glycerol, and succinic acid are quoted from Burton, K., and H. A. Krebs, *Biochem. J.,* **54**:94, 1953. Original sources are given by the authors. Other data are quoted from Rossini, F. D., *et al.,* National Bureau of Standards Circular No. 500, *Selected Values of Chemical Thermodynamic Properties*, 1952, and from Rossini, F. D., *et al., Selected Values of Physical and Thermodynamic Properties of Hydrocarbons and Related Compounds*, American Petroleum Institute Research Project 44, 1953. Data for organic compounds are based on graphite as the standard state of carbon.

*Note:* Subscripts accompanying the names of compounds refer to the physical state of the substance: (g) gas, (l) liquid, (s) crystalline solid, (aq) aqueous solution.

A more traditional method for determining $\Delta H_{reaction}$ is to calculate it from equilibrium constants determined at two temperatures. This is the method of van't Hoff and will be discussed in Chapter 3, along with other thermodynamic calculations based on equilibrium constants.

The *heat of combustion,* cited (the second example) on page 59 is the enthalpy change for a reaction in which one mole of the substance is completely oxidized. Heats of combustion are available for a great many compounds.* Chemical equations for the oxidation of organic compounds give $CO_2$ and $H_2O$ as the products of combustion. For or-

---

* For these data, as well as for heats of formation, etc., the student is referred to Rossini, F. D., *et al.,* the National Bureau of Standards Circular No. 500, *Selected Values of Chemical Thermodynamic Properties*, 1952.

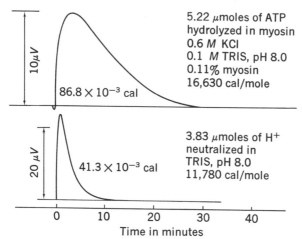

**5.22 μmoles of ATP hydrolyzed in myosin**
**0.6 M KCl**
**0.1 M TRIS, pH 8.0**
**0.11% myosin**
**16,630 cal/mole**

10μV

86.8 × 10⁻³ cal

20 μV

41.3 × 10⁻³ cal

**3.83 μmoles of H⁺ neutralized in TRIS, pH 8.0**
**11,780 cal/mole**

0   10   20   30   40

**Time in minutes**

**Figure 2.7**  Experimental determination of $\Delta H$ of hydrolysis of ATP at pH 8.0. Upper curve: heat change of enzymatic ATP hydrolysis including subsequent neutralization of one proton per ATP molecule. Lower curve: proton neutralization in the same buffer. Difference is heat of reaction of ATP hydrolysis proper at pH 8.0. [Reproduced by permission from Benzinger, T. H., "Heatburst Microcalorimetry," in *Fractions*, No. 2, 1965, published by Spinco Division of Beckman Instruments, Inc.]

ganic compounds containing nitrogen, it is conventional to correct all nitrogen reaction products to nitrogen gas. For example, the equation for the heat of combustion of ethyl amine is written

$$C_2H_5NH_{2(g)} + 3\tfrac{3}{4}O_{2(g)} \rightarrow 2CO_{2(g)} + \tfrac{1}{2}N_{2(g)} + 3\tfrac{1}{2}H_2O_{(l)}.$$

Heats of combustion are determined in the bomb calorimeter.

The caloric values of food, familiar to every weight-watcher, are corrected heats of combustion expressed as *kilocalories per gram*. In the areas of nutrition and dietetics it is less cumbersome to use the kilocalorie, or as it is sometimes called, the "large calorie." Thus, when we say that sugar has a caloric value of 4 calories per gram, we actually mean 4 kilocalories per gram. On the average, the physiological fuel values for the three principal classes of foodstuffs are: carbohydrates, 4 kcal per gram; proteins, 4 kcal per gram; and fats, 9 kcal per gram.

The maximum caloric value of food mixtures for which chemical formulas or exact chemical composition cannot be given can nevertheless be determined by oxidation. Heats of combustion of this type do not, however, represent the physiologically available energy, because the products of oxidation in the body will not necessarily be $CO_2$, $H_2O$, and $N_2$. Most of the nitrogenous waste of the body is excreted in the

**TABLE 2.2** RELATIONSHIP BETWEEN HEAT OF REACTION AT CONSTANT VOLUME ($\Delta E$) AND HEAT OF REACTION AT CONSTANT PRESSURE ($\Delta H$)

| Example | $\Delta n$ | $\Delta E$ and $\Delta H$ relationship |
|---|---|---|
| $2CO_{(g)} + O_{2(g)} \rightarrow 2CO_{2(g)}$ | $-1$ | $\Delta H = \Delta E - RT$ |
| | | $\Delta H = \Delta E - 592$ cal. (25°C) |
| $2HI_{(g)} \rightarrow H_{2(g)} + I_{2(g)}$ | $0$ | $\Delta H = \Delta E$ |
| $C_6H_{12}O_{6(s)} \rightarrow 2C_2H_5OH_{(l)} + 2CO_{2(g)}$ | $+2$ | $\Delta H = \Delta E + 2RT$ |
| | | $\Delta H = \Delta E + 1184$ cal. (25°C) |
| $C_6H_{12}O_{6(s)} + 6O_{2(g)} \rightarrow 6CO_{2(g)} + 6H_2O_{(l)}$ | $0$ | $\Delta H = \Delta E$ |

form of urea, and other incompletely oxidized compounds appear in the urine and feces.

An exact thermochemical analysis of physiological oxidation in living organisms must take many factors into account, and the calculations are complicated.[*] Calorimeters large enough to hold human beings and larger animals were first built at Pennsylvania State University and were used in pioneering experiments designed to determine with exactness the heat lost from living bodies during various types of normal activity. Information of this kind is vital to a precise knowledge of animal bodies as converters of matter and energy. Much of the available energy in food is trapped in organisms as chemical energy rather than released as heat energy. The energetics of living cells will be discussed further in connection with $\Delta G$, the free energy, in Chapter 3.

We have already seen that the heat of reaction at constant pressure is $q_P$ or $\Delta H$, and that the heat of reaction at constant volume is $q_V$ or $\Delta E$. Since at constant pressure

$$\Delta H = \Delta E + P\,\Delta V = \Delta E + \Delta n\,RT,$$

heats of reaction obtained in a bomb calorimeter ($\Delta E$) can be converted to heats of reaction at constant pressure ($\Delta H$). A number of examples are given in Table 2.2. Bear in mind that $\Delta n =$ (number of moles of gaseous products − number of moles of gaseous reactants). At 25°C, $\Delta n\,RT$ for $\Delta n = 1$, will be 592 calories.

## 2.3 The Second Law of Thermodynamics

**The entropy concept and a search for a new state function** The Second Law is a statement of man's recognition that all systems spontaneously change in such a manner as to decrease their capacity for

---

[*] Brody, S., *Bioenergetics and Growth*. Reinhold, New York, 1945. Kleiber, M., *The Fire of Life*. John Wiley and Sons, New York, 1961.

change, *i.e.*, to approach a condition of equilibrium. Our experience contains countless examples of this phenomenon: water runs downhill through resistive channels or in some instances cascading off of rocks and other physical impediments, heat flows from a higher temperature to a lower one, gas spontaneously expands from a higher pressure against a lower external pressure, electric charge moves through resistance from a higher to a lower potential, solute distributes itself uniformly throughout a solution, chemical reactions proceed to a position of minimum change in thermodynamic properties, and living organisms (after reaching a certain age) begin spontaneously to undergo the process of aging with its accompanying degeneration.[*]

The *entropy* (Gr. *entrope*; from *en* in + *trope* turning) of a system is a measure of the degree to which it is removed from equilibrium, but in an inverse sense. That is, a coiled spring has a low entropy; the same spring uncoiled has a high entropy. In this instance, it is the coiled state which is at issue, not the increased potential energy gained through compression of the spring. If a system at constant energy has a high capacity for change, its entropy is low; whereas if a system is at equilibrium, its entropy is relatively high. We will next consider the entropy concept in the symbolism of thermodynamics.

In our attempt to find a mathematical definition of the entropy change in terms of thermodynamic quantities, let us examine the several examples just cited as spontaneous processes. Significantly, some of the processes could occur without the performance of any work whatsoever; therefore, we conclude that, in general, the doing of work is not a suitable criterion for spontaneous change. Of equal importance is the observation that spontaneous processes may occur with a positive, negative, or zero change in internal energy ($\Delta E$). For the isothermal mixing of gases, $\Delta E$ is zero. On the other hand, the spontaneous dissolving of ammonium sulfate in water is a process for which $\Delta E$ is positive. The important point is that the sign of $\Delta E$ is no measure of spontaneity or of the entropy change.

We intuitively feel that entropy should be a state function, that is, the capacity of a system to undergo change should depend only on its properties and not on its previous history. After all these considerations we are left with only $q$ to examine as a possible criterion for spontaneity. We now recall with some concern that $q$ is not a state function and that $dq$ is not a perfect differential. In this apparent dilemma we turn to a modern statement of the Principle of Clausius: "Heat cannot spontaneously flow from a lower to a higher temperature without the expenditure of work on the system." Here we encounter the idea that the temperature may play a role in the thermodynamic criterion of spontaneity.

---

[*] For more details about the last example see Brillouin, L., *Amer. Sci.*, **37**:554, 1949.

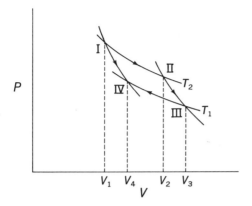

**Figure 2.8**  Carnot cycle.

Heat partially can be transformed into work by an engine operating between two heat reservoirs at different temperatures. In the early part of the nineteenth century scientists and engineers alike were plagued by the question of whether there existed a maximum efficiency for an engine, and if so, what that efficiency might be. A young engineer in Napoleon's army, Sadi Carnot, presented the first successful and valid analysis of the efficiency of an engine working between a high temperature reservoir and a low temperature sink. The entire process is treated as being reversible, for only in this case can maximum work be obtained. We must keep in mind that the process is an abstraction, that is, an imaginary process. An illustration of Carnot's "engine" follows. A sample of gas in the engine constitutes the system; the rest of the engine and everything else around constitutes the environment.

Let us begin by examining Fig. 2.8. We see a diagram composed of four curved sections joined together. If we go around the figure from I to IV and then return to our starting point we will have completed a cyclic process consisting of the following steps: (1) an isothermal expansion, (2) an adiabatic expansion, (3) an isothermal compression, and (4) an adiabatic compression. Our goal is to calculate the net work which is done when the process is carried out reversibly and to relate this information to the maximum efficiency of the engine. Since we have already considered both the isothermal and adiabatic processes earlier in this chapter, we need not develop any new equations for our mathematical analysis. We will use Eqs. 2.4 and 2.7 at the appropriate times.

The working substance of Fig. 2.8 is one mole of an ideal gas. The four steps of the process considered in detail are as follows: (1) the gas, initially in state I, is expanded isothermally and reversibly from I to II at $T_2$ with the absorption of an amount of heat, $q_2$; the work done *on the surroundings* is defined according to our conventions as $-w_1$; since $\Delta E = 0$ for the isothermal, reversible expansion of a perfect gas, $q_2 = -w_1$; (2) the gas is then expanded reversibly and adiabatically from II to III;

the work done on the surroundings is $-w_2$, $q = 0$; and $-\Delta E_1 = -w_2 = \bar{C}_V(T_2 - T_1)$; (3) as the first step in the restoration the gas is compressed from III to IV reversibly and isothermally; the work done by the surroundings on the gas is $w_3$; since $\Delta E = 0$ for this step also, $w_3 = -q_1$, where $-q_1$ is the heat given off to the surroundings; (4) the last stage of the cycle is the adiabatic reversible compression from IV to I; the work done by the surroundings on the gas is $w_4$, $q = 0$, and $\Delta E_2 = w_4 = \bar{C}_V(T_2 - T_1)$.

Now the *net* work done by the gas on the surroundings is the gross work done on the surroundings less the gross work done by the surroundings on the gas, or

$$-w_{\text{net}} = [(-w_1) + (-w_2)] - [(w_3) + (w_4)].$$

Since $-\Delta E_1 = \Delta E_2$, then $-w_2 = w_4$, and therefore,

$$-w_{\text{net}} = (-w_1 - w_3) = (q_2) - (-q_1) = (q_2 + q_1).$$

The efficiency of the engine is, by definition

$$\frac{-w_{\text{net}}}{q_2} = \frac{q_2 + q_1}{q_2}. \tag{2.18}$$

We wish to reconsider the Carnot cycle in terms of the behavior of the ideal gas:

for step (1), $-w_1 = RT_2 \ln(V_2/V_1) = q_2$     (this is Eq. 2.4);

for step (2), $-w_2 = \bar{C}_V(T_2 - T_1)$     (this is Eq. 2.7);

for step (3), $w_3 = -RT_1 \ln(V_4/V_3) = -q_1$     (Eq. 2.4 again);

for step (4), $w_4 = \bar{C}_V(T_2 - T_1)$     (Eq. 2.7 again);

and

$$-w_{\text{net}} = RT_2 \ln(V_2/V_1) + RT_1 \ln(V_4/V_3).$$

From Eq. 2.8 it can be shown that $(V_2/V_1) = (V_3/V_4)$ and therefore,

$$-w_{\text{net}} = R(T_2 - T_1) \ln(V_2/V_1),$$

and

$$\text{efficiency} = \frac{-w_{\text{net}}}{q_2} = \frac{R(T_2 - T_1) \ln(V_2/V_1)}{RT_2 \ln(V_2/V_1)} = \frac{T_2 - T_1}{T_2}. \tag{2.19}$$

The Carnot cycle is of considerable importance to engineers and all who are concerned with heat engines or refrigerators (the latter is simply the cycle run in reverse). From the viewpoint of other science disciplines the Equation of Carnot (2.19) is of interest because of its historic role in the development of thermodynamics and its identification of the thermodynamic zero of temperature. Inspection of Eq. 2.19 reveals that an efficiency of unity could result only if $T_1$ is zero. Lord Kelvin is credited with demonstrating that the thermodynamic temperature scale with zero so defined is indeed the same as the absolute temperature scale defined through the behavior of ideal gases. In recognition of his contribution, the absolute temperature scale is now called the Kelvin scale.

Comparison of Equations 2.18 and 2.19 yields

$$\frac{q_2 + q_1}{q_2} = \frac{T_2 - T_1}{T_2},$$ (2.20)

or

$$\frac{q_1}{T_1} + \frac{q_2}{T_2} = 0.$$ (2.21)

We have not forgotten our suggestion that perhaps there is some function involving $q$ and $T$ that will properly express the entropy in thermodynamic terms. From Eq. 2.21 we conclude that the ratio of $q_{rev}$ to the absolute temperature must be a state function, since $\Sigma(q_{rev}/T) = 0$ for a cyclic process. Remember that one of the criteria for a state function is that $\oint d(\text{function}) = 0$.

Our problem now is to determine whether a change in the function, $q_{rev}/T$, is a measure of the extent to which the system has undergone spontaneous change.

Clausius showed that $\oint dq/T$ is less than zero if irreversibility enters into the cycle in any manner (note that $q$ here is not $q_{rev}$). The argument arises from the fact that in cases that have been observed irreversible processes have lower efficiency than do reversible ones, *i.e.*,

(a real process, an   $\dfrac{q_2 + q_1}{q_2} < \dfrac{T_2 - T_1}{T_2}$   (an ideal process,
irreversible one)                                           a reversible one)

and

$$\frac{q_2}{T_2} + \frac{q_1}{T_1} < 0.$$ (2.22)

Therefore:

$$\oint dq/T < 0.$$

If it is assumed that in general irreversible processes are less efficient than reversible ones, then the above inequality may be used to prove that the function, $dq_{rev}/T$, always increases for an irreversible process occurring in an isolated system, and one is led to the second law of thermodynamics. We do so by a reasoning process which now follows.

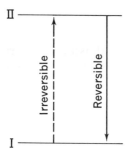

**Figure 2.9** Diagram for imaginary reversible and irreversible processes.

Consider two different states of a system, I and II, which differ in their capacities for spontaneous change. Let an irreversible change occur in the system while isolated from its surroundings such that it moves from I to II by the broken line shown in Fig. 2.9. Then, let the system return from II to I by a reversible path (solid line). During this second change the system interacts with its surroundings and there is a finite value of $\int_{II}^{I} dq_{rev}/T$. For the cyclic process, since irreversibility is a factor, we apply Eq. 2.22 and obtain

$$\int_{I}^{II} dq_{irrev}/T + \int_{II}^{I} dq_{rev}/T < 0. \tag{2.23}$$

Since the system was isolated from its surroundings during the irreversible process, the first term of Eq. 2.23 is zero and the second term is, consequently, less than zero. This can be true only if there is a *decrease* in the value of this function $q_{rev}/T$, in going from state II to state I— which means that the value of $q_{rev}/T$ is higher at II (after the irreversible process) than it is at I (before the irreversible process). It is important to realize that the real processes we observe around us are spontaneous under the circumstances of their occurrence, and that all spontaneous processes are irreversible. Although we cannot completely isolate any system from its surroundings, we often can effectively do so within experimental error and, in the general sense, we presume our universe to be an isolated system. We therefore establish that the thermodynamic criterion for a spontaneous change must be that $\int dq_{rev}/T$ approaches a maximum and define the entropy change for the change of state as equal to this quantity.

The symbol chosen for the entropy is $S$. Unlike $E$ and $H$, $S$ does not have the dimensions of energy, but is energy per degree, since it is the heat energy dissipated at a particular temperature, or is the sum of the increments lost per degree over a temperature range. Thus, $S$ is energy per degree; to obtain an energy term we must multiply it by $T$, *i.e.*, $TS$ has the units of energy. Entropy is, therefore, the capacity factor of heat energy; temperature is the intensity factor.

**Statements of the Second Law of Thermodynamics** The mathematical statement of the Second Law of Thermodynamics is

$$\text{(differential form)} \quad dS = dq_{rev}/T,$$

and

$$\text{(integrated form)} \quad \Delta S = q_{rev}/T. \tag{2.24}$$

Considerations along the lines of thought presented above lead to various verbal statements of the Second Law of Thermodynamics:

1. Efficiencies greater than those provided by a reversible Carnot engine cannot be obtained and Carnot efficiency may only be approached in the limit.
2. It is impossible to achieve perpetual motion of the second kind, *i.e.*, to construct a machine that will convert heat into work isothermally. (Note that such a process is not forbidden by the First Law.)
3. Heat cannot spontaneously flow from a lower to a higher temperature without the expenditure of work on the system. This is the Principle of Clausius, previously stated. Clausius' original postulates, which became known as statements of the First and Second Laws of Thermodynamics, appear under the chapter heading on page 40. Translated, these are: "The energy of the world is constant. The entropy of the world moves toward a maximum."
4. Heat cannot by a cyclic process be taken from a reservoir and converted into work without at the same time delivering heat from the higher temperature to a lower temperature reservoir (William Thomson, Lord Kelvin).

Although the First Law forbids that energy be created from nothing, it is noncommital about the spontaneous direction of energy transformation. The prediction of this direction is the unique contribution of the Second Law. A particularly vivid illustration of this difference is given by Nash.*

If a speeding lead bullet is stopped by an unyielding (and thermally insulated) sheet of armor, all of the *kinetic energy* of the bullet is converted into *internal energy* that manifests itself in a rise of temperature. But we never find that equal bits of lead, heated to the same

*Nash, L. K., *Elements of Chemical Thermodynamics*, p. 29. Addison-Wesley, Reading, Mass., 1962.

temperature, suddenly cool down and move off with the velocity of bullets, although such a development would be perfectly compatible with the first principle of thermodynamics.

The Second Law tells us that the universe must also be "running down" through natural spontaneous processes that make its energy increasingly unavailable for useful work. When these concepts of "entropic doom" were propounded in the latter part of the nineteenth century, they provoked a scientific and philosophical uproar that was nearly a century subsiding. The First Law reassures us of the constancy of "something" we call energy; the Second Law warns us of the ever-increasing unavailability of that energy because of "something else" we call entropy.

**Entropy and probability** The relationship between entropy and probability can readily be seen. Spontaneous processes are those which we observe to happen around us, *i.e.*, they are the events most likely to occur, those with which there is associated a high probability. The entropy expressed as a probability function is

$$S = A \ln W + B, \qquad (2.25)$$

where $W$ is the probability of the event happening and $A$ and $B$ are constants. We can appreciate the logarithmic nature of the equation by recognizing that entropies are additive, but that isolated single probabilities are multiplied together to yield the probability of a final event. For example, let us spread out face-downward a deck of 52 playing cards. What is the chance of drawing a heart? One out of four. What is the chance of drawing an ace? One out of 13. What is the chance of drawing the ace of hearts? $1/4 + 1/13$? No, the chance of drawing the ace of hearts is $1/4 \times 1/13$, or $1/52$. Our experience confirms the correctness of this result. Since entropies are additive, the probability terms, $1/4$ and $1/13$ can be handled in an additive sense if we express them as logarithms: $\log (1/4 \times 1/13) = \log 1/4 + \log 1/13$.

The Second Law is a statistical law and holds only if we observe a large number of events. If we have only half a dozen gas molecules in a box, it is quite possible that at some instant, all of them might be near the same wall in the course of their bounding back and forth. If, however, we have a more normal population in the container, the distribution will be completely random and homogeneous. A group seated in a theater is not the least apprehensive that all of the oxygen molecules might at some instant move simultaneously to the left and leave the right half of the audience gasping for breath. Experience has instilled confidence that such an occurrence, though possible, is so improbable that it is not worth worrying about. Perhaps the point is more clearly illustrated in the tossing of pennies. If we toss 10 pennies 500 times, the distribution of 5 heads and 5 tails is predicted to occur 123 times, or 24.6%, whereas the probability of a 9:1 distribution is 5 out of 500 tosses, or 1.00%. If the number of pennies is increased to 100, the chance that the distribution will be 50-50 is $10^{16}$ times greater than the probability that it will be 90-10. If we extend this idea to a roomful of air containing $10^{26}$ molecules, our expectation that the distribution will be uniform (most probable) is statistically justified.

James Clerk Maxwell (1831–1879) was one of the first to see clearly that the principle embodied in the Carnot engine (a principle that would later be called the Second Law of Thermodynamics) was not of universal validity. Whereas others strove to prove that the principle held with the same sweeping generality as the Law of Conservation of Energy, Maxwell insisted that it was valid only on a statistical basis. The only obstacle to violation of the Second Law was, in Maxwell's words, that man is not clever enough to devise a means of doing so. He conceived of a finite being of miniscule size who could operate a sliding panel in a diaphragm separating two gases, so as to exchange fast-moving molecules for slow ones. Swapping these on a one-for-one basis, the little "demon" would eventually collect the higher energy molecules in one compartment and the lower energy molecules in the other with the result that the hotter system would get hotter and the colder system colder. Figure 2.10 suggests how a Maxwellian demon could air-condition one room while heating another simply by separating molecules of different speed. This would be a perpetual motion machine of the second kind, since the work of air-conditioning would be done simply by utilizing the heat energy of air at constant temperature. The essential point made by Maxwell was that no external work had been done, only the intelligence of a very observant and neat-fingered being had been employed.

For nearly a century the "demon" envoked by Maxwell for the purpose of contradicting the Second Law of Thermodynamics has intrigued both the physical and biological areas of the scientific world. According to W. Ehrenberg[*] in a highly recommended article on this subject,[†] William Thomson postulated that the demon should have some or all of the properties of animation, atomic dimensions, and intelligence; M. Smoluchowski presented the case that the original demon could not be automated but that an intelligent being could operate a perpetual-motion machine of the second kind; Leo Szilard maintained that the intelligence which the demon must possess was a kind of memory and as such, Szilard opened the way leading to information theory; sometime later Leon Brillouin concluded that Maxwell's demon could not operate without producing a net increase in entropy. So the arguments continue.

---

[*] Ehrenberg, W., "Maxwell's Demon." *Scientific American,* November 1967, p. 103.
[†] See also, Klein, Martin J., "Maxwell, His Demon and the Second Law of Thermodynamics." *American Scientist,* Vol. 58, Jan.–Feb., 1970, p. 84.

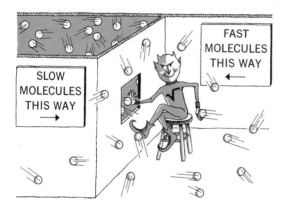

**Figure 2.10** Maxwell's demon separates molecules to air-condition his quarters. Fast and slow moving molecules travel in both directions, but the demon only permits certain ones to go through the hole.

Whatever the results and whatever man's ingenuities, Maxwell's demon has endeared itself to students of thermodynamics because it points up in a humorous and imaginative way the limitations of macroscopic beings in dealing with a world of submicroscopic molecules.

The Second Law is a more general statement of the well-known Laws of Chance. Those individuals who profit from the operation of roulette wheels and dice tables know with certainty that a successful winning streak is the improbable and not the probable event. They are betting on the side of the Second Law, which predicts that the numbers on the wheel or spots on the dice will appear randomly in a statistical number of tries.

**Calculating the entropy change** The mathematical expression of the Second Law was given in Eq. 2.24:

$$S_2 - S_1 = \Delta S = \frac{q_{rev}}{T},$$

which tells us that $\Delta S$, the heat energy per degree that is unavailable for useful work at constant temperature, is equal to the heat absorbed (or evolved) reversibly, divided by the absolute temperature at which the process occurs.

*First case: a change of state at constant temperature* In this instance $\Delta S$ is the molar heat of fusion or vaporization divided by the melting point or boiling point respectively.

*Example 2.6* Calculate the entropy change occurring when one mole of water is converted into one mole of steam reversibly at its boiling point.

$$\Delta S = \frac{q_{rev}}{T} = \frac{(18)(539)}{373} = 26 \text{ cal per degree.}$$

**Second case: a perfect gas changing volume at constant temperature**
We must first recall that $\Delta E = 0$ for an isothermal volume change of this type. We have already established that $q = -w$ and, provided the process is carried out reversibly, we obtain for one mole of gas,

$$-w_{max} = 2.303 \; RT \; \log\!\left(\frac{V_2}{V_1}\right) \qquad \text{(using Eq. 2.4).}$$

Therefore we may write,

$$\Delta S = \frac{q_{rev}}{T} = \frac{-w_{max}}{T} = 2.303 \; R \; \log\!\left(\frac{V_2}{V_1}\right) = 2.303 \; R \; \log\!\left(\frac{P_1}{P_2}\right). \quad (2.26)$$

*Example 2.7* Calculate the entropy change for one mole of a perfect gas reversibly expanding from a volume of 10.0 liters to 20.0 liters at constant $T$.

$$\Delta S = (2.303)(1.987)\left(\log\frac{20.0}{10.0}\right) = 1.38 \text{ cal per degree}$$

In order to emphasize the relationship between probability and entropy, let us obtain the entropy change accompanying the isothermal change in volume for an ideal gas from Eq. 2.25. To begin, we can partition the volumes $V_1$ and $V_2$ into very tiny imaginary cubes, all of identical size. There will be $n_1$ imaginary cubes in $V_1$, and $n_2$ in $V_2$. Since the imaginary cubes are all identical, it follows that

$$\frac{n_2}{n_1} = \frac{V_2}{V_1}. \quad (2.27)$$

Our first objective is to evaluate the number of ways we can arrange the $N$ ideal gas molecules in the $n_1$ imaginary cubes in the initial volume. The first ideal gas molecule which we place in $V_1$ can enter any of the $n_1$ imaginary cubes, and therefore can be placed in this volume $n_1$ ways. Since an ideal gas molecule is assumed to occupy a negligible volume, the second (and all succeeding) ideal gas molecules can also be placed in any of the $n_1$ imaginary cubes. There will be no preference either for or against placing two molecules close together because ideal gas molecules do not exert forces on each other. The total distinguishable number of ways in which the $N$ ideal gas molecules can arrange themselves in the $n_1$ cubes is obtained by multiplying together the number of ways each molecule can exist in these cubes and dividing by $N!$. The division by $N!$ is required because the ideal gas molecules are

$$W_1 = \frac{n_1^N}{N!} \quad (2.28)$$

all identical. By similar reasoning we can obtain the number of ways in which the $N$ ideal gas molecules can arrange themselves in the $n_2$ imaginary cubes in the final volume.

$$W_2 = \frac{n_2^N}{N!}. \tag{2.29}$$

The entropy change can now be evaluated from Eq. 2.25:

$$\Delta S = A \ln W_2 + B - A \ln W_1 - B = A \ln \frac{W_2}{W_1}. \tag{2.30}$$

Substitution from Eqs. 2.28 and 2.29, with cancellation of $N!$, yields

$$\Delta S = A \cdot \ln \frac{n_2^N}{n_1^N} = A \ln \left( \frac{n_2}{n_1} \right)^N. \tag{2.31}$$

Utilization of Eq. 2.27 allows us to introduce the volumes:

$$\Delta S = A \ln \left( \frac{V_2}{V_1} \right)^N. \tag{2.32}$$

In order to facilitate comparison with Eq. 2.26, we can finally bring the exponent before the logarithm and convert from ln to log.

$$\Delta S = 2.303 \, NA \, \log \left( \frac{V_2}{V_1} \right). \tag{2.33}$$

Equation 2.26 is the entropy change which occurs in the isothermal volume change of one mole of ideal gas. It is identical to Eq. 2.33 since $R = NA$ when one mole of an ideal gas is involved, *i.e.*,

$$A = \frac{R}{N}. \tag{2.34}$$

Therefore the constant $A$ in Eq. 2.26 is simply the gas constant per molecule. This constant is called the Boltzmann constant, and is often signified by the symbol $k$.

*Third case: a substance being heated or cooled over a temperature range*   Since the change is not isothermal, we will have to treat it as a series of minute equilibria and integrate all the $dq/T$ terms to obtain the desired mathematical result. As previously, it is necessary to find a way of expressing $q$ in terms of $T$ to perform the integration. For a process in which a substance is heated or cooled over a temperature range, the heat exchange is equal to the heat capacity of the system multiplied by the temperature change. For one mole of substance (as we have previously discussed in connection with $\Delta H$, the enthalpy change)

$$dq = \bar{C}_V \, dT \quad \text{(for a constant volume process)},$$

and

$$dq = \bar{C}_P \, dT \quad \text{(for a constant pressure process)}.$$

These are the differential forms of the relationships presented in Eqs. 2.7 and 2.16. In the integrated forms used earlier we wrote

$$\Delta E = \bar{C}_V \, \Delta T$$

and

$$\Delta H = \bar{C}_P \, \Delta T,$$

subject to the limitation that $C$, the molar heat capacity, was constant over the temperature range $\Delta T$.

Now, however, our mathematical problem is to integrate (at constant $P$)

$$dS = \frac{dq}{T} = \bar{C}_P \frac{dT}{T}.$$

On integration we obtain

$$\Delta S = \bar{C}_P \, \ln\!\left(\frac{T_2}{T_1}\right) \tag{2.35}$$

for one mole of substance, and for $n$ moles of substance

$$\Delta S = n\bar{C}_P \, \ln\!\left(\frac{T_2}{T_1}\right).$$

If $\bar{C}_P$ is of the usual temperature dependent form, $\bar{C}_P = a + bT + cT^2 + \cdots$, the integration will yield a correspondingly complex expression for $\Delta S$.

Let us now consider an example which requires the use of the three kinds of changes presented.

*Example 2.8* Calculate the entropy change for the following process, all steps being carried out reversibly. One mole of ice at 0°C is melted to water and then heated to 100°C, at which temperature it is converted into steam at constant pressure. The steam is then expanded at constant temperature to a volume of 154.0 liters. Assume that the gas behaves ideally and that the pressure is one atm.

Step 1. Melting the ice at 273°K.

$$\Delta S = \frac{q_{\text{rev}}}{T} = \frac{(18)(79)}{273} = 5.2 \text{ cal per degree.}$$

Step 2. Heating the water between 273°–373°K.

$$\Delta S = 2.303 \ \bar{C}_P \ \log\!\left(\frac{T_2}{T_1}\right) = (2.303)\,(18)\!\left(\log\frac{373}{273}\right) = 5.6 \text{ cal per degree.}$$

Step 3. Vaporizing the water at 373°K.

$$\Delta S = \frac{q_{rev}}{T} = \frac{(18)\,(539)}{373} = 26.0 \text{ cal per degree.}$$

Step 4. Expanding the steam isothermally (temperature is held constant at 373°).

$$\Delta S = 2.303 \ R \ \log\!\left(\frac{V_2}{V_1}\right) = (2.303)\,(1.987)\!\left(\log\frac{154}{30.6}\right) = 3.2 \text{ cal per degree.}$$

Note: $V_1$ is obtained by correcting the volume of a mole of gas at one atm and 273°K to the expanded volume at 373°K, *i.e.*, $(22.4)\!\left(\frac{373}{273}\right) = 30.6$ liters.

Step 5. Summation.

$$\Delta S_{process} = 5.2 + 5.6 + 26.0 + 3.2 = 40 \text{ cal per degree.}$$

**The entropy change and predicting the direction of spontaneity** To be able to predict whether a chemical reaction or physical process may proceed spontaneously is of importance and, at times, of economic value to the chemist and chemical engineer. If we know the entropy change for the system and its surroundings, this prediction can be made. In other words, we must be able to measure $\Delta S_{system}$ and $\Delta S_{surroundings}$. We need to know $q$ and $w$ in terms of reversible processes, since they (unlike the state functions) depend on the pathway. If all of these requirements are met, then

if $\Delta S_{system} + \Delta S_{surroundings} = 0$, the system is at equilibrium;

if $\Delta S_{system} + \Delta S_{surroundings} = $ a positive number, the change may occur spontaneously;

if $\Delta S_{system} + \Delta S_{surroundings} = $ a negative number, the change cannot occur spontaneously.

Despite the entropy function's great potential usefulness in this way, the practical difficulties of fulfilling the requirements, *i.e.*, making measurements of reversible changes in the system *and* in the surroundings, make it highly inconvenient to use. For example, plants and animals spontaneously develop into complex organized structures, seemingly in violation of the Second Law of Thermodynamics. However, whereas randomness is decreasing in growing terrestrial organisms, the sun — the ultimate source of energy on earth — is consuming itself with a prodigality that drives it inexorably toward an ultimate entropic doom. Only a fraction of this energy is trapped in the biosphere; the rest radiates to the far reaches of space where, so far as we know, it is forever wasted. Thus, using the entropy function as a criterion of spontaneity often becomes a sheer impossibility because of the requirement that we measure changes in the surroundings. Early workers in the field of thermodynamics soon defined other thermodynamic functions that serve as criteria of spontaneity for changes which may occur in systems under certain restrictions. The most useful of these functions, the free energy, was proposed by J. Willard Gibbs. The free energy function serves as the criterion of spontaneity for changes occurring at constant temperature and pressure and is conveniently evaluated by measuring changes in the properties of the *system alone*. This function, the free energy, will be the subject of Section 2.5.

### 2.4 Absolute Entropies and the Third Law of Thermodynamics

Although it is not possible to know absolute enthalpies for reasons previously discussed, it is possible to calculate absolute entropies. If all atomic motion ceases at $0°K$, all perfect crystalline substances should have zero entropy at this temperature. It is an accepted principle of thermodynamics that absolute zero is absolutely unattainable, although temperatures within $2 \times 10^{-5}$ of a degree of the absolute zero have been achieved. The unattainability of the absolute zero is a consequence of the fact that there is no sink at a lower temperature to which the last minute amounts of heat energy may flow. The principle of the zero entropy of any perfect crystalline substance at absolute zero of temperature is called the Third Law of Thermodynamics and was first suggested by W. Nernst. The Third Law was first clearly stated by G. N. Lewis and M. Randall:

> If the entropy of each element in some crystalline state be taken as zero at the absolute zero of temperature, every substance has a finite positive entropy; but at the absolute zero of temperature the entropy may become zero, and does so become in the case of perfect crystalline substances.

A practical use of the principle is found in the calculation of $S°$ values, standard entropies (25°C), by extrapolating heat capacity data to $0°K$. The molar heat capacity, $\bar{C}$, will not be constant over such a large temperature range. The summation, therefore, will involve integrating and adding together the appropriate $\bar{C}_P \, dT/T$ terms from absolute zero to the desired temperature. By this process we obtain the absolute entropies or "Third Law" entropies for various gases, liquids, and solids. A few of these data have been collected together in Table 2.3. These values represent the entropy change from absolute zero to 298°K. Many difficulties accompany measurements of this type, not the least of them being the anomalous behavior of certain substances as their temperatures approach the absolute zero.

Absolute entropies are used to calculate standard entropies of formation ($\Delta S_f^\circ$) for various compounds of interest. The entropy of formation of a compound is equal to the difference between the absolute entropy of the product and the sum of the absolute entropies of the reactants appearing in the formation equation:

$$\Delta S_f^\circ = \Sigma \ S_{product}^\circ - \Sigma \ S_{reactants}^\circ. \tag{2.36}$$

Note from Table 2.3 that $S^\circ$ values for the elements listed are not equal to zero, as are their heats of formation. In Table 2.4 are given $\Delta S_f^\circ$ for some selected compounds of biological interest.

*Example 2.9* From the data in Table 2.3, prove that $S_f^\circ$ for ethanol is $-82.4$ calories per degree per mole, as given in Table 2.4.

The equation for the formation of ethanol from the elements is

$$2C + 3H_2 + \tfrac{1}{2}O_2 \rightarrow C_2H_5OH.$$

Substituting $S^\circ$ values from Table 2.3 into Eq. 2.36 yields

$$\Delta S_f^\circ = \underset{\text{ethanol}}{38.4} - [\underset{\text{carbon}}{(2)\,(1.361)} + \underset{\text{hydrogen}}{(3)\,(31.21)} + \underset{\text{oxygen}}{(1/2)\,(49.00)}]$$

$$= -82.5 \text{ cal deg}^{-1} \text{ mol}^{-1}$$

## 2.5  The Free Energy

The concept of a free energy function (also called net work or isothermally useful work) was arrived at independently by Gibbs and Helmholtz. The Gibbs free energy, $G$, and the Helmholtz free energy, $A$, each finds its unique usefulness along a different path. The special attractiveness of the Gibbs concept is that it gives us a workable criterion of spontaneity for conditions under which the chemist normally works. As will be shown below, if we know the free energy change at constant temperature and pressure, we can predict whether a reaction will proceed spontaneously, *and,* we can obtain this information from readily measurable quantities.

We will use the symbol $G$ for the Gibbs free energy, although the student should be aware that other symbols are currently used, notably $F$.

The definition of $G$ is

$$G = H - TS. \tag{2.37}$$

**TABLE 2.3** ABSOLUTE ENTROPIES OF SELECTED COMPOUNDS AT 25°C

| Substance | $S°$ in calories per degree per mole | Substance | $S°$ in calories per degree per mole |
|---|---|---|---|
| Acetaldehyde$_{(g)}$ | 63.5 | Propylene$_{(g)}$ | 63.80 |
| Acetic acid$_{(l)}$ | 38.2 | Urea$_{(s)}$ | 25.00 |
| Acetylene$_{(g)}$ | 48.0 | CO$_{(g)}$ | 47.30 |
| Benzene$_{(l)}$ | 41.9 | CO$_{2(g)}$ | 51.06 |
| n-Butane$_{(g)}$ | 74.12 | H$_2$O$_{(l)}$ | 16.72 |
| 1-Butene$_{(g)}$ | 73.04 | H$_2$O$_{(g)}$ | 45.11 |
| Cyclohexane$_{(l)}$ | 48.84 | NH$_{3(g)}$ | 46.01 |
| Ethane$_{(g)}$ | 54.85 | SO$_{2(g)}$ | 59.40 |
| Ethanol$_{(l)}$ | 38.4 | NaCl$_{(s)}$ | 17.30 |
| Ethylene$_{(g)}$ | 52.45 | HCl$_{(g)}$ | 44.62 |
| n-Hexane$_{(l)}$ | 70.72 | C (graphite) | 1.361 |
| Methane$_{(g)}$ | 44.50 | O$_{2(g)}$ | 49.00 |
| Methanol$_{(l)}$ | 30.32 | H$_{2(g)}$ | 31.21 |
| Propane$_{(g)}$ | 64.51 | | |

*Sources:* Data are quoted from Rossini, F. D., *et al.*, National Bureau of Standards Circular No. 500, *Selected Values of Chemical Thermodynamic Properties*, 1952, and from Rossini, F. D., *et al.*, *Selected Values of Physical and Thermodynamic Properties of Hydrocarbons and Related Compounds*, American Petroleum Institute Research Project No. 44, 1953.

*Note:* Subscripts refer to the physical state of the substance: (g) gas, (l) liquid, (s) crystalline solid.

**TABLE 2.4** STANDARD ENTROPIES OF FORMATION OF SELECTED COMPOUNDS AT 25°C

| Substance | $\Delta S_f°$ in calories per degree per mole | Substance | $\Delta S_f°$ in calories per degree per mole |
|---|---|---|---|
| Acetic acid$_{(l)}$ | −76.1 | Ethanol$_{(l)}$ | −82.4 |
| Acetone$_{(l)}$ | −74.32 | Furmaric acid$_{(s)}$ | −126.2 |
| L-Alanine$_{(s)}$ | −153.7 | α-D-Glucose$_{(s)}$ | −291.7 |
| L-Aspartic acid$_{(s)}$ | −194.1 | L-Glutamic acid$_{(s)}$ | −222.4 |
| n-Butanol$_{(l)}$ | −131.5 | Glycerol$_{(l)}$ | −153.6 |
| n-Butyric acid$_{(l)}$ | −125.2 | Succinic acid$_{(s)}$ | −155.1 |

*Source:* Data are quoted from Burton, K., and H. A. Krebs, *Biochem. J.*, **54**:94, 1953. Original sources are given by these authors.

*Note:* Subscripts refer to the physical state of the substance: (l) liquid, (s) crystalline solid.

At constant temperature, a change in the free energy will be

$$\Delta G = \Delta H - T\Delta S. \tag{2.38}$$

What does this equation tell us? At constant $T$, the Gibbs free energy change, which we have already predicted to be the chemist's most useful thermodynamic concept, is dependent upon the enthalpy change, $\Delta H$, and the entropy change, $\Delta S$. In biological systems, both volume and pressure are usually constant, so that $\Delta H = \Delta E$. A *decrease* in enthalpy (or internal energy) plus an *increase* in entropy will both contribute to a decrease in the free energy, which, as we will see, serves to drive a reaction forward in the spontaneous direction.

The free energy defined in this way is the function that can be used to make calculations for systems at constant temperature and pressure because it has the property of a *potential* under these conditions. By a potential function we mean a function of the system which, when differentiated with respect to displacement, provides the force tending to return or drive the system to equilibrium. The free energy is not unique in having the property of a potential. Other state functions are also potential functions under certain specified conditions (paths) along which the system may change. For example, for changes occurring at constant entropy and constant volume, $E$ is the criterion of spontaneity because it has the property of a potential under these conditions. These concepts are most clearly illustrated with open systems (mass and energy may both leave or enter), as will be shown at the beginning of Chapter 3. For the present, we will confine ourselves to very simple closed systems.

We will derive the basic equations which apply to closed systems limited to pressure-volume work. We begin with Eq. 2.37.

$$G = H - TS,$$

and substitute the definition, $H = E + PV$, to obtain

$$G = E + PV - TS,$$

which we differentiate to obtain

$$dG = dE + d(PV) - d(TS),$$

and expand as

$$dG = dE + P\,dV + V\,dP - T\,dS - S\,dT.$$

Substituting the differential form of the First Law, $dE = dq + dw$, we write

$$dG = dq + dw + P\,dV + V\,dP - T\,dS - S\,dT.$$

If the change is reversible, by the Second Law, $dq - T\,dS = 0$, and restricting the system to the work of expansion allows us to say that $dw + P\,dV = 0$. Substituting these relationships simplifies the equation to

$$dG = V\,dP - S\,dT. \tag{2.39}$$

If the system is held at constant temperature, $dT$ is zero, Eq. 2.39 reduces to $dG = V\,dP$, and accordingly to the expression given in Eq. 2.40, in which the subscript $T$ is used to indicate that the temperature is held constant.

$$(\partial G/\partial P)_T = V. \tag{2.40}$$

Similarly, if the pressure is held constant the quantity $dP$ is zero, and we are led to Eq. 2.41, in which the subscript $P$ indicates that the pressure is constant.

$$(\partial G/\partial T)_P = -S. \tag{2.41}$$

These fundamental equations are the basis for useful relationships from which the free energy change may be calculated. Before proceeding, however, let us note other equations similar to Eq. 2.39 which define the conditions under which $H$ and $E$ have the properties of potential functions:

$$dE = T\,dS - P\,dV. \tag{2.42}$$

$$dH = T\,dS + V\,dP. \tag{2.43}$$

For a system at constant entropy and constant volume, $dE = 0$ defines the condition of equilibrium; for a system at constant entropy and constant pressure, $dH = 0$ defines the condition of equilibrium. Similarly, we see from Eq. 2.39 that $dG = 0$ (or, for a macroscopic change, $\Delta G = 0$), will be the condition for equilibrium in a system at constant pressure and constant temperature. The above equations are for closed systems limited to work of expansion.

The sign of $\Delta G$, negative or positive, predicts whether the reaction can proceed spontaneously. The "$\Delta$" quantities such as $\Delta G$, $\Delta H$, $\Delta E$, etc. always refer to the value of the term in the *final* state minus its value in the *initial* state. The approach to equilibrium at constant temperature and pressure always means that $\Delta G$ of the system is approaching zero, that is, decreasing. If a quantity is decreasing, over any stage of the process, its value in a final state minus its value in an initial state will be negative. Since all systems spontaneously approach equilibrium within the restrictions of any imposed conditions (Second Law), the sign of $\Delta G$ at constant temperature and pressure will be negative for a spontaneous process. The terms *endergonic* and *exergonic* have been coined for reactions in which a system gains ($+\Delta G$) or loses ($-\Delta G$) free energy, respectively. A memory device to help the student remember these terms and not confuse them with *endothermic* and *exothermic*, is that the terms related to free energy contain the word "erg" whereas those related to heat of reaction contain the term "therm."

A generalization which should be memorized is given in the following summary: For a system at constant temperature and pressure

when $\Delta G = $ a negative number, the reaction may proceed spontaneously;

when $\Delta G = $ a positive number, the reaction cannot proceed spontaneously;

when $\Delta G = 0$, the system is at equilibrium and can neither gain nor lose free energy.

Now we will proceed to show the usefulness of the free energy concept under conditions of constant temperature and pressure and under the condition that the system is allowed to do all forms of work. Let us write again the differential equation which immediately preceded Eq. 2.39:

$$dG = dq + dw + P\,dV + V\,dP - T\,dS - S\,dT.$$

If we restrict the change in the system to constant $T$ and $P$ and reversibility, we have

$$dG = dq + dw + P\,dV - T\,dS.$$

Introducing the Second Law, $dq_{rev} = T\,dS$, gives

$$dG = dw + P\,dV.$$

By integrating and expressing the change as a *decrease* in free energy, we obtain

$$-\Delta G = -w_{max} - P\,\Delta V. \tag{2.44}$$

Equation 2.44 states that the free energy decrease at constant temperature and pressure is equal to the total work the system can do less $P\Delta V$ work. Work symbolized by $-\Delta G$ is frequently called *net work* or *useful work* and can be chemical, electrical, photochemical, or osmotic work, to cite a few examples. Specifically excluded is the expansion of a gas against constant pressure when this type of work is not harnessed in a *useful* way. An example of such unproductive $P\Delta V$ work is the expansion of air caused by a fire burning in a fireplace. In living systems, the

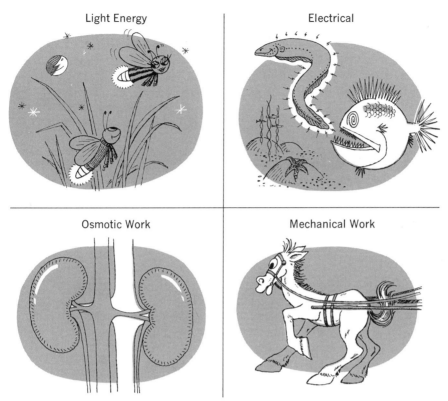

Light Energy

Electrical

Osmotic Work

Mechanical Work

**Figure 2.11**  Some results of useful work ($\Delta G$) in living systems. In (a) the firefly emits light to attract a mate. In (2) the electric eel wards off predators with a lethal electric discharge. Kidneys concentrate solutes against an unfavorable concentration gradient in (3); and in (4) muscle contraction moves heavy loads.

useful work aspect of the free energy change is particularly meaningful, because plant and animal cells do not absorb heat from their surroundings to do mechanical work and thereby to carry on the business of the life processes. In homeothermic animals, particularly, the absorption of heat to do work at constant temperature is forbidden by the Second Law. The work of the cell depends on the chemical energy obtained from absorbed nutrients; this kind of work is to be found in the concept of $G$, the free energy. Figure 2.11 illustrates some of the many types of useful work encountered in living creatures.

Leaving the complexities of living systems for the time being, let us return to our simple imaginary system, the ideal gas. We may readily derive from the basic differential equation we have just been using that for a mole of ideal gas undergoing a volume change reversibly at constant temperature

$$-\Delta G = -w_{\max} -\Delta (PV) . \tag{2.45}$$

Since $\Delta(PV) = 0$ for a gas behaving ideally in an isothermal change,

$$-\Delta G = -w_{max}. \tag{2.46}$$

The system we are discussing is a familiar one by now, and we know that both $\Delta E$ and $\Delta H$ are zero. Therefore, $q_{rev} = -w_{max}$, and since $q_{rev} = T\,\Delta S$,

$$-\Delta G = T\,\Delta S. \tag{2.47}$$

This equation tells us that the free energy decrease accompanying the isothermal expansion of an ideal gas results solely from the increase in *randomness* of the system. Recalling that we have already established for this system that

$$\Delta S = 2.303\ R\ \log(V_2/V_1) \qquad \text{(Eq. 2.26)},$$

we know

$$-\Delta G = 2.303\ RT\ \log(V_2/V_1) = 2.303\ RT\ \log(P_1/P_2). \tag{2.48}$$

Thus, in an isothermal system, we can measure the free energy change of an ideal gas from its pressure change alone. The biologist or biochemist, however, is rarely concerned with measuring $\Delta G$ from pressure changes in living systems. Rather, he is interested in free energy changes in solutions ranging in complexity from physiological saline to protoplasm. It is imperative, therefore, to express the free energy change in terms of the activities of solutes. However, we will defer this problem to Chapter 3, following a consideration of phase equilibria and the laws governing the behavior of solutions.

To illustrate the use of Eqs. 2.47 and 2.48, the following example is included.

> **Example 2.10** Calculate $\Delta E$, $\Delta H$, $\Delta G$, and $\Delta S$ for the reversible isothermal expansion of a perfect gas. Let 56 grams of nitrogen, at 27°C and 10 atm pressure, expand isothermally and reversibly until the pressure is 2 atm. Assume that the gas behaves ideally.
>
> First of all, for an ideal gas changing pressure and volume reciprocally at constant temperature, both $\Delta E$ and $\Delta H$ are zero. $\Delta G$ is calculated from Eqs. 2.47 and 2.48. $\Delta S$ is obtained by the same calculation since
>
> $$\Delta S = -\Delta G/T.$$
>
> By Eq. 2.48
>
> $$\Delta G = 2.303\ RT\ \log(P_2/P_1) = (2.303)\,(1.987)\,(300)\,(\log 2/10)$$
>
> $$= -959\ \text{cal per mole.}$$
>
> However, 56 g of nitrogen is 2 moles, therefore
>
> $$\Delta G_{system} = -1.92\ \text{kcal.}$$

Since this is equal to $-T \Delta S$,

$$\Delta S_{\text{system}} = 1{,}920/300 = 6.4 \text{ calories per degree.}$$

**The Gibbs-Helmholtz Equation** The way in which the free energy depends on temperature at constant pressure has been given in Eq. 2.41.

$$(\partial G/\partial T)_P = -S.$$

Gibbs and Helmholtz independently discovered that this simple relationship led to a useful expression involving the enthalpy. First, Eq. 2.37 is substituted into Eq. 2.41 above:

$$(\partial G/\partial T)_P = \frac{G - H}{T}. \tag{2.49}$$

Taking the values of $G$ and $H$ between their initial and final states, we may also write,

$$(\partial \Delta G/\partial T)_P = \frac{\Delta G - \Delta H}{T}. \tag{2.50}$$

Eq. 2.50 is one of several forms of the Gibbs-Helmholz Equation. Others which are sometimes more conveniently used can be obtained. If $\Delta G/T$ is differentiated with respect to temperature at constant pressure, we get

$$\left[\frac{\partial(\Delta G/T)}{\partial T}\right]_P = -\frac{\Delta G}{T^2} + \frac{1}{T}\left(\frac{\partial \Delta G}{\partial T}\right)_P. \tag{2.51}$$

We can substitute $(\Delta G - \Delta H)/T$ for $(\partial \Delta G/\partial T)_P$ according to Eq. 2.50 and simplify Eq. 2.51 to the following:

$$\left[\frac{\partial(\Delta G/T)}{\partial T}\right]_P = -\frac{\Delta H}{T^2}. \tag{2.52}$$

Since $\partial(1/T)/\partial T = -T^{-2}$, we can arrive at an additional version of the Gibbs-Helmholtz Equation by making this substitution in Eq. 2.52

$$\left[\frac{\partial(\Delta G/T)}{\partial T}\right]_P = \frac{\partial(1/T)}{\partial T}\Delta H, \tag{2.53}$$

which rearranges to give,

$$\left[\frac{\partial(\Delta G/T)}{\partial(1/T)}\right]_P = \Delta H. \tag{2.54}$$

It is apparent that this relationship enables us to calculate $\Delta H$ for a reaction at constant pressure by plotting $\Delta G/T$ versus $1/T$. How do we

**TABLE 2.5** STANDARD FREE ENERGIES OF FORMATION OF SELECTED COMPOUNDS AT 25°C

| Compound | $\Delta G_f^\circ$ in kilocalories per mole | Compound | $\Delta G_f^\circ$ in kilocalories per mole |
|---|---|---|---|
| Acetaldehyde$_{(g)}$ | −31.96 | $\alpha$-D-Glucose$_{(s)}$ | −217.56 |
| Acetic acid$_{(l)}$ | −93.8 | L-Glutamic acid$_{(s)}$ | −174.05 |
| Acetone$_{(l)}$ | −37.18 | Glycerol$_{(l)}$ | −114.02 |
| Acetylene$_{(g)}$ | 50.00 | $n$-Hexane$_{(l)}$ | −1.03 |
| L-Alanine$_{(s)}$ | −88.34 | Methane$_{(g)}$ | −12.14 |
| L-Aspartic acid$_{(s)}$ | −174.88 | Methanol$_{(l)}$ | −39.73 |
| Benzene$_{(l)}$ | 30.99 | Propane$_{(g)}$ | −5.61 |
| Benzoic acid$_{(s)}$ | −61.1 | Propylene$_{(g)}$ | 14.99 |
| $n$-Butane$_{(g)}$ | −4.10 | Urea$_{(s)}$ | −47.12 |
| $n$-Butanol$_{(l)}$ | −40.39 | CO$_{(g)}$ | −32.81 |
| 1-Butene$_{(g)}$ | 17.09 | CO$_{2(g)}$ | −94.26 |
| $n$-Butyric acid$_{(l)}$ | −90.65 | H$_2$O$_{(l)}$ | −56.69 |
| Cyclohexane$_{(l)}$ | 6.37 | H$_2$O$_{(g)}$ | −54.64 |
| Ethane$_{(g)}$ | −7.86 | NH$_{3(g)}$ | −3.98 |
| Ethanol$_{(l)}$ | −41.77 | SO$_{2(g)}$ | −71.79 |
| Ethylene$_{(g)}$ | 16.28 | NaCl$_{(s)}$ | −91.79 |
| Formaldehyde$_{(g)}$ | −26.30 | HCl$_{(g)}$ | −22.77 |
| Fumaric acid$_{(s)}$ | −156.49 | | |

*Sources:* Data for L-alanine, L-aspartic acid, fumaric acid, D-glucose, L-glutamic acid, and glycerol are quoted from Burton, K., and H. A. Krebs in *Biochem. J.*, **54**:94, 1953. Original sources are given by the authors. Other data are quoted from Rossini, F. D., *et al.*, National Bureau of Standards Circular No. 500, *Selected Values of Chemical Thermodynamic Properties*, 1952, and Rossini, F. D., *et al.*, *Selected Values of Physical and Thermodynamic Properties of Hydrocarbons and Related Compounds*, American Petroleum Institute Research Project 44, 1953. Data for organic compounds are based on graphite as the standard state of carbon.

*Note:* Subscripts refer to the physical state of the substance: (g) gas, (l) liquid, (s) crystalline solid.

obtain $\Delta G$ for a series of different constant temperatures? This can be done by measuring the equilibrium constant of the reaction, but we will defer this topic until Chapter 3.

**The free energy of formation** In Section 2.2 the Laws of Thermochemistry were stated and applied to the problem of determining heats of formation and heats of reaction. For $\Delta G$ of formation and $\Delta G$ of reaction a completely analogous situation exists. We may write

$$\Delta G_{reaction} = \Sigma \, \Delta G_{formation\ of\ products} - \Sigma \, \Delta G_{formation\ of\ reactants}, \quad (2.55)$$

which we compare to Eq. 2.17. The free energy of formation of any element in its standard state is taken as zero. Data for the standard free energies of formation of a number of compounds are given in Tables

**TABLE 2.6** STANDARD FREE ENERGIES OF FORMATION OF SELECTED COMPOUNDS AND IONS IN AQUEOUS SOLUTION AT 1 M ACTIVITY AT 25°C

| Substance | $\Delta G_f^\circ$ in kilocalories per mole | Substance | $\Delta G_f^\circ$ in kilocalories per mole |
|---|---|---|---|
| Acetaldehyde | −33.38 | Glyoxylate⁻ | −112.0 |
| Acetic acid | −95.38 | $\beta$-Hydroxybutyric acid | −127.00 |
| Acetate⁻ | −88.99 | $\beta$-Hydroxybutyrate⁻ | −121.00 |
| Acetoacetate⁻ | −118.00 | Isocitrate³⁻ | −277.65 |
| Acetone | −38.52 | $\alpha$-Ketoglutarate²⁻ | −190.62 |
| cis-Aconitate³⁻ | −220.51 | Lactate⁻ | −123.76 |
| L-Alanine | −88.75 | $\alpha$-Lactose | −362.15 |
| DL-Alanylglycine | −114.57 | $\beta$-Lactose | −375.26 |
| L-Aspartic acid | −172.31 | L-Leucine | −81.68 |
| L-Aspartate⁺⁻⁻ | −166.99 | DL-Leucylglycine | −110.90 |
| n-Butanol | −41.07 | Mannitol | −225.29 |
| n-Butyric acid | −90.86 | Malate²⁻ | −201.98 |
| n-Butyrate⁻ | −84.28 | $\beta$-Maltose | −357.80 |
| Citrate³⁻ | −279.24 | Methanol | −41.88 |
| Cysteine | −81.21 | Oxalacetate²⁻ | −190.53 |
| Cystine | −159.00 | Oxalate²⁻ | −161.3 |
| Ethanol | −43.39 | n-Propanol | −42.02 |
| Formaldehyde | −31.2 | 2-Propanol | −44.44 |
| Formic acid | −85.1 | Pyruvate⁻ | −113.44 |
| Formate⁻ | −80.0 | Succinic acid | −178.39 |
| Fructose | −218.78 | Succinate²⁻ | −164.97 |
| Fumaric acid | −154.67 | Sucrose | −370.90 |
| Fumarate²⁻ | −144.41 | L-Threonine | −123.0 |
| $\alpha$-D-Galactose | −220.73 | L-Tyrosine | −92.55 |
| $\alpha$-D-Glucose | −219.22 | Urea | −48.72 |
| L-Glutamic acid | −171.76 | $CO_2$ | −92.31 |
| L-Glutamate⁺⁻⁻ | −165.87 | $HCO_3^-$ | −140.31 |
| Glycerol | −116.76 | $NH_4^+$ | −19.00 |
| Glycine | −89.26 | $OH^-$ | −37.60 |
| Glycogen (per glucose unit) | −158.3 | $H^+$ | 0.00 |
| Glycolate⁻ | −126.9 | | |

Sources: Data for DL-Alanylglycine and DL-Leucylglycine are quoted from Edsall, J. T., and J. Wyman, *Biophysical Chemistry I*, pp. 236–237. Academic Press, New York, 1958. Original sources are given by the authors. All other data are quoted from K. Burton, in Krebs, H. A., and H. L. Kornberg, *Ergeb. Physiol. biol. Chem. u. exptl. Pharmakol.*, **49**:212, 1957. Additional data are found in this reference.

2.5 and 2.6. Using these data, we can calculate free energy changes for reactions that cannot be studied experimentally or do not lend themselves to other methods of obtaining $\Delta G$. The use of Tables 2.5 and 2.6 is illustrated in Example 2.11. In Chapter 3 we will learn how to calculate $\Delta G$ from the equilibrium constant of a reaction. Sometimes, however, the equilibrium position of a reaction lies so far in one direction that $K_{equil}$ is not known with certainty and $\Delta G$ calculated from such a $K_{equil}$ is correspondingly uncertain.

> *Example 2.11* Calculation of $\Delta G°$ for a reaction from $\Delta G_f°$ data. Acetobacter are used commercially for the synthesis of dilute acetic acid from ethanol. A dilute solution of ethanol is allowed to trickle down over beechwood shavings that have been inoculated with a culture of the bacteria. Air is forced through the vat countercurrent to the alcohol flow. The net chemical reaction is
>
> $$C_2H_5OH_{(aq)} + O_2 \rightarrow CH_3COOH_{(aq)} + H_2O_{(l)}.$$
>
> Applying Eq. 2.48 to the data of Tables 2.5 and 2.6 gives
>
> $$\Delta G° = [-56.69 + (-95.38)] - [-43.39 + 0] = -108.7 \text{ kcal/mole}.$$
>
> (Remember that $\Delta G_f°$ of any element in its standard state is zero.)

## 2.6 Thermodynamic Studies on the Structure of Water

Several distinctive properties of water allow it to play an important role in biological processes. First, the dielectric constant ($D$) of water is one of the highest known. Since the interionic force of attraction varies inversely with $D$ (as explained in Section 1.1), the attraction between ions is diminished when they are dissolved in water, and, as a result, many ionic compounds are very water-soluble. Second, water has an unusually high heat capacity, a major factor in temperature control in homeothermic animals, as well as being of importance in protecting plants from adverse effects of fluctuating temperature. Third, the high heat of vaporization of water enables man and certain other animals to avoid overheating by evaporation of water from the skin. Animals which do not sweat appreciably can evaporate water from the surface of the tongue by panting. With few exception, the evaporation of water from plants during transpiration is essential for maintaining reasonable leaf temperatures. Other properties of water such as high surface tension, low viscosity, high melting and boiling points, and the higher density of its liquid state as compared with its solid state are also worthy of mention.

While these important properties of water are essential to biological processes, they serve to complicate attempts to treat quantitatively the thermodynamics of the liquid state. A second major handicap is the lack of a satisfactory general equation for the liquid state. Although several workable equations of state have been developed for gases (the Perfect Gas Equation, the van der Waals Equation, etc.), no comparable generalizations have been devised for liquids.

An initial turn in the right direction was taken by the theorists who recognized that the molecules of liquid water are associated in some

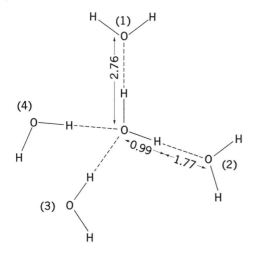

**Figure 2.12** Tetrahedral structure of fully coordinated water. Molecules (1) and (2) as well as the central $H_2O$ molecule lie entirely in the plane of the paper. Molecule (3) lies above this plane, molecule (4) below it, so that oxygens (1), (2), (3), and (4) lie at the corners of a regular tetrahedron. Distances are in angstroms. [From Edsall, J. T., and J. Wyman, *Biophysical Chemistry*, Vol. I, p. 31, Copyright 1958. Reproduced by permission of Academic Press, New York.]

way, possibly in the form of low molecular weight polymers. Modern development of these ideas has led to the postulation of a highly hydrogen-bonded structure full of regions that are ice-like in their ordered network of chains of linked water molecules.

The structure of ice at ordinary pressures is a good basis from which to begin an examination of liquid water. The water molecules in ice are tetrahedrally coordinated by hydrogen bonding as shown in Fig. 2.12. The crystalline array is the same as the arrangment of silicon atoms in the tridymite form of $SiO_2$. A sketch of the ice structure that shows the position of hydrogen and oxygen atoms both as lattice points and in terms of their van der Waals radii is presented in Fig. 2.13. The overall arrangement is a very spacious one and accounts for the density of ice being lower than that of water.

Many models have been proposed to account for the properties of liquid water. These have postualted various degrees of polymerization, hydrogen-bonding, "ice-likeness" — including penetration of non-hydrogen-bonded molecules into the centers of hexagonally packed water molecules. To summarize several decades of effort let us say that all of these models possessed merit and some measure of usefulness. The most recent summary of the state of our knowledge about the structure of water is a completely fascinating article by Frank.* He has concluded that it seems likely that cold water is structured primarily "of hydrogen-bonded, four-coordinated, framework regions, with interstitial monomers occupying some fraction of the cavities the framework encloses. The precise geometry of the framework has not been specified, but some

---

° Frank, H. S., *Science*, **169**, p. 635–641 (1970).

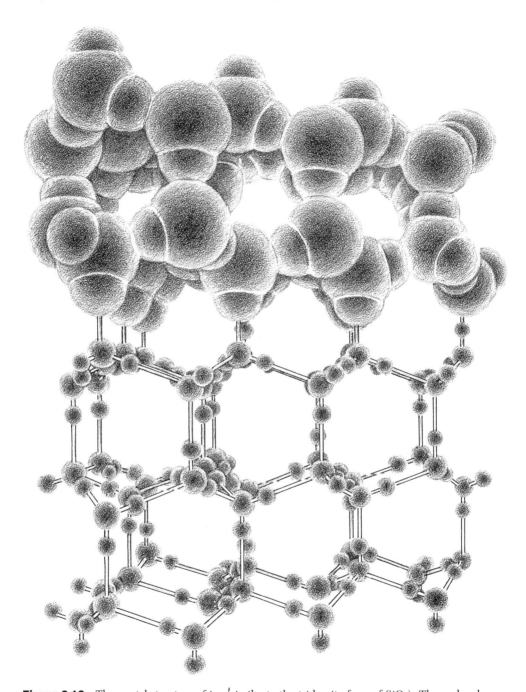

**Figure 2.13** The crystal structure of ice (similar to the tridymite form of $SiO_2$). The molecules are shown with approximately their correct size (relative to the interatomic distances). Note hydrogen bonds and open structure that gives ice its low density. [From *College Chemistry*, 3rd Ed. by Pauling, Linus. W. H. Freeman and Company. Copyright © 1964.]

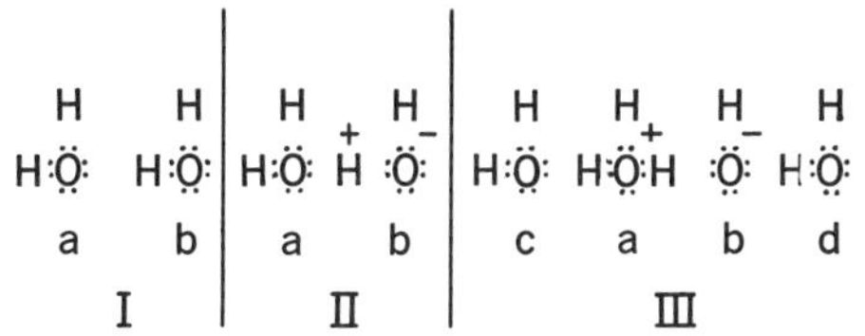

H   H | H ₊ H_ | H   H₊  H_  H
H:Ö:  H:Ö:|H:Ö: H :Ö:|H:Ö: H:Ö:H  :Ö:  H:Ö:
a    b | a    b | c    a    b    d
   I   |   II  |      III

**Figure 2.14** Resonance in hydrogen-bonded water molecules leading to cooperative interaction in hydrogen bonding. [From Frank, H. S., *Proc. Roy. Soc. (London)*, **247A**:482, 1958. Reproduced by permission.]

evidence suggests that it is rather regular at low temperatures and becomes more random as the water gets warmer."

To indicate in some detail how one of the models has proven useful in that it leads to a satisfactory prediction of many of the experimentally established properties of water, we will look at a proposal by Nemethy and Scheraga.* These workers presented a quantitative statistical treatment of the structure of water based on the model proposed by Frank and Wen.† As Frank proposed in the model that is reproduced here as Fig. 2.14, hydrogen bonding in water molecules (I) leads to a partial charge separation (II), which permits the molecules *a* and *b* to form additional hydrogen bonds with their neighbors more easily (III). This leads to bond formation and bond breaking, which involves groups of water molecules cooperatively and produces short-lived "flickering clusters," as Frank and Wen have called them. The "flickering cluster" aspect of the model permits a flexible structure that has the low viscosity characteristics of liquid water.

In addition to the pseudo-ice structure of the clusters of hydrogen-bonded water molecules, one must keep in mind the dipole-dipole and London interactions of the nonbonded water molecules that fill up the spaces between the clusters. A sketch of this arrangement has been prepared by Némethy and Scheraga and is shown in Fig. 2.15. The clusters are believed to have an average lifetime as short as $10^{-10}$ to $10^{-11}$ sec, based on other physical measurements. This dynamic state is the result of local energy fluctuations in the liquid. The whole system tends toward a state of equilibrium in which its free energy will be a minimum. The water molecules will form as many hydrogen bonds as it is possible to form without their bending unduly away from the tetrahedral angle of 109° 28'. Dimers and other small aggregates as well as extended or isolated chains are regarded as being energetically unfavored. In the absence of a more quantitative description of Frank's later model referred to above we note that both involve regions of hydrogen-bonded molecules confused with the presence of interstitial monomers. For a full discussion of these ideas, the student should examine the interesting and highly readable descriptive sections of the references given in the

*Nemethy, G., and H. A. Scheraga, *J. Chem. Phys.*, **36**:3382, 1962.
†Frank, H. S., and W. Y. Wen, *Discussions Faraday Soc.*, **24**:133, 1957.

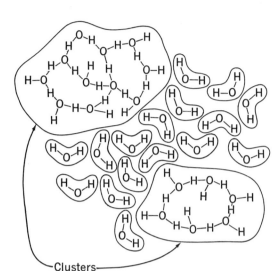

**Figure 2.15** Two "flickering clusters" separated by unbonded water molecules. The molecules in the interior of the clusters are tetracoordinated but not drawn as such in this two-dimensional diagram. [From Némethy, G., and H. A. Scheraga, *J. Chem. Phys.*, **36**:3387, 1962. Reprinted by permission of the American Institute of Physics.]

footnotes below. The mathematical treatment consists of summing up all the energetic interactions of all the species postulated to exist in liquid water so as to arrive at equations which will predict the internal energy, free energy, entropy, heat capacity, molar volume, etc., of liquid water. This is the approach of statistical thermodynamics mentioned in the introductory section of the chapter.

The statistical model successfully accounts for many of the known properties of water and is in agreement with much spectral data. (This model also has some deficiencies, however, which have been carefully pointed out by its authors.) At 20°C the percentage of unbonded water molecules per mole was 29.48, the remaining molecules being distributed among four species having different degrees of bonding. At this temperature 46.2 percent of the hydrogen bonds were calculated to be unbroken. As temperature increased, the average cluster size decreased and the percentage of unbonded molecules increased.

Following their thermodynamic treatment of the structure of water, Némethy and Scheraga used the model to examine the properties of aqueous solutions of hydrocarbons and extrapolated these findings to alkyl sidechain-water interactions in proteins.* A different model system has been proposed by Klotz† for the study of solvent-sidechain interactions in proteins. We will return to this topic in the next chapter following a development of the fundamental equations of heterogeneous equilibrium.

---

*Némethy, G., and H. A. Scheraga, *J. Phys. Chem.*, **66**:1773, 1962; and Scheraga, H. A., in *Forms of Water in Biologic Systems*, edited by Whipple, H. E. Annals of the New York Academy of Sciences, **125**:253–276, 1965.

†Klotz, I. M., "Water," in *Horizons in Biochemistry*, edited by Kasha, M., and B. Pullman, Academic Press, New York, 1962.

The energy of the hydrogen bond in ice is 4.8 kcal mole$^{-1}$, and the value for water is only slightly less. In water some of the hydrogen bonds are broken or bent, and there is additional van der Waals stabilization of water with broken hydrogen bonds because of its closer packing.

## Suggested Additional Reading

### General: Thermodynamics

Bransom, S. H., *Applied Thermodynamics*. D. Van Nostrand, New York, 1961.

Daniels, F., and R. A. Alberty, *Physical Chemistry*, 3rd Ed. John Wiley and Sons, New York, 1966.

Lewis G. N., and M. Randall, *Thermodynamics*, 2nd Ed., revised by Pitzer, K. S., and L. Brewer. McGraw-Hill, New York, 1961.

Mahan, B. H., *Elementary Chemical Thermodynamics*. W. A. Benjamin, New York 1963.

Moore, W. J., *Physical Chemistry*, 4th Ed. Prentice-Hall, Englewood Cliffs, N.J., 1972.

Morowitz, H. J., *Entropy for Biologists*. Academic Press, New York, 1970.

Nash, L. K., *Elements of Chemical Thermodynamics*. Addison-Wesley, Reading, Mass., 1962.

Pimentel, G. C., and R. D. Spratley, *Understanding Chemical Thermodynamics*. Holden-Day, San Francisco, 1969.

Rossini, F. D., *Chemical Thermodynamics*. John Wiley and Sons, New York, 1950.

Wall, F. T., *Chemical Thermodynamics*, 2nd Ed. W. H. Freeman and Company, San Francisco, 1965.

### Biochemical Applications

Bray, H. G., and K. White, *Kinetics and Thermodynamics in Biochemistry*. Academic Press, New York, 1957.

Edsall, J. T., and J. Wyman, *Biophysical Chemistry I*. Academic Press, New York, 1958.

Klotz, I. M., *Energy Changes in Biochemical Reactions*. Academic Press, New York, 1967.

Lehninger, A. L., *Bioenergetics*, 2nd Ed. W. A. Benjamin, New York, 1971.

Patton, A. R., *Biochemical Energetics and Kinetics*. W. B. Saunders, Philadelphia, 1965.

### History of Thermodynamics

Wightmen, W. P. D., *The Growth of Scientific Ideas*. Yale University Press, New Haven, 1953.

### Water

Eisenberg, D., and W. Kauzmann, *The Structure and Properties of Water*. Oxford Univ. Press, 1969.

---

*Pauling, L., *College Chemistry*, Freeman, 1970, p. 432 (see also the Eisenberg and Kauzmann reference cited at the end of the chapter).

## Problems

**2.1** For each of the following systems tell whether $\Delta E$, $q$, and $w$ have values or are equal to zero. Tell whether $q = -w$.
(a) A perfect gas changing volume at constant temperature.
(b) A perfect gas expanding into a vacuum isothermally.
(c) One mole of liquid being converted into one mole of gas at the boiling point of the liquid (isothermal process).

*Answer.*
(a) $\Delta E = 0$; $q = -w \neq 0$.
(b) $\Delta E = q = -w = 0$.
(c) $\Delta E \neq 0$; $q \neq -w$.

**2.2** Calculate $q$, $w$, and $\Delta E$ for the process described by the following steps. First 410 calories of heat are added to a system, then the system does 210 calories of work, and finally 200 calories of heat are given off to the surroundings.

**2.3** Calculate the work done when 2 moles of water are vaporized against a constant pressure of 3 atm. The temperature is 25°C.

*Answer.* $w = -1186$ cal.

**2.4** How much heat must be added to 100 g of ice at $-10°C$ to convert it to water at 75°C?

**2.5** In Section 2.1, Example 2.3 is concerned with an irreversible adiabatic expansion.
(a) Calculate the pressure of the gas at the end of the irreversible expansion.
(b) What will be the final pressure and temperature if the expansion is carried out reversibly to the same final volume?

*Answer.*
(a) 1.5 atm.
(b) 1.0 atm and 183°K.

**2.6** (a) Calculate the standard heat of formation of acetic acid at 25°C:

$$2C_{(s)} + 2H_{2(g)} + O_{2(g)} \rightarrow CH_3COOH_{(l)} \qquad \Delta H_f^\circ = ?$$
from $\hspace{8cm} \Delta H^\circ$

$$CH_3COOH_{(l)} + 2O_{2(g)} \rightarrow$$
$$2CO_{2(g)} + 2H_2O_{(l)} - 208{,}340 \text{ cal./mole } CH_4COOH_{(l)}$$
$$C_{(s)} + O_{2(g)} \rightarrow CO_{2(g)} -94{,}050 \text{ cal./mole } C_{(s)}$$
$$H_{2(g)} + \tfrac{1}{2}O_{2(g)} \rightarrow H_2O_{(l)} - 68{,}320 \text{ cal./mole } H_{2(g)}$$

(b) Name and state the Laws of Thermochemistry you used to solve the problem.

*Answer.* (a) $-116.4$ kcal./mole $CH_3COOH_{(l)}$.

**2.7** On page 64 it is stated that, on the average, carbohydrates have a physiological food value of 4 kcal/g. What is the heat of combustion of $\alpha$-D-glucose, expressed as kcal/g?

*Answer.* $-3.72$ kcal/g.

**2.8** What would be the free energy change, in kcal/g glycogen, for the following process at 25°C?

$$6O_{2(g)} + \text{glycogen (per glucose unit)}_{(1\,M\,\text{activity, aq})} \rightarrow 6CO_{2(g)} + 5H_2O_{(l)}.$$

*Answer.* $-4.26$ kcal/g glycogen.

**2.9** Two possible metabolic fates of $\alpha$-D-glucose are given below. Compare the standard free energy changes at 25°C for these processes. Which process represents the more efficient utilization of $\alpha$-D-glucose?

(a) $\alpha$-D-glucose$_{(1\,M\,\text{activity, aq})} \rightarrow 2$ ethanol$_{(1\,M\,\text{activity, aq})} + 2CO_{2(g)}$
(b) $6O_{2(g)} + \alpha$-D-glucose$_{(1\,M\,\text{activity, aq})} \rightarrow 6CO_{2(g)} + 6H_2O_{(l)}$.

*Answer.* (b) is more efficient since $\Delta G° = -686.5$ kcal/mole $\alpha$-D-glucose in (b) and $-56.08$ kcal/mole $\alpha$-D-glucose in (a).

**2.10** The disaccharide maltose can be hydrolyzed to two molecules of the monosaccharide glucose according to the equation

$$C_{12}H_{22}O_{11(s)} + H_2O_{(l)} = 2C_6H_{12}O_{6(s)}.$$

The following data are available from the "Handbook of Chemistry and Physics," 38th ed., 1956, pp. 1777, 1779.

$$C_6H_{12}O_{6(s)} + 6O_{2(g)} = 6CO_{2(g)} + 6H_2O_{(l)}$$
$$\Delta H° = -673.0 \text{ kcal/mole glucose.}$$

$$C_{12}H_{22}O_{11(s)} + 12O_{2(g)} = 12CO_{2(g)} + 11H_2O_{(l)}$$
$$\Delta H° = -1350.2 \text{ kcal/mole maltose.}$$

(a) Calculate the heat of reaction at constant pressure.

*Answer.* $-4.2$ kcal/mole maltose.

(b) If the hydrolysis were carried out at constant volume, would the heat evolved be greater than, less than, or about the same as the answer to part (a)? Why?

*Answer.* About the same since no gas is involved.

**2.11** Calculate $\Delta H°$ for each of the following reactions at 25°:

(a) $C_2H_{4(g)} + H_2O_{(l)} \rightarrow C_2H_5OH_{(l)}$.

(b) $C_6H_{6(l)} + 3H_{2(g)} \rightarrow C_6H_{12(l)}$.

(c) $2C_2H_{4(g)} \rightarrow C_4H_{8(g)}$ (1-butene).

(d) $CH_{4(g)} + \frac{3}{2}O_{2(g)} \rightarrow CO_{(g)} + 2H_2O_{(l)}$.

(e) $CO_{2(g)} + 2NH_{3(g)} \rightarrow \underset{\text{urea}}{(NH_2)_2CO_{(s)}} + H_2O_{(l)}$.

**2.12** Derive the following expressions:

(a) $\Delta H_{reaction} - \Delta E_{reaction} = \Delta n \, RT$

(b) $\Delta H = 0$ for the isothermal expansion of an ideal gas.

**2.13** Calculate the entropy change when one mole of water is changed from ice at $-5°C$ to steam at $105°C$ and 1 atm pressure.

*Answer.* 37.0 cal per degree.

**2.14** The molar heat capacity at constant pressure of carbon dioxide in cal deg$^{-1}$ mole$^{-1}$ is $6.3957 + 10.1933 \times 10^{-3}T - 35.333 \times 10^{-7}T^2$ over the temperature range 300–1400°K. Calculate the entropy change when carbon dioxide is heated at constant pressure from 500–1000°C.

**2.15** (a) From the data given in Tables 2.1 and 2.5 calculate $\Delta S_f°$ of acetic acid at 25°C.

(b) Calculate the same quantity from the absolute entropies of acetic acid, oxygen, hydrogen and carbon, as given in Table 2.3.

**2.16** Given the chemical reaction

$$HC\equiv CH_{(g)} + H_2O_{(l)} \rightarrow CH_3CHO_{(g)},$$

calculate $\Delta H°$ and $\Delta G°$ at 25° from the data given in Tables 2.1 and 2.5.

*Answer.* $\Delta H° = -25.63$ kcal./mole.
$\Delta G° = -25.27$ kcal./mole.

**2.17** Recent measurements of $\Delta H°$ and $\Delta G°$ for the hydrolysis of ATP have shown the changes in these functions to be $-4800$ and $-7000$ calories/mole respectively at 36°C and physiological pH. Calculate $\Delta S°$ for the same conditions. What is significant about the positive sign of $\Delta S°$?

**2.18** (a) What is the free energy change when 10 g of liquid water is vaporized at 100°C and 1 atm?

(b) What is the sign of the enthalpy change in the process in part (a)?

(c) What is the relationship between the enthalpy and entropy changes in part (a)?

*Answer.*
(a) 0.
(b) positive.
(c) $\Delta H° = (373°)(\Delta S°)$.

**2.19** Suppose that we have the following change taking place in a biological polymer

$$\text{Native State} \rightleftarrows \text{Denatured State},$$

and find that the equilibrium shifts to the right as the reaction temperature is elevated.
(a) According to the principle of Le Chatelier, what would you conclude about the sign of $\Delta H°$?
(b) If, at a temperature of 60°C, $\Delta G°$ is negative and $\Delta H°$ is positive, what must be true of the sign and magnitude of $\Delta S°$? What does this mean in terms of the structure of the polymer?

**2.20** The freezing of liquid water has the following enthalpy changes at 1 atm pressure (values are quoted by I. M. Klotz in "Energy Changes in Biochemical Reactions," Academic, 1967, p. 22):

| T, °C | $\Delta H°$, cal/mole |
|---|---|
| −10 | −1343 |
| 0 | −1436 |
| +10 | −1529 |

(a) What is $\Delta S°$ for the process $H_2O_{(l)} = H_2O_{(s)}$ at 0°C?
(b) Does $\Delta S°$ for the process $H_2O_{(l)} = H_2O_{(s)}$ become more negative or more positive as the temperature increases from −10° to +10°C?
(c) Based on the discussion of water in section 2.6, given an explanation for the sign of $\Delta S°$ and for the temperature dependence of $\Delta S°$.

*Answer.*
(a) $\Delta S° = -5.25$ cal/mole/deg.
(b) $\Delta S°$ becomes more negative as temperature increases.
(c) Formation of a H-bonded lattice on freezing restricts the motion of the molecules, giving a negative $\Delta S°$.

**2.21** For which of the following two reactions should $\Delta S$ be more positive? Why?

(a)

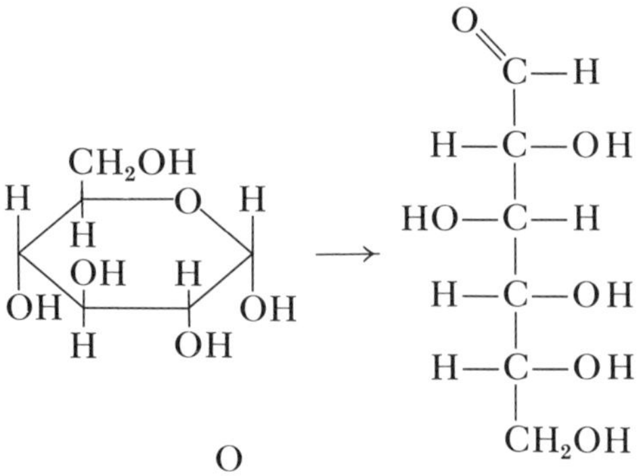

$$
\begin{array}{c}
\text{CH}_2\text{OH} \\
\text{H} \quad \text{O} \quad \text{H} \\
\text{H} \\
\text{OH} \quad \text{H} \\
\text{OH} \qquad \text{OH} \\
\text{H} \quad \text{OH}
\end{array}
\longrightarrow
\begin{array}{c}
\text{O} \\
\text{C}-\text{H} \\
\text{H}-\text{C}-\text{OH} \\
\text{HO}-\text{C}-\text{H} \\
\text{H}-\text{C}-\text{OH} \\
\text{H}-\text{C}-\text{OH} \\
\text{CH}_2\text{OH}
\end{array}
$$

(b) 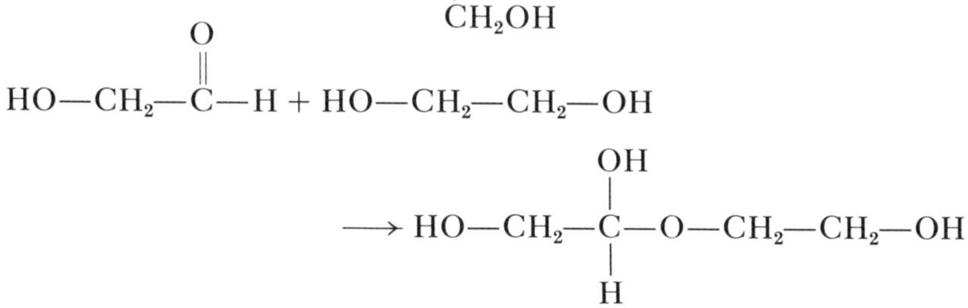
$$
\text{HO}-\text{CH}_2-\overset{\displaystyle \text{O}}{\overset{\|}{\text{C}}}-\text{H} + \text{HO}-\text{CH}_2-\text{CH}_2-\text{OH}
$$

$$
\longrightarrow \text{HO}-\text{CH}_2-\overset{\displaystyle \text{OH}}{\underset{\displaystyle \text{H}}{\text{C}}}-\text{O}-\text{CH}_2-\text{CH}_2-\text{OH}
$$

*Answer.* $\Delta S$ for (a) is positive owing to the loss of rigidity and $\Delta S$ for (b) is negative owing to the loss of independent translational motion of the reactants.

**2.22** The heat of solution of glutamine was studied in the microcalorimeter. In one experiment the reaction was carried out in three successive steps in which 69.0 micromoles of glutamine was diluted to a final concentration of 0.0046 $M$. The heat absorbed was found to be 330 millicalories. In a second experiment the reaction was carried out in only one step: 36.2 micromoles was diluted to a final concentration of 0.0024 $M$ and the heat absorbed was found to be 174 millicalories. Calculate the molar heat of solution for each of the two experiments. What is significant about the data?

**2.23** Using the data given in Table 2.6, calculate $\Delta G°$ for the following reaction at 25°C

$$\text{L-aspartate}^{+--} \rightleftarrows \text{Fumarate}^{2-} + \text{NH}_4^+.$$

*Answer.* +3.58 kcal./mole.

# 3 CHEMICAL EQUILIBRIUM IN HETEROGENEOUS AND HOMOGENEOUS SYSTEMS

The evil demon disappears like the sudden ceasing of the basso parts in music, which hitherto wildly permeated the piece; what before seemed beyond control is now ordered as by magic. . . .

*L. Boltzmann*

Early in Chapter 1 (Table 1.3), we considered various manifestations of energy, some of which are commonplace—for example, heat energy. Others, notably surface energy or chemical energy, are not readily associated with our day-to-day experience and therefore seem more obscure. However, we must now direct our attention to this chemical energy of systems, because it is the very heart of chemical thermodynamics. Furthermore, the useful work that can be obtained from specific combinations of chemical reactants drives the life processes, and these biochemical aspects of thermodynamics fall within the purview of that mushrooming discipline known as molecular biology.

## 3.1 The Chemical Potential

Considered thermodynamically, living cells are *open systems*. As defined earlier, an open system is one in which matter and energy may leave or enter. The fundamental thermodynamic equations which we wrote for closed systems (see page 45) must now be modified by the addition of terms related to the changes in the mass of a system. For the thermodynamic functions, $E$, $H$, and $G$, we write the following differential equations:

$$dE = T\ dS - P\ dV + \mu_1\ dn_1 + \mu_2\ dn_2 + \mu_3\ dn_3 + \cdots, \qquad (3.1)$$
$$dH = T\ dS + V\ dP + \mu_1\ dn_1 + \mu_2\ dn_2 + \mu_3\ dn_3 + \cdots, \qquad (3.2)$$
$$dG = -S\ dT + V\ dP + \mu_1\ dn_1 + \mu_2\ dn_2 + \mu_3\ dn_3 + \cdots, \qquad (3.3)$$

where $\mu$ is called the *chemical potential* and $n$ is the number of moles of an individual component. For each component in the system there will be a $\mu \, dn$ term. Thus, in a system of three components, the number of moles of each would be designated by the symbols $n_1$, $n_2$, and $n_3$. The corresponding chemical potentials would be identified by the notation $\mu_1$, $\mu_2$, and $\mu_3$. Any single component is traditionally referred to as the *i*th component. If we hold constant, in each equation, all variables on the right except $n_1$, we obtain the following set of relationships

$$\mu_1 = \left(\frac{\partial E}{\partial n_1}\right)_{S,V,n_2,n_3,\ldots} = \left(\frac{\partial H}{\partial n_1}\right)_{S,P,n_2,n_3,\ldots} = \left(\frac{\partial G}{\partial n_1}\right)_{T,P,n_2,n_3,\ldots} \tag{3.4}$$

From Eq. 3.1 we can also obtain

$$\mu_1 = -T\left(\frac{\partial S}{\partial n_1}\right)_{E,V,n_2,n_3,\ldots} \tag{3.5}$$

We may arrive at the same relationship in a different way. First we state that the three thermodynamic functions, $E$, $H$, and $G$ are functions of certain variables:

$$E = f(S, V, n_i).$$
$$H = f(S, P, n_i).$$
$$G = f(T, P, n_i).$$

Taking the total derivative of each of these functions, we obtain

$$dE = \left(\frac{\partial E}{\partial S}\right)_{V,n_i} dS + \left(\frac{\partial E}{\partial V}\right)_{S,n_i} dV + \sum_i \left(\frac{\partial E}{\partial n_i}\right)_{S,V,n_j} dn_i, \tag{3.6}$$

$$dH - \left(\frac{\partial H}{\partial S}\right)_{P,n_i} dS + \left(\frac{\partial H}{\partial P}\right)_{S,n_i} dP + \sum_i \left(\frac{\partial H}{\partial n_i}\right)_{S,P,n_j} dn_i, \tag{3.7}$$

$$dG = \left(\frac{\partial G}{\partial T}\right)_{P,n_i} dT + \left(\frac{\partial G}{\partial P}\right)_{T,n_i} dP + \sum_i \left(\frac{\partial G}{\partial n_i}\right)_{T,P,n_j} dn_i. \tag{3.8}$$

We now define a new term, $\mu_i$, the chemical potential, as being

$$\mu_i = \left(\frac{\partial E}{\partial n_i}\right)_{S,V,n_j}$$

Finally, we state that it can be shown

$$\mu_i = \left(\frac{\partial E}{\partial n_i}\right)_{S,V,n_j} = \left(\frac{\partial H}{\partial n_i}\right)_{S,P,n_j} = \left(\frac{\partial G}{\partial n_i}\right)_{T,P,n_j}$$

This is identical with Eq. 3.4.

These equations show that the chemical potential is the partial molar internal energy or enthalpy or free energy, depending on which variables are held constant. Since chemical processes in living cells and under most laboratory conditions occur at constant temperature and pressure, the definition of the chemical potential as the partial molar free energy is the most useful. We may think of the chemical potential as the intensity factor of chemical energy, as it was defined in Table 1.3, or we may think of it as a rate, as shown in Eqs. 3.4 and 3.5. Both concepts are important. For a macroscopic change we write

$$\Delta\mu = \mu_2 - \mu_1 = \Delta G \text{ per mole at constant } P \text{ and } T. \tag{3.9}$$

In our initial description of Gibb's free energy, emphasis was given to its importance as a criterion of equilibrium in systems at constant $T$ and $P$. For such systems, equilibrium is achieved when $\Delta G = 0$. Thus, when the free energy is no longer changing, the chemical potential of any component can no longer be changing. So, we may write as a corollary to our statement concerning equilibrium at constant $T$ and $P$, that $\Delta\mu_i = 0$. The chemical potential of a given component must be equal in all parts of a system at equilibrium. The relationships shown in Eq. 3.4 and Eq. 3.5 should help clarify the point made in Chapter 2 regarding criteria for equilibria. The criterion of equilibrium depends on how the system is defined, that is, on which of its various properties are held invariant. For systems at constant $S$, $V$, and $N$ (total number of moles), $\Delta E = 0$ is the criterion of equilibrium, and it must follow that $\Delta\mu_i = 0$. For systems at constant $S$, $P$, and $N$, $\Delta H = 0$ is the criterion of equilibrium with the corollary that $\Delta\mu_i = 0$. As we will discuss in Section 3.3, the total number of intensive variables required to define a system is fixed by the number of components (chemical individuals) and phases (homogeneous regions within the system). However, choice is available in the selection of variables.

Just as we defined standard changes for the other thermodynamic functions and obtained $\Delta H°$, $\Delta S°$, and $\Delta G°$, we speak of the standard chemical potential, $\mu°$, as the free energy change per mole of substance formed, consumed, or transferred from one phase to another in its standard state—at one atm pressure and a specified temperature and in its standard reference form.

## 3.2  Definitions

**The mole fraction**   The concentration of a solute in a solution can be expressed in various ways, the *molarity, molality,* and *normality* being familiar examples. In many instances it is useful to describe the composition in terms of a percentage. Thus, we have *percent by weight, percent by volume,* and *mole percent.* The mole percent is equal to the number of moles of a particular component per 100 moles of all components. In dealing with the physical chemistry of solutions, the *mole* *fraction* has proved to be equally useful. The mole fraction of any component is its decimal fraction of all the moles present; the sum of all mole fractions must equal unity.

To illustrate: if we have 3 moles of $A$, 2 moles of $B$ and 6 moles of $C$ combined in a homogeneous solution, the mole fraction $A$ will be

$$X_A = \frac{3}{3 + 2 + 6} \quad \text{or} \quad \frac{n_A}{n_A + n_B + n_C} \tag{3.10}$$

where $X$ stands for the mole fraction of a component and $n$ equals the number of moles.

**The escaping tendency**   In describing the behavior of a heterogeneous system such as a liquid in contact with a gas phase, or a solid in the presence of its vapor, we use the term *escaping tendency* to denote the degree to which the molecules of the substance can pass through the interface separating the phases and reach the vapor phase. The word "escaping" invokes a mental picture of the liquid or solid state molecules being imprisoned by their neighbors and held within the boundaries of the phase by some kind of force. On the whole, this is not such a bad analogy, since the forces of attraction between the molecules constitute one of the major determinants of a components' escaping tendency. Other determinants are the kinetic energy of the substance, its shape, size, etc. One of the best measures of the escaping tendency of a given substance is its vapor pressure. The greater the vapor pressure, the higher the escaping tendency of a particular species. If more than one substance is present in the vapor phase, the respective equilibrium vapor pressures are called *partial* vapor pressures, as defined earlier in the discussion of Dalton's Law of Partial Pressures.

**The ideal solution**   Earlier we found it convenient to use the concept of the ideal gas, and it is now advantageous to employ a similar device: the ideal solution. Just as real gases approach ideality under certain conditions of temperature and pressure, so real solutions approach ideality under certain conditions of dilution. The ideal gas is defined as one wherein the molecules possess negligible volume and display no forces of intermolecular attraction. Obviously, we cannot use these same criteria to define the ideal solution, because the molecules of a liquid must have appreciable relative volume and must attract one another for the liquid to exist. The ideal solution is one in which the inherent properties of the solvent and solute are not changed by the presence of new neighbors except as these properties are affected by the dilution involved. No new forces arise and no old ones are canceled. The escaping tendency of the solvent (and the solute, too, if it is volatile) is reduced only to the extent that the solvent molecules are spatially hindered or blocked by the presence of the other component. Properties such as volumes are additive. The temperature does not rise or fall on mixing. Thus, if we have two components, A and B, the forces between all the molecules are the same whether we are considering A and A, B and B, or A and B.

With real solutions, the solvent and solute can markedly affect the escaping tendency of each other by altering the forces of attraction between an individual molecule and its neighbors. If a particular solvent, such as alcohol, has a low escaping tendency because of strong dipole-dipole attraction between its molecules, separating these dipoles from one another by mixing in a nonpolar solute like benzene will permit the polar alcohol molecules to escape from the surface more freely. If we calculate what the vapor pressure of the polar solvent should be, simply on the basis of dilution, we find that the experimentally determined value is higher, *i.e.*, the solvent is behaving nonideally. If we reverse the situation and dissolve a highly polar solute in a nonpolar solvent, the induced attractive forces between solvent and solute will restrict the movement of the nonpolar molecules and reduce their vapor pressure below the calculated value.

We can employ the principle of the limiting law to make useful applications of the laws governing the behavior of ideal solutions. If there are relatively few solute molecules, interactions between solute molecules will be minimal and the effect of solute molecules on the behavior of the solvent will be slight. Hence, the prime requisite in applying the laws for *ideal* solutions to *real* solutions is to restrict application to *dilute* solutions. Mathematically complex equations have been developed for calculating the properties of real solutions of higher solute concentration. These treatments are not entirely satisfactory, however, and are beyond the scope of this presentation.

### 3.3 Principles Governing Heterogeneous Equilibria

Before embarking on an exploration of some relatively simple equations for calculating vapor pressures, freezing points, boiling points, and osmotic pressures of solutions, we must first acquaint ourselves with the two fundamental theoretical equations that provide the foundation on which the limiting equations rest. These two basic relationships are the Phase Rule and the Clausius-Clapeyron Equation.

**The Phase Rule**   In Fig. 3.1 we see what is commonly referred to as a "phase diagram" because it quantitatively describes the behavior of a system which is not homogeneous, but which contains homogeneous regions, known as phases, separated from one another by observable boundaries, sometimes called surfaces of discontinuity. Any single phase is homogeneous, but it does not have to be continuous. For example, ice constitutes a phase whether it is a solid block or in pieces; a finely divided precipitate is also a single phase. The number of phases is the number of separate physical states in a system. There may be, however, more than one liquid phase (*e.g.*, water and benzene) or more than one solid phase (*e.g.*, ice and solid salt). The number of independent chemical individuals which must be specified in order to describe the chemical nature of the system is called the number of *components*. Figure 3.1 is the familiar phase diagram for water: it describes a one-component system. If salt were dissolved in the water, we would obtain a two-component system. A solution of two proteins in a buffer would constitute a five-component system: protein $A$, protein $B$, water, acid, and the salt of the acid. Figure 3.1 is diagrammatic in nature and not drawn to scale.

The study of such heterogeneous systems was greatly simplified by a generalization developed in 1874 by Gibbs. The generalization is known as the Phase Rule and may be stated in a simple way:

$$f = c - p + 2. \tag{3.11}$$

where $c$ is the number of components, $p$ is the number of phases, and $f$ is the number of *degrees of freedom*, that is, the number of independent intensive variables such as temperature, pressure and concentration which must be specified in order to describe the system completely. An alternate statement is that the number of degrees of freedom is the number of independent intensive variables which can be changed without changing the number of phases. The rule applies only to systems in equilibrium, but is valuable in solving practical problems dealing with systems containing many variables. Examples of such complex systems are (1) the various solutions and compounds that result when iron is heated with carbon, (2) a natural deposit such as the Stassfurt salt deposits, and (3) the solid system known as concrete.[*]

The Phase Rule may be derived by a deductive process in which we determine the *minimum* number of variables that must be specified to describe a system. We have already shown that for a system at constant temperature and pressure

$$\mu_i = \left( \frac{\partial G_i}{\partial n_i} \right)_{T,P,n_j}$$

---

[*]A discussion of these examples is beyond the scope of this presentation, but the student who pursues the topic will find it intensely interesting. See the book by Findlay *et al* in the listing in *Suggested Additional Reading*.

If the system is at equilibrium, $\Delta\mu_i = 0$, $\Delta G_i = 0$, and $\Delta n_i = 0$. Therefore, if we specify the temperature, pressure, and composition of each phase, we will obtain the *total* number of variables that describe the system at equilibrium. If there are $p$ phases and $c$ components, there are $pc + 2$ variables. (The 2 is added to include the temperature and pressure.) Our goal now is to reduce the total number to the minimum number. The composition of each phase can be described in terms of the mole fractions of the various components present. Since the sum of the mole fractions in any phase must equal unity, the composition of the phase can be determined if we know the values of all the mole fractions save one. (That one can be calculated by subtracting the others from unity.) If there are $p$ phases, then there are $p$ mole fractions which need not be specified to describe the system. The total number of independent variables is thereby reduced to $(pc + 2) - p$, or $p(c - 1) + 2$.

We can further reduce the total number of independent variables by remembering that at equilibrium the chemical potential of any component must be uniform throughout the entire system. If we designate the various phases by the superscripts $\alpha$, $\beta$, and $\gamma$, then for components 1, 2, and 3, we can write

$$\mu_1^\alpha = \mu_1^\beta = \mu_1^\gamma = \ldots,$$
$$\mu_2^\alpha = \mu_2^\beta = \mu_2^\gamma = \ldots,$$
$$\mu_3^\alpha = \mu_3^\beta = \mu_3^\gamma = \ldots,$$

For each equality sign, we may reduce the number of independent variables by one. We can see from inspection that in any single equation of the type just written, the number of equality signs is equal to $(p - 1)$. For $c$ components, then, we will have $c(p - 1)$ equality signs. If we subtract this term from the total number of independent variables given at the end of the preceding paragraph, we obtain

Minimum number of variables required
$$\text{to describe the system} = p(c - 1) + 2 - c(p - 1)$$

or

$f$, the number of degrees of freedom, $= c - p + 2$
This is the Phase Rule     (Eq. 3.11).

Returning to Fig. 3.1, we may analyze it in the light of the Phase Rule. Since water is the only chemical individual present, $c = 1$. Water can exist in three phases – solid, liquid, and vapor – and the number of these which can coexist in equilibrium depends on the temperature and pressure, as we shall see.

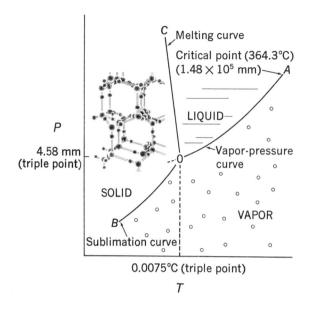

**Figure 3.1** Phase diagram for water (not drawn to scale).

The line $OA$ represents the vapor pressure curve of water. Along $OA$ liquid water and vapor exist in equilibrium. Such a curve is determind experimentally by careful measurements of the vapor pressure of pure water at various temperatures. The line $OB$ is the sublimation curve of ice. Ice lies above the curve and vapor below it. Ice and vapor exist together in equilibrium only along the line $OB$ where the two phases meet. The line $OC$ describes the effect of pressure on the melting point of ice; along the line ice and water exist in equilibrium. Since $OC$ inclines toward the ordinate, we can see that increasing the pressure lowers the melting point of ice. The three curves possess a common intersection, $O$, known as a *triple point*. The student should bear in mind that phase diagrams like Fig. 3.1 are projections of a three-dimensional system (see Fig. 1.6) onto a pressure-temperature plane. The projection is made along the volume axis.

The diagram contains areas, lines, and a triple point. We can make a tabulation to show the corresponding degrees of freedom:

| Regions of the diagram | c | p | f |
|---|---|---|---|
| Areas | 1 | 1 | 2 |
| Lines | 1 | 2 | 1 |
| Triple point | 1 | 3 | 0 |

How do we interpret the various degrees of freedom? In the areas, $f = 2$ means that we must specify both temperature and pressure to describe the system, or that both variables can change without changing the number of phases. In other words, liquid water alone can exist at a variety of temperatures and pressures. Along the curves, $f = 1$ means that we can establish a point on a line (*i.e.*, describe the system) by specifying either temperature or pressure. At any one temperature there is only one possible vapor pressure along either the solid-vapor or liquid-vapor lines; at any one pressure there is only one possible melting point for ice (line *OC*). At the triple point, no degrees of freedom exist, which is to say that the system is invariant and the three phases can exist in quilibrium at only one possible temperature and pressure. The temperature and pressure of the triple point were determined by allowing ice and water to come to equilibrium with vapor in a previously evacuated space. The vapor pressure was found to be 4.58 mm and the temperature, 0.0075°C. If the system is placed under air at one atmosphere pressure, the melting point drops to 0°C, the temperature which is defined by this phenomenon.

The sublimation curve *OB* theoretically extends to absolute zero; the vapor pressure curve *OA* terminates at the critical temperature (364.3°C) and a pressure of 194.6 atmospheres. The upper end of the melting point curve *OC* is complicated by the appearance of various new crystalline modifications of ice which, unlike ordinary ice, are denser than water. These new phases are not shown in Fig. 3.1. We will consider the phase diagram of a two-component system after introducing the second basic equation, the Clausius-Clapeyron Equation.

**The Clausius-Clapeyron Equation**  We can see in Fig. 3.1 that the slopes of the lines *BO* and *OA* in the *P* versus *T* plot must be given by $dP/dT$, which is the rate at which the vapor pressure is changing with the absolute temperature, and the slope of curve *OC*, $dP/dT$, is the reciprocal of the rate at which the melting point changes with pressure. An equation that relates these changes to other measurable properties of the system was proposed by Clapeyron in 1834 and later modified by Clausius. We will derive the equation from the thermodynamic relationships which we have already established for the chemical potential, however, it will be helpful to examine the concept of the *partial molar quantity*. Any property, such as volume, enthalpy, free energy, etc., that depends upon the quantity of substance being considered, is called an *extensive property*. When given per mole of a substance under clearly defined conditions it becomes a characteristic property of that substance under those conditions, and is designated by placing a bar over the particular function or property of the system. The meaning of the volume effectively occupied by one mole of a solute in a solution, called the *partial molar volume* of the solute, may be visualized as the change in total volume of a very large amount of the solution when one additional mole of solute is added. An infinite amount of initial solution would be required to permit this addition without changing the concentration. Mathematically, the partial molar volume $\bar{V}_1$, of substance 1 present in solvent $s$, which may also contain additional solutes 2, 3, ... is given by

$$\bar{V}_1 = \left(\frac{\partial V}{\partial n_1}\right)_{T,P,n_s,n_2,n_3,\ldots}$$

where $V$ is total volume, $T$ is temperature, $P$ is pressure, and $n$ represents the number of moles of the subscript substance present in volume $V$ of solution. $\bar{V}_1$ will be a function of $T$, $P$, and the composition of the solution.

If we have one component distributed between two phases, $\alpha$ and $\beta$, we already know that at equilibrium

$$\mu^\alpha = \mu^\beta.$$

Also, if we have a one-component system, the chemical potential will be equal to the free energy per mole

$$\mu^\alpha = \bar{G}^\alpha \quad \text{and} \quad \mu^\beta = \bar{G}^\beta.$$

The symbol $\bar{G}_a^\alpha$ refers to the Gibbs free energy of component $a$ in phase $\alpha$ and means the "partial molar free energy of $a$ in phase $\alpha$."

If, because of a small change in temperature and pressure, there is a slight change in $\mu^\alpha$, it follows that there will be the same change in $\mu^\beta$ (when phases $\alpha$ and $\beta$ are in contact and at equilibrium under the new conditions), so that we can write

$$d\bar{G}^\alpha = d\bar{G}^\beta.$$

The relationship between $dG$ and changes in temperature and pressure was given much earlier in Section 2.5:

$$dG = V\,dP - S\,dT \qquad \text{(Eq. 2.39)}.$$

Restricting the equation to partial molar quantities, we can substitute the above equation in the one immediately preceding it:

$$\bar{V}^\alpha\,dP - \bar{S}^\alpha\,dT = \bar{V}^\beta\,dP - \bar{S}^\beta\,dT.$$

Rearranging the preceding equation to give the $dP/dT$ term, yields

$$\frac{dP}{dT} = \frac{\bar{S}^\beta - \bar{S}^\alpha}{\bar{V}^\beta - \bar{V}^\alpha} = \frac{\Delta\bar{S}}{\Delta\bar{V}}. \tag{3.12}$$

In Section 2.3, we made calculations for $\Delta S$ associated with phase changes. The value of $\Delta S$ was given by the Second Law, $\Delta S = q_{rev}/T$, in which case $q_{rev}$ was the molar heat of vaporization or molar heat of fusion, according to the system described. This substitution gives

$$\frac{dP}{dT} = \frac{q}{T\,\Delta\bar{V}}. \tag{3.13}$$

So far we have made no simplifying assumptions, and the equation cannot be integrated in its present form. Equation 3.13 is essentially the Clapeyron Equation. The equation can be made considerably more useful, however, if certain assumptions are introduced, as was first done by Clausius.

Restricting ourselves for the moment to liquid-vapor equilibrium, we see that $\Delta V$ is $(\bar{V}_g - \bar{V}_l)$. For water, $\bar{V}_l$ is roughly 0.1 percent of $\bar{V}_g$, and may be neglected in the term $(\bar{V}_g - \bar{V}_l)$. Equation 3.13 then becomes

$$\frac{dP}{dT} = \frac{q}{T\bar{V}_g}. \tag{3.14}$$

Assuming, in addition, that the gas obeys the Perfect Gas Law, we may substitute $RT/P$ for $\bar{V}_g$ to obtain

$$\frac{dP}{dT} = \frac{qP}{RT^2}. \tag{3.15}$$

Whereas the Clapeyron Equation contains three variables, $P$, $V$, and $T$, the Clausius-Clapeyron Equation (3.15) contains only two, $P$ and $T$. We introduce the traditional notation, $\Delta H_V$, for $q$ in Eq. 3.15 to obtain the usual differential form of the Clausius-Clapeyron Equation

$$\frac{dP}{dT} = \frac{\Delta H_V(P)}{RT^2}.$$

The equation may now be integrated if it is assumed that the molar heat of vaporization, $\Delta H_V$, is constant, *i.e.*, is temperature independent. The result of collecting variables and integrating between the limits $(P_1, T_1)$ and $(P_2, T_2)$ is

$$\ln\frac{P_2}{P_1} = -\frac{\Delta H_V}{R}\left(\frac{1}{T_2} - \frac{1}{T_1}\right),$$

which is usually rearranged in the form

$$\ln\frac{P_2}{P_1} = \frac{\Delta H_V(T_2 - T_1)}{RT_2 T_1}. \tag{3.16}$$

This final form of the Clausius-Clapeyron Equation is very much a limiting law since we have imposed all the restrictions of the Ideal

Gas Equation in addition to assuming that $\Delta H_V$ will be constant over the range of $P$ and $T$ chosen.

Now that we have derived the two basic relationships, the Phase Rule and the Clausius-Clapeyron Equation, we are ready to consider some heterogeneous equilibria associated with certain phenomena that are collectively known as the *colligative properties* of solutions.

### 3.4 The Colligative Properties of Solutions

Restricting ourselves to solid solutes with no appreciable vapor pressure, let us examine what happens to a solvent when such solutes are dissolved in it. Certain of its properties are affected and, perhaps oddly, in a way that depends strictly on the number of solute particles added. It would seem that the inherent nature of the solute should have an influence on these changes, that is, that urea, for example, should affect the solvent in a way different from glucose. However, certain properties change in proportion to the concentration of the solute, and are unrelated to its chemical structure, provided that the solute does not react with the solvent and does not dissociate or associate and thereby produce a change in the number of solute molecules or particles. These properties of a solution are called the *colligative* properties. The word colligative means "bound-together" and serves as a class name for these phenomena. Four of these are traditionally discussed as a group, and are: the vapor pressure lowering, the boiling point elevation, the freezing point depression, and the osmotic pressure. The fundamental phenomenon is the vapor pressure lowering and the other three are its direct consequences.

**Raoult's Law** François-Marie Raoult (1830–1901), of the University of Grenoble, was the first experimental scientist to make sufficiently precise measurements to permit a mathematical treatment of the effects of solutes on the physical properties of the solvent. Raoult discovered that the vapor pressure of a solution containing a nonvolatile solute is directly proportional to the concentration of the solvent, *i.e.*, the larger the mole fraction of solvent, the larger the vapor pressure. This generalization has been named Raoult's Law in his honor. We can state it mathematically by saying

$$P = kX_1, \tag{3.17}$$

where $P$ is the vapor pressure of the solution, $k$ is a proportionality constant, and $X_1$ is the mole fraction of solvent. If we consider the situation in which no solute is present and $X_1$, therefore, is equal to unity, the identity of $k$ is revealed: it is the vapor pressure of the pure solvent, $P_0$. We now write

$$P = P_0 X_1. \qquad (3.18)$$

Since the sum of all the mole fractions must equal 1.0, we can write that $X_1 + X_2 = 1.0$, where $X_2$ is the mole fraction of solute. Substituting $(1 - X_2)$ for $X_1$ in the preceding equation and rearranging gives

$$\frac{P_0 - P}{P_0} = X_2. \qquad (3.19)$$

Expressing the equations in words, we can say that the relative lowering of the solvent vapor pressure is equal to the mole fraction of solute, and the vapor pressure of the solution is proportional to the mole fraction of solvent. Equations 3.18 and 3.19 are both rearranged statements of Raoult's Law. Consider the following problem.

*Example 3.1* The vapor pressure of a solution containing a nonvolatile solute at 28°C is 27.371 mm. The vapor pressure of the pure solvent (water) at this temperature is 28.065 mm. What is the mole fraction of solvent? Mole fraction of solute?
Substituting the data into Eq. 3.19 gives

$$X_2 = \frac{0.694}{28.07} = 0.025.$$

Or, from Eq. 3.18:

$$X_1 = \frac{27.37}{28.07} = 0.975.$$

Only one calculation need be made since the sum of the mole fractions is unity.

Raoult's Law holds exactly only for ideal solutions, and holds approximately for dilute solutions. The greater the dilution, the closer the approach to ideality. For dilute solutions we can make a further simplification in Eq. 3.19.
By definition, $X_2 = n_2/(n_1 + n_2)$ where $n_2$ is the number of moles of solute and $n_1$ is the number of moles of solvent. If the solution is dilute, $n_2$ will be small in comparison with $n_1$ (about 1 percent of $n_1$ or less), and $n_2$ can be dropped from the denominator term (not from the numerator!). This leads to

$$\frac{P_0 - P}{P_0} = X_2 = \frac{n_2}{n_1 + n_2} \cong \frac{n_2}{n_1}. \qquad (3.20)$$

The number of moles, $n$, is the weight of a component divided by its gram molecular weight, $w/M$; and $n_2/n_1$ is $w_2/M_2$ divided by $w_1/M_1$, or $w_2M_1/w_1M_2$. Making this substitution yields

$$\frac{P_0 - P}{P_0} = \frac{w_2M_1}{w_1M_2}. \tag{3.21}$$

Now the molality of a solution is defined as

$$m = \frac{w_2/M_2}{w_1/1000} = \frac{1000w_2}{w_1M_2}, \tag{3.22}$$

where $w_1/1000$ is the number of thousand grams of solvent present. We can now substitute $m/1000$ for $w_2/w_1M_2$ in Eq. 3.21 and obtain

$$\frac{P_0 - P}{P_0} = \frac{mM_1}{1000}. \tag{3.23}$$

If the solvent is water, as it always is in physiological solutions, $M_1 = 18$, and the final equation is

$$\frac{P_0 - P}{P_0} = 0.018 \ m. \tag{3.24}$$

Equation 3.24 is a useful simplification of Raoult's Law because concentration of solute is given as a molality rather than as a mole fraction. Molalities are used in the equations for the other colligative properties, and therefore, the latter may all be calculated rapidly from the vapor pressure lowering alone. It should be reemphasized that the use of Eq. 3.24 is restricted to dilute, aqueous solutions.

**Freezing point lowering and boiling point elevation**  We have previously stated that the freezing point lowering and boiling point elevation were the direct result of the vapor pressure lowering. This relationship is best illustrated by a phase diagram.

A two-component system must be considered to explain the colligative properties of solutions in terms of the Phase Rule, since the addition of solute increases the number of chemical individuals to 2. The Phase Rule then predicts

$$f = c - p + 2 = 2 - p + 2 = 4 - p.$$

When only one phase is present, there will be three degrees of freedom: temperature, pressure, and concentration. Since a phase diagram is a two-dimensional projection of a space model along the volume axis, four-dimensional geometry would be required to show concentration as an additional degree of freedom. The problem is solved by combining

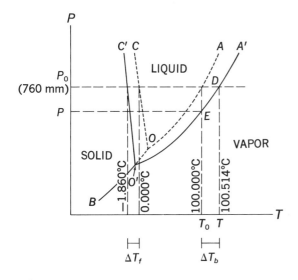

**Figure 3.2** Phase diagram for water and for a hypothetical 1 *m* ideal solution (not drawn to scale).

in one figure the two-dimensional projections for systems in which concentration is fixed, *i.e.*, it is no longer a degree of freedom. Figure 3.2 shows two of these: the dotted lines correspond to the pure solvent (concentration of solute is zero) and the solid lines correspond to a solute concentration of 1.0 *m*. The solution is regarded as behaving ideally. The figure is not drawn to scale. Comparison with Fig. 3.1 will identify the dotted lines as the phase diagram for water.

The possibilities indicated by the Phase Rule may be summarized thus:

| Regions of the diagram | $c$ | $p$ | $f$ |
|---|---|---|---|
| Areas in the $A'O'C'$ plane | 2 | 1 | 3 |
| Lines in the $A'O'C'$ plane | 2 | 2 | 2 |
| The triple point, $O'$ | 2 | 2 | 1 |

No invariant points exist, therefore one variable must always be specified to describe the system. Along the lines, $A'O$ and $O'C'$, concentration is already fixed at 1.0 *m* and either temperature or pressure must also be specified. In the areas in the $A'O'C'$ plane all three variables must be specified to describe the system. An invariant triple point exists on the phase diagram for a *saturated* solution. The pure solute appears as a fourth phase, and when all phases coexist,

$$f = c - p + 2 = 2 - 4 + 2 = 0.$$

Such a cross section has not been shown in Fig. 3.2, but may readily be visualized by imagining a third v-shaped segment lying below $A'O'C'$ and roughly parallel to it.

**TABLE 3.1** FREEZING POINT AND BOILING POINT CONSTANTS

| | °C per molality unit | |
|---|---|---|
| Substance | $K_f$ | $K_b$ |
| Acetic acid | 3.90 | 2.93 |
| Benzene | 5.12 | 2.64 |
| Camphor | 40.0 | . . . |
| Ethanol | . . . | 1.23 |
| Naphthalene | 6.8 | 5.65 |
| Toluene | . . . | 3.37 |
| Water | 1.86 | 0.514 |

Across the top of Fig. 3.2 has been drawn an isobar corresponding to one atmosphere pressure. The intersection of $OC$ with the isobar occurs at 0°C, the freezing point of water; the intersection with $OA$ occurs at 100°C, the boiling point of water. The corresponding intersections with $O'C'$ and $O'A'$ occur at −1.86°C and 100.514°C: Every point on $A'O'C'$ corresponding to a point at the same temperature on $AOC$, lies at a lower pressure. In other words, the lowering of the freezing point of the solution and the elevation of the boiling point are a consequence of the vapor pressure lowering. The point $O'$ lies at the intersection of the vapor pressure curve of the solution with the sublimation curve of pure ice. The temperature at this point will correspond to the melting point of the solution under the equilibrium vapor pressure of the solvent. Since experimental measurements are made at one atmosphere pressure, the observed melting point will correspond to the intersection of $O'C'$ with the one atmosphere isobar, as already noted.

The molal freezing point depression and boiling point elevation for the ideal 1 molal solution are also called the cryoscopic and ebullioscopic constants. They vary with the solvent, as may be seen by inspection of Table 3.1.

The mathematical expressions for the freezing point depression and boiling point rise are

$$\Delta T_f = mK_f, \tag{3.25}$$

where $K_f$ is the freezing point constant, and

$$\Delta T_b = mK_b, \tag{3.26}$$

where $K_b$ is the boiling point constant. As before, $m$ is molality. If we know the vapor pressure lowering, we can readily obtain the freezing point lowering and boiling point rise by calculating the molality from a simplified form of Raoult's Law (Eq. 3.24). This sequence is actually the reverse of the experimental approach. Vapor pressures are exceedingly difficult to determine with accuracy, whereas boiling points and freezing points prove more tractable. In any event, if one of these colligative properties is known, the other two may be calculated.

The freezing point and boiling point constants are determined experimentally, but they can be verified by a theoretical derivation. The following derivation is for the boiling point rise.

We begin with the Clausius-Clapeyron Equation (Eq. 3.16) since it is the fundamental relationship relating vapor pressure to temperature, as depicted by the lines of a phase diagram. In Fig. 3.2 line $O'A'$ is the vapor pressure curve of interest. We will select one point $(D)$ at the intersection of the one atm pressure isobar; the coordinates of this point as shown on the figure are $(T, P_0)$. The second point $(E)$ is chosen at the intersection of the solution curve $(O'A')$ with the $T_0$ line, that is, at the boiling point of the pure solvent. The coordinates of this point are $(T_0, P)$. The notation has been so chosen that $P_0$ and $T_0$ correspond to the vapor pressure and temperature of the pure solvent at its boiling point. Inserting the upper values, $P_0$ and $T$, and the lower values, $P$ and $T_0$, into Eq. 3.16, we obtain

$$\ln\frac{P_0}{P} = \frac{\Delta H(T - T_0)}{RTT_0}.$$

The following simplifications are made:

(a) According to Raoult's Law, for ideal solutions of nonvolatile solutes $P = P_0 X_1$ (Eq. 3.18). Taking the logarithm of both sides of this equation gives $\ln P = \ln P_0 + \ln X_1$; rearranging, $\ln (P_0/P) = -\ln X_1 = -\ln (1 - X_2)$. According to Maclaurin's theorem, $-\ln (1 - X_2) = X_2 + \frac{1}{2}X_2^2 + \frac{1}{3}X_2^3 + \cdots$. Since $X_2$ is much smaller than unity for dilute solutions, we can discard the exponential terms and say that $-\ln (1 - X_2)$ is approximately equal to $X_2$.

(b) The mole fraction of solute, $X_2$, is equal to $n_2/(n_1 + n_2)$. As we have done before the dilute solutions, we may neglect $n_2$ in the denominator and let $X_2$ be approximately equal to $n_2/n_1$ (see Eq. 3.20).

(c) The change that we will be measuring, the boiling point rise, occurs over such a narrow temperature range that $T$ is practically equal to $T_0$. Therefore, no appreciable error will be incurred if we substitute $T_0^2$ for $T_0 T$. In the numerator we recognize that $(T - T_0)$ is $\Delta T_b$.

Making the substitutions according to (a), (b), and (c), we have

$$\ln\frac{P_0}{P} = -\ln X_1 = -\ln (1 - X_2) = X_2 = \frac{n_2}{n_1} = \frac{\Delta H \, \Delta T_b}{RT_0^2}. \qquad (3.27)$$

For aqueous solutions and 1000 g of solvent, $n_1 = 1000/18 = 55.6$, and $n_2$ is the molality of the solution. We make the substitution $(n_2/n_1)$

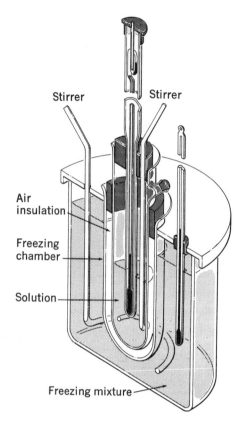

**Figure 3.3** Apparatus for the determination of freezing point lowering.

$= 0.018$ $m$ just as we did in Eqs. 3.20–3.24. Solving for $\Delta T_b$, the temperature elevation, yields

$$\Delta T_b = 0.018\ m\frac{(RT_0^2)}{\Delta H}. \tag{3.28}$$

The boiling point of water at one atmosphere is $373.15°K$; the molar heat of vaporization is $18 \times 539$ calories at this temperature and one atm pressure. The quantity $\Delta H$ is expressed in calories per mole, and $R$ must agree dimensionally, *i.e.*, be expressed as $1.987$ calories/mole-degree. Substituting these values gives

$$\Delta T_b = \frac{(0.018\ m)\,(1.987)\,(373.15)^2}{(18)\,(539)} = m\,(0.513°). \tag{3.29}$$

This is the correct constant for the boiling point elevation for water. By a similar derivation we may obtain the corresponding expression for the freezing point constant. Experimental values obtained with molal solutions do not correspond to the molal boiling point and freezing point constants, because such solutions are too concentrated to behave ideally. Experimental measurements are made on dilute solutions and the molal constants calculated accordingly. An apparatus for the experimental measurement of a freezing point is shown in Fig. 3.3.

Plants that are adapted to arid climates* often have thick cuticles and exterior layers of wax, which impose mechanical barriers to the escape of water vapor. In addition, however, species which display drought resistance or winter hardiness have higher concentrations of solutes in the cell protoplasm and consequently lower protoplasm vapor pressure than less resistant species.

The older scientific literature of plant physiology had much to say about "bound water," a term that is no longer fashionable, principally because of its vagueness. Although there were for many years persuasive indications that living tissue did contain more than one kind of water, the subject did not prove amenable to definitive experimentation. Recently, however, nuclear magnetic resonance studies of living tissue have yielded quantitative estimations of the extent of immobilized water present at various temperatures.† Much of the water in cells is presumably highly structured when it occurs in apposition to the surfaces of macromolecules. Water layers around macromolecules in animal protoplasm are also believed to be highly structured. The published papers on the various forms of water in biological systems—the subject of a conference sponsored by the New York Academy of Sciences—are recommended to those having further interest in this topic.‡

Illustrations of several types of calculations involving the colligative properties follow.

*Example 3.2* Calculation of molecular weight. A certain nondissociating solute is subjected to carbon-hydrogen analysis. The empirical formula is shown to be $(CH_2O)_x$. A solution of the compound containing 1.000 g dissolved in 100.00 g of water freezes at $-0.103°C$. What is the molecular weight and the correct molecular formula?

First the molality of the solution is calculated.

$$\Delta T_f = m K_f \qquad \text{(Eq. 3.25)},$$
$$0.103 = m(1.86),$$
$$m = 0.0554.$$

Since a solution of 1,000 g per 100.00 g of water is found to be $0.0554\,m$, $M_2$, the molecular weight of the solute, is calculated from the definition of the molality.

---

*Hadley, Neil F., "Desert Species and Adaptation." *American Scientist*, May-June, 1972, p. 338.

†Bratton, C. B., A. L. Hopkins, and J. W. Weinberg, *Science*, **147**:738, 1965. Sussman, M. V., and L. Chin, *Science*, **151**:324, 1966.

‡See Whipple, H. E., in Suggested Additional Reading.

$$m = \frac{w_2/M_2}{w_1/1000} \qquad \text{(Eq. 3.22)},$$

$$0.0554 = \frac{1.000/M_2}{100.00/1000},$$

$$M_2 = 181.$$

Since the formula weight of $CH_2O$ is 30.03, the molecular formula must be $(CH_2O)_6$, or $C_6H_{12}O_6$.

*Example 3.3*  Calculation of freezing and boiling points from vapor pressure data. The vapor pressure of an aqueous solution at 28°C is 27.995 mm. The vapor pressure of water at the same temperature is 28.065 mm. What will be the approximate freezing point and boiling point of this solution?

Since only an approximate answer is desired, the molality is calculated from the limiting equation 3.24.

$$\frac{P_0 - P}{P_0} = 0.018 \ m,$$

$$\frac{28.065 - 27.995}{28.065} = 0.018 \ m,$$

$$m = 0.139.$$

Now the freezing point depression and boiling point elevation may be calculated.

$\Delta T_f = (0.138)(1.86) = 0.257°.$
The freezing point is $-0.257°C.$
$\Delta T_b = (0.138)(0.514) = 0.071;$
The boiling point is $100.071°C.$

In this example the error resulting from the use of the approximate equation (3.24) is around 0.2 percent.

*Example 3.4*  Determination of the freezing point constant for a nonaqueous solvent. Naphthalene melts at 80.1°C. When 0.1106 g of anthranilic acid is dissolved in 20.000 g of naphthalene, the melting point is lowered by 0.278°. Taking the molecular weight of the solute to be 137.12 and assuming that no association occurs on solution, calculate the molal freezing point constant.

First the molality of the solution is calculated

$$m = \frac{w_2/M_2}{w_1/1000} = \frac{0.1106/137.12}{20.000/1000} = 0.0404.$$

Now calculate $K_f$.

$$\Delta T_f = mK_f,$$

$$K_f = \frac{0.278}{0.0403} = 6.88°/\text{molal}.$$

**TABLE 3.2** PHYSICOCHEMICAL MECHANISMS FOR SOLVENT OR SOLUTE MOVEMENT IN BIOLOGICAL SYSTEMS

| Mechanism | Essential features | Principal moving components |
|---|---|---|
| Osmosis | Escaping tendency of solvent differs on the two sides of a membrane. | Water and small ions, if the membrane is permeable. |
| Gibbs-Donnan effect | Large charged species is trapped on one side of a membrane and causes unequal distribution of small ions at equilibrium. | Small ions |
| Imbibition | Water is trapped by organic macromolecules, which bind it in an ice-like structure that resists desiccation. | Water |
| Active transport | Solutes are transported across a living membrane into a region of higher free energy as a consequence of exergonic processes occurring within the membrane. A trapping mechanism may be involved. | Solutes (ions and nonelectrolytes) |

**Osmotic pressure** Living systems possess various physicochemical mechanisms for transporting solvent and solutes. Some of these mechanisms are readily explained in terms of the escaping tendency of solvents and the selective permeability of membranes. Others are imperfectly understood at present. One type of mechanism called "active transport" is displayed only by living membranes. When an organism dies and can no longer supply energy to a membrane, active transport ceases. A summary of this and other common mechanisms is given in Table 3.2. We will consider only the first two.

In 1748, Abbé Nollet first observed that solvent passes through a membrane from a dilute solution into a more concentrated one. If pressure is applied to the more concentrated solution, the flow of solvent can be slowed, halted, or reversed, depending on the amount of pressure applied. The *osmotic pressure* of a solution is the minimum pressure that must be applied to the solution, in excess of the pressure on the solvent, to prevent the flow of solvent from pure solvent into the solution, when the two are separated by a membrane *not* permeable to the solute particles (semipermeable). The tendency of solvent to move from a more dilute solution to a more concentrated one can be demonstrated even when the only connecting surface between the two solutions is a layer of vapor. If two solutions of different concentrations are placed in

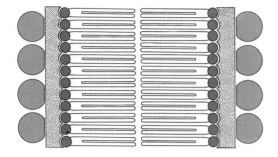

**Figure 3.4**  Schematic representation of the "butter sandwich" arrangement of the living membranes. Two layers of lipids, their hydrocarbon tails pointing in and their water-soluble heads pointing out, comprise the middle section. They lie between two thin sheets of protein (medium gray bands). These sheets of protein are thought to be coated with globular proteins (light gray circles). [From "The Chemistry of Cell Membranes" by Hokin, L. E., and M. R. Hokin. Copyright © October 1965 by Scientific American, Inc. All rights reserved.]

a sealed desiccator which is kept at constant temperature, the liquid level of the more concentrated solution will rise while that of the more dilute solution will fall. Solvent will distill from one beaker to the other until the two solutions are at the same concentration if enough solvent is present. This will be the position of rest, or equilibrium, with the entire system at its lowest energy level. We recall that the tendency of natural systems to go to equilibrium is embodied in the Second Law of Thermodynamics.

Membranes differ in their composition, structure, pore size, and response to different solvents. Some permit only the solvent to pass through. Presumably this occurs by a sort of vapor distillation through membrane, or the solvent may actually dissolve in the membrane as it passes through it. Some membranes permit the passage of ions and small molecules. The passage of ions through a membrane is complicated by electrostatic interaction between charged groups on the membrane and the solute ions. It is certainly not the pore size alone that determines which molecules will be admitted and which excluded.

A well-known model for living membranes suggests that they are composed of lipid interspersed between protein in a "butter sandwich" arrangement (Fig. 3.4). An alternative model has recently been suggested to be more appropriate for many membranes.*† The lipid molecules are still believed to exist in a bilayer, with their polar heads directed toward the surface of the membrane, just as in Fig. 3.4. According to this model the protein does not form a layer on the surface. Instead the protein molecules are positioned at discrete intervals on the membrane. In some instances they are partially exposed on a surface of the membrane and partially buried, whereas in others they may completely penetrate the membrane, so that part of a protein molecule may appear on both surfaces of a membrane. Thus the surface of the membrane

---

*Singer, S. J. and G. L. Nicolson, *Science*, 175, 720 (1972).
†Fox, C. F., in *Scientific American*, February 1972, p. 31.

would be a background of polar lipid heads studded at various positions by the protruding portions of globular proteins. It has been suggested that active transport is accomplished as a result of conformational changes, some of which require the expenditure of metabolic energy, in some of the proteins protruding through the membrane.*

Despite these properties of membranes, the osmotic effects must be independent of the nature of the membrane used in measuring them. Otherwise, we could build a perpetual motion machine by taking advantage of two different osmotic pressures developing from the same solvent passing through two different membranes into a common solution. Solvent would be continually forced out through the membrane producing the lower osmotic pressure and could be made to turn a turbine and do mechanical work, while being recycled through the membrane producing the higher osmotic pressure. This hypothetical machine is shown schematically in Fig. 3.5. The possibility of such a perpetual motion machine was first analyzed by Ostwald, the distinguished German physical chemist of the latter part of the nineteenth century. Two different membranes may show initially different osmotic pressures with which each membrane equilibrates with the solution. If enough time is allowed, however, the same pressure will be attained with both membranes. Were this not the case, we would have a clear-cut exception to the Second Law of Thermodynamics.

Pfeffer, in 1877, measured the osmotic pressures ($\Pi$) of solutions containing the same weight of solute dissolved in different volumes of solvent. He showed that the $\Pi V$ products were essentially constant at constant temperature. Further, he showed that for a given solution the osmotic pressure increased regularly with temperature, the $\Pi/T$ ratio being essentially constant. The Dutch chemist, van't Hoff, perceived that a combination of these results would yield an empirical osmotic pressure equation for solutions. The equation is

$$\Pi = CRT, \tag{3.30}$$

where $\Pi$ is the osmotic pressure and $C$ is the concentration of the solution. The equation is satisfactory for dilute solutions only; $C$ is expressed as molality or molarity. True linearity is attained only with ideal solutions and therefore, the van't Hoff Equation is another illustration of a limiting law.

An osmotic pressure equation can be derived in an exact way from theoretical considerations. The osmotic pressure of a solution is also a consequence of the fact that the vapor pressure of a solution of a nonvolatile solute is lower than the vapor pressure of the pure solvent. With the assumption that the solvent vapor behaves ideally, the equation for osmotic pressure is readily derived. When such a solution is separated

---

*Fox, C. F., in *Scientific American*, February 1972, p. 31.

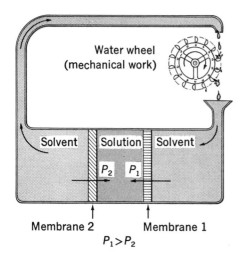

**Figure 3.5** Hypothetic perpetual motion machine, which could be built if there were a difference in osmotic pressure caused by membrane characteristics. This imaginary machine was proposed by Ostwald and would be a perpetual motion machine of the *second kind*. (See Item 2 under "Statements of the Second Law, page 71.)

from the pure solvent by a semipermeable membrane and this system brought to a condition of equilibrium by increasing the total pressure on the solution, the free energy per mole of solvent must be the same on both sides of the membrane. The decrease in free energy per mole of solvent in the solution (from its free energy in the standard state) is given by Eq. 2.48: $\Delta \bar{G} = RT \ln P/P_0$. The *increase* in pressure that is placed on the solution must be just sufficient to restore the free energy of the solvent in the solution to the free energy of the pure solvent. This increase in free energy is given by Eq. 2.40, written in a slightly different notation. That is, at constant temperature,

$$dG = V \, dP$$

and therefore, since the liquid is only slightly compressible,

$$\Delta G = V \, \Delta P.$$

The increase in pressure in the case we are considering is the osmotic pressure, *i.e.*, $\Delta P = \Pi$. The molar volume of the solvent is denoted by $\bar{V}$. Hence, as an alternate form of Eq. 2.40 we write, for liquids,

$$\Delta \bar{G} = \bar{V} \Pi.$$

At equilibrium,

$$\text{Displacing Force} = -\text{Restoring Force},$$

or

$$\Pi \bar{V} = -RT \ln \frac{P}{P_0} = RT \ln \frac{P_0}{P}.$$

Therefore,

$$\Pi = \frac{RT}{\bar{V}} \ln \frac{P_0}{P}$$

$$= \frac{RT}{\bar{V}} (-\ln X_1). \tag{3.31}$$

We have already shown that $\ln \dfrac{P_0}{P} = X_2$ for dilute solutions (Eq. 3.27). Substituting this identity in 3.31 gives

$$\Pi = \frac{RTX_2}{\bar{V}}. \tag{3.32}$$

Since $X_2$ is approximately equal to $n_2/n_1$ for dilute solutions,

$$\Pi = \frac{RTn_2}{\bar{V}n_1}.$$

The term $(\bar{V}n_1)$ is equal to the volume of the solvent since it is the volume of one mole $(\bar{V})$ multiplied by the number of moles, $n_1$. In a dilute solution, this will be approximately equal to the volume of the solution, $V$. Making this substitution yields

$$\Pi = \frac{n_2RT}{V}. \tag{3.33}$$

When $V$ is one liter, $n_2$ is the molarity. Compare this final result with the van't Hoff empirical equation (3.30) and it will be apparent that they are identical. Equation 3.33 is a useful and special case of the ideal equation for osmotic pressure (3.31). The latter is not restricted to dilute solutions, but does require that the solvent vapor behave ideally.

*Example 3.5* Osmotic pressure measurements were made on sucrose solutions at 20°C.

For a 0.1 $m$ (0.098 $M$) solution, the *observed* osmotic pressure was 2.59 atm. For a 1.0 $m$ (0.825 $M$) solution, the *observed* osmotic pressure was 26.64 atm.

Calculate the osmotic pressure for these two solutions using Eqs. 3.31, and 3.32, and 3.30. Compare the results.

(a) From Eq. 3.31, the most nonlimiting form,

$$\Pi = \frac{RT}{\bar{V}} (-\ln X_1),$$

for the 0.1 $m$ solution,

$$\Pi = \frac{(0.082)(293°)}{0.018} \left( -\ln \frac{55.56}{55.66} \right) = 2.40 \text{ atm};$$

for the 1.0 $m$ solution,

$$\Pi = \frac{(0.082)(293°)}{0.018}\left(-\ln\frac{55.56}{56.56}\right) = 23.8 \text{ atm.}$$

(b) From Eq. 3.32

$$\Pi = \frac{RTX_2}{\bar{V}}$$

for the 0.1 $m$ solution,

$$\Pi = \frac{(0.082)(293°)(0.1/55.66)}{0.018} = 2.40 \text{ atm;}$$

for the 1.0 $m$ solution,

$$\Pi = \frac{(0.082)(293°)(1.0/56.56)}{0.018} = 23.6 \text{ atm.}$$

(c) From Eq. 3.30

$$\Pi = CRT$$

for the 0.098 $M$ solution (0.1 $m$),

$$\Pi = (0.098)(0.082)(293°) = 2.35 \text{ atm;}$$

for the 0.825 $M$ solution (1.0 $m$),

$$\Pi = (0.825)(0.082)(293°) = 19.8 \text{ atm.}$$

The student should not be misled by the form of Eq. 3.33 and its resemblance to the Perfect Gas Law. There is no implication or suggestion that the osmotic pressure arises from the bombardment of the vessel walls by the solute molecules. Such thinking is erroneous. Equation 3.33 (also 3.30) predicts that one mole of solute contained in one liter of solvent will give rise to an osmotic pressure of 22.41 atmospheres at 0°C in the ideal case. Real solutions show discernible deviations at 1 $m$ concentrations from the predicted osmotic pressure, just as they do from predicted freezing points and boiling points.

Equation 3.33 can be rearranged in terms of M, the molecular weight. Substituting g/M for $n_2$, the number of moles of solute, gives

$$\Pi = \left(\frac{\text{g}}{\text{V}}\right)\frac{RT}{\text{M}},$$

or

$$\Pi = \frac{CRT}{\text{M}}$$

where $C$ is the concentration in grams per liter. Dividing both sides of the equation by $C$ gives a term, $\Pi/C$, which is called the *reduced osmotic pressure.* For a solute that behaves ideally, $\Pi/C$ will be a constant that is independent of $C$. With certain solutes, such as linear polymers, however, $\Pi/C$ changes with $C$. The general equation which holds for these cases is

$$\frac{\Pi}{CRT} = A_1 + A_2C + A_3C^2 + \cdots$$

where $A_1 = 1/M_N$ and $M_N$ is the number average molecular weight. In highly dilute solutions and at experimentally attainable concentrations the small terms can be dropped and the equation simplified to

$$\frac{\Pi}{CRT} = \frac{1}{M_N} + A_2C \qquad (3.34)$$

only if (1) $A_2$ is not too large (because $A_3$ depends approximately on $A_2^2$) and (2) $M_N$ is not too large. Equation 3.34 is the equation of a straight line of slope $A_2$ called the "second virial coefficient." Its intercept on the $\Pi/CRT$ axis is the reciprocal of $M_N$. This method is used with high molecular weight solutes and is most accurate for values of M between 10,000 and 200,000.

**Measurement of the colligative properties**   The four colligative properties are not measured with equal facility. The freezing point lowering can usually be determined with the greatest ease and accuracy. As we have already noted in Table 3.1, different solvents have different molal constants. For a particular experiment, the solvent possessing the largest molal constant will be the one of choice, provided the solute will dissolve and behave reasonably ideally. In certain solvents, some solutes display abnormal molecular weights because of dimerization and thereby present a difficulty. Advantages and disadvantages must be weighed before a decision is made as to which colligative property to measure. Another aspect of the problem is the relative magnitude of the change in the colligative properties for any individual solute concentration. A $0.01\ m$ solution will give a freezing point depression (in water) of only $0.0186°$, whereas the osmotic pressure will be 170.24 mm at 0°C. After a solution is diluted 50-fold, the osmotic pressure may still be measurable, but the freezing point depression will be too small to be determined. For this reason, the osmotic pressure is the only colligative property that has been useful in the problem of molecular weight determinations for proteins, polysaccharides, and other high polymers. Additional methods for the determination of molecular weights of macromolecules will be described in Chapter 7.

   In Fig. 3.6 is shown a diagram of an apparatus for the precise measurement of osmotic pressure. Just enough external pressure is applied to the solution to equal the osmotic pressure. One of the modern osmometers (the Mechrolab high-speed membrane servo-osmometer) has an optical

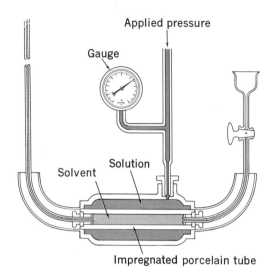

**Figure 3.6** The osmotic pressure apparatus of Berkeley and Hartley. An external pressure is applied to the solution so that its osmotic pressure is exactly balanced and no water flows from the cell into the solution. The level of liquid in the capillary tube is used as a criterion of equilibrium. If any water flows out of the cell, the liquid level will drop in the capillary. [After the original sketch of the Earl of Berkeley and E. G. J. Hartley, *Trans. Roy. Soc. (London)*, A **206**:486, 1906.]

detector, and a servo-operated pressure system that automatically establishes osmotic equilibrium. An attachment is available for continuous strip chart readout. This particular instrument is used for molecular weights in the 20,000 to 1,000,000 range. For molecular weights in the 100 to 20,000 range, the same manufacturers supply a vapor pressure osmometer that works on a different principle: Molecular weights are determined thermoelectrically by measuring temperature differences caused by vapor pressure lowering of solvent by solute.[*]

The absence of reliable techniques for the determination of molecular weights of proteins abetted the state of confusion that existed until the 1920's over the nature of the molecular species. The disparity among the reported molecular weights for the common proteins strengthened the argument that proteins existed as heterogeneous aggregates. Refinements in the techniques of protein purification and osmometry eventually eliminated various sources of error so that reproducible results could be obtained. Sørensen's experiments with albumin[†] and Adair's studies on hemoglobin[‡] proved to be the turning points in the dispute. Sørensen showed that large deviations in osmotic pressure resulted from varying the pH of the albumin solutions and concluded that it was the ionization of the protein and not its heterogeneity that was responsible for the deviations. Adair obtained the first accurate molecular weight of hemoglobin on a highly purified, salt-free preparation and showed that the

---

[*]The vapor pressure osmometer is based on the original work of Brady, A. P., H. Huff, and J. W. McBain, *J. Phys. and Colloid Chem.*, **55**:304, 1951.

[†]Sørensen, S. P. L., *Comptes rendus des travaux du Laboratoire Carlsburg*, **12**, 1917.

[‡]Adair, G. S., *Proc. Roy. Soc. (London)*, **A109**:292, 1925.

lower weights estimated earlier resulted from salt impurities or the use of highly impermeable membranes, some of which required as long as eleven days to equilibrate. The fact that we now admit that some proteins do exist simultaneously in several polymeric forms is another illustration of dialecticism in scientific ideas. Yesterday's dogma is today's heresy and will be tomorrow's dogma—revised. In many instances scientific theories spiral onward through cycles of acceptance, rejection, and reacceptance.

**The Gibbs-Donnan effect**   With a pseudo-physiological solution, such as a saline solution of serum albumin, the osmotic pressure at higher albumin concentrations seriously deviates from the values predicted by the van't Hoff Equation. There is, however, a further complication in this system. In addition to small ions, very large charged particles are present —the protein molecules. If the experiment is performed at pH 5.4, the pH at which the serum albumin molecules have equal numbers of positive and negative charges, the ionization of the protein makes no difference because its net charge is zero. At this pH, departures from the van't Hoff Law can be charged to the nonideality of the solution. On the other hand, if the experiment is performed at pH 7.4, larger deviations occur, resulting in higher osmotic pressures. These additional deviations are the consequence of the Gibbs-Donnan effect. The failure to recognize this pH-related phenomenon contributed to the experimental difficulties (mentioned in the preceding paragraph) that plagued protein chemists during the early part of this century. Experimental curves illustrating the effect are shown in Fig. 3.7.

To describe the Gibbs-Donnan effect, we must consider a membrane permeable to both solvent and small ions present in buffers. Such a membrane corresponds to the membranes making up the vascular systems of the body, since inorganic salts diffuse through them into the interstitial fluids, but proteins do not, under normal conditions. We can see from the experimental results (Fig. 3.7) that the effect is pH-dependent and must be related to the charge on the protein.

Let us consider a system in which pH is 7.4 and the protein may be represented by $R^- Na^+$, since this protein will be an anion at a pH above its isoelectric point (pH of electrical neutrality of the molecule). The protein is all contained in solution, inside a membrane and cannot pass through it. As a second solute we select $Na^+ Cl^-$ and stipulate that, initially, all of the $Na^+ Cl^-$ is outside of the membrane. This would be the case if we were to place an aqueous solution of the protein inside an osmometer and immerse the apparatus in a solution of dilute sodium chloride. The system is shown diagramatically below, where [ ] represents the concentration of the ion so bracketed and the subscripts $i$ and $o$ refer to initial concentrations inside and outside, respectively.

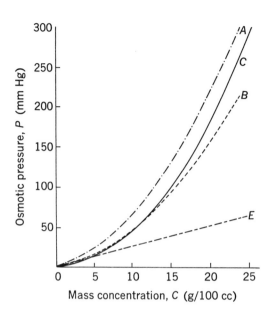

**Figure 3.7**  The effect of concentration on the osmotic pressure of protein solutions as related to charge. Curve $E$ is the hypothetical plot for a solute of molecular weight 60,000 which obeys the van't Hoff osmotic pressure equation. Curve $B$ is the experimental plot for serum albumin ($M = 60,000$) at its isoelectric point, pH 5.4. Differences between $E$ and $B$ are a reflection of the nonideality of the solution at higher solute concentrations. Curve $A$ is the experimental plot for serum albumin at pH 7.4. The difference between $A$ and $B$ is due to the net negative charge on the protein molecules at pH 7.4 and the consequent Gibbs-Donnan Effect. Curve $C$ is the experimental plot for human plasma at pH 7.4. Although the average molecular weight of the solute in Curve $C$ is larger than in Curve $A$, the general shape of the curve is the same. [From Scatchard, G., C. Batchelder and A. Brown, *J. Clin. Investigation*, **23**:459, 1944.]

INITIAL STATE OF THE SYSTEM

| Inside | MEMBRANE | Outside |
|---|---|---|
| $[R^-]_i$ | | $[Cl^-]_o$ |
| $[Na^+]_i$ | | $[Na^+]_o$ |
| $[R^-]_i = [Na^+]_i$ | | $[Cl^-]_o = [Na^+]_o$ |

The membrane is permeable to small ions, and chloride ion will move to the left, as ions tend to distribute themselves uniformly. If only chloride ion moved, the solution outside the membrane would possess an excess of positive charges; therefore, each chloride ion that moves across the membrane must be accompanied by a sodium ion. Let $x$ represent the concentration of chloride ions that have moved inside the membrane by the time equilibrium has been reached. The protein cannot pass through the membrane and remains inside it. At equilibrium, the redistribution will result in the following state:

FINAL STATE OF THE SYSTEM

| Inside | MEMBRANE | Outside |
|---|---|---|
| $[R^-]_i$ | | $[Cl^-]_o - x$ |
| $[Na^+]_i + x$ | | $[Na^+]_o - x$ |
| $[Cl^-] = x$ | | $[Cl^-]_o - x = [Na^+]_o - x$ |

We must also assume that the volumes of the solutions are held constant: if we do, the concentration relationships may be handled in a simple way. With the diagrams in mind, we may use subscripts "left" and "right" to denote final concentrations inside and outside the membrane. We note without making application of the condition at this time that electrical neutrality must apply on both sides of the membrane so that—for univalent ions—the total concentration of positive ions is equal to that of both the left and right negative ions. A more interesting condition is the thermodynamic requirement that at equilibrium the chemical potentials of the diffusible substance—sodium chloride—must be equal on both sides of the membrane. From this latter consideration and by a derivation beyond the scope of this text, we can write the following equation for dilute solutions:

$$[Na^+]_{left} \cdot [Cl^-]_{left} = [Na^+]_{right} \cdot [Cl^-]_{right},$$

and therefore,

$$([Na^+]_i + x)(x) = ([Na^+]_o - x)^2.$$

Expanding and solving for $x$, we have

$$x = \frac{[Na^+]_o^2}{[Na^+]_i + 2[Na^+]_o}. \tag{3.35}$$

Whatever the initial concentrations may be, we now have a general equation that will permit us to calculate the equilibrium concentrations. For example, if the initial sodium proteinate is 0.1 $M$ and the initial external sodium chloride is 0.1 $M$, $x = 0.033$ $M$, i.e., one-third of the sodium chloride moves inside the membrane. This redistribution will result in an appreciable increase in the osmotic pressure of the solution within the membrane. In Table 3.3, equilibrium concentrations and osmotic pressures are given for various initial concentrations; calculations were based on the simple equation derived above (3.35). In each example, it was assumed that water did not move.

We have so far considered only singly charged penetrating and non-penetrating ions. If we develop equations for multiply charged ions and also allow for incomplete dissociation of the protein, the algebra becomes more complicated. If, in addition, we develop equations that are strictly applicable and possess none of the defects of simple limiting laws, the mathematical difficulties extend beyond algebraic complications. These equations exist but are beyond the scope of this presentation.

The unequal distribution of ions across living membranes gives rise to electrical potentials, which have important physiological consequences.

**TABLE 3.3** INITIAL CONCENTRATIONS AND FINAL EQUILIBRIUM STATE IN GIBBS-DONNAN EFFECT

| Initial Concentrations (molarities) | | Final Concentrations (molarities) | | | Ideal osmotic pressure at 25°C for final state (atmospheres)* |
|---|---|---|---|---|---|
| Inside | Outside | Inside | | Outside | |
| $[Na^+]_i$ | $[Na^+]_o$ | $[Cl^-]$ or $x$ | $([Na^+]_i + x)$ | $([Na^+]_o - x)$ | |
| 0.01 | 1.0 | 0.498 | 0.508 | 0.502 | 0.29 |
| 1.0 | 1.0 | 0.333 | 1.333 | 0.667 | 32.8 |
| 1.0 | 0.5 | 0.125 | 1.125 | 0.375 | 36.7 |
| 1.0 | 0.01 | 0.000098 | 1.000098 | 0.0099 | 48.7 |

*Ideal osmotic pressure calculated for final state inside the membrane is based on the sum of the concentrations of all contributing ions: $R^-$, $Cl^-$, and $Na^+$. The assumption that solvent does not move has been made in the interest of simplifying the calculation.

Hydrostatic pressures may or may not be equal on both sides of a membrane. If they are, there will be no impedance to the migration of water across the membrane. The ability of certain membranes to concentrate ions is astounding. For example, in the nasal salt glands of the albatross, the petrel, and certain other marine birds there are membranes that transport such high concentrations of sodium chloride from the internal tissues to the exterior of the gland that a 5 percent salt brine drips from the tip of the bird's beak. This specialized adaptation permits the birds to drink sea water and survive in an environment devoid of fresh water. The petrel, which spends most of its life in flight and seldom rests on the water, has a special method of ejecting the fluid through a pair of tubes atop its beak. The brine is expelled by a jet of exhaled air in "water pistol" fashion.* The biochemical mechanism by which the salt transport is effected by these glands is not fully understood because of the complexities of membrane structure and active transport processes.

**Dialysis** The migration of solvent and small ions through membranes is used to advantage in a laboratory technique known as dialysis. A solution of protein or some other polymer is placed in a cellophane bag that is securely closed and immersed in distilled water, dilute buffer, or whatever solvent is suitable to the experiment. The large molecules are retained within the membrane and the small ions and molecules migrate through the membrane until equilibrium is attained. By frequent or continual changes of the dialysis bath, a protein can be freed of all salt contamination except for electrically bound cations or anions. Proteins

*Schmidt-Nielsen, K., "Salt Glands." *Scientific American*, January 1959, p. 109.

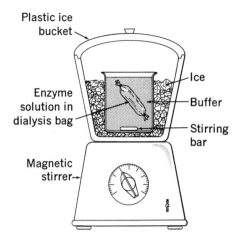

Plastic ice
bucket

Enzyme
solution in
dialysis bag

Magnetic
stirrer→

Ice

Buffer

Stirring
bar

**Figure 3.8** A simple method for purifying an enzyme by dialysis.

are frequently dialyzed against dilute buffer, rather than distilled water, to prevent precipitation or inactivation resulting from solvent attack on the tertiary structure of the protein. The purification of an enzyme by dialysis is shown in Fig. 3.8.

### 3.5 Biological Applications of Phase Diagrams and the Phase Rule

Phase changes peculiar to biological systems are rather limited in number. Genuinely crystalline macromolecules, *i.e.*, proteins and viruses, do not display true melting points since they are irreversibly denatured by heating. The polymerization of protein subunits and the "melting" of DNA may be cited as phase changes, although they are of the type designated as second- or third-order.

Fruitful application of the Phase Rule and phase diagrams has been achieved in the area of protein solubility. In particular, solubility curves have proved highly useful in determining the purity of protein preparations. Of the various physical criteria of purity available to the protein chemist, including homogeneity in the ultracentrifuge or under electrophoresis, the criterion of constant solubility of the protein remains the ultimate reliable index, provided the protein does not polymerize or form solid solutions with other components of the mixture. The solubility of a protein in a solvent is influenced by a variety of factors, among which are ionic strength, temperature, and pH of the solvent.*

---

*See the book by Cohn, E. J., and J. T. Edsall listed in *Suggested Additional Reading* at the end of this chapter.

### 3.6 Free Energy and Chemical Equilibrium

**The general equilibrium expression** The concentration of an ideal solution is related to the vapor pressure of the solvent by Raoult's Law, as we have already discussed:

$$P = P_o X_1 \quad \text{(Equation 3.18)},$$

where $P$ is the partial pressure of the solvent, $P_o$ is the vapor pressure of the pure solvent and $X_1$ is mole fraction of solvent. A similar relationship was discovered for volatile solutes and the empirical statement is called Henry's Law. Henry's Law states that the concentration of a gas dissolved in a liquid is directly proportional to the pressure of the gas in equilibrium with the liquid.

$$P = kX \tag{3.36}$$

where $P$ is the vapor pressure of the solute (gas), $X$ is the mole fraction of the gas in the liquid, and $k$ is a proportionality constant, called Henry's Law constant. (If the volatile component is the solvent and $X$ refers to mole fraction of solvent, the Henry's Law constant becomes the vapor pressure of the pure solvent and the relationship becomes identical with Raoult's Law.)

From Henry's Law it is apparent that the vapor pressure of a component in solution can be expressed in terms of the concentration of that component, an approach that is advantageous to the chemist. Suppose that we have two solutions of the same solute and solvent but at different concentrations. The solute is volatile, and in each solution the solvent and solute are in equilibrium with their vapor.

From Eq. 2.38 we write

$$\Delta \bar{G} = RT \ln \frac{P_2}{P_1},$$

where $\Delta \bar{G}$ is the difference in the molar free energy of the solute in one solution as compared with the other, and $P_2$ and $P_1$ are the respective equilibrium vapor pressures of the *solute* in the two solutions.

Substituting from Henry's Law gives

$$\Delta \bar{G} = RT \ln \frac{kX_2(P_2)}{kX_1(P_1)} = RT \ln \frac{X_2(P_2)}{X_1(P_1)} \tag{3.37}$$

The equation may be written

$$\bar{G}_2 - \bar{G}_1 = RT \ln \frac{X_2(P_2)}{X_1(P_1)} \tag{3.38}$$

Note that in this derivation, $X_2$ and $X_1$ *both refer to solute.*

It is often desirable to define a system in which the mole fraction $X_1$ is unity and the corresponding $\bar{G}$ is written as $\bar{G}°$. By $\bar{G}°$ is meant the molar free energy of the solute in its standard state—pure solute under one atmosphere pressure at $T°K$, but in such a state of separation that solute molecules do not interact, *i.e.*, in a hypothetical state of infinite dilution at a mole fraction of unity.

For the general case the equation* is

$$\bar{G} - \bar{G}° = RT \ln X. \tag{3.39}$$

Our purpose, however, is to arrange this equation so that a concentration term such as molarity or molality can be used in place of mole fraction. Putting in the necessary conversion constant gives

$$\bar{G} - \bar{G}° = RT \ln aC,$$

where $a$ is approximately constant at low concentrations. Rearranging and incorporating $RT \ln a$ into the standard free-energy term yields

$$\bar{G} - \bar{G}° = RT \ln C, \tag{3.40}$$

or

$$\bar{G} = \bar{G}° + RT \ln C,$$

which says that the free energy of a component in solution is equal to the standard free energy of the component, $\bar{G}°$, plus the term $RT \ln C$, where $C$ is the molar concentration of the component in solution. The standard state of the solute is now redefined as being at a concentration of one mole per liter at one atmosphere pressure and $T°K$. To determine the free energy change associated with a change in concentration, we write the appropriate equations for $\bar{G}_1$ corresponding to $C_1$ and $\bar{G}_2$ corresponding to $C_2$ and subtract them:

$$\bar{G}_2 = \bar{G}° + RT \ln C_2$$
$$\bar{G}_1 = \bar{G}° + RT \ln C_1$$
$$\overline{\Delta G = \bar{G}_2 - \bar{G}_1 = RT \ln \frac{C_2}{C_1}.} \tag{3.41}$$

This equation gives the molar free energy *change* corresponding to a change in the concentration of one component (solute). For $n$ moles, $\Delta G = nRT \ln C_2/C_1$. If $C_2$ is the final concentration and $C_1$ the initial

---

*Many will prefer to write this equation as $\mu - \mu° = RT \ln X$ for ideal solutions where $\mu°$ is the chemical potential of the solute in its standard state and $X$ is its mole fraction in its solution. Most standard physical chemistry texts will probably handle derivations like those to follow from the chemical potential point of view and will introduce additional concepts in order to handle non-ideal solutions. The authors prefer the treatment given here for a simplified introduction to these ideas.

concentration, we can see from the form of the equation that an increase in concentration will cause the solution to gain free energy with respect to that particular component ($\Delta G$ will be positive), and that dilution with respect to the same component will cause the solution to lose free energy ($\Delta G$ will be negative). A loss in free energy represents a loss in the capacity to do work.

Equation 3.41 can be used to calculate the osmotic work done by living systems in transferring solutes from regions of lower concentration to those of higher. For example, the osmotic work done by the kidney in forming urine may be calculated from the concentrations of the various solutes present in both plasma and urine. The osmotic work for each individual substance is calculated separately; the total osmotic work is the sum of the individual terms.

> **Example 3.6**   Calculate the osmotic work done by the kidneys in secreting 0.158 moles of $Cl^-$ in a liter of urine water at 37°C when the concentration of $Cl^-$ in the plasma is 0.104 $M$, and in urine 0.158 $M$. According to Eq. 3.41, after converting $\ln(C_2/C_1)$ to 2.303 $\log(C_2/C_1)$,
>
> Osmotic Work $= \Delta G = (2.303)(0.158)(1.987)(310) \log \dfrac{0.158}{0.104} = 40.7$ calories.

So far, the discussion has been restricted to a consideration of the free energy change associated with one solute in solution. In a chemical reaction at least two solutes (reactant and product) will be present, and usually more than two, depending on the stoichiometry. In the general case of two products and two reactants, the chemical reaction is expressed by the familiar general equation

$$aA + bB = cC + dD.$$

For the same reaction we can write a general free energy equation,

$$\Delta G = \Sigma\, G_{products} - \Sigma\, G_{reactants},$$

which is analogous to Eq. 2.39. When the system is in a state of equilibrium,

$$\Sigma\, G_{products} = \Sigma\, G_{reactants} \quad \text{and} \quad \Delta G = 0.$$

No useful work can be done. We are interested in the chemical reaction that *can* do useful work, and if we write free energy expressions for all the reactants and products and subtract the former from the latter, we will be able to calculate $\Delta G$. In other words, we will write the expressions given in Eq. 3.40 (second form of it) for the general chemical reaction just cited. Each equation will have to be written for the correct number of moles of component (the coefficients of $A$, $B$, $C$, and $D$). This is done as follows:

(1)  $G_A = aG_A^\circ + a(RT \ln C_A)$.
(2)  $G_B = bG_B^\circ + b(RT \ln C_B)$.
(3)  $G_C = cG_C^\circ + c(RT \ln C_C)$.
(4)  $G_D = dG_D^\circ + d(RT \ln C_D)$.

$$\Delta G = [(3) + (4)] - [(1) + (2)] \quad \text{or} \quad (G_C + G_D) - (G_A + G_B).$$

Therefore,

$$\Delta G = (cG_C^\circ + dG_D^\circ - aG_A^\circ - bG_B^\circ) + RT \ln \frac{(C_C)^c (C_D)^d}{(C_A)^a (C_B)^b}$$

(The student is advised to work through the algebra to assure himself that the final result is correct.) The parentheses that contain the $G^\circ$ terms are now replaced by $\Delta G^\circ$, which is called the standard free energy change for the reaction since

$$\Delta G^\circ = \Sigma\, G^\circ_{\text{products}} - \Sigma\, G^\circ_{\text{reactants}}.$$

This reduces the final form of the equation to

$$\Delta G = \Delta G^\circ + RT \ln \frac{(C_C)^c (C_D)^d}{(C_A)^a (C_B)^b} = \Delta G^\circ + RT \ln Q. \qquad (3.42)$$

The log term in the above equation is called, for convenience, log $Q$, and $Q$ here has the same *form* as an equilibrium constant. Concentrations that appear in the $Q$ term are the actual concentrations regardless of their "displacements" from equilibrium.

If it should happen that reactants and products actually are present in equilibrium concentrations, or at equilibrium pressures, then $\Delta G = 0$ and $Q = K$ where $K$ is the equilibrium constant of the reaction. For this special case we write

$$\Delta G = 0 = \Delta G^\circ + RT \ln K.$$

Rearranging gives

$$\Delta G^\circ = -RT \ln K. \qquad (3.43)$$

This equation for the standard free energy change is one of the most useful of all thermodynamic expressions.

**$K_P$ and $K_C$**   The numerical value of $\Delta G^\circ$ depends on whether the expression for the equilibrium constant contains the pressures of components, in which case $K$ is usually designated as $K_P$, or whether it is composed of concentration terms and called $K_C$. The *sign* of the standard free energy change will be independent of the exact form of $K$. For chemical reactions involving gases, $K_C$ will be equal to $K_P$ if the number of moles of gaseous reactants is equal to the number of moles of gaseous products; an example in which $K_C = K_P$ is

$$H_{2(g)} + I_{2(g)} \rightarrow 2HI_{(g)}.$$

The general rule is

$$K_P = K_C (RT)^{\Delta n}, \tag{3.44}$$

where $\Delta n$ = (moles of gaseous products) − (moles of gaseous reactants) as described in the stoichiometry of the reaction. As an illustration of the validity of the general rule, consider the example

$$3H_{2(g)} + N_{2(g)} = 2NH_{3(g)}.$$

For this reaction $\Delta n = -2$, and $K_P = K_C (RT)^{-2}$. The latter relationship may be proved by writing the $K_P$ and $K_C$ expressions.

$$K_P = \frac{(P_{NH_3})^2}{(P_{H_2})^3 (P_{N_2})}.$$

$$K_C = \frac{(C_{NH_3})^2}{(C_{H_2})^3 (C_{N_2})}.$$

According to the Ideal Gas Law,

$$C = n/V = P/RT,$$

then,

$$K_C = \frac{(P_{NH_3}/RT)^2}{(P_{H_2}/RT)^3 (P_{N_2}/RT)} = \frac{(P_{NH_3})^2}{(P_{H_2})^3 (P_{N_2})} \cdot (RT)^2 = K_P (RT)^2.$$

Therefore,

$$K_P = K_C (RT)^{-2}.$$

When calculations involve several equilibrium constants (which occurs in the computation of $\Delta G°$ for several reactions) the $K$'s chosen must be of uniform type. Similar care must be used in the selection of $K$ values for the van't Hoff Equation to be considered in the next section.

If the dissociating system is maintained at constant temperatures and pressure, its volume increases as the number of moles of molecules increases, although there is no change in the total mass of the system. This means that the density of the gas will decrease since density = mass/volume. Under conditions of constant $P$ and $T$ an inverse proportionality exists therefore between $d$, the density of the system, and the number of moles present for a given weight of gas, so that

$$\frac{d_{\text{final}}}{d_{\text{initial}}} = \frac{1}{1 - \alpha(1 - n)}$$

and

$$\alpha = \frac{d_i - d_f}{d_f(n - 1)}. \tag{3.45}$$

We may substitute for the densities in terms of the corresponding molecular weights, since $d_i = g/V_i \propto M_i$ and $d_f = g/V_f \propto M_f$. The molecular weight of the undissociated gas is given by $M_i$ whereas $M_f$ is the average molecular weight of the equilibrium mixture of the gas and its dissociation products. This leads to

$$\alpha = \frac{M_i - M_f}{M_f(n - 1)}. \tag{3.46}$$

Always bear in mind that equilibrium constants apply to reactions as written. For example, in the following reaction, $K_C$ and $K_P$ are for the *formation* of water, not its *decomposition* to hydrogen and oxygen.

$$H_{2(g)} + \tfrac{1}{2}O_{2(g)} \rightleftharpoons H_2O_{(g)}$$

The relationship between the forward and reverse reaction equilibrium constants is: $K_{\text{forward reaction}} = 1/K_{\text{reverse reaction}}$. An equilibrium constant for a particular reaction is written in a specific way. $K_P$ for the formation of water is for the above reaction and *not* for

$$2H_{2(g)} + O_{2(g)} \rightleftharpoons 2H_2O_{(g)}.$$

$K_P$ for the second reaction is equal to $(K_P)^2$ for the first.

**Equilibrium calculations** In making calculations involving equilibrium constants the student will conserve time and energy by taking several simple precautions. First, inspect the reaction equation carefully, noting the physical state of each substance, so that no error will be made in writing the equilibrium constant expression. For example, in the formation of hydrogen iodide either solid or gaseous iodine may be used in the reaction. Thus for

$$H_{2(g)} + I_{2(g)} \rightleftharpoons 2HI_{(g)}, \quad K_P = \frac{(P_{\text{HI}})^2}{(P_{\text{H}_2})(P_{\text{I}_2})}.$$

However, for

$$H_{2(g)} + I_{2(s)} \rightleftharpoons 2HI_{(g)}, \quad K_P = \frac{(P_{HI})^2}{P_{H_2}}.$$

The numerical values of the two preceding $K_P$'s are related, and one can be calculated from the other if the vapor pressure of the solid phase, iodine, is known at the appropriate temperature.

Another frequently cited example is the thermal decomposition of calcium carbonate to yield carbon dioxide and calcium oxide:

$$CaCO_{3(s)} \rightleftharpoons CO_{2(g)} + CaO_{(s)}.$$

The equilibrium constant for this reaction is simply

$$K_P = P_{CO_2},$$

as long as all three components of the system are present.

A second admonition concerns a careful reading of the problem to be solved. Before plunging in to write the equilibrium constant expression, the student should thoughtfully consider the data given and he should determine exactly how the pressure is to be expressed. For an individual gas in a mixture, the pressure may be given in one of several ways:

$$\text{Pressure of } A = P_A = \frac{n_A RT}{V_{total}} = P_{total}(X_A)$$

where $X_A$ is the mole fraction of $A$. If pressures are to be calculated in terms of either $nRT/V$ or $P_t(X)$, the equilibrium constant expression can usually be simplified by cancelling out like terms in the various pressure expressions. Example 3.8 illustrates this point.

Finally, if the pressure or concentration term includes relative or fractional numbers of moles expressed algebraically, the student must be certain that these terms correctly express the stoichiometry. For example, suppose that the reaction under consideration is

$$N_{2(g)} + 3H_{2(g)} \rightleftharpoons 2NH_{3(g)}$$

and suppose that the problem is to calculate the percent conversion of nitrogen and hydrogen to ammonia when three volumes of hydrogen and one volume of nitrogen are mixed and allowed to come to equilibrium at some fixed total pressure. The $K_P$ expression is easy enough:

$$K_P = \frac{[(X_{NH_3})P_{total}]^2}{[(X_{N_2})P_{total}][(X_{H_2})P_{total}]^3}$$

but the difficulty arises in deciding how to write the mole fractions. One approach is to let the number of moles of $NH_3$ at equilibrium $= x$. Then, let the number of moles of nitrogen initially present $= y$. The number of moles of hydrogen initially present will equal $3y$, since equal volumes of gases contain equal numbers of moles at the same temperature and pressure. The number of moles of nitrogen at equilibrium will be $(y - \frac{1}{2}x)$, and the number of moles of hydrogen at equilibrium will be $(3y - \frac{3}{2}x)$. The total number of moles at equilibrium is the sum of the three quantities:

$$\text{Total moles} = x + (y - \tfrac{1}{2}x) + (3y - \tfrac{3}{2}x) = 4y - x.$$

The solution of the quadratic gives directly the fractional number of moles of $y$ converted to $x$.

*Example 3.7*  Calculate the percentage conversion of nitrogen to ammonia when 3 volumes of hydrogen and one volume of nitrogen are mixed at 450°C and 10 atm pressure. $K_P$ for the formation of ammonia at this temperature is $4.34 \times 10^{-5}$ atm$^{-2}$.

Using the mole fraction expressions just derived above,

$$K_P = 4.34 \times 10^{-5} = \frac{\left[\left(\dfrac{2x}{4x - 2x}\right)10\right]^2}{\left[\left(\dfrac{y - x}{4y - 2x}\right)10\right]\left[\left(\dfrac{3y - 3x}{4y - 2x}\right)10\right]^3}$$

$$= \frac{\dfrac{4x^2}{(4y - 2x)^2}}{(y - x)(3y - 3x)^3} \cdot \frac{1}{10^2}$$

$$= \frac{\dfrac{4x^2}{(4y - 2x)^2}}{27(y - x)^4} \cdot \frac{1}{10^2}.$$

Taking the square root of both sides,

$$6.59 \times 10^{-3} = \frac{4x(2y - x)}{27(y - x)^2 10}$$

which reduces to the quadratic equation,

$$4.34x^2 - 8.69y + 0.34y^2 = 0$$

from which we obtain, $x = 0.067y$ or the conversion of $N_2$ to $NH_3$ is 6.7%. From this answer we calculate the percentage ammonia at equilibrium to be 3.47%. Experiment yields about 2% $NH_3$ at equilibrium and there is a difference because it is not really valid to treat these gases as being ideal at 10 atm and 450°C.

Frequently in gas dissociation calculations data for the density of the system initially and at equilibrium are employed. A simple relationship exists between the density of a partially dissociated gas system and its degree of dissociation. One of the classical examples of gaseous dissociation is

$$N_2O_{4(g)} \rightleftarrows 2NO_{2(g)}.$$

The symbol $\alpha$ is a common usage for the fraction of the number of moles initially present which dissociate, although $x$ or $y$ would be equally meaningful. For each mole of $N_2O_4$ initially present, the amount present at equilibrium is $(1-\alpha)$. Correspondingly, the number of moles of $NO_2$ formed will be $2\alpha$. The total number of moles at equilibrium is: $(1-\alpha) + 2\alpha = (1 + \alpha)$. For the general case where $n$ moles of products are formed from one mole of reactant, the total number of moles would be: $(1 - \alpha) + n\alpha = 1 - \alpha(1 - n)$.

*Example 3.8* For the reaction $N_2O_{4(g)} \rightleftarrows 2NO_{2(g)3}$ the equilibrium constant $K_P$ at 25°C is 0.120.

(a) Calculate the degree of dissociation of $N_2O_4$ at this temperature and at a total pressure of 2 atm. (b) Calculate the average molecular weight and density of the final equilibrium mixture. We have just pointed out that for each mole of $N_2O_4$ initially, the number of moles of $N_2O_4$ at equilibrium is $(1 - \alpha)$, the number of moles of $NO_2$ is $2\alpha$ and the total number of moles is $(1 + \alpha)$.

$$X_{NO_2} = \frac{2\alpha}{(1 + \alpha)}$$

$$X_{N_2O_4} = \frac{(1 - \alpha)}{(1 + \alpha)}$$

$$K_P = 0.120 = \frac{\left[\left(\frac{2\alpha}{1 + \alpha}\right)2\right]^2}{\frac{(1 - \alpha)}{(1 + \alpha)}2} = \frac{8\alpha^2}{(1 - \alpha^2)}$$

$$\alpha^2 = \frac{0.12}{8.12}$$

$$\alpha = 0.121.$$

The average molecular weight of the final equilibrium mixture is $M_f$ and can now be calculated from Eq. 3.46.

$$\alpha = \frac{M_i - M_f}{M_f(n - 1)} = \frac{92.0 - M_f}{M_f(2 - 1)}$$

$M_f = 82.1$ grams/molecular unit.

Eq. 1.24 states that $M = dRT/P$, so that

$$d_f = \frac{PM_f}{RT} = \frac{(2)(82.1)}{(0.082)(298)} = 6.72 \text{ g/l.}$$

**Spontaneous change in biochemical systems**  Many important bio-chemical reactions take place *in vivo* under conditions that make it impossible to measure actual concentrations. We can obtain a partial insight into the energetic potentialities involved if we measure the equilibrium constant for the reaction of interest and calculate the cor-responding $\Delta G°$. The *sign* of $\Delta G°$ determines the favored direction of the standard free energy change and the *magnitude* of the term dis-tinguishes the high energy reactions from the low energy ones.

To reiterate, $\Delta G°$ is the difference between the free energies of prod-ucts and reactants in their standard states, and it may be calculated if the equilibrium constant for the reaction is known. When products and reactants are in their standard states ($P = 1$ atm, $C = 1$ mole per liter), the $Q$ term becomes unity and $\Delta G = \Delta G° + RT \ln 1$, *i.e.*, $\Delta G = \Delta G°$.

In Chapter 2 consideration was given to a model system, a perfect gas changing volume and pressure at constant temperature. For this system, both $\Delta E$ and $\Delta H$ are zero. When the principles derived from this simplified system are applied to solutions, the molar concentration, rather than the pressure, is the parameter that changes. The $\Delta H$ is usually not equal to zero and must be considered along with $\Delta G$. For the perfect gas changing $V$ and $P$ at constant $T$ ($\Delta H = 0$), $\Delta G$ depends only on $-T \Delta S$. An increase in the entropy of the system will always result in $\Delta G$ being a negative number, and the change will proceed spontane-ously. With non-zero values of $\Delta H$, however, $\Delta G$ depends on both the entropy and enthalpy changes, since

$$\Delta G = \Delta H - T \Delta S \qquad \text{(This is Eq. 2.38 – for calculating energy changes at constant } T.)$$

Even if the entropy of the system is increasing, $\Delta G$ will be negative only if $\Delta H$ is negative, or is positive but smaller than $T \Delta S$. To return to a point emphasized in Chapter 2, we see again from this analysis that the sign of $\Delta H$ is not the deciding factor in determining the spontaneity of a reac-tion. At one time it was thought that all exothermic reactions (those for which $\Delta H$ is negative) were spontaneous, and that all endothermic reactions ($\Delta H$ is positive) were not. A moment's reflection will call to mind that many salts, such as KCl and $NH_4Cl$, spontaneously dissolve in water ($\Delta G$ is negative) although the beaker containing the solution grows cold to the hands and the solution must absorb heat from the room to return to its original temperature ($\Delta H$ is positive). The enthalpy change alone cannot, therefore, be used to determine the spontaneity of chemical reactions.

### 3.7 The Effect of Temperature on Chemical Equilibrium

One of the first principles mastered in general chemistry is that temperature affects the position of equilibrium in a chemical reaction. The principle of Le Chatelier predicts that raising the temperature of an endothermic reaction will favor the formation of products, whereas raising the temperature of an exothermic reaction will shift the equilibrium in favor of reactants. Earlier in this chapter we derived the mathematical relationship between the absolute temperature and the vapor pressure of a substance. The form of the Clausius-Clapeyron Equation that was developed at the time was

$$\ln \frac{P_2}{P_1} = \frac{\Delta H (T_2 - T_1)}{R T_2 T_1} \qquad \text{(Eq. 3.16)}.$$

An equivalent form of this equation, called the van't Hoff Equation, applies when we wish to express the dependence of the equilibrium constant on temperature. The van't Hoff Equation is

$$\ln \frac{K_2}{K_1} = \frac{\Delta H^\circ (T_2 - T_1)}{R T_2 T_1}, \qquad (3.47)$$

where $\Delta H^\circ$ is the standard heat of reaction (enthalpy change) and is assumed to be constant over the range of temperature involved. The van't Hoff Equation permits $\Delta H^\circ$ to be calculated if the equilibrium constant of a reaction can be measured at two temperatures and if it is valid to assume the temperature independence of $\Delta H^\circ$ as stated. The preferred procedure would be to measure the equilibrium constant at a minimum of three temperatures. It is then possible to take advantage of the differential form of Eq. 3.47,

$$\frac{d(\ln K)}{d(1/T)} = \frac{-\Delta H^\circ}{R} \qquad (3.48)$$

and plot $\ln K$ vs. the reciprocal of the absolute temperature. If $\Delta H^\circ$ is indeed independent of temperature over the range studied, $-\Delta H^\circ / R$ will be a constant, and the plot will yield a straight line (see Fig. 3.9). If a curve is obtained, the slope (and therefore $\Delta H^\circ$) is changing in the temperature range.

The heat of reaction, $\Delta H$, is the difference between the enthalpies of the final and initial and final states, *i.e.*,

$$\Delta H_{\text{Reaction}} = H_{\text{Products}} - H_{\text{Reactants}}$$

and, at constant pressure,

$$\left[ \frac{\partial (\Delta H)}{\partial T} \right]_P = \left[ \frac{\partial H_{\text{Prod.}}}{\partial T} \right]_P - \left[ \frac{\partial H_{\text{React.}}}{\partial T} \right]_P.$$

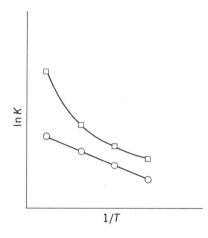

**Figure 3.9** Diagrammatic plots of ln $K$ vs. $1/T$ where $\Delta H^\circ$ is independent of temperature (circles) and varies with those temperatures (squares) in the temperature range studied.

From Eq. 2.16

$$\left[\frac{\partial(\Delta H)}{\partial T}\right] = C_{P_\text{Prod.}} - C_{P_\text{React.}} = \Delta C_p. \tag{3.49}$$

If we integrate at a constant pressure of one atm, $\Delta H$ is $\Delta H^\circ$ and

$$\Delta H^\circ_{T_2} - \Delta H^\circ_{T_1} = \int_{T_1}^{T_2} \Delta C_P dT. \tag{3.50}$$

If $\Delta C_P$ is temperature independent, (a constant, not zero) Eq. 3.50 becomes

$$\Delta H^\circ_{T_2} - \Delta H^\circ_{T_1} = \Delta C_P[T_2 - T_1]$$

and $\Delta H^\circ$ is clearly temperature dependent. If $\Delta C_P = 0$, *i.e.*, the products and reactants have the same heat capacities, $\Delta H^\circ$ is temperature independent. If the heat capacities of products and reactants are temperature dependent and are of the mathematical form shown following Eq. 2.16 on page 58, the variation of $\Delta H^\circ$ with temperature will reflect this heat capacity dependence on temperature. Equation 3.47 could legitimately be utilized to evaluate $\Delta H^\circ$ from the data represented by the circles in Fig. 3.9, but it would be wrong to use this equation in an attempt to analyze the data represented by the squares. In the latter it would be necessary to perform the difficult feat of measuring the slope at a point or using the temperature dependent expression for $\Delta C_P$, develop an integrable expression for $\Delta H^\circ$ in terms of $T$, substitute in Eq. 3.48 and integrate the resulting equation.

The life scientist will frequently be interested in the study of those reactions that occur in an aqueous solution. Reactions in water are often accompanied by enough differences in $C_P$ for reactants and products that a change in $\Delta H°$ with temperature will result so that Eq. 3.47 does not apply. One mechanism that can lead to this situation can be introduced by an examination of the transfer of simple aliphatic hydrocarbons from the liquid (or nonpolar solute) to water. The pertinent data for several hydrocarbons has been summarized by Némethy and Scheraga.[*] With butane, $\Delta G°$ for the transfer is +6 kcal/mole at 25°. Based upon the low solubility of hydrocarbons in water, $\Delta G°$ should be positive. However, the low solubility of butane in water is not caused by unfavorable changes in enthalpy. At 25° $\Delta H°$ is small, about $-0.8$ kcal/mole, and actually favors the transfer of butane to the aqueous solution. The transfer is accompanied by a large decrease in entropy ($\Delta S° = -22$ cal$^{-1}$deg$^{-1}$ mole$^{-1}$), which more than compensates for the small negative $\Delta H°$ and leads to a positive $\Delta G°$.

The sign of $\Delta H°$ and $\Delta S°$ can be understood if we assume a hydrocarbon, such as butane, induces the water molecules in its vicinity to act somewhat as though they had been chilled.[†] The water surrounding the hydrocarbon is frequently referred to as "ice-like," or as forming "icebergs," but this does not mean that the water necessarily resembles the familiar form of ice.[‡] Instead, we see that the water has become more structured, possibly by the formation of additional hydrogen bonds between water molecules or by increasing the strength of existing hydrogen bonds, or even by a combination of these two effects. The formation of new and/or stronger hydrogen bonds would lead to the evolution of heat and a negative $\Delta H°$. The formation of the "ice-like" structure requires a decrease in the freedom of movement of the water molecules, and the consequence is a negative $\Delta S°$.

The concept of the hydrophobic bond[**] can be readily visualized from the data presented above for butane. Many biological molecules contain hydrocarbon substituents. Examples are proteins containing hydrophobic amino acids such as L-leucine and L-isoleucine, and triglycerides, which have long chain fatty acids in ester linkage. The assumption is that water reacts to exposure to these hydrocarbon substituents in a manner similar to exposure to butane. The biological molecule can exist in a number of conformations which differ in the degree to which the hydrocarbon portions are exposed to the solvent. Exposure of the hydrocarbon portions to the solvent causes the nearby water to become more "ice-like," accompanied by a large negative entropy change and a positive contribution to the free energy. The more stable (lower free energy) conformations of the biological molecule will

[*]Némethy, G. and H. A. Scheraga, *J. Chem. Phys.*, **36**, 3401 (1962).

[†]Frank, H. S. and M. J. Evans, *J. Chem. Phys.*, **13**, 507 (1945).

[‡]The reader should re-examine Section 2.6 *Thermodynamic Studies on the Structure of Water* of Chapter 2 in this text.

[**]Kauzmann, W., *Advan. Protein Chem.*, **14**, 1 (1959).

minimize the contact of water with its hydrocarbon substituents. Globular proteins accomplish this by folding into compact structures with the majority of their hydrophobic amino acid side chains in the interior, away from contact with the solvent.[*] The formation of the "ice-like" structure, and the accompanying large decrease in entropy, is thereby avoided.

Let us now direct our attention to the effect of temperature upon the butane-water system. An aqueous solution of butane has a much higher value of $C_P$ than do the pure liquids separately. The partial molal heat capacity at constant pressure is +80 cal $deg^{-1}$ $mole^{-1}$ for butane in water.[†] The heat capacity of liquid butane is about 30 cal $deg^{-1}$ $mole^{-1}$. Why should more heat be required to raise the temperature of butane in water than is required for raising the temperature of butane and water separately? The effect can be explained by assuming that the water surrounding the dissolved butane becomes more "ice-like." When raising the temperature of an aqueous solution of butane, we must supply not only the heat which would be required for heating the butane and water separately, but also additional heat to "melt" the "ice-like" structures surrounding the dissolved butane molecules. It is this additional heat which accounts for the positive $\Delta C_P$ which is a consequence of the solution of butane in water. If we assume that the hydrocarbon substituents of biological molecules affect water in a manner similar to butane, we would anticipate a difference in the heat capacities of products and reactants for any reaction that involves a net change in the exposure of hydrocarbon substituents to water, *i.e.*, that which involves a change in hydrophobic interactions. The thermal denaturation of ribonuclease proceeds with $\Delta C_P = +2$ kcal/deg mole at 30°. This large value of $\Delta C_P$ would be consistent with the transfer of several hydrophobic amino acid side chains from the interior of the protein to contact with the solvent upon denaturation. The thermal denaturation of ribonuclease and of many other proteins proceeds with a temperature dependence of $\Delta H°$ of denaturation.[‡]

In the examples that follow, the various ways in which Eqs. 3.42–3.47 may be used are illustrated. For ease in calculation all logarithmic terms are expressed to the base 10.

*Example 3.9* Calculate $\Delta G$ for a chemical reaction in which reactants and products are not at equilibrium concentrations. Consider the reaction

$$H_{2(g)} + I_{2(g)} \rightarrow 2HI_{(g)}.$$

[*]Dickerson, R. E., and I. Geis, *The Structure and Action of Proteins*, Harper & Row, 1969, Chapters 3 and 4.

[†]Némethy, G. and H. A. Scheraga, *J. Chem. Phys.*, **36**, 3401 (1962).

[‡]Brandts, J. F., in *Structure and Stability of Biological Macmolecules*, S. N. Timasheff and G. D. Fasman, Eds., Marcel Dekker, N.Y., 1969, p. 213.

The $K_P$ for this reaction is 53 at 444°C. What will be the $\Delta G$ of the reaction if one mole of iodine vapor at a pressure of 0.4 atm reacts with one mole of hydrogen at 0.2 atm to form two moles of HI at a pressure of 0.8 atm, all components being at a temperature of 444°C?

Equation 3.42 (in the second form) and Eq. 3.43 are used to solve the problem. The two equations are combined to give the form needed, *i.e.*,

$$\Delta G = \Delta G° + 2.303 \ RT \log Q \qquad \text{(Eq. 3.42)}$$

and

$$\Delta G° = -2.303 \ RT \log K_P \qquad \text{(Eq. 3.43)},$$

are combined to give

$$\Delta G = -2.303 \ RT \log K_P + 2.303 \ RT \log Q$$

$$= -2.303 \,(1.987)\,(717)\,(\log 53) + 2.303\,(1.987)\,(717)\left(\log\frac{(0.8)^2}{(0.2)\,(0.4)}\right)$$

$$= -2694 \text{ calories.}$$

***Example 3.10*** Calculate $\Delta H°$ from equilibrium constants obtained at two temperatures. In the enzymatic conversion of L-aspartate to fumarate and ammonium ion, the equilibrium constant was found to be $1.60 \times 10^{-2}$ at 39°C, and $0.74 \times 10^{-2}$ at 29°C. What is $\Delta H°$ of the reaction over this temperature range? The appropriate equation is 2.47 if we assume $\Delta H°$ independent of $T$.

$$2.303 \log \frac{K_2}{K_1} = \frac{\Delta H°(T_2 - T_1)}{RT_2T_1} \qquad (K \text{ is } K_C).$$

$K_2$ and $T_2$ refer to the system at 39°C and $K_1$ and $T_1$ refer to the system at 29°C. The data are substituted accordingly:

$$2.303 \log \frac{1.60 \times 10^{-2}}{0.74 \times 10^{-2}} = \frac{\Delta H°(10)}{(1.987)\,(312)\,(302)}$$

$$\Delta H° = +14,400 \text{ calories per mole.}$$

***Example 3.11*** Calculate $\Delta G°$ at 39°C for the reaction described in Example 3.10.

$$\Delta G°_{312} = -2.303 \ RT \log K_C = (-2.303)\,(1.987)\,(312)\,(\log 1.60 \times 10^{-2})$$

$$= +2,560 \text{ calories per mole.}$$

Conversely, if the $\Delta G°$ of a reaction is known or can be calculated from standard free energies of formation, the equilibrium constant is readily calculated. $K$ obtained in this way will be for the temperature of the $\Delta G°$ value used.

***Example 3.12*** Calculate $\Delta S°$ for the enzymatic reaction described in Example 3.10. In this example, $\Delta H°$ was found to be 14,400 calories for the reaction when written

$$\text{L-aspartate} \longrightarrow \text{fumarate} + \text{NH}_4^+.$$

At 39° (312°K) the equilibrium constant is $1.60 \times 10^{-2}$. In Example 3.11 these data were used to calculate $\Delta G°$ at the temperature specified; it was found to be 2,560 calories per mole.

The definition of $G$ is

$$G = H - TS.$$

For a change in the standard free energy at constant temperature,

$$\Delta G° = \Delta H° - T \, \Delta S°,$$

$$2{,}560 = 14{,}400 - (312)\,(\Delta S°),$$

$$\Delta S°_{312} = +38.0 \text{ calories per mole per degree.}$$

## 3.8  Living Systems and $\Delta G$

The principal energy currency of living systems is a molecule of moderate complexity known as adenosine-5'-triphosphate, usually abbreviated to ATP. In solutions at pH 7 and above, the principal ionic form of the molecule is

Adenosine-5'-triphosphate (ATP)

ATP is the most ubiquitous and important member of a group of organic compounds that participate in metabolic processes involving energy storage, utilization, and transfer. Chemically, these compounds are principally organic phosphates or thioesters, and are commonly described as having "high-energy bonds."[*] To a certain degree this term is misleading, since it does not refer in the usual sense to the energy attendant upon bond formation or bond rupture. Most physical chemistry textbooks contain tables giving average bond strengths in terms of $\Delta H°$, the heat absorbed on bond rupture. The bonds are presumed to break

---

[*]In this context, the "high energy bond" refers to the energy distribution of electrons in all of the bonds around the phosphorous atoms.

symmetrically with all reactants and products in the gas phase. This type of bond energy is appropriately called "heat of atomization." The reference temperature is 25°C. From these data we can calculate a heat of reaction by adding together all the types of bonds formed and subtracting from that total the sum of the various bonds broken, each bond evaluated in terms of bond strength in kilocalories per mole at 25°C. For example, take the chemical reaction catalyzed by the enzyme aspartase:

$$^-OOC-CH_2-CH-COO^- \longrightarrow \ ^-OOC-CH=CH-COO^- + NH_4^+$$
$$| $$
$$NH_3^+$$

L-Aspartate                                                      Fumarate

When aspartate is converted to fumarate and ammonium ion, the bonds broken are N—C and C—H. In addition a C—C bond is changed into a C=C bond, and an N—H bond is formed. Based on the average values of bond energies given by T. L. Cottrell,[*] $\Delta H°$ for the reaction is found to be 19 kcal/mole. Considering that this calculation is based on energies of homolytic bond scission in gas phase, it agrees surprisingly well with experimentally determined values for $\Delta H°$ based on equilibrium constants for the reaction in dilute neutral solution. Wilkinson and Williams[†] found $\Delta H°$ to be 14.4 kcal per mole whereas Bada and Miller[‡] obtained somewhat lower values, around 6–7 kcal per mole. When $\Delta H°$ was calculated for the reverse reaction, that is, from equilibrium constants obtained by combining fumarate and ammonium salt in the presence of enzyme and allowing the system to go to equilibrium, the enthalpy change was found to be $-15.5$ kcal/mole, in excellent agreement with the value from the forward reaction.

For a group of compounds of which ATP is the most prominent member, it has been found to be of value when considering changes associated with bond formation to discuss free energy instead of enthalpy. The resulting quantities are energies of *bond transfer* and are free energies. Calculations are based on the decrease in free energy accompanying the transfer of a specific group, such as phosphate, to water — that is, the free energy of hydrolysis. Although the isolated hydrolysis of ATP is energetically wasteful and of uncertain significance in metabolism, the free energy of hydrolysis is a standard of reference by which we may compare ATP with other members of the group. The symbol ~ is used to designate the particular bond about which the energy is localized. Thus the ATP molecule is written as Ad-Rib-P ~ P ~ P in a shorthand notation designed to differentiate between the high-energy pyrophosphate linkages and the ester phosphate bond connecting ribose to the first phos-

[*]Cottrell, T. L., *The Strength of Chemical Bonds.* Butterworth Scientific Publications, London, 1958.
   [†]Wilkinson, J. S. and V. R. Williams, *Arch. Biochem. Biophys.*, **93**:80, 1961.
   [‡]Bada, J. L. and S. L. Miller, *Biochemistry*, **7** 3403 (1968).

phate group. The transfer of a terminal phosphate group of ATP to water would result in the release of approximately twice the free energy that would accompany the hydrolysis of an ordinary phosphate ester bond like that occurring in adenosine monophosphate (AMP). For example, in the abbreviated notation for these compounds,

$$\text{Ad-Rib-P} \sim \text{P} \sim \text{P}^{4-} + \text{H}_2\text{O} \rightarrow \text{Ad-Rib-P} \sim \text{P}^{3-} + \text{HPO}^{2-} + \text{H}^+$$
$$\text{(ATP)} \qquad\qquad\qquad \text{(ADP)}$$

$$\Delta G' = -7 \text{ kcal/mole} \qquad (3.51)$$

$$\text{Ad-Rib-P}^{2-} + \text{H}_2\text{O} \rightarrow \text{Ad-Rib} + \text{HPO}_4^{2-}$$
$$\text{(AMP)}$$

$$\Delta G' = -3 \text{ kcal/mole} \qquad (3.52)$$

The notation $\Delta G'$ means that $\Delta G°$ is calculated for the reaction at physiological pH, usually 7.0, and moderate temperature, usually 25°C. The value of $-\Delta G'$ in kilocalories for these hydrolyses is called the *phosphate transfer potential*. Free energies of hydrolysis and phosphate transfer potentials for a number of organic phosphates are given in Table 3.4.

Compounds displaying large decreases in $\Delta G'$ on the hydrolysis of phosphate groups fall into several categories: (a) polyphosphates, such as ATP, (b) enol-phosphates, (c) acyl-phosphates, and (d) guanidino-phosphates. The student is referred to a standard biochemistry textbook for the structural formulas of these compounds.

The large free energy of hydrolysis of high-energy phosphate compounds is attributed to several factors, special prominence being given to the increase in resonance stabilization and the decrease in electrostatic repulsion resulting from hydrolysis. Counting the various resonance forms of ATP$^{4-}$ and comparing that number with the sum of the various resonance forms of ADP$^{3-}$ and HPO$_4^{2-}$ will verify that hydrolysis leads to resonance stabilization. In addition, the removal of the terminal phosphate of ATP$^{4-}$ relieves the electrostatic repulsion existing between its four closely situated negatively charged oxygens. The two products, ADP$^{3-}$ and HPO$_4^{2-}$, are free to move apart in water and thereby reduce the repulsive forces between their like charges.

It must be emphasized that Eq. 3.51 is an oversimplification which holds true only in solutions on the alkaline side of pH 7. The three phosphate compounds, ATP, ADP, and inorganic phosphate, all undergo a dissociation step near neutral pH (see Table 4.6). Therefore, to consider every major ionic species present at neutrality, we must include ATP$^{4-}$, ATP$^{3-}$, ADP$^{3-}$, ADP$^{2-}$, HPO$_4^{2-}$, and H$_2$PO$_4^-$. If, in addition, we include various physiologically important calcium and magnesium complexes of these compounds which exist in the range pH 4–10, we will have altogether seven species of ATP, seven species of ADP, and four species of inorganic phosphate.[*] Only then can the pH dependence of the

---

[*]Alberty, R. A., *J. Biol. Chem.*, **243**, 1337, 1968.

**TABLE 3.4**  FREE ENERGY OF HYDROLYSIS
OF PHOSPHATE COMPOUNDS*

| Compound | $\Delta G'$ at 25°C kcal/mole |
|---|---|
| Acetyl adenylate | −13.3 |
| Phosphoenolpyruvate | −12.8 |
| 1,3-Diphosphoglycerate (acyl phosphate only) | −11.8 |
| Creatine phosphate | −10.3 |
| Acetyl phosphate | −10.1 |
| Arginine phosphate | −8.0 |
| ATP | −7.0 |
| Inorganic pyrophosphate | −6.6 |
| ADP | −6.4 |
| Glucose-1-phosphate | −5.0 |
| 2-Phosphoglycerate | −4.1 |
| Fructose-6-phosphate | −3.8 |
| Glucose-6-phosphate | −3.3 |
| Fructose-1-phosphate | −3.1 |
| 3-Phosphoglycerate | −3.0 |
| Glycerol-1-phosphate | −2.3 |

*Source:* Data are quoted from Atkinson, M. R., and R. K. Morton in Chapter 1 of *Comparative Biochemistry*, Vol. II, edited by Mason, H. S., and M. Florkin. Academic Press, New York, 1960. Original sources are given by the authors.

*The phosphate transfer potential is obtained by changing the sign of $\Delta G'$ at 25°C.

hydrolysis of ATP be viewed in an unequivocal way. Since this topic is more appropriately considered along with the other ionic equilibria in Chapter 4, we have deferred further discussion until then. In the meanwhile we will continue to use Eq. 3.51 and similar formulations while being aware of their limitations.

In living systems the energy of phosphate group transfer is not wasted in fruitless hydrolysis reactions. Rather, it is conserved for future use by being transferred with little or no loss of free energy, or is utilized to effect the synthesis of needed metabolites. Let us discuss these two processes separately.

**Transfer reactions in which the energy of ATP is conserved**  In the oxidation of foodstuffs in cells, the energy of oxidation is trapped in the form of ATP molecules or reduced pyridine nucleotides. Whereas these processes are common to nearly all living systems, there exists in man a means for storing much of the chemical energy of ATP in another form.

This particular form, creatine phosphate, is stable in the muscle cells where it is stored, but readily gives up its high-energy phosphate group to resynthesize ATP. The equation for this reaction is:

$$\underset{\text{(ATP)}}{\text{Ad-Rib-P} \sim \text{P} \sim \text{P}^{4-}} + \text{Creatine} \overset{\underset{\text{kinase}}{\overset{\text{phospho-}}{\text{creatine}}}}{\rightleftarrows} \underset{\text{(ADP)}}{\text{Ad-Rib-P} \sim \text{P}^{3-}} + \text{Creatine} \sim \text{P}^{2-} + \text{H}^+$$

$\Delta G'$ at pH 7.4, 30°C, and 0.01 $M$ $Mg^{2+}$ is reported to be +2.8 kcal/mole. Although the direction of the free energy change does not favor the synthesis of creatine phosphate at standard conditions, the accumulation of ATP in resting muscle will cause the reaction to go to the right, thereby storing high energy phosphate until needed. Creatine phosphokinase was discovered in 1935 by K. Lohmann, who had earlier discovered ATP.

A second example of an energy-conservation transfer occurs in the enzymatic phosphorylation of Coenzyme A by ATP.

$$\text{ATP}^{4-} + \text{CoA—SH} \rightleftarrows \text{CoA—S} \sim \text{PO}_3^{2-} + \text{ADP}^{3-} + \text{H}^+$$

$\Delta G°$ for this reaction at pH 7.4, 37°C, is reported to be −0.2 kcal/mole.

**Reactions involving utilization of the free energy of ATP**   In addition to the mechanical work of muscle, the chemical energy of ATP is diverted by living systems into the production of electrical energy, light, and chemical work. Of these various forms of useful work, the last is the most common and will be used as an illustration. An important characteristic of many of the chemical processes which involve ATP as a reactant is the near-irreversibility of these reactions under biological conditions. Equilibrium constants range from $10^3$ upward for these reactions, meaning that very large concentrations of products will be required to effect any measure of reversibility. Thus, ATP is a phosphorylating agent *par excellence.*\*

The synthesis of glycerol-1-phosphate is an example of a reaction involving a major decrease in free energy. Living cells are unable to phosphorylate the primary alcohol group of glycerol by esterifying it with inorganic phosphate. The ester can be synthesized enzymatically, however, if ATP is the phosphorylating agent. The reaction is:

$$\text{Glycerol} + \text{ATP}^{4-} \xrightarrow{\quad \overset{\alpha\text{-glycerolkinase}}{\quad\quad\quad\quad} \quad} \text{Glycerol-1-P}^{2-} + \text{ADP}^{3-} + \text{H}^+.$$

---

\*On the other hand, a few reactions are known which favor the formation of ATP from ADP and a suitable organic phosphate compound. The twelfth reaction of Table 3.5 is one of this type. These reactions, known as *substrate-level phosphorylations*, are few in number but of great metabolic importance. We will return to them in Chapter 5.

The free energy decrease for this reaction is calculated by combining equations and data (Table 3.4) for the hydrolysis of glycerol-1-P and ATP (an application of Hess' Law):

$$\begin{aligned} \text{ATP}^{4-} + \text{H}_2\text{O} &\rightarrow \text{ADP}^{3-} + \text{P}_i^{2-} + \text{H}^+ && \Delta G' = -7.0 \text{ kcal/mole} \\ \text{Glycerol} + \text{P}_i^{2-} &\rightarrow \text{Glycerol-1-P}^{2-} + \text{H}_2\text{O} && \Delta G' = +2.3 \text{ kcal/mole} \\ \hline \text{Glycerol} + \text{ATP}^{4-} &\rightarrow \text{Glycerol-1-P}^{2-} + \text{ADP}^{3-} + \text{H}^+ && \Delta G' = -4.7 \text{ kcal/mole} \end{aligned}$$

Viewed in this light, the high negative free energy of hydrolysis of ATP may be regarded as the thermodynamic driving force of the reaction, although the biochemical mechanism does not involve two consecutive hydrolysis steps as written above. Many biochemical syntheses proceed, however, by virtue of an exergonic reaction driving an endergonic reaction which is coupled to it. Let us consider these *tandem reactions* in more detail.

**Tandem reactions**   Some of the sequential or tandem reactions that occur in living systems couple together the free energy changes of two or more reactions so as to favor the formation of a desired end product. That $\Delta G°$ of a chemical reaction is positive does not mean that no product will be formed from reactants. It means that the $K_{equil}$ is a number less than unity. The sign of $\Delta G°$ tells us the direction of spontaneous change for reactants in standard states and is not an "all or none" index. If a reaction for which $\Delta G° = +$calories and therefore $K_{equil} < 1$ is followed by a reaction for which $\Delta G° = -$calories and therefore $K_{equil} > 1$, the exergonic second reaction can "drive" the endergonic first reaction and permit the desired product to be formed. Several examples of these coupled reactions occur in glycolysis. One of them involves the formation of pyruvate from phosphoglycerate. The isomerization of 3-phosphoglycerate to 2-phosphoglycerate is not favored energetically:

(1)   3-phosphoglycerate$^{3-}$ $\rightarrow$ 2-phosphoglycerate$^{3-}$

$$\Delta G' = 1.06 \text{ kcal/mole.}$$

However, the tandem reaction, which is the conversion of 2-phosphoglycerate into phosphoenolpyruvate, goes spontaneously to the right, as does the next reaction, the conversion of phosphoenolpyruvate into pyruvate and ATP.

(2)   2-phosphoglycerate$^{3-}$ $\rightarrow$ 2-phosphoenolpyruvate$^{3-}$ + $\text{H}_2\text{O}$

$$\Delta G' = -0.64 \text{ kcal/mole.}$$

(3)   2-phosphoenolpyruvate$^{3-}$ + ADP$^{3-}$ + $\text{H}^+$ $\rightarrow$ pyruvate$^-$ + ATP$^{4-}$

$$\Delta G° = -6.1 \text{ kcal/mole.}$$

The sum of reactions 1, 2, and 3 is:

$$\text{3-phosphoglycerate}^{3-} + \text{ADP}^{3-} + \text{H}^+ \rightarrow \text{pyruvate}^- + \text{ATP}^{4-} + \text{H}_2\text{O}$$

$$\Delta G' = -5.68 \text{ kcal/mole.}$$

Thus, 3-phosphoglycerate is continuously removed from the system and glycolysis proceeds in a forward direction. Free energy data for a number of reactions of biochemical interest have been compiled by Burton and Krebs. These have been grouped according to metabolic pathways and presented in Table 3.5.

**Calculations involving** $\Delta G^\circ$, $\Delta G'$, **and** $\Delta G''$   To insure that we understand the relationship between $\Delta G^\circ$, $\Delta G'$, and $\Delta G''$ as shown in Table 3.5, let us calculate $\Delta G'$ and $\Delta G''$ for the first reaction shown under the heading, Miscellaneous Reactions.

*Example 3.13*   For the reaction

$$\tfrac{1}{2} \text{ Butyrate}^- + \tfrac{1}{2}\text{O}_2 + \text{CoA} + \tfrac{1}{2}\text{H}^+ \rightarrow \text{Acetyl CoA} + \text{H}_2\text{O}$$

$\Delta G^\circ$ is given as $-42.5$ kcal in Table 3.5. $\Delta G'$ is given as $-37.7$ kcal, and $\Delta G''$ is $-36$ kcal.
The appropriate equation for the calculation of $\Delta G'$ is

$$\Delta G' = \Delta G^\circ + 2.303 \; RT \log Q,$$

where the $Q$ term contains only the activity of hydrogen ion. Since $\text{H}^+$ is a reactant, $[\text{H}^+]^{1/2}$ occurs in the denominator of $Q$. The calculation follows:

$$\Delta G' = \Delta G^\circ + 2.303 \; RT \log \frac{1}{(\text{H}^+)^{1/2}}$$

$$= -42.5 + (2.303)(1.987 \times 10^{-3})(298) \log \frac{1}{(10^{-7})^{1/2}}$$

$$= -37.7 \text{ kcal.}$$

$\Delta G''$ is of a still more general nature, in that only water is in its standard state. The usual conditions for other reactants are: pH, 7.0, solutes at an activity of 0.01 $M$, $\text{CO}_2$ at 0.05 atm pressure, and $\text{O}_2$ at 0.2 atm pressure. The equation is

$$\Delta G'' = \Delta G^\circ + 2.303 \; RT \log Q$$

For the reaction given at the beginning of this example

$$Q = \frac{(\text{Acetyl CoA})(\text{H}_2\text{O})}{(\text{Butyrate})^{1/2}(\text{O}_2)^{1/2}(\text{CoA})(\text{H}^+)^{1/2}} = \frac{(0.01)(1)}{(0.01)^{1/2}(0.2)^{1/2}(0.01)(10^{-7})^{1/2}}$$

$$= 7.1 \times 10^4.$$

**TABLE 3.5** FREE-ENERGY DATA FOR REACTIONS OCCURRING IN CERTAIN METABOLIC PATHWAYS. ($\Delta G'$ values are for unit activities of all components except $H^+$, which is taken as $10^{-7}$ moles per liter. $\Delta G''$ values are for 0.2 atm $O_2$, 0.05 atm $CO_2$, and 0.01 M concentrations of reactants except $H_2O$, which is always taken at unit activity, and $H^+$, which is taken at $10^{-7}$ M.)

| Reaction | $\Delta G°$ kcal/ mole | $\Delta G'$ kcal/ mole | $\Delta G''$ kcal/ mole |
|---|---|---|---|
| GLYCOLYSIS AND ALCOHOLIC FERMENTATION | | | |
| Glycogen + $HPO_4^{2-}$ → G-1-$P^{2-}$ + $H_2O$ | +0.55 | +0.55 | +3.28 |
| G-1-$P^{2-}$ → G-6-$P^{2-}$ | −1.72 | −1.72 | −1.72 |
| Glucose + $ATP^{4-}$ → G-6-$P^{2-}$ + $ADP^{3-}$ + $H^+$ | +3.9 | −5.1 | −5.7 |
| G-6-$P^{2-}$ → F-6-$P^{2-}$ | +0.50 | +0.50 | +0.50 |
| F-6-$P^{2-}$ + $ATP^{4-}$ → $HDP^{4-}$ + $ADP^{3-}$ + $H^+$ | +5.2 | −4.2 | −4.2 |
| $HDP^{4-}$ → glyceraldehyde-3-$P^{2-}$ + dihydroxyacetone $P^{2-}$ | +5.51 | +5.51 | +2.78 |
| Dihydroxyacetone $P^{2-}$ → glyceraldehyde-3-$P^{2-}$ | +1.83 | +1.83 | +1.83 |
| Glyceraldehyde-3-$P^{2-}$ + $NAD^+$ + $HPO_4^{2-}$ → 1,3-di-P-glycerate$^{4-}$ + NADH + $H^+$ | +11.05 | +1.50 | +4.2 |
| 1,3-di-P-glycerate$^{4-}$ + $ADP^{3-}$ → glycerate-3-$P^{3-}$ + $ATP^{4-}$ | −4.75 | −4.75 | −4.75 |
| Glycerate-3-$P^{3-}$ → glycerate-2-$P^{3-}$ | +1.06 | +1.06 | +1.06 |
| Glycerate-2-$P^{3-}$ → enolpyruvate-2-$P^{3-}$ + $H_2O$ | −0.64 | −0.64 | −0.64 |
| Enolpyruvate-2-$P^{3-}$ + $ADP^{3-}$ + $H^+$ → pyruvate$^-$ + $ATP^{4-}$ | −14.5 | −6.1 | −5.0 |
| Pyruvate$^-$ + NADH + $H^+$ → lactate$^-$ + $NAD^+$ | −15.54 | −5.1 | −6.0 |
| Glycogen (1 glucose unit) + $H_2O$ → 2 lactate$^-$ + 2$H^+$ | −32.2 | −51.6 | −54.0 |
| Glucose → 2 lactate$^-$ + 2$H^+$ | −27.9 | −47.4 | −49.7 |
| Pyruvate$^-$ + $H^+$ → acetaldehyde + $CO_2$ | −14.65 | −5.1 | −6.9 |
| Acetaldehyde + NADH + $H^+$ → ethanol + $NAD^+$ | −14.90 | −5.4 | −5.4 |

CITRIC ACID CYCLE

| Reaction | | | |
|---|---|---|---|
| Pyruvate⁻ + $\frac{1}{2}$O₂ + CoA + H⁺ → acetyl CoA + CO₂ + H₂O | −66 | −61.8 | −55 |
| Oxaloacetate²⁻ + acetyl CoA + H₂O → citrate³⁻ + CoA + H⁺ | +1.8 | −7.5 | −7.8 |
| Citrate³⁻ → isocitrate³⁻ | +1.59 | +1.59 | +1.59 |
| Isocitrate³⁻ + $\frac{1}{2}$O₂ + H⁺ → α-ketoglutarate²⁻ + H₂O + CO₂ | −63.9 | −54.4 | −55.6 |
| α-ketoglutarate²⁻ + $\frac{1}{2}$O₂ + CoA + H⁺ → succinyl CoA + H₂O + CO₂ | −69.1 | −59.6 | −58.0 |
| Succinyl CoA + GDP³⁻ + HPO₄²⁻ → succinate²⁻ + GTP⁴⁻ + CoA | −0.77 | −0.77 | −0.77 |
| Succinate²⁻ + $\frac{1}{2}$O₂ → fumarate²⁻ + H₂O | −36.13 | −36.1 | −35.7 |
| Fumarate²⁻ + H₂O → malate²⁻ | −0.88 | −0.88 | −0.88 |
| Malate²⁻ + $\frac{1}{2}$O₂ → oxaloacetate²⁻ + H₂O | −45.24 | −45.3 | −44.8 |

MISCELLANEOUS REACTIONS

| Reaction | | | |
|---|---|---|---|
| $\frac{1}{2}$Butyrate⁻ + $\frac{1}{2}$O₂ + CoA + $\frac{1}{2}$H⁺ → acetyl CoA + H₂O | −42.5 | −37.7 | −36 |
| Acetate⁻ + CoA + H⁺ → acetyl CoA + H₂O | +3.7 | +13.2 | +16 |
| 2 Acetyl CoA + H₂O → acetoacetate⁻ + 2CoA + H⁺ | −0.7 | −10.2 | −13 |
| Pyruvate⁻ + 2$\frac{1}{2}$O₂ + H⁺ → 3CO₂ + 2H₂O | −282.84 | −273.5 | −273.5 |
| NADH + $\frac{1}{2}$O₂ + H⁺ → NAD⁺ + H₂O | −61.91 | −52.4 | −51.9 |
| Glucose + 6O₂ → 6CO₂ + 6H₂O | −686.5 | −686.5 | −688.5 |
| Glutamate⁺²⁻ + $\frac{1}{2}$O₂ → α-ketoglutarate²⁻ + NH₄⁺ | −43.63 | −43.63 | −45.9 |
| Aspartate⁺²⁻ + $\frac{1}{2}$O₂ → oxaloacetate²⁻ + NH₄⁺ | −42.42 | −42.42 | −44.7 |
| Alanine⁺⁻ + $\frac{1}{2}$O₂ → pyruvate⁻ + NH₄⁺ | −43.57 | −43.57 | −45.8 |
| Glutamate⁺²⁻ + NAD⁺ + H₂O → α-ketoglutarate²⁻ + NH₄⁺ + NADH + H⁺ | +18.28 | +8.78 | +6.0 |
| Malate²⁻ + NADP⁺ → pyruvate⁻ + NADPH + CO₂ | −0.18 | −0.18 | −2.0 |
| Oxaloacetate²⁻ + H₂O → pyruvate⁻ + HCO₃⁻ | −6.39 | −6.39 | −9.12 |

*Sources:* $\Delta G'$ data are quoted from K. Burton, H. A. Krebs, and H. L. Kornberg, *Ergeb. Physiol. Biol. Chem. u. exptl. Pharmakol.*, **49**:212, 1957. This extensive review article is in English. Other data are quoted from K. Burton and H. A. Krebs, *Biochem. J.*, **54**:94, 1953. Some of the 1957 $\Delta G'$ data do not correspond to the 1953 $\Delta G°$ data cited for the same enzymatic reaction. A few data have been calculated to make the table complete.

Therefore

$$\Delta G'' = \Delta G^\circ + (2.303)(1.987 \times 10^{-3})(298) \log 7.1 \times 10^4$$

$$= -42.5 + 6.6 = -35.9 \text{ kcal.}$$

To make a rapid estimate of $\Delta G'$ from $\Delta G^\circ$ data the student should note that for $H^+$ as a *product*, the term $(2.303\ RT \log H^+)$ is equal to approximately $-9.5$ kcal for pH 7.0 and, correspondingly, when $H^+$ is a reactant, the term $(2.303\ RT \log 1/H^+)$ is equal to approximately $+9.5$ kcal. Inspection of the data of Table 3.5 will verify this generalization.

For reactions in which gases are neither products nor reactants and the number of moles of reactants equals the number of moles of products in the stoichiometry, $\Delta G' = \Delta G''$. If all of these conditions hold and, in addition, $H^+$ is neither a product nor reactant, $\Delta G^\circ = \Delta G' = \Delta G''$. See, for example, in Table 3.5 the triosephosphate isomerase reaction (the tenth reaction under Glycolysis).

**The use of $K_{app}$**   In addition to the conventions just described in the preceding paragraphs, another convention is commonly used in the thermodynamic treatment of biochemical free energy data. Since $\Delta G'$ is the standard free energy change at pH 7.0, it is convenient to speak of the *apparent* equilibrium constant, $K_{app}$, which is directly calculated from $\Delta G'$:

$$\Delta G' = -2.303\ RT \log K_{app}.$$

The use of $K_{app}$ in this sense should not be confused with $K'$, another kind of apparent equilibirum constant, that designates an equilibrium constant calculated from concentration data rather than activity values. In Chapter 4 extensive use is made of the $K'$ notation. What is to be understood here is that $K_{app}$ is a term that includes the true thermodynamic equilibrium constant $K$ and the term $Q$ (see Eqs. 3.42 and 3.43). That is,

$$K_{app} = K/Q \tag{3.53}$$

since

$$\Delta G' = -2.303\ RT \log K_{app} = -2.303\ RT \log K - 2.303\ RT \log Q.$$

Therefore, if an enzymatic reaction at pH 7.0 and 25°C has a $\Delta G' = -8,000$ cal, $K_{app}(\text{pH7})$ would be

$$K_{app}(\text{pH7}) = \text{antilog} \frac{-8,000}{-2.303 \times 1.99 \times 298} = \text{antilog } 5.82$$

$$K_{app}(\text{pH7}) = 6.6 \times 10^5.$$

Suppose that we want to know $K_{app}$ at pH 6.0 instead of pH 7.0? If we evaluate $K$, we can then calculate $K_{app}$ at any pH from Eq. 3.53, since the $Q$ term contains only $[H^+]$ or its reciprocal, all other reactants and products being at unit activity. However, it is not necessary to calculate $K$ in order to determine $K_{app}$ at pH 6.0 if $K_{app}$ is known for pH 7.0 or any other pH. Assuming that the reaction under consideration is of the type $A + B = C + D + H^+$. It follows that $Q = [H^+]$. For the general case of pH $= y$,

$$K_{app}(pHy) = K/10^{-y}$$

and for pH $= x$,

$$K_{app}(pHx) = K/10^{-x}.$$

Therefore

$$K_{app}(pHy)/K_{app}(pHx) = 10^{-x}/10^{-y}.$$

For the specific data given above, that is, $K_{app}(pH7) = 6.6 \times 10^5$, $K_{app}(pH6)$ $= (6.6 \times 10^5)(10^{-7}/10^{-6}) = 6.6 \times 10^4$.

If the reaction equation contains $H^+$ as a reactant rather than a product, $Q = 1/[H^+]$, the general equation becomes

$$K_{app}(pHy)/K_{app}(pHx) = 10^{-y}/10^{-x}.$$

If $\Delta G'_y$ is known at a certain pH, pHy, and it is desirable to know $\Delta G'_x$ at another pH, pHx, the calculation can be made directly from the free energy and pH data. It is not necessary to calculate either $K_{app}$ or $K$. For $H^+$ as a *product*, $Q = [H^+]$ and

$$\Delta G'_{pHy} - \Delta G'_{pHx} = 2.303\ RT\ (pHx - pHy).$$

For $H^+$ as a *reactant*, i.e., $Q = 1/[H^+]$

$$\Delta G'_{pHy} - \Delta G'_{pHx} = 2.303\ RT\ (pHy - pHx).$$

The student should find it easy to verify both these relationships.

Finally, the student should be cautioned that although the use of $\Delta G'$ values is indispensable in providing a standard of reference for studying the energetics of living cells, the $\Delta G'$ is the difference between the free energy of products and reactants in their standard states at pH 7. Physiological concentrations are believed to be much lower, although no one can say positively what concentrations may be present in the vicinity of the active sites of enzymes. The complex internal structure of the living cell enables it to channel metabolites in the most beneficial way. We are presently compelled to measure the concentrations of these metabolites in cell lysates or pressed juices in which most of the cellular structure has been destroyed. In other words, $\Delta G'$ is principally an informed estimate that does not tell us the exact magnitude of the true $\Delta G$ of the various biochemical reactions of interest. Correction of $\Delta G'$ values to $\Delta G''$ as is done in Table 3.5 gives a more realistic free energy change for physiological conditions.

An additional word of explanation must be offered with regard to the frequent revisions made in the value of $\Delta G'$ of hydrolysis of ATP and other high-energy biological metabolites. For $\Delta G'$ values in the range of $-10{,}000$ calories per mole, great experimental difficulties attend the direct measurement of the equilibrium constant. Calculations are usually made by an indirect route, for example, from enthalpy and entropy changes calculated from thermal data, from oxidation-reduction data, or by applying Hess' Law to reactions that can be added to give the final desired hydrolysis equation and $\Delta G'$ of hydrolysis.

Recent microcalorimetric measurements of the heat of reaction and equilibrium point of the hydrolysis of glutamine have yielded thermodynamic data which have been combined with reliable data from the glutamine synthetase reaction to achieve a new evaluation of the free energy of hydrolysis of ATP.[*] The two reactions and their sum follow. Glutamine has been abbreviated to Ge and Glutamate to Ga. The superscript, $+2-$, *e.g.*, means a zwitterion having one positive group and two negatively charged groups.

$$Ge^{+-} + H_2O \rightarrow Ga^{+2-} + NH_4^+, \qquad\qquad \Delta G' = -3750 \text{ cal}$$
$$Ga^{+2-} + ATP^{4-} + NH_4^+ \rightarrow Ge^{+-} + ADP^{3-} + H^+ + HPO_4^{2-}, \Delta G' = -3670 \text{ cal}$$
$$\overline{ATP^{4-} + H_2O \rightarrow ADP^{3-} + H^+ + HPO_4^{2-}, \qquad\qquad \Delta G' = -7420 \text{ cal}}$$

$\Delta G'$ was found to be $-7{,}420$ calories/mole at pH 7.0 and 37°C in the presence of magnesium ions. This value agrees well with other estimates, which have ranged from $-7$ to $-10$ kcal/mole. The ATP hydrolysis equilibrium is discussed more fully in Chapter 4, Section 4.5, after a development of the principles of multiple equilibria.

### Suggested Additional Reading

#### General: Chemical Equilibrium

Daniels, F., and R. A. Alberty, *Physical Chemistry*, 3rd Ed. John Wiley and Sons, New York, 1966.

Moore, W. J., *Physical Chemistry*, 4th Ed. Prentice-Hall, Englewood Cliffs, N.J., 1972.

---

[*] Benzinger, T. H., C. Kitzinger, R. Hems, and K. Burton, *Biochem. J.*, **71**: 400, 1959.

### The Phase Rule

Findlay, A., A. N. Campbell, and N. O. Smith, *The Phase Rule and Its Applications*, 9th Ed. Dover Publications, New York, 1951.

Veis, A., in *Biological Polyelectrolytes*, Veis, A., Ed. Marcel Dekker, New York, 1970, p. 211.

### Colligative Properties: Medical and Biochemical Applications

Clark, W. M., *Topics in Physical Chemistry*, 2nd Ed. Williams and Wilkins, Baltimore, 1952.

Cohn, E. J., and J. T. Edsall, *Protein, Amino Acids, and Peptides*. Reinhold, New York, 1943. (Reprinted by Hafner Publishing Company, New York, 1965.)

Johlin, J. M., *Introduction to Physical Biochemistry*, 2nd Ed. Paul B. Hoeber, New York, 1949.

Kupke, D. W., in *Advances in Protein Chemistry*, Vol. 15. Academic Press, New York, 1960.

### Membranes

Chance, B., C. Lee and J. Blasie, Editors, *Probes of Structure and Function of Macromolecules and Membranes*, Vol. 1, Academic Press, New York, 1971.

Henn, F. A., and T. E. Thompson, *Annual Review of Biochemistry*, **38**, 241 (1969).

Korn, E. D., *Annual Review of Biochemistry*, **38**, 263 (1969).

Rothfield, L., and A. Finkelstein, *Annual Review of Biochemistry*, **37**, 463 (1968).

### ATP and Living Systems

Lehninger, A. L., *Bioenergetics*, 2nd Ed. W. A. Benjamin, New York, 1971.

### Structure of Water

Eisenberg, D. and W. Kauzmann, *The Structure and Properties of Water*. Oxford Univ. Press, 1969.

Solomon, A. K., *Sci. Am.*, Feb. 1971, p. 89. Available as Scientific American Offprint No. 1213 from W. H. Freeman and Company. "The State of Water In Red Cells."

Whipple, H. E., editor, *Forms of Water in Biologic Systems*. Annals of the New York Academy of Sciences, **125** (Article 2):249–772, 1965.

## Problems

**3.1** In the Krebs Cycle, citrate is converted to isocitrate. (The cycle is also known as Citric Acid Cycle.) From the algebraic sign of $\Delta G°$ for the isomerization, what is the favored or spontaneous direction of the reaction? Calculate $\Delta G°$ for the following reaction and explain what bearing this reaction will have on the isomerization of citrate and the operation of the Krebs Cycle:

$$\text{Isocitrate}^{3-} + \tfrac{1}{2}O_{2(g)} + H^+ \rightleftharpoons \alpha\text{-ketoglutarate}^{2-} + H_2O_{(l)} + CO_{2(g)}.$$

Assume the standard state to be an activity $= 1\ M$.

*Answer.* $\Delta G°$ for isomeriztion $= +1.59$ kcal. Isomerization to isocitrate is nonspontaneous for reactants and products in their standard states. $\Delta G°$ for the oxidation and decarboxylation of isocitrate is $-63.9$ kcal. This large free energy decrease will favor the constant removal of isocitrate and permit the cycle to move forward despite the nonspontaneity of the isomerization.

**3.2** Consider mixing a solvent and a solute (nonvolatile) to form an ideal solution. What is the enthalpy of mixing? What is the sign of the entropy of mixing and the free energy of mixing? Explain in words the reason for the sign of the entropy of mixing.

**3.3** Calculate the concentration of a solution prepared by dissolving 1.000 g ethyl alcohol in 150.0 g water in terms of (a) molarity, (b) molality, (c) normality, and (d) mole fraction. Assume the density to be that of pure water.

**3.4** A sample of blood has a freezing point of $-0.560°C$. What is the effective molal concentration of the solute in the serum? According to this concentration, what is the osmotic pressure in the blood serum at body temperature ($37°C$)?

**3.5** Assuming that glucose and water form an ideal solution, what is the vapor pressure at $20°C$ of a solution of 1.000 g of glucose in 100.0 g of water? The vapor pressure of pure water is 17.535 mm at the temperature specified.

*Answer.* 17.517 mm.

**3.6** Calculate the vapor pressure of water at $150°C$. Discuss the assumptions made in deriving the Clausius-Clapeyron Equation.

**3.7** Draw the phase diagram for water. Label all the regions, the triple point, and the critical point. Comment on the slope of the lines separating the phase regions.

**3.8** A system contains four components: water, salt, acid, and a protein. How many degrees of freedom will there be if the system is (a) a homogeneous solution, (b) a saturated solution containing one solid, (c) a saturated solution containing two separate solid phases?

How many degrees of freedom will there be if temperature, pressure, concentration of salt, and concentration of acid (*i.e.*, a buffer) are held constant?

**3.9** A solution of sucrose ($C_{12}H_{22}O_{11}$) is made by dissolving 34.2g of sugar in 1000g of water.

(a) Calculate the vapor pressure of the solution at 20°C.

(b) From the value of $\ln P_0/P$ calculate the osmotic pressure at 20°C, the freezing point, and the boiling point of the solution. Do not use the approximate relationship given in Eq. 3.28.

The density of the solution at 20°C is 1.024. The vapor pressure of water at this temperature is 17.535 mm of Hg. The heat of vaporization of water is 539 cal per g, and the heat of fusion is 79.7 cal per g.

**3.10** Using the Clausius-Clapeyron Equation as it appears in Eq. 3.16 or 3.27, solve the following problems:

(a) Calculate the freezing point of an aqueous solution of glycine for which the mole fraction of glycine is 0.0058, using the fact that the heat of fusion of water is 79.7 cal per g.

(b) The heat of vaporization of benzene is 94.3 cal per g and its boiling point is 80.2°C. A solution containing 11.94 g of solute per 1000 g of benzene raised the boiling point 0.152°. What is the mole fraction solute? What is the molecular weight of the solute?

**3.11** Calculate how many grams of glycerine must be added to 1000 g of water to reduce the vapor pressure by 2.000 mm at 25°C. The vapor pressure of pure water at this temperature is 23.756 mm.

*Answer.* 470 g by Eq. 3.19; 430 g by Eq. 3.21.

**3.12** Two g of an unknown substance is added to 100.0 g of benzene. The vapor pressure of the benzene is reduced 1.00 mm. If the temperature is 20°C, and the vapor pressure of pure benzene is 76.5 mm, what is the molecular weight of the substance?

**3.13** Calculate the freezing point of an aqueous solution which contains 50.0 g of ethylene glycol per 500.0 g of water.

*Answer.* −2.99°C.

**3.14** A sample of a nonvolatile fatty acid weighing 1.250 g is dissolved in 500 g of carbon tetrachloride. The boiling point of the solvent is raised 0.06°. The boiling point of pure $CCl_4$ is 76.75°C, and the molal boiling point constant is 4.88. What is the molecular weight of the fatty acid?

**3.15** It is necessary to determine the molecular weight of an unknown lipid by boiling point elevation. The lipid can be dissolved in either chloroform or methanol. If the accuracy of the molecular weight determination depends on the size of the boiling point elevation, which solvent should be used? Chloroform boils at 61.5°C with a heat of vaporization of 59.0 cal/g, and methanol boils at 64.7°C with a heat of vaporization of 262.8 cal/g.

*Answer.*
$K_b = 3.76$ for chloroform
$K_b = 0.86$ for methanol
Choose chloroform

**3.16** The fluid in certain protoplasts has an osmotic pressure of 5 atm. What must be the molal concentration of an aqueous sucrose solution if it is to be isoosmotic with the fluid in these cells at 30°C?

*Answer. 0.2 m.*

**3.17** The following osmotic pressures were observed with isoelectric solutions of denatured aldolase in 6 M guanidinium chloride + 0.1 M mercaptoethanol at 25°C. (Based on S. Lapanje and C. Tanford, *J. Amer. Chem. Soc.*, **89**, 5030 (1967)).

| Aldolase Conc., g/cm³ | Osmotic Pressure, cm. of solvent |
|---|---|
| .0012 | 0.66 |
| .0018 | 1.01 |
| .0027 | 1.56 |
| .0037 | 2.22 |
| .0051 | 3.19 |

(a) What is the molecular weight of denatured aldolase? The density of the solvent is 1.14 g/cm³.
(b) Native aldolase has been found to have a molecular weight of 158,000 (K. Kawahara and C. Tanford, *Biochemistry*, **5**, 1578 (1966)). If the native aldolase molecule contains several polypeptide chains, which dissociate upon denaturation, how can you account for this result?

*Answer.*
(a) 42,000.
(b) Native aldolase contains four polypeptide chains.

**3.18** The following osmotic pressures were observed with the synthetic polyamino acid poly-L-proline in water at 30°C (W. Mattice and L. Mandelkern, *J. Amer. Chem. Soc.*, **93**, 1769 (1971)).

| Conc., mg/cm³ | Osmotic pressure, cm solvent |
|---|---|
| 3.71 | 2.18 |
| 5.56 | 3.58 |
| 8.34 | 6.13 |
| 11.12 | 9.22 |
| 13.90 | 13.38 |

What is the molecular weight of this sample of poly-L-proline?

*Answer.* 54,000.

**3.19** The following osmotic pressures were observed with isoelectric solutions of bovine serum albumin at 25°C:

Native bovine serum albumin in 0.15 M NaCl (based on G. Scatchard et al., *J. Amer. Chem. Soc.*, **68**, 2320 (1946)).

| BSA Conc., g/cm³ | Osmotic pressure, cm solvent |
|---|---|
| 0.002 | 0.73 |
| 0.003 | 0.10 |
| 0.004 | 1.47 |
| 0.005 | 2.22 |

Denatured bovine serum albumin in 6M guanidinium chloride + 0.1 M mercaptoethanol (based on S. Lapanje and C. Tanford, *J. Amer. Chem. Soc.*, **89**, 5030 (1967)).

| BSA Conc., g/cm³ | Osmotic pressure, cm solvent |
|---|---|
| 0.002 | 0.69 |
| 0.003 | 1.08 |
| 0.004 | 1.50 |
| 0.006 | 2.45 |

The density of 0.15 M NaCl can be taken as 1.00, and that of 6M guanidinium chloride + 0.1 M mercaptoethanol as 1.14 g/cm³.
(a) What is the molecular weight of native bovine serum albumin?
(b) What is the molecular weight of the denatured bovine serum albumin?
(c) What would the error in the molecular weights of native and denatured bovine serum albumin be if they were estimated single measurements at a concentration of 0.006 g/cm³? Does native or denatured bovine serum albumin form a nearly ideal solution?

*Answer.*
(a) 69,000.
(b) 71,000.
(c) Estimated molecular weights would be 68,000 in 0.15 M NaCl and 54,000 in 6 M guanidinium chloride + 0.1 M mercaptoethanol.

**3.20** Seventy-five g of a substance is dissolved in 1000 g of water. This solution at 25°C has a density of 1.02 g per cc and an osmotic pressure of 2.1 atm. Calculate the molecular weight of the substance. Use Eq. 3.33.

*Answer.* 829.

**3.21** Find an expression, in terms of the density of the solvent, for the cm-gm-sec value of the gas constant as it should be used in Eq. 3.34 if C is to be expressed in g/cm³ and $\pi$ in cm of solvent.

*Answer.*

$$R = \left(\frac{0.082\ 1 \cdot atm}{mole°K}\right)\left(\frac{76.0cmHg}{atm}\right)\left(\frac{1000cm^3}{1}\right)\left(\frac{13.542g\ Hg/cm^3}{density\ of\ solvent}\right)$$

$$= 84{,}500/density\ of\ solvent.$$

**3.22** Five g of substance A is added to 150.0 g of water. The boiling point of the solution is raised 0.01°C. What is the molecular weight of A?

**3.23** A membrane separates a solution of sodium proteinate (0.001 M) from a solution of 0.01 M sodium chloride. If the membrane is permeable to sodium chloride, but impermeable to the protein anion, how many moles of salt (per liter) will move across the membrane, and what will be the final external concentration of salt? Assume that water does not move.

*Answer.*
0.0048 moles/liter;
0.0052 moles/liter.

**3.24** A mixture of proteins is analyzed by the method of solubility curves. A plot of "grams of protein dissolved per unit volume of solvent" (*y* axis) versus "grams of protein added per unit volume of solvent" (*x* axis) shows four straight line segments. The *x*, *y* coordinates of the points of intersection of the segments are: (2.5, 2.5), (5.0, 4.0), (10.0, 6.5), and (20.0, 8.5). Identifying the pure solutes as *A*, *B*, *C*, and *D*, in order of separation, what are the solubilities of each of these, and what is the percentage composition of the mixture?

*Answer.*

| Component | Percent of Mixture | Solubility in g per unit vol of solvent |
|---|---|---|
| *A* | 40 | 1.0 |
| *B* | 10 | 0.5 |
| *C* | 30 | 3.0 |
| *D* | 20 | 4.0 |

**3.25** Consider the following reaction in which all components are gases:

$$A + 2B = C$$

(a) What is the relationship of $K_C$ to $K_P$ for this system?
(b) If $\Delta H^\circ$ at 25°C is 17,500 cal, what is the value of $\Delta E^\circ$?
(c) If one mole of $C$ is contained in 10 liters volume, and $x$ moles dissociate to give $B$ and $A$, what will be the $K_C$ expression for calculating the value of $x$?

*Answer.*
(a) $K_P = K_C(RT)^{-2}$.
(b) $\Delta E^\circ = +18,684$ cal.
(c) $K_C = \dfrac{100(1-x)}{4x^3}$.

**3.26** Calculate the change in the free energy per one mole of solute when 500 ml of a 0.1 $M$ solution of glucose in water at 30°C is diluted to one liter.

*Answer.* $-417$ cal.

**3.27** The standard free energy change for the reaction catalyzed by fumarase (fumarate$^{2-}$ + $H_2O$ → malate$^{2-}$) is given in Table 3.5. Verify the result using the standard free energies of formation in aqueous solution at 1 $M$ activity at 25°C.

**3.28** Imagine that you are a cell whose purpose is to accumulate a store of compound $B$ from compound $A$, which is supplied by the blood at a concentration of 0.001 M. $\Delta G^\circ$ is +500 cal for the reaction

$$A_{(aq)} = B_{(aq)} + C_{(aq)}$$

$A$, $B$, and $C$ are all nonvolatile. Describe two mechanisms by which you might accumulate a store of compound $B$. Explain the thermodynamic basis of your mechanisms.

**3.29** (a) Show that $K_P = K_C(RT)^{\Delta n}$.
(b) Show the mathematical relationship between $K_C$ and $K_P$ for these reactions:

$$HCl_{(g)} + NH_{3(g)} \qquad \rightarrow NH_4Cl_{(s)}.$$
$$CH_2 = CH_{2(g)} + HBr_{(g)} \rightarrow CH_3CH_2Br_{(g)}.$$

**3.30** The *Handbook of Chemistry and Physics* (38th ed., 1956, p. 1643) gives the $K_W$ for water as $\log K_W = -14.000$ at 24°C and $\log K_W = -14.7338$ at 5°C.
(a) What would the standard enthalpy of ionization of water be if it could be assumed that it were independent of temperature?
(b) Based on the assumption in (a), predict $\log K_W$ at 60°C.

(c) The experimental value of log $K_W$ at 60° is −13.0171. Compare this with the result in (b) and explain the difference.

*Answer.*
(a) 14,700 cal.
(b) −12.83.
(c) $\Delta H°$ is not constant over the range 5° to 60°.

**3.31** For an enzymatic reaction the equilibrium constants at 39°C and 29°C are 0.0160 and 0.0074, respectively. Calculate $\Delta G°$ and $\Delta S°$ for a temperature of 29°C.

**3.32** What will be the minimum osmotic work required at 37°C for the kidneys to secrete 0.333 mole of urea per kilogram of water if the concentration of urea in plasma is 0.005 molar and the concentration in urine is 0.333 molar?

*Answer.* 861 cal.

**3.33** In Table 3.5, $\Delta G'$ for the condensation of oxaloacetate and acetyl CoA to form citrate, CoA and hydrogen ion is given as −7.5 kcal.
(a) What is the apparent equilibrium constant at pH 7?
(b) Calculation $K_{app}$(pH 5.5).

**3.34** Bada and Miller(*Biochemistry*, 7, 3403 (1968)), report the equilibrium constant for the following reaction to be $5.1 \times 10^{-3}$ at 300.5°K:

$$\text{Aspartate}^{--+} \rightleftharpoons \text{Fumarate}^{--} + \text{NH}_4^+$$

If this system is used for the synthesis of aspartic acid (by incubating potassium fumarate and ammonium chloride with enzyme in buffer at appropriate temperature and pH), will a better percentage yield of aspartic acid be obtained by starting with 0.5 M fumarate and 0.5 M ammonium chloride or with 0.05 M concentrations? What is the percentage conversion of fumarate to aspartate in each case? If, on the other hand, aspartate is used for the synthesis of fumarate, is a high percentage conversion of reactant to product favored by high or low starting concentrations of aspartate?

**3.35** Use Eq. 3.42 to verify the $\Delta G''$ values given in Table 3.5. See how many of the $\Delta G''$ values listed under Miscellaneous Reactions you can calculate correctly. The pressures and concentrations to be used in the log $Q$ term are given in the heading to the table.

**3.36** If the degree of dissociation, $\alpha$, of 50 g of nitrogen tetroxide ($N_2O_4$) into nitrogen dioxide ($NO_2$) at 25°C and 1 atm is 0.185, what is the density of the partially dissociated gas under these conditions? What volume does it occupy?

# 4 | SOLUTIONS OF ELECTROLYTES

After all, what have we said except that the ionic theory is not complete? But perfection is rare in the science of chemistry. Our scientific theories do not as a rule spring full-armed from the brow of their creator. They are subject to slow and gradual growth, and we must candidly admit that the ionic theory in its growth has reached the "awkward age." Instead, however, of judging it according to the standard of perfection, let us simply ask what it has accomplished and what it may accomplish in scientific service.

*G. N. Lewis*

The first four sections of this chapter contain a reasonably complete general treatment of acid-base equilibria, offered principally for the student whose preparation in this area is deficient. Other students will find it more profitable to begin with Section 4.4.

### 4.1 General Considerations of Strong and Weak Electrolytes

The concept of electrolytic dissociation is today a secure tenet of our scientific dogma. We are incredulous on learning that Arrhenius, on presenting this theory in his doctoral dissertation in 1884, received the lowest possible passing mark from the University of Uppsala. In the years that followed, his ideas were ignored more than opposed. Finally, through the joint efforts of van't Hoff and Ostwald, the basic theory of Arrhenius won acceptance.

Before the development of Arrhenius' theory, van't Hoff had observed that most salts as well as acids and bases gave aqueous solutions having osmotic pressures much higher than those predicted by his equation

$$\Pi = CRT \quad \text{(Eq. 3.30).}$$

**TABLE 4.1** THE EFFECT OF MOLALITY ON THE EXPERIMENTALLY OBSERVED FREEZING-POINT DEPRESSION

| Compound | | Molalities | | | | | | | |
|---|---|---|---|---|---|---|---|---|---|
| | | 0.005 | 0.006 | 0.01 | 0.02 | 0.05 | 0.1 | 0.2 | 0.5 |
| KCl | $\Delta T_f/m$ | 3.648 | 3.640 | 3.610 | 3.564 | 3.502 | 3.451 | 3.394 | 3.314 |
| | $i$ | 1.963 | 1.959 | 1.943 | 1.918 | 1.885 | 1.861 | 1.833 | 1.800 |
| $K_2SO_4$ | $\Delta T_f/m$ | 5.308 | 5.282 | 5.198 | 5.040 | 4.776 | 4.568 | 4.324 | 3.948 |
| | $i$ | 2.857 | 2.843 | 2.798 | 2.713 | 2.570 | 2.459 | 2.333 | 2.316 |
| $MgSO_4$ | $\Delta T_f/m$ | 3.148 | 3.112 | 3.006 | 2.854 | 2.638 | 2.460 | 2.270 | 2.008 |
| | $i$ | 1.694 | 1.675 | 1.618 | 1.536 | 1.420 | 1.324 | 1.223 | 1.084 |
| $K_3Fe(CN)_6$ | $\Delta T_f/m$ | 6.840 | 6.810 | 6.696 | 6.192 | 5.60 | 5.30 | 5.00 | 4.55 |
| | $i$ | 3.681 | 3.665 | 3.604 | 3.333 | 3.02 | 2.86 | 2.70 | 2.45 |

*Source:* Noyes, A. A. and K. G. Falk, *J. Am. Chem. Soc.*, **32**:1011, 1910.

He introduced an empirical correction factor, $i$, which would permit the equation to hold:

$$\Pi = iCRT. \tag{4.1}$$

Van't Hoff initially believed $i$ to be a constant for a given solute. The correction factor was determined by dividing the observed osmotic pressure by the osmotic pressure for a "normal" solute at the same molality. These "normal" solutes were all nondissociating substances like sucrose and urea. Any of the colligative properties could be used analogously:

$$i = \frac{\Pi_{obs}}{\Pi_{theor}} = \frac{\Delta T_{f\,obs}}{\Delta T_{f\,theor}} = \frac{\Delta T_{b\,obs}}{\Delta T_{b\,theor}} = \frac{(P_0 - P)_{obs}}{(P_0 - P)_{theor}}. \tag{4.2}$$

This empirical correction took care of the data but did not take care of the explanation. Arrhenius showed that $i$ was not a constant, but was dependent on concentration. In Table 4.1 we can make this observation for ourselves by inspecting the classical data of Noyes and Falk. Arrhenius' theory proposed that acids, bases, and salts underwent dissociation in water to varying degrees, thereby producing more particles and increasing the osmotic pressure, freezing point lowering, and other colligative properties proportionately. He made no distinction between strong electrolytes such as the alkali metal halides, now known to be fully ionized in solution, and weak electrolytes such as the lower fatty acids. Subsequent experience has proved that the strong and weak electrolytes are most satisfactorily handled by different mathematical treatments, but this development does not detract from the brilliance of Arrhenius' original concepts.

Major opposition to Arrhenius' theory was given by the argument that the separation of molecules into fully charged particles would require more energy than appeared to be available. The argument failed to appreciate the effect of the high dielectric constant of water in reducing the work of separation of ions. In the opening paragraphs of Chapter 1, we showed that the force of attraction between ions is given by Coulomb's Law and is equal to the force that must be overcome if ions are to remain apart in solution:

$$\text{force} = \frac{Q_1 Q_2}{D r^2},$$

where $Q_1$ and $Q_2$ are the charges on the ions, $r$ is the distance between them, and $D$ is the dielectric constant of the solvent.

As we have also noted, the expression for the energy (or work) of separation is

$$\text{energy} = \frac{Q_1 Q_2}{D r}.$$

Since $D$, the dielectric constant, appears in the denominator of both equations, it is apparent that a large dielectric constant such as that of water (78.6 at 25°C) reduces appreciably both the energy and the force. For example, the energy of separation in water will be considerably smaller than that in practically all organic solvents because of the relatively smaller dielectric constants of the latter. The dielectric constant—sometimes called the specific inductive capacity—is a characteristic of the substance that also depends on the temperature and wavelength at which it is measured. Under comparable conditions, at approximately room temperature and at a wavelength of infinity, water has a dielectric constant of about eighty and the dielectric constant values of more common organic liquids vary from close to two up to about twenty-five. Once the ions are separated, water molecules orient themselves around the charged particles because of ion-dipole attractive forces and thus afford the ions an insulating sheath of solvent and hinder the contact of oppositely charged ions. Solvation of ions by oriented water molecules is shown diagrammatically in Fig. 4.1.

A further obstacle to the acceptance of Arrhenius' theory was the widely held belief that conductivity of electrolytes was a consequence of the molecules being pulled apart by the electric potential. Arrhenius showed that similar values for the degree of dissociation of solutes were obtained from measurements of conductivity and colligative properties. Since the colligative properties were measured in the absence of an applied electric potential, dissociation had to be the consequence of the process of solution and not of the external field.

Arrhenius believed that an equilibrium was established between the molecules and the ions of an electrolyte in solution. The difference between strong and weak electrolytes was attributed to a difference in degree of ionization. Dilution of a solution was supposed to shift the equilibrium in the direction of dissociation and ionization. Actually, the

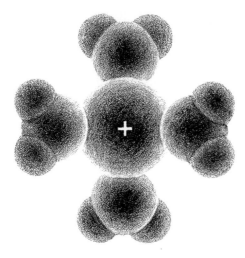

**Figure 4.1** Solvation of ions by oriented water molecules.

Arrhenius theory works quite well for weak electrolytes, although it is quantitatively unsatisfactory for strong electrolytes. We shall proceed, therefore, to consider the two separately.

**Weak electrolytes** Covalent compounds that undergo partial dissociation in water are classified as weak electrolytes. The polarization of the covalent bond involved in the dissociation will determine the extent to which ions are produced. The chloroacetic acids illustrate this point very well. As one, two, and finally three chlorine atoms are substituted on the *alpha* carbon of acetic acid, the additive electron-withdrawing effect of the chlorines increasingly polarizes the O—H bond of the carboxyl group. Finally, in trichloroacetic acid, the electron pair is shifted so much in the direction of oxygen that the acid is strongly dissociated in water. Compare the equilibrium constants for the dissociations: whereas $K$ for trichloroacetic acid is 0.2 moles/liter; $K$ for acetic acid is $1.86 \times 10^{-5}$ moles/liter.

With weak electrolytes, the degree of dissociation is the parameter in which we are interested. The symbol $\alpha$ represents the degree of dissociation, which is always less than unity. When multiplied by 100, $\alpha$ becomes the percentage dissociation. The degree of dissociation may be calculated from $i$, the van't Hoff factor mentioned earlier. Since $i$ represents the ratio of the observed colligative property to the theoretical value of that property if there were no dissociation, *i.e.*, $\Delta T_{f\,\text{obs}}/\Delta T_{f\,\text{theor}}$, it is equal to the ratio of the number of particles present in the solution of weak electrolyte to the number of particles that would be present were there no dissociation:

$$i = \frac{\Delta T_{f\,\text{obs}}}{\Delta T_{f\,\text{theor}}} = \frac{\text{total number of particles actually present}}{\text{total number of particles in absence of dissociation}}.$$

If we start with one mole of compound in a liter of solution and $\alpha$ moles dissociate, $(1 - \alpha)$ moles per liter is the concentration of the undissociated molecules. If $n$ is the number of ions formed per molecule, the concentration of ions is equal to $n\alpha$. The total number of moles of particles per liter is $(1 - \alpha) + n\alpha$. Substituting in the above expression gives

$$i = \frac{(1 - \alpha) + n\alpha}{1}.$$

Therefore,

$$\alpha = \frac{i - 1}{n - 1}. \tag{4.3}$$

At infinite dilution, $i$ approaches $n$. If $i$ were actually to become equal to $n$, $\alpha$ would be unity and dissociation would be 100 percent. This never happens to the weak electrolyte. The calculation of $i$ from colligative property data is illustrated in the following example.

*Example 4.1* Calculate the degree of dissociation of an 0.08 $m$ solution of chloroacetic acid which has a freezing point of $-0.168°$.
First, calculate $i$.

$$i = \frac{\Delta T_{f\,\text{obs}}}{\Delta T_{f\,\text{theor}}} = \frac{-0.168}{(0.08)(-1.86)} = 1.13.$$

Now calculate $\alpha$ from Eq. 4.3.

$$\alpha = \frac{1.13 - 1}{2 - 1} = 0.13 \qquad (13\% \text{ dissociation}).$$

In this example, $n = 2$ because chloroacetic acid dissociation gives only two ions per molecule.

**Strong electrolytes** Experimental evidence acquired since Arrhenius' time has proved that many strong electrolytes, such as potassium chloride, contain fully developed ions even in the solid state, and it is therefore nonsense to write equilibrium equations and equilibrium constants for their dissociation. On the other hand, we must account for the fact that solutions of potassium chloride do not show exactly twice the theoretical freezing point lowering (Table 4.1) and that the van't Hoff $i$ value approaches 2.0 only if the data are extrapolated to infinite dilution ($m = 0.0$ in Fig. 4.2).

In 1923 the following explanation was proposed by Debye and Hückel. The ions in solution, although separated from one another by shielding layers of solvent, nevertheless attract one another weakly. The more concentrated the solution, the closer an ion will be to its neighboring ions. Thus the solute particles, instead of moving about freely and

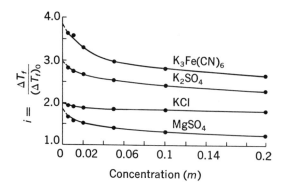

**Figure 4.2** Dependency of $i$, the van't Hoff factor, on solute concentration. [From Noyes, A. A., and K. G. Falk, *J. Am. Chem. Soc.*, **32**:1026, 1910.]

independently, assume a certain degree of orientation. Each ion is surrounded by an "ion atmosphere" of opposite charge so that its movement is slightly restricted. The restriction of movement of the solute interferes with its effect on the escaping tendency of the solvent. Therefore, instead of the vapor pressure of the solvent being reduced exactly in proportion to the number of ions present, it is reduced slightly less than this proportion would predict.

*Activity coefficient*   The effective concentration of solutes in solution is expressed by their *activity,* in contrast to their absolute concentration —the latter being based on the weight of solute in moles per liter. Activities are related to concentrations by a proportionality factor called the activity coefficient, for which the symbol $\gamma$ is used. The equation for the relationship is

$$a = \gamma C, \tag{4.4}$$

where $a$ is the activity and $C$ is the concentration. Both $a$ and $C$ are expressed in moles per liter. The activity coefficient approaches unity as the solution approaches infinite dilution. For an electrolyte, the activities of the individual ions will generally be unequal. Hence, when we refer to the activity of the compound, we are referring to the mean activity of its ions. For a uni-univalent electrolyte, the mean activity is formally designated as $a\pm$ and the appropriate equation is

$$a_\pm = \sqrt{a_+ \cdot a_-}.$$

Similarly, the activity coefficient is generally taken as the mean activity coefficient for the ions of the electrolyte:

$$\gamma_\pm = \sqrt{\gamma_+ \cdot \gamma_-}.$$

*Debye-Hückel limiting law and the concept of ionic strength*   A method for calculating $\gamma$ was developed by Debye and Hückel. For our purposes we shall present the Debye-Hückel Equation in a limiting form which restricts it to dilute aqueous solutions at a temperature of 25°C. The more general form of the equation can be used with other solvents and at other temperatures by making the proper substitutions in the constant term. The limiting form of the equation is

$$-\log_{10} \gamma = 0.51 \; Z^2 \sqrt{I} \qquad (4.5)$$

where $Z$ is the number of electronic charges on the ion for which $\gamma$ is determined, and $I$ is the *ionic strength* of the solution with respect to all ionizing solutes.[*] The ionic strength is a special parameter, devised by G. N. Lewis, which expresses the concentration of all the solutes in terms of both their molality (or molarity) and their charges. If the behavior of an ion is a reflection of its ion atmosphere, it is logical to consider all the ions in solution. In biological systems we are never so fortunate as to have only one solute present. However, the concentrations of some of the electrolytes are so small that they may safely be ignored. For example, the ions produced by the dissociation of water are usually ignored within the pH range of 4 to 10 because the concentrations involved are so low compared with ions from the solutes.

The ionic strength, $I$, is calculated as one-half the sum of the concentrations of all the ions, each concentration multiplied by the square of the charge of the ion concerned.

$$I = \tfrac{1}{2} \sum_i C_i Z_i^2. \qquad (4.6)$$

The subscript $i$ refers to the $i$th ion, or final ion in the summation. In older literature the symbol $\Gamma/2$ is frequently used instead of $I$, and the statement is made that with $\Gamma/2$, concentrations are given in molalities. In dilute aqueous solutions, it makes very little difference which units of concentration are employed. Some authors prefer the symbol $\omega$ for ionic strength.

---

[*]The general equation is usually given in the form

$$-\ln \gamma = Z^2 \left( \frac{A\sqrt{I}}{1 + Br\sqrt{I}} \right)$$

where

$$A = \frac{e^2}{DkT} \left( \frac{2\pi \mathfrak{N} e^2}{1000 \; DkT} \right)^{1/2}$$

$$B = 2 \left( \frac{2\pi \mathfrak{N} e^2}{1000 \; DkT} \right)^{1/2}$$

The symbols are defined as follows: $e$ is the charge on the electron, $\mathfrak{N}$ is Avogadro's number, $D$ is the dielectric constant of the solvent, $k$ is the Boltzmann constant (the molecular gas constant $R/\mathfrak{N}$), $T$ is the absolute temperature, and $r$ is the distance of closest approach between the centers of the ions.

*Example 4.2* Calculate the ionic strength of a solution that is 0.1 $M$ with respect to $KH_2PO_4$ and 0.05 $M$ with respect to $K_2HPO_4$.

As a first step, write down the various species of ions which are present in the solution: $K^+$, $H_2PO_4^-$, and $HPO_4^{2-}$. These salts give solutions which fall within the 4 to 10 pH range, and therefore $H^+$ and $OH^-$ may be neglected.

The second step is to determine the total concentration of each ion:

$$[K^+] = 0.1 + (2 \times 0.05)\,M;$$

$$[H_2PO_4^-] = 0.1\,M;$$

$$[HPO_4^{2-}] = 0.05\,M.$$

Finally the concentration and charge terms are substituted into the ionic strength expression:

$$I = \tfrac{1}{2}([K^+] \cdot 1^2 + [H_2PO^-] \cdot 1^2 + [HPO^{2-}] \cdot 2^2)$$
$$= \tfrac{1}{2}(0.2 + 0.1 + 0.05 \cdot 4) = 0.25.$$

A summary comparison of the properties of strong and weak electrolytes is presented in Table 4.2.

## 4.2 Acid-Base Equilibria: Practical Considerations and Calculations

By this time, the student has undoubtedly been exposed to a historical account of our changing concepts regarding acids and bases. The Brönsted-Lowry definition of acids and bases as proton donors and acceptors, respectively, will best serve here. Recognizing the fact that free protons ($H^+$) do not exist as such in aqueous solution but are always solvated ($H_3O^+$ or some higher order of solvation), we will nevertheless take the liberty of writing the hydrogen ion as $H^+$ in the interest of simplicity. As a final restriction, the discussion in this section will be limited to acids and bases sufficiently weak that their dissociation constants lie between 1 and $10^{-14}$.

In any dissociation which can be represented in the following way,

$$HA \underset{\longleftarrow}{\overset{H_2O}{\rightleftharpoons}} A^- + H^+, \tag{4.7}$$

the species HA and $A^-$ are referred to as a conjugate pair. HA is the conjugate acid (proton donor) and $A^-$ is the conjugate base (proton acceptor). In precisely the same sense one may write,

$$BH^+ \underset{\longleftarrow}{\overset{H_2O}{\rightleftharpoons}} B + H^+, \tag{4.8}$$

**TABLE 4.2** A COMPARISON OF THE PROPERTIES OF ELECTROLYTES

| Descriptive Property | Strong Electrolyte | Weak Electrolyte |
|---|---|---|
| Dissociation equation | $M^+X^- \xrightarrow{H_2O} M^+ + X^-$ | $AH \underset{H_2O}{\rightleftarrows} A^- + H^+$ |
| Equilibrium constant | extremely large | $K' = \dfrac{[A^-][H^+]}{[AH]}$ |
| Causes of deviation of observed colligative properties from predicted values | Solvated ions are surrounded by ionic atmosphere | The apparent equilibrium constant, $K'$, is dependent on the ionic strength. |
| Parameter which is usually measured | $\gamma$, the activity coefficient | $\alpha$, the degree of dissociation |
| Appropriate equation (limiting law) | $\log_{10} \gamma = -0.51\, Z^2 \sqrt{I}$ (in water at 25°C) | $\alpha = \dfrac{i-1}{n-1}$ |

*Note:* $K'$ is called the apparent equilibrium constant. It is calculated from concentrations; the true equilibrium constant is calculated from activities.

where $BH^+$ is the conjugate acid and B is the conjugate base. These conjugate pairs do not have to be acids and bases in the traditional sense; *e.g.*, in the dissociation

$$CH_3COOH \underset{H_2O}{\rightleftarrows} CH_3COO^- + H^+, \qquad (4.9)$$

the acetate anion is the base. Analogously, in the equation

$$NH_4^+ \underset{H_2O}{\rightleftarrows} NH_3 + H^+, \qquad (4.10)$$

the ammonium ion is the acid. Biochemists prefer to handle the dissociation of organic bases in the manner illustrated with ammonia, rather than to write formulas for hydrated species like $NH_4OH$, $RNH_3OH$, etc. Mathematically, it is often a simplification to handle all the dissociations as proton-producing or proton-consuming and to avoid some of the traditional notation such as the $K_b$ (see p. 184). For the sake of completeness in this presentation, however, the $K_b$ and its related term, the $pK_b$, will be explained. As long as the terms remain in the literature and current reference works, they must of necessity be perpetuated in the textbooks.

**The titration curve**   A visual concreteness is given to the concept of the acid dissociation constant, $K_a$, by plotting the titration curve of a weak acid. A titration curve is obtained experimentally by adding small increments of standard alkali to a solution of a weak acid and determining the pH after each addition. The same volumes of alkali are added to the solvent alone, and the solvent correction curve so obtained is subtracted from the titration curve for acid and solvent. When the net result is plotted as volume of standard alkali versus pH, a titration curve for a single acid group looks like the idealized curve shown in Figure 4.3.

The curve has a sigmoid or S-shape, which tells us that the pH does not change at a constant rate as alkali is added in uniform increments. There is an initial rapid change represented by the flat lower part of the curve, followed by a steep rise during which the pH is changing very little. Finally, the curve flattens off at the top and the pH again changes rapidly for each small addition of alkali. The chemical event having the major influence on the shape of the plot is the dissociation of the weak acid. As the concentration of hydroxyl ions increases in the solution, the acid begins to surrender its protons. The pH at which this event occurs depends on various electrical forces operating within the structure of the molecule and involves concepts that will not be discussed here. Suffice it to say that the location of the dissociation phenomenon on the pH scale is an inherent characteristic of each proton-releasing group. As the acid releases protons, they combine with hydroxyl ions so as to keep the pH fairly constant during the dissociation. When most of the acid has been deprotonated, and is in the conjugate base form, the top of the titration curve is reached. Few protons remain to neutralize hydroxyl ions and the pH rises. We will return to a quantitative discussion of titration curves after learning how to make calculations from dissociation constants.

**First case: an acid or a base in solution**  Let us take the acid first. As we have already written, the dissociation equation is

$$HA \underset{\longleftarrow}{\overset{H_2O}{\rightleftarrows}} H^+ + A^- \qquad (Eq.\ 4.7).$$

The double arrows, denoting the equilibrium, identify HA as a weak acid. An alternate statement of the dissociation is

$$HA + H_2O \rightleftarrows H_3O^+ + A^-. \qquad (4.11)$$

This equation, though possibly a more accurate description of events, possesses several disadvantages as compared with Eq. 4.7: the proton is represented as $H_3O^+$ which is difficult to distinguish from $H_2O$ in rapid reading, and furthermore, may not be as chemically correct as $H_9O_4^+$ or some other species; and although water appears as a reactant in the equation (4.11), it is invariably discarded in the equilibrium constant expression (incorporated into $K_a$) since the concentration of water is constant. If we write the dissociation of HA according to Eq. 4.11, then consistency dictates that we write the dissociation of water as

$$2H_2O \rightleftarrows H_3O^+ + OH^-. \qquad (4.12)$$

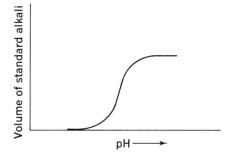

**Figure 4.3** Idealized titration curve.

For the sake of simplicity, the dissociation of HA will be written according to Eq. 4.7 and the dissociation of water as

$$H_2O \rightleftharpoons H^+ + OH^-. \qquad (4.13)$$

The equilibrium constant for the dissociation of an acid, $K_a$, relates the reactant and products of Eq. 4.7 in accordance with the Law of Mass Action:

$$K_a = \frac{[H^+][A^-]}{[HA]}, \qquad (4.14)$$

where the bracketed terms represent the equilibrium activities of the components of the dissociation.* Such a $K_a$ is called a *thermodynamic* equilibrium constant and holds for ideal solutions or those at infinite dilution. When chemical concentrations, rather than activities, are used in the brackets, the constant is written $K_a'$ and is called the *apparent* equilibrium constant. In the following discussion and examples, apparent equilibrium constants will be used wherever applicable and available.

The equilibrium constant for the dissociation of water at 25°C is

$$K = \frac{[H^+][OH^-]}{[H_2O]} = 1.8 \times 10^{-16} \text{ mole/liter.} \qquad (4.15)$$

Since the water is present in great excess, its concentration is constant at 55.56 moles per liter. Incorporating this concentration term into the

---

*The bracketing of these symbols should not necessarily be taken as the meaning of brackets elsewhere. We have used them sometimes to indicate activity and other times to indicate concentration. No great difficulty should result in this elementary treatment of the subject because activity coefficients are generally assumed to be 1.00 (to shorten the discussion and therefore activity to equal concentration). However, the student should realize that the ratio of activity coefficients required to convert $K$ (the equilibrium activity ratio) to $K'$ (the equilibrium concentration ratio) rarely even approximates being equal to unity, when ions are involved in the chemical change.

equilibrium constant yields a new expression

$$K_w = [H^+][OH^-] = 1.01 \times 10^{-14} \text{ mole}^2/\text{liter}^2, \qquad (4.16)$$

which is usually rounded off to $1 \times 10^{-14} \text{ mole}^2/\text{liter}^2$. As with all equilibrium constants, $K_w$ is temperature dependent. For example, at 40°C $K_w \cong 4 \times 10^{-14}$.

In pure water $H^+$ and $OH^-$ are the only ions present, and

$$[H^+] = [OH^-] = 1 \times 10^{-7} \text{ moles per liter.} \qquad (4.17)$$

It is a convenience to write this expression in the "$p$" form devised by Sørensen. Taking the negative logarithm of both sides gives

$$-\log [H^+] = -\log [OH^-] = 7.0,$$
$$pH = pOH = 7.0 \quad \text{(in pure water at 25°C).} \qquad (4.18)$$

Since the equilibrium constant, $K_w$, must be satisfied in any aqueous solution, we will find it useful to write Eq. 4.16 in the "$p$" form:

$$-\log K_w = (-\log [H^+]) + (-\log [OH^-]) = 14.0,$$

$$pK_w = pH + pOH = 14.0. \qquad (4.19)$$

Since the dissociation constants for weak acids and bases are small numbers of the order of $10^{-2}$ to $10^{-12}$, it is a great convenience to handle them in a form analogous to that used for hydrogen ion concentration: they may be expressed in the "$p$" form. Thus $-\log K$ becomes $pK$, just as $-\log [H^+]$ is called pH. If the dissociation constant is for a proton-releasing reaction, the designation is $pK_a$; for a reaction releasing $OH^-$, the designation is $pK_b$.

In the example we are considering, only one compound is present in the solution: the weak acid. It dissociates as designated in Eq. 4.7. For each $H^+$ produced, one $A^-$ is also formed, and therefore, $[H^+] = [A^-]$. We can now write

$$K_a' = \frac{[H^+]^2}{C_a - [H^+]}, \qquad (4.20)$$

where $[H^+]$ is the concentration of $H^+$ or $A^-$, and $C_a$ is the initial concentration of HA. The value of $[H^+]$ is obtained by expanding Eq. 4.20 and applying the quadratic formula

$$[H^+] = y = \frac{-b \pm (b^2 - 4ac)^{1/2}}{2a}, \qquad (4.21)$$

where $a$ and $b$ are the coefficients of $[H^+]^2$ and $[H^+]$ respectively, and $c$ is the constant term.

If, however, $K'_a$ is small enough that the concentration of ions produced is negligible in comparison with the original concentration of acid, we can discard $[H^+]$ in the denominator term (not in the numerator!) and write

$$K'_a = \frac{[H^+]^2}{C_a},$$  (4.22)

and

$$[H^+] = \sqrt{K'_a \cdot C_a}.$$  (4.23)

If $[H^+]$ is less than 5 percent of $C_a$ the simplified expression (4.22) may be used. Let us compare the two calculations for an appropriate example such as a 0.1 $M$ solution of a weak acid having a $K'_a$ equal to $1.8 \times 10^{-4}$ mole/liter. Using Eq. 4.20 and solving by the quadratic formula (4.21) gives $[H^+] = 4.15 \times 10^{-3}$ $M$. Using the simplified version (Eq. 4.23), $[H^+] = 4.24 \times 10^{-3}$ $M$. When rounded off to two significant figures, the answers obtained by both methods are the same.

Since $[H^+]$ is the unknown in the calculation, one cannot estimate prior to the operation whether it will be less than 5 percent of $C_a$. An alternate generalization is that one may use the simplified version of the expression if $K_a$ is of the order of $10^{-5}$ mole/liter or smaller, provided $C_a$ is of sufficient magnitude. If $C_a$ is 0.01 $M$ or larger, $[H^+]$ is almost certain to be negligibly small in the denominator. As an illustration of a situation in which $[H^+]$ may *not* be neglected, consider the following problem: calculate $[H^+]$ in a 0.005 $M$ solution of acetic acid. $K_a$ for acetic acid is $1.86 \times 10^{-5}$ mole/liter. Calculating $[H^+]$ by Eq. 4.20 shows it to be equal to $2.99 \times 10^{-4}$ $M$, which is larger than 5 percent of 0.005.

Since one is frequently justified in using the simplified expression (Eq. 4.22), it is advantageous to write it in the logarithmic form, which leads directly to pH. Writing again

$$[H^+] = \sqrt{K'_a \cdot C_a}$$

and taking the negative logarithm of both sides, we have

$$-\log [H^+] = \tfrac{1}{2}(-\log K'_a - \log C_a),$$

which is

$$pH = \tfrac{1}{2}(pK'_a - \log C_a).$$  (4.24)

In Table 4.3 are given the thermodynamic dissociation constants and corresponding $pK_a$ values for various weak acids at 25°C. Additional constants can be found in recent editions of *Lange's Handbook* or *The Handbook of Chemistry and Physics*. Some of the more useful $pK'_a$ values and the corresponding ionic strengths are presented in Table

**TABLE 4.3** DISSOCIATION CONSTANTS OF WEAK ACIDS AT 25°C

| Acid | $K_a$ mole/liter | $pK_a$ |
|---|---|---|
| Urea | 0.67 | 0.18 |
| Oxalic acid $(K_1)$ | $7.5 \times 10^{-2}$ | 1.27 |
| Phosphoric acid $(K_1)$ | $7.5 \times 10^{-3}$ | 2.12 |
| Malonic acid $(K_1)$ | $1.49 \times 10^{-3}$ | 2.83 |
| Monochloroacetic acid | $1.35 \times 10^{-3}$ | 2.87 |
| Phthalic acid $(K_1)$ | $1.3 \times 10^{-3}$ | 2.88 |
| Tartaric acid $(K_1)$ | $1.1 \times 10^{-3}$ | 2.96 |
| Nitrous acid | $5.1 \times 10^{-4}$ | 3.29 |
| Formic acid | $1.8 \times 10^{-4}$ | 3.75 |
| Acetoacetic acid | $1.6 \times 10^{-4}$ | 3.8 |
| Lactic acid | $1.37 \times 10^{-4}$ | 3.86 |
| Tartaric acid $(K_2)$ | $6.9 \times 10^{-5}$ | 4.16 |
| Succinic acid $(K_1)$ | $6.31 \times 10^{-5}$ | 4.2 |
| Benzoic acid | $6.2 \times 10^{-5}$ | 4.21 |
| Oxalic acid $(K_2)$ | $5.4 \times 10^{-5}$ | 4.27 |
| $\beta$-Hydroxybutyric acid | $3.98 \times 10^{-5}$ | 4.4 |
| Anilinium ion | $2.5 \times 10^{-5}$ | 4.60 |
| Acetic acid | $1.86 \times 10^{-5}$ | 4.73 |
| Butyric acid | $1.48 \times 10^{-5}$ | 4.83 |
| Propionic acid | $1.34 \times 10^{-5}$ | 4.87 |
| Pyridinium ion | $6.0 \times 10^{-6}$ | 5.22 |
| Phthalic acid $(K_2)$ | $3.9 \times 10^{-6}$ | 5.41 |
| Succinic acid $(K_2)$ | $2.35 \times 10^{-6}$ | 5.63 |
| Malonic acid $(K_2)$ | $2.0 \times 10^{-6}$ | 5.70 |
| Carbonic acid $(K_1)$ | $4.52 \times 10^{-7}$ | 6.34 |
| Hydrogen sulfide $(K_1)$ | $1.0 \times 10^{-7}$ | 7.00 |
| Phosphoric acid $(K_2)$ | $6.31 \times 10^{-8}$ | 7.20 |
| Barbital (veronal) | $3.7 \times 10^{-8}$ | 7.43 |
| Tris (tris-hydroxymethyl-methylammonium ion) | $8.32 \times 10^{-9}$ | 8.08 |
| Hydrazinium ion | $3.3 \times 10^{-9}$ | 8.48 |
| Boric acid $(K_1)$ | $5.8 \times 10^{-10}$ | 9.23 |
| Ammonium ion | $5.7 \times 10^{-10}$ | 9.24 |
| Phenol | $1.0 \times 10^{-10}$ | 10.0 |
| Carbonic acid $(K_2)$ | $4.68 \times 10^{-11}$ | 10.33 |
| Methylammonium ion | $2.5 \times 10^{-11}$ | 10.60 |
| Phosphoric acid $(K_3)$ | $3.98 \times 10^{-13}$ | 12.4 |
| Hydrogen sulfide $(K_2)$ | $1.26 \times 10^{-13}$ | 12.9 |

**TABLE 4.4** APPARENT $pK_a'$ VALUES FOR
CERTAIN WEAK ACIDS AT 25°C

| Acid | $pK_a'$ mole/liter | Ionic Strength |
|------|---------|----------------|
| Acetic acid | 4.66 | 0.1 |
| Malonic acid ($pK_2$) | 5.37 | 0.1 |
| Imidazolium ion | 7.07 | 0.1 |
| Phosphoric acid ($pK_2$) | 6.86 | 0.1 |
| Ammonium ion | 9.4 | 0.1 |
| Carbonic acid ($pK_2$) | 10.02 | 0.1 |

*Source:* Selected from data in Clark, W. M., *Topics in Physical Chemistry,* pp. 282, 287. Williams and Wilkins, Baltimore, 1952.

4.4. A comparison of the data of Tables 4.3 and 4.4 should emphasize to the student the effect of ionic strength on the dissociation constant. It is unfortunate that we do not have available more data of the type presented in Table 4.4, since they are eminently more useful in the laboratory. When one must actually prepare a 0.1 $M$ or 0.01 $M$ buffer, the constants from Table 4.4 should be used rather than those from Table 4.3.

We shall now consider the weak base. Since any weak base exists in solution in equilibrium with its conjugate acid, the $pK_a$ is widely used in dissociation calculations. The logarithmic form of the dissociation expression involving $K_a$ (or $K_a'$) contains a pH term, and since our thinking is pH-oriented rather than pOH-oriented, such a development is a logical one. We cannot, however, use Eq. 4.24 because it refers to the dissociation of a weak acid to give an acidic solution; the weak base gives a basic solution. As we shall proceed to derive, the correct expression for the pH of a solution of a weak base is

$$\mathrm{pH} = \tfrac{1}{2}(pK_w + pK_a' + \log C_b), \qquad (4.25)$$

where $C_b$ is the initial concentration of the base and $pK_a'$ refers to the conjugate acid. This is a limiting expression obtained through simplifications like those employed in obtaining Eq. 4.24.

When a weak base such as ammonia or an organic amine dissolves in water, it ionizes according to the equilibrium expression

$$\mathrm{B} + \mathrm{H_2O} \rightleftharpoons \mathrm{BH^+} + \mathrm{OH^-}. \qquad (4.26)$$

Recalling the notation used in Eq. 4.7, we can write a similar expression for the base $\mathrm{A^-}$:

$$\mathrm{A^-} + \mathrm{H_2O} \rightleftharpoons \mathrm{HA} + \mathrm{OH^-}. \qquad (4.27)$$

The choice of a general notation is rather immaterial as long as the chemical event is correctly depicted. Specific examples are

$$NH_3 + H_2O \rightleftarrows NH_4^+ + OH^-; \tag{4.28}$$

$$CH_3COO^- + H_2O \rightleftarrows CH_3COOH + OH^-. \tag{4.29}$$

In the examples, $NH_4^+$-$NH_3$ and $CH_3COOH$-$CH_3COO^-$ are recognized as conjugate acid-base pairs.

Returning to the general notation used in Eq. 4.26, we write the equilibrium constant

$$K = \frac{[OH^-][BH^+]}{[B][H_2O]}.$$

Since the concentration of water is constant, $[H_2O]$ is eliminated from the denominator just as it was for the $K_w$ expression, leaving

$$K_b = \frac{[OH^-][BH^+]}{[B]}, \tag{4.30}$$

where the subscript $b$ identifies the dissociation as one which produces hydroxyl ions. The $K_a$ for the conjugate acid is (see Eq. 4.8)

$$K_a = \frac{[H^+][B]}{[BH^+]}. \tag{4.31}$$

A convenient and simple relationship exists between the $K_a$ and $K_b$ of a conjugate acid-base pair:

$$K_a \cdot K_b = K_w = 10^{-14} \text{ mole}^2/\text{liter}^2 \text{ at } 25°. \tag{4.32}$$

In the negative logarithmic form,

$$pK_a + pK_b = pK_w = 14.0. \tag{4.33}$$

Equation 4.30 can now be simplified for limiting cases. Since $[OH^-]$ is equal to $[BH^+]$, $[OH^-]$ denotes the concentration of either ion, and $C_b$ is the initial concentration of base. We write

$$K_b' = \frac{[OH^-][BH^+]}{[B]} = \frac{[OH^-]^2}{C_b - [OH^-]}. \tag{4.34}$$

When $[OH^-]$ is less than 5 percent of $C_b$ we may further simplify to

$$K_b' = \frac{[OH^-]^2}{C_b}. \tag{4.35}$$

Proceeding as before, we obtain

$$[OH^-] = \sqrt{K_b' \cdot C_b}, \tag{4.36}$$

and in the negative logarithmic form,

$$pOH = \tfrac{1}{2}(pK'_b - \log C_b). \tag{4.37}$$

To convert Eq. 4.37 to pH and $pK'_a$ form, we substitute for pOH according to Eq. 4.19 and for $pK'_b$ according to Eq. 4.33 (using $pK_w$ rather than 14.0). Collecting terms and rearranging gives

$$pH = \tfrac{1}{2}(pK_w + pK'_a + \log C_b),$$

which is Eq. 4.25 that we wished to derive. The student should make the substitutions himself to be satisfied that Eq. 4.37 is equivalent to Eq. 4.25. Data for the conjugate acids of a number of weak bases have been included in Table 4.3. The use of Eq. 4.25 and Eq. 4.37 is illustrated in the following example.

> *Example 4.3*   Calculate the pH of a 0.01 $M$ solution of ammonium hydroxide. The $pK'_a$ for ammonium ion is 9.4. The corresponding $pK'_b$ will be $14.0 - 9.4$, or 4.6. The corresponding $K'_b$ must be of the order of $10^{-5}$ mole/liter; therefore, according to the general rules previously stated, the simplified versions of the pH and pOH expressions may be used. According to Eq. 4.25,
>
> $$pH = \tfrac{1}{2}(pK_w + pK'_a + \log C_b)$$
> $$= \tfrac{1}{2}(14.0 + 9.4 + -2.0) = 10.7.$$
>
> According to Eq. 4.37,
>
> $$pOH = \tfrac{1}{2}(pK'_b - \log C_b)$$
> $$= \tfrac{1}{2}(4.6 - -2.0) = 3.3.$$
>
> To obtain the pH, now subtract the pOH from 14.0:
>
> $$pH = 14.0 - 3.3 = 10.7.$$
>
> The same answer is obtained by either method, but an extra step is required if Eq. 4.37 is used.

**Second case: the equivalence point in a titration**   Quite often we need a rapid method for calculating the pH of a solution of the salt of a weak base and a strong acid (*e.g.*, ammonium sulfate), or the salt of a weak acid and a strong base (*e.g.*, sodium acetate). When using concentrated ammonium sulfate solutions in protein fractionation, we must be able to calculate the pH of the solution so as to decide on the concentration of buffer to be included in the medium. A second practical problem occurs in the titration of a weak acid or a weak base. The end point of the titration will be the pH given by a solution of the salt, *e.g.*, $Na^+A^-$ or $BH^+Cl^-$. Since this pH will almost never be 7.0, it must be calculated in advance so that a suitable indicator can be chosen or an appropriate pH meter setting selected.

The reaction of $Na^+A^-$ or $BH^+Cl^-$ with water is commonly referred to as a *hydrolysis*, but this particular term should not obscure the fact that the chemical process is merely an acid-base reaction of the type discussed in the preceding section. The $Na^+$ and $Cl^-$ ions produced in the respective reactions are hydrated by water but do not alter the concentrations of $H^+$ and $OH^-$ ions, and, therefore, will not be included in the hydrolysis equations.

When a weak acid, HA, is exactly neutralized with NaOH, the products are $Na^+A^-$ and $H_2O$. The anion $A^-$, which is the conjugate base of HA, reacts with water, according to Eq. 4.27:

$$A^- + H_2O \rightleftharpoons HA + OH^-;$$

$$(\text{base}) + (\text{acid}) = (\text{acid}) + (\text{base}).$$

The equilibrium constant for the reaction is

$$K = \frac{[HA][OH^-]}{[A^-][H_2O]}. \tag{4.38}$$

Proceeding as before to incorporate $[H_2O]$ into $K$ and to employ concentrations gives

$$K' = \frac{[HA][OH^-]}{[A^-]}. \tag{4.39}$$

Since $[HA] = [OH^-]$,

$$K' = \frac{[OH^-]^2}{[A^-]}. \tag{4.40}$$

Looking back a few paragraphs, we see that these expressions are identical with the equations leading to Eq. 4.34. If the same restriction applies, *i.e.*, $[OH^-]$ is less than 5 percent of $[A^-]$, then Eqs. 4.37 and 4.25 are directly applicable:

$$pOH = \tfrac{1}{2}(pK_b' - \log C_b) \qquad (\text{Eq. 4.37}),$$

and

$$pH = \tfrac{1}{2}(pK_w + pK_a' + \log C_b) \qquad (\text{Eq. 4.25}),$$

where $C_b$ is the initial concentration of $Na^+A^-$, $pK_a'$ refers to HA, and $pK_b'$ to its conjugate base, $A^-$.

Equation 4.25 is a general equation for the pH of a solution of a salt of a weak acid and a strong base. The $pK_w$ is always 14 and may be written as such. The equation shows very clearly the effect of dilution on the pH of these solutions. As $C_b$ becomes smaller and approaches $K_a'$, the pH will approach 7.0.

The salt of the weak base and the strong acid is handled in a similar way, but the pH expression is obtained directly. When a salt of this type dissolves in water or when it is formed at the equivalence point of a titration, hydrolysis of the conjugate weak acid occurs. For example, in ammonium chloride, the ammonium ion reacts with water:

$$NH_4^+ + H_2O \rightleftharpoons NH_3 + H_3O^+, \tag{4.41}$$

or we may write an equivalent expression,

$$NH_4^+ \underset{\xleftarrow{H_2O}}{\rightleftharpoons} NH_3 + H^+. \tag{4.42}$$

The apparent equilibrium constant is

$$K_a' = \frac{[NH_3][H^+]}{[NH_4^+]}. \tag{4.43}$$

Since, in the hydrolysis, $NH_3$ and $H^+$ are formed in equal amounts and, if the contribution to $[H^+]$ by dissociation of water is negligible,

$$K_a' = \frac{[H^+]^2}{[NH_4^+]}, \tag{4.44}$$

and since the initial and equilibrium concentrations of ammonium ion are very nearly the same,

$$K_a' = \frac{[H^+]^2}{C_a},$$

which is identical with Eq. 4.22. The negative logarithmic form is

$$pH = \tfrac{1}{2}(pK_a' - \log C_a),$$

which is identical with Eq. 4.24. This is the general equation for the pH of a solution of a salt of a weak base and a strong acid or the pH of the end point in a titration involving these species. As the solution becomes more dilute, the pH approaches 7.0 as $C_a$ approaches $K_a'$.

---

*Example 4.4*  Calculate the pH of a 0.05 $M$ solution of sodium acetate (salt of a weak acid and a strong base).

Since acetate anion is the conjugate base of a weak acid, the solution will be basic at equilibrium. The correct approximate equation is

$$pH = \tfrac{1}{2}(pK_a' + pK_w + \log C_b).$$

The $pK_a'$ for acetic acid is 4.66; $pK_w$ is 14.0; $C_b = 0.05$. These values give

$$pH = \tfrac{1}{2}(4.66 + 14.0 - 1.30) = 8.68.$$

*Example 4.5* Calculate the pH of a 0.05 $M$ solution of ammonium chloride (salt of a weak base and a strong acid).

Since ammonium ion is the conjugate acid of a weak base, the final pH will be less than 7.0. The correct approximate equation is

$$pH = \tfrac{1}{2}(pK_a' - \log C_a).$$

The $pK_a'$ is 9.4 and the log of 0.05 is $-1.30$. Therefore

$$pH = \tfrac{1}{2}(9.4 + 1.30) = 5.35.$$

**Third case: buffers—weak acids or weak bases in the presence of their salts** A buffer is defined as a solution that will maintain a reasonably constant pH when relatively large amounts of either acid or base are added. Of course, no solution can buffer indefinitely, and the pH will change markedly if the capacity of the buffer is exhausted. One of the most delicately buffered systems is whole blood. Although hemoglobin, other proteins, and phosphate all make a contribution, the work horse is the sodium bicarbonate-carbonic acid buffer system. When this buffer is exhausted, as it may be in severe acidosis, the pH changes drastically and the consequences may be fatal.

With a buffer composed of a weak acid and its salt, we proceed as we have for the weak acid or weak base alone and write an equation describing the pertinent equilibrium. For the acid, an equation previously written applies:

$$HA \rightleftharpoons H^+ + A^- \qquad \text{(Eq. 4.7),}$$

which leads to Eq. 4.14:

$$K_a = \frac{[H^+][A^-]}{[HA]}.$$

The salt of the weak acid, $Na^+A^-$, is assumed to be fully dissociated and therefore no dissociation equilibrium expression is written. Since the anion, $A^-$, is the conjugate base of the acid, it will tend to hydrolyze according to Eq. 4.27, but the equilibrium is greatly shifted to the left and, the few hydroxyl ions produced are neutralized by the protons released by the acid. Therefore, so little $A^-$ is lost through hydrolysis that the hydrolysis process can be ignored. An exact calculation must also consider the contribution of HA dissociation to the total concentration of $A^-$. In an approximate calculation involving monobasic acids with dissociation constants of the order of $1 \times 10^{-5}$ mole/liter or smaller, ionization of the conjugate acid is substantially reduced through the presence of the salt. For this reason, $[A^-]$ is taken as equal to the $[A^-]$ from the salt alone. Similarly so little HA is lost through dissociation and so little gained through the hydrolysis of $A^-$ that $[HA]$ is taken to be equal to the undissociated acid. Although the $[H^+]$ term is small, it appears in a product rather than a sum (or difference) and cannot be dropped. The apparent equilibrium constant is

$$K'_a = \frac{[H^+][A^-]}{[HA]},$$

where $[A^-]$ is the equilibrium concentration of salt and $[HA]$ is the equilibrium concentration of acid. In a buffer, $[H^+]$ and $[A^-]$ are not equal, and the negative logarithmic form is

$$-\log K'_a = -\log [H^+] - \log \frac{[A^-]}{[HA]},$$

and the "$p$" form,

$$pK'_a = pH - \log \frac{[A^-]}{[HA]},$$

which rearranges to

$$pH = pK'_a + \log \frac{[A^-]}{[HA]}, \tag{4.45}$$

or as it is generally stated

$$pH = pK'_a + \log \frac{[\text{conjugate base}]}{[\text{conjugate acid}]}. \tag{4.46}$$

Equation 4.46, frequently called the Henderson-Hasselbalch Equation, is a limiting law useful in making buffer calculations. From the equation we can see the relationship between pH, $pK'_a$, and the fraction of acid neutralized in a titration. Examination of Fig. 4.4,A should clarify the relationships. The solid line shows the titration curve for acetic acid. Exactly half-way up the steep portion of the curve, the composition of the solution will be a 1:1 ratio of acid to salt. According to the Henderson-Hasselbalch Equation the pH at this point is

$$pH = pK'_a + \log 1$$

$$= pK'_a.$$

This relationship is the basis of the experimental determination of the apparent dissociation constants of weak acids and the corresponding $pK'$ values. When partially neutralizing an acid with base to form the salt *in situ*, we must remember that it is the concentration of the *un-neutralized* acid, and not the initial concentration, that is substituted into the HA term. Only when the buffer is formulated of an acid and its

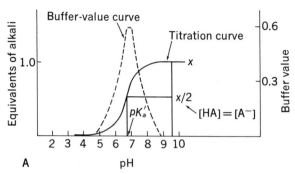

**Figure 4.4**   A: Relationship of the titration curve to the $pK_a'$ and to the buffer value.

salt, and base is *not* used, is initial acid concentration substituted into the HA term.

The midpoint of the titration curve, the $pK_a'$, is the pH at which the buffer displays its greatest ability to resist pH change. Inspection of Fig. 4.4,A reveals that this is the region on the titration curve where its slope is greatest. These considerations prompted Van Slyke to define the buffer value of a system in terms of the slope of the titration curve, that is,

$$\text{Buffer Value} = \frac{d(\text{Equivalents of Base Added})}{d(\text{pH})}. \qquad (4.47)$$

If we write the Henderson-Hasselbalch Equation in the form

$$\text{pH} = pK_a' + \log\frac{\alpha}{1 - \alpha}, \qquad (4.48)$$

where $\alpha$ is the ratio of the moles of conjugate base to the total moles of conjugate acid plus conjugate base, we may readily derive the relationship

$$\text{Buffer Value} = 2.303 \, C\alpha(1 - \alpha), \qquad (4.49)$$

where $C$ is the initial concentration of acid (the concentration before any base is added). At the midpoint of the titration, the $pK_a'$, $\alpha = 0.5$; the maximum buffer value of any system is therefore $0.575 \, C$. In Fig. 4.4,A, the buffer value curve is shown as a dashed line. This curve corresponds to $C = 1 \, M$.

Figure 4.4,A illustrates one additional point of importance: the useful range of a buffer is 1.5 pH units on either side of its $pK_a'$. Beyond these limits the buffer value rapidly approaches zero.

The form of the Henderson-Hasselbalch Equation appearing as Eq.

4.48 shows the pH of a buffer to be independent of the concentrations of acid and conjugate base, and dependent solely on their relative proportions. Ideally, this would be true. Since, however, for real solutions $K_a'$ is a concentration-dependent term, Eq. 4.48 will hold only for dilute solutions.

The following approximate relationships may be helpful in explaining the importance of restricting the working range of a buffer to $pK' \pm 1.5$ units:

| pH | Percent of Conjugate Acid Titrated | $\dfrac{\alpha}{1-\alpha}$ |
|---|---|---|
| $(pK' - 3)$ | ~ 0.1 | $\sim\dfrac{.001}{.999}$ |
| $(pK' - 2)$ | ~ 1 | $\sim\dfrac{.01}{.99}$ |
| $(pK' - 1)$ | ~10 | $\sim\dfrac{.10}{.90}$ |
| $pK'$ | 50 | $\dfrac{.5}{.5}$ |
| $(pK' + 1)$ | ~90 | $\sim\dfrac{.90}{.10}$ |
| $(pK' + 2)$ | ~99 | $\sim\dfrac{.99}{.01}$ |
| $(pK' + 3)$ | ~99.9 | $\sim\dfrac{.999}{.001}$ |

If the acid and anion forms of the substance titrated absorb light of different wavelengths, a *spectrophotometric titration* may be used to determine the $pK_a'$. A series of solutions is prepared having the same concentration of solute but differing in pH. The pH range is chosen to coincide with the dissociation to be studied. The pH values are recorded and the spectra of the solutions are taken with a recording spectrophotometer. From the absorbancy shift we may determine the pH at which the acid is half neutralized. Figure 4.4,B shows three curves from a spectrophotometric titration of pyridoxamine-5-phosphate. The single acid peak with a maximum near 295 m$\mu$ is replaced by peaks to either side as the pH rises. At pH 3.7 the group is half-titrated and pH $= pK_a'$.

*Example 4.6* Determine the concentration ratio of sodium acetate to acetic acid necessary to give a buffer of pH 5.0. For acetic acid, the $pK_a'$ is 4.66. These values are substituted into the Henderson-Hasselbalch Equation:

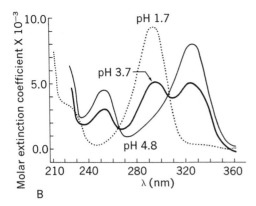

B

**Figure 4.4**   B: Spectrophotometric titration of the phenolic hydroxyl of pyridoxamine-5-phosphate. Group is half-titrated at pH 3.7. [From Williams, V. R., and J. B. Neilands, *Arch. Biochem. Biophys.*, **53**:59, 1954.]

$$pH = pK'_a + \log\frac{[\text{salt}]}{[\text{acid}]},$$

$$5.00 = 4.66 + \log\frac{[\text{salt}]}{[\text{acid}]},$$

$$\log\frac{[\text{salt}]}{[\text{acid}]} = 0.34,$$

Ratio of salt to acid $= 2.19$.

If the buffer is made from acetic acid and sodium hydroxide, the moles of sodium acetate formed are equal to the moles of base used, but enough acetic acid must be used to neutralize the sodium hydroxide and leave an excess of acid so that the ratio of salt to acid is 2.19. In other words,

$$\log\frac{[\text{salt}]}{[\text{acid}]} = \log\frac{[\text{NaOH}]}{[\text{HOAc} - \text{NaOH}]} = \log\frac{2.19}{3.19 - 2.19}.$$

At Equilibrium      Initial Concentrations

Buffers may also be composed of weak bases and their salts. The Henderson-Hasselbalch Equation may be used for these since it is completely general:

$$pH = pK'_a + \log\frac{[\text{proton acceptor}]}{[\text{proton donor}]},$$

or similarly,

$$pH = pK'_a + \log\frac{[\text{conjugate base}]}{[\text{conjugate acid}]} \qquad (\text{Eq. 4.46}).$$

The $pK'_a$ will be for the conjugate acid of the weak base, *e.g.*, $NH_4^+$ in an ammonia-ammonium salt buffer. The Henderson-Hasselbalch Equation here is

$$pH = 9.4 + \log\frac{[\text{NH}_3]}{[\text{NH}_4^+]}.$$

A parallel form of the above equation in terms of pOH may be derived by starting with the equilibrium for the dissociation of the base in water, but the Henderson-Hasselbalch form is more convenient since it gives pH directly.

*Example 4.7* A solution of ammonia in water is neutralized with dilute HCl to a pH of 8.7. What are the relative concentrations of free base and salt?

$$pH = pK_a' + \log\frac{[NH_3]}{[NH_4^+]},$$

$$8.7 = 9.4 + \log\frac{[NH_3]}{[NH_4^+]},$$

$$\log\frac{[NH_3]}{[NH_4^+]} = -0.7,$$

$$\frac{[NH_3]}{[NH_4^+]} = 0.20.$$

Frequently it is necessary to prepare a buffer of a certain pH and ionic strength. This calculation will be discussed following a consideration of the dissociations of polyprotic acids.

**Fourth case: polyprotic acids** When phosphoric acid dissociates in successive steps to yield three protons per molecule, the equilibrium constants for the individual steps are designated as $K_1$, $K_2$, and $K_3$ corresponding to $pK_1$, $pK_2$, and $pK_3$ of the titration curve (Fig. 4.5). In solutions of polyprotic acids (no salts present) the first dissociation is usually the only one making a significant contribution to the pH, since $K_2$ is ordinarily 1/100 (or less) of $K_1$, and $K_3$ is correspondingly smaller than $K_2$. In $H_3PO_4$,

$$H_3PO_4 \rightleftharpoons H^+ + H_2PO_4^-; \quad K_1 = 7.5 \times 10^{-3} \text{ mole/liter;}$$
$$H_2PO_4^- \rightleftharpoons H^+ + HPO_4^{2-}; \quad K_2 = 6.3 \times 10^{-8} \text{ mole/liter;}$$
$$HPO_4^{2-} \rightleftharpoons H^+ + PO_4^{3-}; \quad K_3 = 4.0 \times 10^{-13} \text{ mole/liter.}$$

If we wish to calculate the concentration of $HPO_4^{2-}$ in a 0.1 $M$ solution of $H_3PO_4$, we consider only the first two equilibria since the amount of $HPO_4^{2-}$ disappearing in the final dissociation step will be negligible ($K_3 \ll K_2$). The above constants are thermodynamic equilibrium constants and we will assume in the following calculation that activities are equal to concentrations. (For a more realistic answer we should use apparent equilibrium constants.) The expression for the first equilibrium constant is

$$K_1 = \frac{[H^+][H_2PO_4^-]}{[H_3PO_4]} = 7.5 \times 10^{-3} \text{ mole/liter.}$$

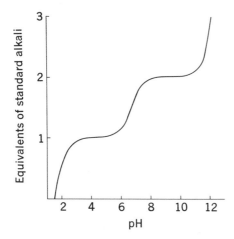

**Figure 4.5** Titration curve for phosphoric acid.

This is a weak acid in the absence of its salts, so $[H^+] \cong [H_2PO_4^-]$ for this step. The second equilibrium constant is

$$K_2 = \frac{[H^+][HPO_4^{2-}]}{[H_2PO_4^-]} = 6.3 \times 10^{-8} \text{ mole/liter.}$$

It is $[HPO_4^{2-}]$ that is desired. Solving the $K_2$ expression for this term yields

$$[HPO_4^{2-}] = \frac{K_2[H_2PO_4^-]}{[H^+]}.$$

$K_2$ is so small compared with $K_1$, that the hydrogen ion concentration can safely be estimated from $K_1$ alone, and the equilibrium concentration of $H_2PO_4^-$ following the second ionization step may be regarded as no different from $[H_2PO_4^-]$ in step one, since the amount lost through further dissociation is negligible. Therefore, in the $K_2$ expression $[H^+]$ = $[H_2PO_4^-]$ leaving

$$[HPO_4^{2-}] = K_2 = 6.3 \times 10^{-8} \ M.$$

Dissociations involving $H_2CO_3$ are handled in a similar way, the dissociation constants differing by a factor of $10^{-4}$.

**Fifth case: equivalence point in the titration of a polyprotic acid**   In the titration of a polyprotic acid like $H_3PO_4$, we may wish to neutralize only one or two of the three protons. What will be the pH of the solution at the desired equivalence point? The simple limiting equations developed earlier (Eqs. 4.24 and 4.25, the latter being for salts of weak bases) cannot be used for systems involving a second dissociation.

Inspection of the titration curve for $H_3PO_4$ (Fig. 4.5) shows that the maximum concentration of a particular salt will occur at a pH midway between the two dissociation steps that produce the anion and consume it. The species $HPO_4^{2-}$, for example, is formed in the second dissociation step and converted to $PO_4^{3-}$ in the third dissociation step. This salt should, therefore, be at maximum concentration exactly halfway between the two $pK_a$ values. That is,

$$pH = \tfrac{1}{2}(pK_1 + pK_2) \tag{4.50}$$

An ion which is both an acid and base, for example, $HPO_4^{2-}$, is called an *amphoteric species*. The pH of a solution of any individual amphoteric species is midway between the two relevant $pK_a$ values. Also, Eq. 4.50 indicates that the hydrogen ion concentration under these circumstances is independent of the concentration of the solute, there being no concentration term in the equation. Equation 4.50 may be proved mathematically. Let $H_2A$ represent a diprotic acid. The species $HA^-$ is the anion of the salt formed by the addition of one equivalent of base. Not only is $HA^-$ the conjugate base of $H_2A$ but it is also an acid. In water it may lose a proton as an acid or gain a proton as a base. These equilibria are

$$HA^- \overset{H_2O}{\rightleftarrows} H^+ + A^{2-} \qquad \text{(as an acid)},$$

$$HA^- + H_2O \rightleftarrows H_2A + OH^- \qquad \text{(as a base)}.$$

The equilibrium constant for the acid dissociation is $K_2$; the equilibrium constant for the basic dissociation is a $K_b$ and is therefore equal to $K_w/K_1$ since $K_a K_b = K_w$ for any conjugate pair. The concentration of $HA^-$ must satisfy both equilibria,

$$(a) \quad K_2 = \frac{[H^+][A^{2-}]}{[HA^-]} \quad \text{and} \quad K_b \text{ (for } HA^-) = \frac{K_w}{K_1} = \frac{[H_2A][OH^-]}{[HA^-]}.$$

Therefore,

$$(b) \quad [HA^-] = \frac{[H^+][A^{2-}]}{K_2} = \frac{[H_2A][OH^-]K_1}{K_w}.$$

However,

$$(c) \quad \frac{[H_2A][OH^-]K_1}{K_w} = \frac{[H_2A]K_1}{[H^+]}.$$

Equating the first fraction in Eq. *b* with the second fraction in Eq. *c*, we obtain

$$(d) \quad \frac{[H^+][A^{2-}]}{K_2} = \frac{[H_2A]K_1}{[H^+]},$$

and, solving for $[H^+]$,

(e) $\quad [H^+]^2 = \dfrac{K_1 K_2 [H_2A]}{[A^{2-}]}$.

This relationship expressed in the negative logarithmic form ($p$ form)

(f) $\quad 2pH = (pK_1 + pK_2) + \log \dfrac{[A^{2-}]}{[H_2A]}$.

Inspection of the last equation reveals that Eq. 4.50 — pH = $\frac{1}{2}(pK_1 + pK_2)$ — will hold when $[A^{2-}] = [H_2A]$, since log 1 is equal to zero.

One of the most satisfactory ways to prove that $\log [A^{2-}]/[H_2A]$ is indeed equal to zero at the midpoint between the two $pK_a$ values is to plot the Michaelis pH functions. The Michaelis functions are based on the definitions of $K_1$ and $K_2$ plus the Conservation Equation for the three species, $H_2A$, $HA^-$, and $A^{2-}$. The Conservation Equation is

(g) $\quad A_{total} = [H_2A] + [HA^-] + [A^{2-}]$

where $A_{total}$ is the initial concentration of acid. Equation g shows that the sum of the equilibrium concentrations of the three species on the right must always be equal to the initial acid concentration.

A Michaelis function for an individual species is the ratio of $A_t$ to the concentration of that species. The functions corresponding to $H_2A$, $HA^-$, and $A^{2-}$ are $f$, $f^-$, and $f^{2-}$. The relationships are:

(h) $\quad f = \dfrac{A_t}{[H_2A]}$

(i) $\quad f^- = \dfrac{A_t}{[HA^-]}$

(j) $\quad f^{2-} = \dfrac{A_t}{[A^{2-}]}$

If we substitute for $A_t$ in accordance with Eq. g, Eq. h becomes

(k) $\quad f = \dfrac{[H_2A]}{[H_2A]} + \dfrac{[HA^-]}{[H_2A]} + \dfrac{[A^{2-}]}{[H_2A]}$.

(The same operation is performed with Eqs. i and j.) The first term on the right-hand side of Eq. k is unity. Substituting for the second and third terms in accordance with the definitions of $K_1$ and $K_2$ converts Eq. k into

(l) $\quad f = 1 + \dfrac{K_1}{[H^+]} + \dfrac{K_1 K_2}{[H^+]^2}$

Similar substitutions in Eqs. *i* and *j* give

$$(m) \quad f^- = 1 + \frac{[H^+]}{K_1} + \frac{K_2}{[H^+]}$$

$$(n) \quad f^{2-} = 1 + \frac{[H^+]}{K_2} + \frac{[H^+]^2}{K_1 K_2}$$

(The student should verify these equations for himself by making the appropriate substitutions.)

The reciprocals of the Michaelis functions are the fractional parts of $A_t$ represented by each species; for example, $1/f = [H_2A]/A_t$. Plotting these reciprocals versus pH gives a family of curves from which we may readily determine the relative concentrations of the three species at any pH. Two of these plots are shown on page 120 of the book by Dixon and Webb that is listed on page 237 of *Suggested Additional Reading*. One plot is for a system having $pK_1$ and $pK_2$ corresponding to 5 and 10, respectively; the second plot is for a system with $pK$'s of 7 and 8. In both of these illustrations, as for any pair of $pK$ values, the reciprocals $1/f$ and $1/f^{2-}$ are equal at a point midway between the two $pK$ values. Therefore, if

$$\frac{1}{f} = \frac{1}{f^{2-}},$$

then,

$$\frac{[H_2A]}{A_t} = \frac{[A^{2-}]}{A_t},$$

and

$$[H_2A] = [A^{2-}],$$

when $pH = \frac{1}{2}(pK_1 + pK_2)$.

Equation $f$ then reduces to Eq. 4.50. This is the relationship we wished to prove.

**Sixth case: a buffer of prescribed pH and ionic strength**   In certain experimental procedures, *e.g.*, electrophoresis studies or enzyme kinetics experiments, buffers must be formulated to a prescribed pH and ionic strength. We can engage in a trial and error approach of successive approximations, or we can adjust to the desired ionic strength by adding KCl. It is a simple matter, however, to calculate the correct concentrations of conjugate acid and base that will give both the desired pH and ionic strength.

There are two unknowns, [HA] and its conjugate base $[A^-]$, and, therefore, two equations are needed. Let $x$ be the molar concentration of acid required and let $y$ be the molar concentration of salt. The two equations needed are the Henderson-Hasselbalch Equation and the expression for the ionic strength.

The simplest example would be that of a uni-univalent salt and a monoprotic acid, *e.g.*, sodium acetate and acetic acid. This is the only type of situation in which the concentration of salt will equal the ionic strength. The ionic strength expression concerns sodium acetate only, since the slight dissociation of the acid may be neglected:

$$[Na^+] = y,$$
$$[CH_3COO^-] = y,$$
$$[CH_3COOH] = x.$$

Substituting in the ionic strength expression, we write

$$I = \tfrac{1}{2}(y \cdot 1^2 + y \cdot 1^2) = y.$$

For acetic acid, $pK_a' = 4.66$. If a buffer of $pH = 5.00$ and $\mu = 0.10$ is to be prepared,

$$y = 0.10.$$

The Henderson-Hasselbalch Equation may be written as

$$pH = pK_a' + \log\frac{y}{x}.$$

Substituting 0.10 for $y$ in the Henderson-Hasselbalch Equation, we find $x$ to be 0.046. The buffer is then made up to be 0.10 $M$ with respect to sodium acetate and 0.046 $M$ with respect to acetic acid.

A more involved situation develops with a polyprotic acid. Since phosphate buffer is so commonly used in the biochemistry laboratory, let us derive the necessary equations for a phosphate buffer of $pH = 7.00$ and $I = 0.1$. With phosphoric acid the pairs of conjugate acids and bases and the corresponding $pK_a$ values are

| Conjugate acid | | Conjugate base | $pK_a$ | $pK_a'$ |
|---|---|---|---|---|
| $H_3PO_4$ | $\rightleftharpoons$ | $H_2PO_4^-$ | 2.12 | |
| $H_2PO_4^-$ | $\rightleftharpoons$ | $HPO_4^{2-}$ | 7.20 | 6.86 |
| $HPO_4^{2-}$ | $\rightleftharpoons$ | $PO_4^{3-}$ | 12.4 | |

A pair is selected for which the $pK_a$ lies within a range $\pm 1.5$ pH units from the desired pH of 7.00. The middle pair satisfies this criterion. The potassium salts are most frequently used in enzymatic studies; in other experiments, sodium salts may be more desirable. The potassium salts, $KH_2PO_4$ and $K_2HPO_4$, do not react with each other. We have three ions to consider: $K^+$, $H_2PO_4^-$, and $HPO_4^{2-}$. If the concentration of the acid species ($KH_2PO_4$) is $x$, it will contribute $x$ moles of $K^+$ per liter and $x$ moles of $H_2PO_4^-$ per liter. If the concentration of $K_2HPO_4$ is $y$, it will contribute $2y$ moles of $K^+$ per liter and $y$ moles per liter of $HPO_4^{2-}$. The ionic strength expression is

$$I = \tfrac{1}{2}( [K^+] \cdot 1^2 + [H_2PO_4^-] \cdot 1^2 + [HPO_4^{2-}] \cdot 2^2) = 0.1.$$

$$I = \tfrac{1}{2}(x + 2y \quad + \quad x \quad + \quad 4y \quad ) = 0.1,$$

$$0.1 = x + 3y,$$

$$x = 0.1 - 3y.$$

The Henderson-Hasselbalch Equation is

$$\mathrm{pH} = pK_a' + \log \frac{y}{x},$$

$$7.00 = 6.86 + \log \frac{y}{x}.$$

Substituting for $x$ from the ionic strength equation, gives

$$7.00 = 6.86 + \log \frac{y}{0.1 - 3y}.$$

Therefore,

$$y = 0.0268\ M.$$

From the relationship $x = 0.1 - 3y$,

$$x = 0.0196\ M.$$

Phosphate buffers containing $KH_2PO_4$ and $K_2HPO_4$ may also be formulated from a nomogram recently published by W. C. Boyd (Fig. 4.6).

If it is necessary to prepare the same buffer from $KH_2PO_4$ and $KOH$, we proceed in a slightly different way, but the principle is the same. There will be a chemical reaction between the acid and alkali, and we need to make the calculations in terms of initial concentrations to yield a buffer of the desired pH and ionic strength. Let $x$ equal the initial acid and $y$ equal the initial alkali. Since excess acid will be required (a buffer is composed of a weak acid and its salt), $x$ must of necessity be larger than $y$. Final acid concentration will be equal to $(x - y)$ and final salt ($HPO_4^{2-}$) will be equal to $y$. No KOH will remain at equilibrium. The concentrations of ions in the final solution will be: $[K^+] = x + y$; $[H_2PO_4^-] = x - y$; and $[HPO_4^{2-}] = y$. It should be reemphasized that, even though $H^+$ and $OH^-$ are present in these buffers, at pH values in the 4 to 10 range their concentrations are insignificant compared with

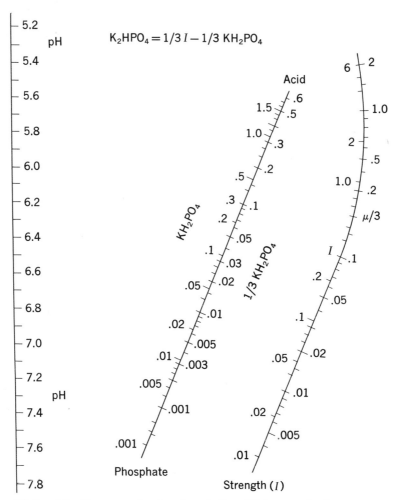

**Figure 4.6**   Nomogram for phosphate buffers. With a straightedge connect the desired points on the pH and ionic strength scales. From the acid phosphate scale read off the corresponding value for $KH_2PO_4$. This will be the correct molar concentration for acid phosphate in the buffer. The acid phosphate scale also gives directly the value of $1/3\ KH_2PO_4$. Enter this concentration into the equation at the top of the nomogram and calculate the molar concentration of secondary phosphate, $K_2HPO_4$. [From Boyd, W. C., *J. Biol. Chem.*, **240**:4097, 1965. Used by permission.]

other ions and they are not included in the ionic strength expression. Following the same steps as before, for this buffer we write

$$I = x + 2y = 0.1;$$

$$x = 0.1 - 2y;$$

$$pH = pK'_a + \log \frac{y}{x - y}.$$

Substituting for $x$ gives

$$pH = pK'_a + \log \frac{y}{0.1 - 3y},$$

and solving the simultaneous equations yields

$$x = 0.0464;$$

$$y = 0.0268.$$

A word of caution should be injected at this point. The Henderson-Hasselbalch Equation is a limiting law. Apparent dissociation constants are used (in the form of $pK'_a$) whenever they are available, and small terms are dropped wherever they are insignificant in a sum.* From the previous discussions it should be apparent that good results will be obtained only when calculations are made for dilute solutions in the pH 4 to 10 range.

**Summary**   The foregoing discussion on weak acids and bases, with and without their salts, has attempted to show a pattern of procedure to be followed. Formulas need not be memorized, if the *modus operandi* is remembered:

1. Write the pertinent equations for the various equilibria.
2. Set up the equilibrium expression for $K_a$ or $K_b$, as preferred. If more than one equilibrium is involved, write the additional equilibrium constant expressions.
3. Make all permissible simplifications, remembering that $[H_2O]$ is incorporated into the constant term. Make any necessary substitutions.
4. Solve the equation for the desired unknown, *i.e.*, $[H^+]$, $[OH^-]$.
5. Put the equation into the negative logarithmic form, and then into the "$p$" form.

The following general equations are some of the more useful approximate ones for the calculation of pH:

*The dissociation of a weak acid alone:*

$$pH = \tfrac{1}{2}(pK'_a - \log C_a). \qquad \text{(Eq. 4.24)}$$

*The dissociation of a weak base alone:*

$$pOH = \tfrac{1}{2}(pK'_b - \log C_b). \qquad \text{(Eq. 4.37)}$$

$$pH = \tfrac{1}{2}(pK_w + pK'_a + \log C_b) \quad \text{(} a \text{ refers to the conjugate}$$
$$\text{acid of the weak base).} \qquad \text{(Eq. 4.25)}$$

---

*For a discussion of these simplifications, the reader should consult the book by W. M. Clark included in the listing in *Suggested Additional Reading*.

*The end point in the titration of a weak base with a strong acid:*

$$\text{pH} = \tfrac{1}{2}(pK'_a - \log C_b) \quad \text{(}a\text{ refers to the conjugate}$$
$$\text{acid of the weak base).} \quad \text{(Eq. 4.24)}$$

*The end point in the titration of a weak acid with a strong base:*

$$\text{pH} = \tfrac{1}{2}(pK_w + pK'_a + \log C_a). \quad \text{(Eq. 4.25)}$$

*An acid in the presence of its conjugate base (or salt):*

$$\text{pH} = pK'_a + \log \frac{[\text{conjugate base}]}{[\text{conjugate acid}]}. \quad \text{(Eq. 4.46)}$$

Whenever they are available and suitable to the concentrations present, $pK'_a$ values should be used instead of $pK_a$ values.

### 4.3 Indicators of pH

The determination of pH is part of the daily routine of most laboratories. Highly satisfactory instruments for this purpose can be purchased from any instrument supplier, so that one relies less and less on the traditional colorimetric methods. Indicator papers which permit the estimation of pH in the 2 to 10 range, however, have become almost indispensable for preliminary pH adjustments or rapid estimations on small volumes. In addition, indicator titrations are still used in many analytical methods, and there continues a proliferation of spectrophotometric methods, involving acid production or oxidation-reduction equilibria in which there is a color change of a suitable indicator in the determinative step. The student must, therefore, fortify himself with a basic knowledge of indicator theory. If the preceding pages of this chapter have been mastered, the theory will already have been learned. All that is required is a slightly different interpretation.

**Theory of indicators** The indicator is a weak acid, which undergoes dissociation in the working pH range, *i.e.*, the $pK_a$ lies between 2 and 12 for the most useful ones. Furthermore, and most distinctively, the acid and its conjugate base (the anion) absorb visible light of different frequencies, so that they are distinctly different colors (or one may be colorless). The Henderson-Hasselbalch Equation in its most general form is

$$\text{pH} = pK_a + \log \frac{[\text{proton acceptor}]}{[\text{proton donor}]}.$$

Similarly we can write for the indicator

$$pH = pK_I + \log \frac{[I^-]}{[HI]},$$

where $pK_I$ is the $pK_a$ of the indicator and HI and $I^-$ refer to the conjugate acid and base respectively of the indicator.

Phenol red, which has a $pK_I$ of 7.8, is yellow in the conjugate acid form and red in the conjugate base form. At pH 6.8 and below, solutions of phenol red are distinctly yellow and cannot be distinguished visually from one another. The parallel situation exists for the red solutions, which exist at pH 8.8 and above. Between pH 6.8 and pH 8.8 there are many intermediate hues ranging from a yellowish orange to reddish orange. Experience is needed to train the eye to recognize the exact shade characteristic of the pH which is desired. The formulas for these two species of phenol red are

Phenol Red yellow acid form      Phenol Red red basic form

Phenol red (phenolsulphonphthalein) is one of a group of triphenyl-methyl compounds that display color changes corresponding to disso-ciative changes in their structures. Phenolphthalein is a familiar member of this group. In some of these compounds, a second color change occurs higher or lower in the pH range. Phenol red actually undergoes three such transitions: the yellow acid of the neutral range becomes red at low pH and the red anion of the neutral range slowly fades until it is colorless in strongly alkaline solutions. This illustration should serve as a warning that the properties of an indicator should be well understood before using it. A red color with phenol red might mean that the solution is slightly alkaline or it might mean that it is strongly acid. A preliminary test with indicator paper of broad range would prevent confusion. Data for some common indicators are given in Table 4.5.

**Sources of error in the use of indicators**   In the process of indicator determination of pH, errors may be incurred for various reasons. The paragraphs that follow list some common sources of difficulty.

**TABLE 4.5** DISSOCIATION CHARACTERISTICS OF SOME USEFUL INDICATORS

| Indicator | $pK_I$ | Color Conjugate Acid | Conjugate Base |
|---|---|---|---|
| Thymol blue (acid range) | 1.7 | red | yellow |
| Methyl yellow | 3.3 | red | yellow |
| Methyl orange | 3.5 | red | yellow |
| Bromphenol blue | 4.0 | yellow | blue |
| Bromcresol green | 4.7 | yellow | blue |
| Methyl red | 5.0 | red | yellow |
| Chlorphenol red | 6.0 | yellow | red |
| Bromcresol purple | 6.2 | yellow | purple |
| Bromthymol blue | 7.1 | yellow | blue |
| Phenol red | 7.8 | yellow | red |
| Cresol red | 8.3 | yellow | red |
| Thymol blue (alkaline range) | 8.9 | yellow | blue |
| Phenolphthalein | 9.7 | colorless | red |

*Source:* Data are quoted from Clark, W. M., *Topics in Physical Chemistry*, p. 291. Williams and Wilkins, Baltimore, 1952.

*Temperature* The $pK_I$ for an indicator is usually stated for 25°C. If the indicator is used at temperatures appreciably above or below 25°, the $pK_I$ will change, the magnitude of the shift depending on the extent to which the equilibrium of the indicator dissociation is affected by the temperature change. If the $pK_I$ is not known for the temperature at which the indicator is used, results must be regarded as approximate.

*Color or turbidity* When indicators are used in solutions which have appreciable color of their own, for example, urine or extracts of vegetable material, some compensation must be made for the interfering color. The reference solution of buffer plus indicator can be placed in a test tube rack with the unknown solution containing indicator in the adjoining space. If a tube of unknown solution containing *no indicator* is now placed behind the reference solution, and a test tube of buffer alone placed behind the unknown-plus-indicator, a rough correction can be made for the color initially present in the unknown. The scheme is illustrated in Fig. 4.7. The rack of tubes is held up to a strong light and the two sets of tubes viewed in tandem. In this way a more accurate matching of unknown and reference solution can be made. Compensation for the turbidity of a solution can be obtained in the same way.

*Salt* The $pK_I$ is dependent upon the salt concentration of the solution because of the effect of ionic strength on the activity coefficient of the various ions in solution (see Eq. 4.5). Since one of the colored forms

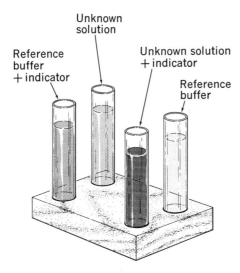

**Figure 4.7** Method for correction for turbidity or color in unknown solution, when pH is being measured by the indicator method.

of the indicator is an ion, its activity or effective concentration will depend on the ionic strength. Therefore, if the unknown solution to be tested or adjusted contains high amounts of salt (for example, sodium chloride from a neutralization), an equal amount of salt should be added to the reference buffer containing the indicator. In the event that this procedure is not feasible, the measurement or adjustment must be regarded as only approximate.

*Complicating reactions*    Proteins are likely to be present in extracts of plant material or biological fluids. Certain indicators combine with proteins, thus introducing serious error into the pH adjustment or measurement. The indicator-protein complex may be a slightly different color from that of the free indicator, and furthermore, the equilibria involving the indicator become complicated. Parenthetically it should be added that proteins and polysaccharides in solution give rise to errors in the use of the glass electrode (pH meter). A film of surface-active material absorbs to the thin glass membrane of the electrode and alters the electrical potential across the membrane. In adjusting the pH of starch solutions one should always check routinely with indicator paper as well. This effect is commonly called "colloid error."

### 4.4   Acid-Base Equilibria in the Dissociation of Amino Acids

Most of the biologically important electrolytes fall into three principal classes; (A) the di- and tricarboxylic acids of the citric cycle and ancillary pathways; (B) the organic phosphates and polyphosphates—that is, the sugar phosphates, certain of the compound and derived lipids, and the

**TABLE 4.6** pK$_a$ VALUES FOR PHOSPHORIC ACID AND SOME METABOLICALLY IMPORTANT DERIVATIVES (ALL DATA FOR 25°C)

| Compound | 1st H$^+$ | pK$_a$ 2nd H$^+$ from Phosphate | 3rd H$^+$ from Phosphate |
|---|---|---|---|
| Phosphoric acid | 2.12 | 7.20 | 12.4 |
| Glucose-1-phosphate | 1.1 | 6.51 | |
| Glucose-6-phosphate | 0.94 | 6.11 | |
| Glycerol-2-phosphate | 1.33 | 6.65 | |
| Adenosine monophosphate | | 6.67 | |
| Guanosine monophosphate | | 6.66 | |
| Inosine monophosphate | | 6.66 | |
| Cytidine monophosphate | | 6.62 | |
| Uridine monophosphate | | 6.63 | |
| Adenosine diphosphate | | 7.20 | |
| Guanosine diphosphate | | 7.19 | |
| Inosine diphosphate | | 7.18 | |
| Cytidine diphosphate | | 7.18 | |
| Uridine diphosphate | | 7.16 | |
| Adenosine triphosphate | | 7.68 | |
| Guanosine triphosphate | | 7.65 | |
| Inosine triphosphate | | 7.68 | |
| Cytidine triphosphate | | 7.65 | |
| Uridine triphosphate | | 7.58 | |
| Pyridoxal-5-phosphate ($pK'_a$) | <2.5 | 6.20 | |
| Pyridoxamine-5-phosphate ($pK'_a$) | <2.5 | 5.76 | |

*Sources:* Data for carbohydrate phosphates from Ashby, J. H., E. M. Crook and S. P. Datta, *Biochem. J.,* **56**:198, 1954, *Ibid ,* **59**.203, 1955, and West, E. S., *Textbook of Biophysical Chemistry,* 3rd Ed., MacMillan, New York, 1963. Data for purine and pyrimidine phosphates from Phillips, R., P. Eisenberg, P. George, and R. J. Rutman, *J. Biol. Chem.,* **240**:4393, 1965. Data for B$_6$ phosphates from Williams, V. R., and J. B. Neilands, *Arch. Biochem. and Biophys.,* **53**:56, 1954.

derivatives of purines and pyrimidines, including nucleic acids; and (C) the amino acids and their derivatives, including proteins. Those in Class (A) are polyprotic acids. We have already given consideration to the practical aspects of their titration and the pertinent calculations. Their dissociation constants have been presented in Table 4.3. Class (B), the organic phosphates, behave in their dissociations much like phosphoric acid itself, with some displacement of the $pK_a$ values. Certain of these have been collected together in Table 4.6, where they may be compared with H$_3$PO$_4$. Class (C), because of the variety of dissociating groups and their metabolic importance, deserves special consideration at this point.

**TABLE 4.7** $pK_a'$ VALUES AND HEATS OF IONIZATION OF DISSOCIATING GROUPS IN AMINO ACIDS AND PROTEINS (ALL DATA AT 25°C)

| Group | $pK_a'$ range | $\Delta H_{ionization}$ (calories per mole) |
|---|---|---|
| Carboxyl (*alpha*) | 1.8–3.6 | ±1,500 |
| Carboxyl (*beta*) | 3.0–4.7 | ±1,500 |
| Carboxyl (*gamma*) | ~4.4 | ±1,500 |
| Phenolic hydroxyl | 9.8–10.4 | 6,000 |
| Sulfhydryl | 8.3–8.6 | ~6,500 |
| Imidazolium | 5.6–7.0 | 6,900–7,500 |
| Ammonium (*alpha*) | 7.9–10.6 | 10,000–13,000 |
| Ammonium (*epsilon*) | 9.4–11.0 | 10,000–12,000 |
| Guanidinium | 11.6–12.6 | 12,000–13,000 |

*Sources:* Data are based on values given by Edsall, J. T., and J. Wyman in *Biophysical Chemistry*, Vol. I, Academic Press, New York, 1958, and by Dixon, M., and E. C. Webb in *Enzymes*, Academic Press, New York, 1964. Data for $\Delta H_{ionization}$ of $-SH$ from Benesch, R. E., and R. Benesch, *J. Am. Chem. Soc.*, **77**:5877, 1955.

**Amino acids as Brønsted acids**  All of the ionizing groups of amino acids can be written in a protonated form and the subsequent dissociations treated as acid dissociations. Because of the convenience of this approach, it is now conventional to refer to the $pK_a$'s of the basic groups as well as those of the carboxyl and other acidic functions. The functional groups which comprise the protonated basic groups are: *alpha* and *epsilon* ammonium, imidazolium, and guanidinium. The acidic groups are: *alpha*, *beta*, and *gamma* carboxyl; sulfhydryl, and phenolic hydroxyl. The $pK_a$ values for these groups will depend on their electrostatic environment, *e.g.*, the *beta* carboxyl in free aspartic acid will not have exactly the same $pK_a$ as it will in the dipeptide, aspartylglycine. For this reason, ranges of $pK_a'$ values are presented in Table 4.7. The ranges have been chosen from representative data and are not all-inclusive.* For example, anomalously high $pK_a'$ values have been found for certain of the phenolic hydroxyl groups of bovine pancreatic ribonuclease† as well as several other proteins.‡ Similarly, in the authors' laboratory, a spectrophotometric difference titration of the sulfhydryl

---

*A highly recommended reference is *Chemistry of the Amino Acids* by J. P. Greenstein and M. Winitz, Wiley and Sons, Inc., New York, 1961. Additional data on apparent dissociation constants for amino acids is to be found in a review by J. Steinhardt and S. Beychok in *The Proteins*, Hans Neurath, Ed., Vol. II, 162–164, 1964.

†Tanford, C., J. D. Hauenstein and D. G. Rands, *J. Amer. Chem. Soc.*, **77**, 6409 (1955).

‡Steinhardt, J. and J. A. Reynolds, *Multiple Equilibria in Proteins*, Academic Press, New York, 1969, Chapter V.

and phenolic hydroxyl groups of $\beta$-methylaspartase accounted for all of the sulfhydryl functions at pH 10, but only 10 percent of the phenolic hydroxyl functions, as previously determined by the Moore-Stein method.

**Titration curves** The titration of any single ionizing group in an amino acid would yield a plot of the type shown earlier in Fig. 4.3. All amino acids, however, possess at least two dissociating groups, the *alpha* amino (or imino) and *alpha* carboxyl. A titration curve corresponding to these dissociations is shown in Fig. 4.8. With glycine, the $pK_a'$ values lie far enough apart that the steep sections of the curve are clearly separated. The isoelectric point is obtained by averaging $pK_1'$ and $pK_2'$, in accordance with Eq. 4.50.

It is of historical interest that a heated dispute was waged over the correct assignment of experimentally determined $pK_a'$ values and the correct formula for the neutral (isoelectric) species. For example, when glycine is dissolved in water, the solution is nearly neutral ($\sim$pH 6). What is the correct formula for the solute: $H_2N-CH_2-COOH$ or $^+H_3N-CH_2-COO^-$? The latter formula is referred to as the *zwitterion* structure (from the German, *zwitter*, hybrid). In the argument over the neutral formulation, all were in agreement that the fully protonated species was exactly the structure, A, shown on the left in the diagram below, and that the species existing at high pH was C. However, depending on the assignment of the $pK_a'$ values, two different intermediate or neutral forms could be postulated: the uncharged molecule, B', and the zwitterion, B. The two possible dissociation sequences were

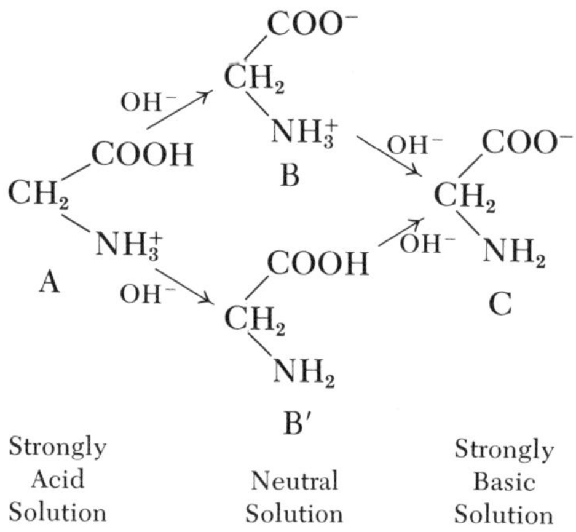

| Strongly Acid Solution | Neutral Solution | Strongly Basic Solution |
|:---:|:---:|:---:|

Overwhelming evidence was marshalled in support of the upper pathway containing the zwitterion. Among the most persuasive data were the heats of ionization (Table 4.7). If a dissociation constant, $K_a'$, is measured at two temperatures and the van't Hoff Equation applied (Eq. 3.44), the $\Delta H$ so obtained is the heat of ionization. Experimentally this amounts to

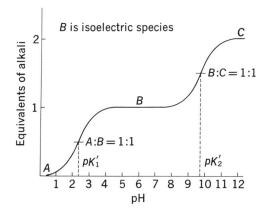

**Figure 4.8** Titration curve for glycine.

little more than carrying out the titration in a refrigerated thermostatted bath set at 5°C for the first run and 40°C for the second run. The $pK_a'$ values are obtained from the plots and inserted directly into the van't Hoff Equation, which can be conveniently rewritten

$$pK_x' - pK_y' = \frac{\Delta H (T_y - T_x)}{2.303 \, RT_y T_x},\tag{4.51}$$

where $pK_x'$ and $pK_y'$ refer to the *same* dissociation step at two temperatures, $T_x$ and $T_y$. For the monoamino-monocarboxylic acids, $\Delta H$ for the first dissociation step was found to be small (of the order of a kilocalorie), whereas $\Delta H$ for the second dissociation step was larger by ten-fold. Comparison of these data with heats of ionization of model compounds like acetic acid (for —COOH) and ethyl ammonium chloride (for —$NH_3^+$) permitted an unambiguous assignment of dissociation constants.

With a dicarboxylic amino acid, for example, aspartic acid, the carboxyl dissociations lie so close together that the steep sections overlap, as shown in Fig. 4.9. The three stages in the ionization are formally assigned to the following species:

| | | | |
|---|---|---|---|
| COOH | COO⁻ | COO⁻ | COO⁻ |
| $\|$ | $\|$ | $\|$ | $\|$ |
| CHNH₃⁺ | CHNH₃⁺ | CHNH₃⁺ | CHNH₂ |
| $\|$ | $\|$ | $\|$ | $\|$ |
| CH₂ | CH₂ | CH₂ | CH₂ |
| $\|$ | $\|$ | $\|$ | $\|$ |
| COOH | COOH | COO⁻ | COO⁻ |
| *A* | *B* | *C* | *D* |

$$\text{COOH} \quad \underset{\text{H+}}{\overset{\text{OH−}}{\rightleftharpoons}} \quad \text{COO}^- \quad \underset{\text{H+}}{\overset{\text{OH−}}{\rightleftharpoons}} \quad \text{COO}^- \quad \underset{\text{H+}}{\overset{\text{OH−}}{\rightleftharpoons}} \quad \text{COO}^-$$

Here, the isoelectric species is not the one most abundant in the zone of neutral pH. The isoelectric point or isoelectric pH will be the average

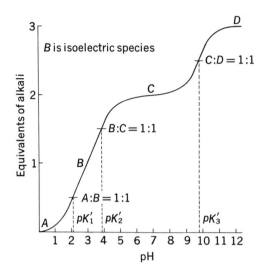

**Figure 4.9**  Titration curve for aspartic acid.

of $pK_1'$ and $pK_2'$, since $B$, the isoelectric species, is in greatest concentration midway between the two. Significant amounts of all three species, $A$, $B$, and $C$, exist at the isoelectric point, 3.0, since $pK_1'$ and $pK_2'$ lie so close together (2.1 and 3.9, respectively). Recalling the Henderson-Hasselbalch Equation (Eq. 4.47), we can immediately see that the ratio of $B$ to $A$ and of $B$ to $C$ must be approximately 10:1 at the isoelectric point.

As a final illustration let us consider an amino acid with two basic groups: lysine. The stages in the titration and the species formally assigned are

$$
\begin{array}{c}
NH_3^+ \\
| \\
(CH_2)_4 \\
| \\
CHNH_3^+ \\
| \\
COOH \\
\\
A
\end{array}
\underset{H^+}{\overset{OH^-}{\rightleftarrows}}
\begin{array}{c}
NH_3^+ \\
| \\
(CH_2)_4 \\
| \\
CHNH_3^+ \\
| \\
COO^- \\
\\
B
\end{array}
\underset{H^+}{\overset{OH^-}{\rightleftarrows}}
\begin{array}{c}
NH_3^+ \\
| \\
(CH_2)_4 \\
| \\
CHNH_2 \\
| \\
COO^- \\
\\
C
\end{array}
\underset{H^+}{\overset{OH^-}{\rightleftarrows}}
\begin{array}{c}
NH_2 \\
| \\
(CH_2)_4 \\
| \\
CHNH_2 \\
| \\
COO^- \\
\\
D
\end{array}
$$

The corresponding titration curve is given in Fig. 4.10. The isoelectric species is $C$, and therefore, the isoelectric point is obtained by averaging $pK_2'$ and $pK_3'$. For practice, the student should write the dissociations and calculate the isoelectric points for histidine and tyrosine. The analysis of cysteine is complicated by the fact that the dissociations which occur after the titration of the carboxyl group appear to proceed simultaneously, rather than sequentially. This interesting phenomenon merits further consideration and will be discussed following a presentation of the theory of polyprotic ionization (Section 4.5).

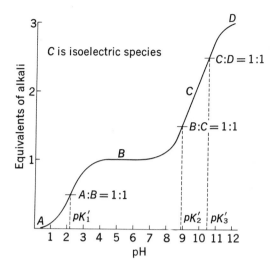

**Figure 4.10** Titration curve for lysine.

**Miscellaneous uses of amino acid $pK'_a$ values** Dissociating groups or proteins, as well as of amino acids, can be identified and/or enumerated by the use of Eq. 4.51. Since titration characteristics of the various functional groups are known, a titration curve of the protein, obtained at two temperatures, will permit us to count the number of carboxyl, imidazolium, and other groups of interest.[*]

Dixon has developed a method for treating enzyme kinetic data in order to identify functional groups at the enzyme active site. If the binding of substrate or catalysis involves dissociating functional groups on the protein or substrate, one may tentatively identify the participating functions by plots of log $V_{max}$, $-\log K_m$, and log $v_0$ versus pH.[†] If the method can be successfully applied to the experimental data, the plots will possess characteristic breaks at pH values corresponding to the $pK'_a$ values of critical functional groups. Since so many enzymes appear to have catalytically essential sulfhydryl and imidazolium groups, it is unfortunate that the heats of ionization for the two are nearly identical. Otherwise Eq. 4.51 could be combined with Dixon's method to both identify and distinguish these two ubiquitous nucleophiles.

### 4.5 Theoretical Considerations Involving Polyprotic Species

**Microscopic and macroscopic dissociation constants** When a polyprotic compound is titrated—let us take a mole of a dibasic acid for a simple example—two equivalents of base are consumed and two $pK'_a$

---

[*] Tanford, C., *Advan. Protein Chem.*, **17**:82, 1962.
[†] See the book by Dixon and Webb included in the listing in *Suggested Additional Reading*.

values may be calculated from the titration curve. These constants may lie relatively far apart as they do for carbonic acid or may be much closer together as for various dicarboxylic acids cited in Table 4.3.

So far we have implied that these $pK_a'$ and the corresponding $K'$ values were the result of an uncomplicated stepwise removal of protons as base was added to the acid. A more careful consideration of all possible events leads us to the inescapable conclusion that the real situation is more complicated. If we designate the dibasic acid by the general notation HAH, there are two pathways by which it may give up protons to become the dianion, $A^{2-}$:

$$
A
\begin{cases}
H_x \xrightleftharpoons{K_{1x}} \underset{H_y}{A^- + H^+} \xrightarrow{K_{2y}} \\[2ex]
H_y \xrightleftharpoons{K_{1y}} \underset{A^- + H^+}{H_x} \xrightarrow{K_{2x}}
\end{cases}
A^{2-} + H^+
$$

We assume that $AH_x^-$ is not the same as $AH_y^-$, although for a symmetrical dibasic acid they may be chemically indistinguishable. The equilibrium constants shown in the diagram are the true *group* ionization constants in contrast with the experimentally determined constants for the two *stages* of titration of the acid. Since the group constants, $K_{1x}, K_{1y}$, etc., involve the idealized microscopically distinct species, they are called *microscopic* dissociation constants (also known as the intrinsic constants). The experimentally evaluated constants are called *macroscopic* dissociation constants or titration stage constants.

The microscopic dissociation constants are related to one another in that $K_{1x}K_{2y} = K_{1y}K_{2x}$. If three of them can be evaluated, the fourth is readily calculated. These four constants are related to the experimentally determined dissociation constants, $K_1$ and $K_2$, in the following way:

$$K_1 = K_{1x} + K_{1y}; \tag{4.52}$$

$$K_2 = \frac{1}{\dfrac{1}{K_{2x}} + \dfrac{1}{K_{2y}}} = \frac{K_{2x}K_{2y}}{K_{2x} + K_{2y}}. \tag{4.53}$$

Let us briefly consider several types of acids and summarize the pertinent relationships:

*A symmetrical dibasic acid with ionizing groups located sufficiently far apart in the molecular structure that they do not influence each other—a hypothetical case*   For this category, $K_{1x} = K_{1y} = K_{2x} = K_{2y}$. Letting $K$ represent any one of these, $K_1 = 2K$, and $K_2 = \frac{1}{2}K$. It then follows that $K_1 = 4K_2$. This is the closest that $K_1$ and $K_2$ can approach each other numerically. The titration curve would be perfectly smooth with no hint of an inflection point, just as if we were titrating two equivalents of a single monobasic acid.

*A symmetrical dibasic acid with interacting ionic groups* For this category, $K_{1x} = K_{1y}$ and $K_{2x} = K_{2y}$ since the two intermediate species, $AH_x^-$ and $AH_y^-$, are chemically indistinguishable. $K_1 = 2K_{1x} = 2K_{1y}$ and $K_2 = \frac{1}{2}K_{2x} = \frac{1}{2}K_{2y}$.

We invariably observe that the second proton is more difficult to remove than the first, so that $K_1 > K_2$. Simple electrostatic repulsion between $AH_x^-$ or $AH_y^-$ and the approaching $OH^-$ is a contributing factor. In addition, increased electronegativity of the now-dissociated group may bind the second proton more tightly.

The validity of the relationship between $K_1$ and $K_{1x}$ for this category may be found in the example of succinic acid. Since $K_1 = 6.9 \times 10^{-5}$ mole/liter, the microscopic constant, $K_{1x}$ or $K_{1y}$, should be approximately equal to $3.5 \times 10^{-5}$ mole/liter. If we titrate the monoester of succinic acid, we observe the dissociation constant to be of this order. The reason for the drop in $K$ is that the concentration of carboxyl groups in the ester is now only half of its value in the free acid, not that the group has lost acid strength.

*An unsymmetrical dibasic acid for which $K_{1x} \gg K_{1y}$* It then follows that $K_{2x} \gg K_{2y}$. The practical consequence is that the concentration of the intermediate species $AH_x^-$ is negligible and essentially only one intermediate form will be present: $AH_y^-$. $K_1$ will be approximately equal to $K_{1x}$, and $K_2$ will be approximately equal to $K_{2y}$. This same line of reasoning has been applied to the acid dissociations of the monoamino-monocarboxylic amino acids, wherein it has been established that the abundant intermediate species is the zwitterion. Evaluation of the microscopic dissociation constants has made it possible to calculate the ratio of zwitterions to uncharged molecules in a solution of glycine:

$$NH_2CH_2COOH \rightleftharpoons NH_3^+CH_2COO^-$$

Edsall and Wyman report the ratio to be $2.23 \times 10^5$.[*]

*An unsymmetrical dibasic acid for which $K_{1x}$ is nearly equal to $K_{1y}$* The titration will proceed along both pathways to nearly the same extent. The concentrations of both intermediate species will rise and fall in a closely parallel fashion as base is added to the system and HAH is converted to $A^{2-}$. If one of the intermediate forms can be distinguished from the other by any characteristic, it will be possible to determine the relative amounts of the two and thereby evaluate the microscopic

---

[*]Edsall, J. T., and J. Wyman, *Biophysical Chemistry*, Vol. I. Academic Press, New York, 1958.

constants. This has been done for the second and third stages of the titration of cysteine by following the formation of the *sulfide* intermediate spectrophotometrically.* The titration diagram may be represented as

$$CH_2\text{—}S^-$$
$$|$$
$$H\text{—}C\text{—}COO^-$$
$$|$$
$$NH_3^+$$

$K_{1x}$      (Sulfide Intermediate)      $K_{2y}$

$$CH_2\text{—}SH \qquad\qquad\qquad\qquad\qquad\qquad CH_2\text{—}S^-$$
$$| \qquad\qquad\qquad\qquad\qquad\qquad\qquad\qquad\qquad\qquad |$$
$$H\text{—}C\text{—}COO^- \qquad\qquad\qquad\qquad\qquad\qquad H\text{—}C\text{—}COO^-$$
$$| \qquad\qquad\qquad\qquad\qquad\qquad\qquad\qquad\qquad\qquad |$$
$$NH_3^+ \qquad\qquad\qquad\qquad\qquad\qquad\qquad\qquad\qquad NH_2$$

$K_{1y}$                   $K_{2x}$

$$CH_2\text{—}SH$$
$$|$$
$$H\text{—}C\text{—}COO^-$$
$$|$$
$$NH_2$$

(Sulfhydryl Intermediate)

The alkyl mercaptides ($RS^-$) absorb ultraviolet light in the wavelength range 230 to 245 m$\mu$ (positions of maxima). This characteristic was exploited to calculate the relative amounts of the two intermediate species as the titration proceeded. The four microscopic dissociation constants were calculated from the data and the $pK$'s found to be: $pK_{1x} = 8.53$; $pK_{1y} = 8.86$; $pK_{2y} = 10.36$; $pK_{2x} = 10.03$. From these constants it was possible to calculate the concentration of each of the four different species as a function of pH. The diagram shown in Fig. 4.11 is reproduced from this research. Theoretical treatments of the type we have been discussing were first published by Michaelis in 1913 and Adams in 1916.†

The problem of the sequence of dissociating groups on certain pyridine derivatives has been of special interest to the authors in connection with the polyfunctional catalyst 2-pyridone (2-hydroxypyridine) and the $B_6$ vitamin group, which are derivatives of 3-pyridone.‡

The protonated forms of these compounds have a phenolic hydroxyl functional group of surprisingly large acid strength. For example, in 2-pyridinium ion the $pK_a'$ of the phenolic functional group is 0.76 as compared with a $pK_a'$ of 10.0 for the same group in phenol. Another interesting aspect of the dissociations of 2-pyridone is that none of the

---

*Benesch, R. E., and R. Benesch, *J. Am. Chem. Soc.*, **77**:5877, 1955.

†For a full presentation of the Michaelis pH functions and their application to various systems, the student is referred to the excellent discussion in the book by Dixon and Webb that is included in the list of *Suggested Additional Reading*.

‡Williams, V. R., and J. B. Neilands, *Arch. Biochem. Biophys.*, **53**:56, 1954. Williams, V. R., and J. G. Traynham, *J. Org. Chem.*, **28**:2883, 1963.

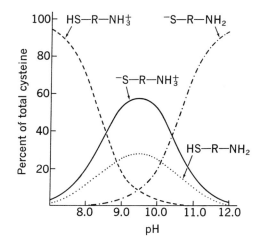

**Figure 4.11**  Concentration of each of the different ionic forms of cysteine as a function of pH, calculated from microscopic dissociation constants. [From Benesch, R. E., and R. Benesch, *J. Am. Chem. Soc.*, **77**:5877, 1955.]

various species which exist in neutral solution corresponds to the formula: 2-hydroxypyridine. Only the zwitterion and the *lactum*, 2-pyridone, are present. These comments are made by way of a warning that many pitfalls await those who assign dissociation constants on the basis of analogy.

**The hydrolysis of ATP**   The equation for the hydrolysis of ATP appears in textbooks in various ways. Two general forms are:

$$ATP + H_2O \rightleftharpoons ADP + P_i \qquad \text{(as to dissociation)} \qquad (4.54)$$

$$(\Sigma ATP) + H_2O \rightleftharpoons (\Sigma ADP) + (\Sigma PO_4) \qquad \begin{array}{l}\text{(inclusive of} \\ \text{all ionic forms)} \end{array} (4.55)$$

Two explicit forms are:

$$ATP^{4-} + H_2O \rightleftharpoons ADP^{3-} + HPO_4^{2-} + H^+ \qquad \begin{array}{l}\text{(limited to} \\ \text{pH} > 7) \end{array} \quad (3.48)$$

$$ATP^{4-} + H_2O \rightleftharpoons ADP^{2-} + HPO_4^{2-} \qquad \begin{array}{l}\text{(similar to 3.48,} \\ \text{but not showing} \\ \text{dissociation of} \\ ADP^{2-} \text{ to } ADP^{3-} + H^+) \end{array} \quad (4.56)$$

As pointed out in Chapter 3, in the range pH 6–8, ATP, ADP, and $P_i$ each undergoes a dissociation, so that $\Sigma ATP = ATP^{4-} + ATP^{3-}$; $\Sigma ADP = ADP^{3-} + ADP^{2-}$; and $\Sigma P_i = HPO_4^{2-} + H_2PO_4^{-}$. In this pH range the equilibrium constant for the hydrolysis of ATP (Eq. 4.55) is

$$K' = \frac{[\Sigma ADP][\Sigma PO_4]}{[\Sigma ATP]} = \frac{[ADP^{3-} + ADP^{2-}][HPO_4^{2-} + H_2PO_4^{-}]}{[ATP^{4-} + ATP^{3-}]} \qquad (4.57)$$

with the activity of water taken as unity. Since the relative amounts of the various species obviously shift with pH, it is highly desirable to rearrange this expression so that the pH dependence of the position of equilibrium and hence the free energy of hydrolysis at a specific pH can be more conveniently handled.

One way to attack the problem is to define the apparent dissociation constants for the three ionizations and to introduce them, along with $H^+$, into Eq. 4.57. The apparent dissociation constants at 37°C and 0.15 ionic strength are:[*]

$$K'_\alpha = [H^+][ATP^{4-}]/[ATP^{3-}] = 10^{-6.50}$$

$$K'_\beta = [H^+][ADP^{3-}]/[ADP^{2-}] = 10^{-6.27} \qquad (4.58)$$

$$K'_\gamma = [H^+][HPO_4^{2-}]/[H_2PO_4^-] = 10^{-6.73}$$

Eq. 4.57 can be rearranged as follows:

$$K'_{eq} = \text{``}K\text{''} \cdot Y = \frac{[ADP^{2-}][HPO_4^{2-}]}{[ATP^{4-}]} \cdot Y \qquad (4.59)$$

where

$$Y = \frac{\left[1 + \dfrac{K'_\beta}{[H^+]}\right]\left[1 + \dfrac{[H^+]}{K'_\gamma}\right]}{1 + \dfrac{[H^+]}{K'_\alpha}}$$

and "$K$" is the apparent equilibrium constant corresponding to the reaction as written in Eq. 4.56. Admittedly, there is nothing *obvious* about this rearrangement, but it turns out to be a convenient form for calculations. The student might gain insight by substituting into $Y$ for $K'_\alpha$, $K'_\beta$, and $K'_\gamma$ in terms of their definitions (see 4.58) and then working back to the expression shown in Eq. 4.57.

Values of $Y$ corresponding to various values of pH can readily be calculated from the apparent dissociation constants given in (4.58). If $K'$ and $Y$ are accurately known at a certain pH, it is then possible to calculate "$K$." For example, Benzinger and Hems found $\Delta G'$ for the hydrolysis of ATP at pH 7.0 and 37°C to be $-7,730$ cal in the presence of magnesium ions.[†] The apparent equilibrium constant (calculated from $\Delta G' = -2.303\ RT \log K'$) is equal to $2.81 \times 10^5$. At pH 7.0 and 37°C, $Y = 7.5$. Therefore "$K$" $= 2.81 \times 10^5/7.5 = 3.75 \times 10^4$. With this pH-independent value of "$K$" and appropriate values of $Y$, $K'$ for the hydrolysis of ATP can be calculated at any pH.

---

[*]Alberty, R. A., M. R. Smith and R. M. Bock, *J. Biol. Chem.*, **193**:425, 1951. For values at 25°C and 0.2 ionic strength see Smith, R. M. and R. A. Alberty, *J. Am. Chem. Soc.*, **78**: 2376, 1956.

[†]Benzinger, T. H. and R. Hems, *Proc. Nat. Acad. Sci. (U.S.)*, **42**:896, 1956.

As complicated as this may appear, it is not so nearly involved as more careful approaches that take into consideration various physiologically important calcium and magnesium complexes of ATP, ADP, and inorganic phosphate. Alberty has recently published a detailed analysis of the effect of pH and metal ion concentration on the equilibrium hydrolysis of ATP to ADP.[*] Seven species of ATP, seven species of ADP and four species of $P_i$ were included. Recalculation of earlier chemical equilibrium data led to $\Delta G' = -9500$ cal per mole at pH 7.0 and 25° in the absence of magnesium ions. In the presence of millimolal magnesium ion and 0.2 ionic strength, $\Delta G'$ was found to be $-8800$ cal per mole. These values are higher than those generally used during the last ten years.

At the end of Chapter 3 we learned how to calculate $\Delta G'$ from $\Delta G°$ by a relationship used in Example 3.13,

$$\Delta G' = \Delta G° + 2.303 \, RT \log Q, \qquad (4.60)$$

where the $Q$ term is either $H^+$ or its reciprocal.

From $\Delta G'$ we readily calculated the corresponding apparent equilibrium constant. Why not use this method in calculating $K'$ for the hydrolysis of ATP at any desired pH? Actually, the method was used by Krebs and Kornberg[†] some years ago to calculate $\Delta G'$ at pH 7.5. The ATP hydrolysis reaction was written

$$ATP^{4-} + H_2O \rightleftharpoons ADP^{3-} + HPO_4^{2-} + H^+ \qquad \text{(same as 3.51)}$$

and $\Delta G°$ was +1.3 kcal/mole at 25°, 1 atm, and with all reactants and products in their standard states of 1 molal activity (pH 0). However, since $ATP^{4-}$, $ADP^{3-}$, and $HPO_4^{2-}$ do not exist in significant concentrations at pH = 0, we need a more flexible approach—one that includes all species pertinent to the pH under consideration. The method of Eq. 4.60 is unsatisfactory here because of the multiple equilibria involved. Values for $K'$ calculated by this approach do not agree with those obtained by using Eq. 4.59 and are not nearly so accurate.

## 4.6 Some Biologically Important Consequences of the Ionic Environment

### "Salting-in" and "salting-out" of polar molecules

Early in his undergraduate training, the student learns that proteins are traditionally grouped into a half dozen or so solubility classes, principally because of the dearth of a better system of classification. He also learns that, of

---

[*]Alberty, R., *J. Biol. Chem.*, **243**:1337, 1968.
[†]Krebs, H. A. and H. L. Kornberg, *Ergeb. Physiol.*, **49**, 212–298 (1957). Appendix by K. Burton.

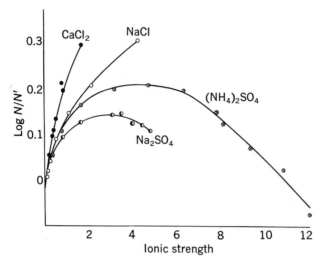

**Figure 4.12**  Solubility of cystine in solutions of different salts at various ionic strengths. [From Cohn, E. J., *Chem. Rev.*, **19**:241, 1936.]

the larger proteins, only the albumins are soluble in distilled water, all other classes requiring some adjustment of the aqueous environment— its salt concentration, pH, or dielectric constant. Less well known is the effect of salt concentration on the solubility of low molecular weight solutes such as the amino acids. However, the increase in solubility of amino acids with increasing ionic strength of the solution was clearly shown over thirty years ago. Two figures are reproduced from the original account: Fig. 4.12, which shows the effect of various inorganic salts on the solubility of cystine in water, and Fig. 4.13, which shows a parallel but more dramatic effect of inorganic salts on aqueous solutions of horse carboxyhemoglobin. The portions of the curves having positive slopes delineate the "salting-in" aspects of increasing ionic strength, whereas those portions of negative slope result from the "salting-out," which occurs most markedly in these figures at higher concentrations of ammonium sulfate and sodium sulfate.

The fundamental mathematical relationship which applies to the salting-in of ions is the Debye-Hückel Equation for dilute solutions, presented in a limiting form in the first part of the chapter:

$$-\log \gamma = 0.51 \ Z^2\sqrt{I} \quad \text{(Eq. 4.5)}.$$

As the ionic strength of a solution increases, the activity coefficient of any particular ion, $\gamma$, decreases. The result is an increase in the solubility of the ion. This relationship may be clarified: The solubility product, $K_{sp}$, is related to the average activity coefficient, $\gamma\pm$, in a way that helps us understand Eq. 4.5.

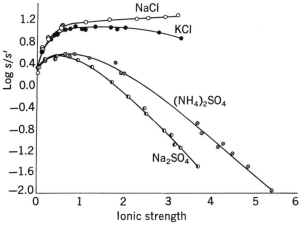

**Figure 4.13** Effect of the ionic strength of various inorganic salts on the solubility of hemoglobin. [From Cohn, E. J., *Chem. Rev.*, **19**:241, 1936.]

A relatively insoluble salt, AgCl, dissolves in water:

$$AgCl_{(s)} \underset{aq}{\overset{}{\rightleftharpoons}} Ag^+ + Cl^-.$$

The thermodynamic constant, $K_{sp}^{\circ}$, is the product of the activities of the ions, since the solid AgCl is in its standard state and therefore at unit activity:

$$K_{sp}^{\circ} = (a_{Ag^+})(a_{Cl^-}).$$

This is equal to (see Eq. 4.4)

$$K_{sp}^{\circ} = (\gamma_{Ag^+} \cdot C_{Ag^+})(\gamma_{Cl^-} \cdot C_{Cl^-}).$$

The apparent solubility product constant, $K_{sp}'$, will be simply $(C_{Ag^+}) \cdot (C_{Cl^-})$. This term may be factored out of the preceding equation so that

$$K_{sp}^{\circ} = (\gamma_{Ag^+} \cdot \gamma_{Cl^-}) K_{sp}'.$$

Taking the logarithm of both sides and rearranging gives

$$-\log (\gamma_{Ag^+} \cdot \gamma_{Cl^-}) = \log \frac{K_{sp}'}{K_{sp}^{\circ}}$$

Now the average activity coefficient, $\gamma \pm$, is $(\gamma_{Ag^+} \cdot \gamma_{Cl^-})^{1/2}$ and therefore,

$$-\log \gamma \pm = \log \left(\frac{K_{sp}'}{K_{sp}^{\circ}}\right)^{1/2} = 0.51 \ Z^2 \sqrt{I}.$$

Since $(K_{sp})^{1/2}$ is the molar solubility of the salt, we can also write

$$-\log \gamma \pm = \log \frac{S}{S^\circ} = 0.51 \ Z^2 \sqrt{I}. \tag{4.61}$$

Instead of the ratio of the molar solubilities, we may use the ratio of the mole fractions or any other appropriate units provided we are consistent. Equation 4.61 shows clearly that the Debye-Hückel Equation relates the solubility properties of the solute to the ionic strength of the solution. The thermodynamic solubility terms, $K^\circ$, $S^\circ$, or $X_0$, may be thought of as representing the solubility of the solute in the pure solvent, where it is essentially freed of an ionic atmosphere.

With further increases in the ionic strength of the solution, Eq. 4.61 may no longer fit the experimental data satisfactorily and the solubility may begin to decrease. A simple extension of Eq. 4.61 contains an additional term, $K_s I$, opposite in sign to the first right-hand term. Where $K_s$ is an empirical "salting-out" constant which tends to increase with the number of large non-polar residues in the dipolar ion.

$$\log \frac{S}{S^\circ} = 0.51 \ Z^2 \sqrt{I} - K_s I. \tag{4.62}$$

In related equations describing the effect of ions on the solubility of dipolar ions (see Figs. 4.12 and 4.13), both right-hand terms are linear in $I$. A limiting equation applicable to the salting-out of dipolar ions is:

$$\log \frac{S}{S^\circ} \cong - K_s I. \tag{4.63}$$

We can rearrange this equation to give

$$\log S = - K_s I + \log S^\circ. \tag{4.64}$$

A plot of $\log S$ versus $I$ will have a slope equal to $-K_s$ and an intercept of $\log S^\circ$. These plots relate only to the salting-out aspects of increasing ionic strength and have no sections of positive slope, since the salting-in term of the general equation has been dropped. Plots of this type have been useful in the study of proteins and some data are shown in Fig. 4.14. Worthy of emphasis is the point that $S^\circ$ is not an actual solubility, but a hypothetical ideal solubility obtained by extrapolating the plot to zero ionic strength.

The salting-out technique has proved invaluable in the purification of proteins, since, as shown in Fig. 4.14, individual proteins may have widely different salting-out characteristics. One of the classical criteria for distinguishing albumins from globulins is their relative solubilities in ammonium sulfate solutions. Since ammonium sulfate is the salt of

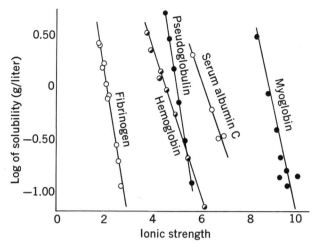

**Figure 4.14** Solubility of proteins in ammonium sulfate
solutions. [From Cohn, E. J., and J. T. Edsall, *Proteins,
Amino Acids and Peptides*. Reinhold, New York, 1943,
p. 602. (Reprinted by Hafner Publishing Co., New York,
1965.)]

a weak base and, as we have already noted, gives rise to slightly acidic
solutions, the pH of which depends on concentration, it is customary to
add buffer to the salt solutions to hold pH constant during fractionation.

As may be observed from Figs. 4.12 and 4.13, expressing the ion
concentration as ionic strength does not smooth out differences between
various salts. The sulfates are more effective than the chlorides in these
plots. The German chemist Franz Hofmeister observed that various inorganic ions could be arranged in a series which predicted their salting-
out effectiveness for a great many different proteins. This particular
arrangement is known as the Hofmeister series and is the consequence
of certain inherent characteristics of the ions, *i.e.*, size, hydration, charge,
etc. In the years that have elapsed since Hofmeister's time empiricism
has given way to a powerful theoretical treatment of his and other ionic
activity series. In the following pages we will briefly summarize the
theory and lay emphasis on its importance in interpreting biological
phenomena.

**The formation of coordination compounds** Thus far the discussion
of electrolytes has been restricted to simple model compounds and to
a proton-oriented treatment of the dissociation equilibria. Many ionic
reactions of biological significance do not involve proton exchange at
all, but should be considered in the light of the broader acid-base theory
of G. N. Lewis. Whereas a Brønsted acid is a proton donor, a Lewis acid

is an electron pair acceptor. By corollary, a Lewis base is an electron pair donor. Both viewpoints apply to a reaction such as the protonation of ammonia,

$$:NH_3 + H^+ \rightleftharpoons (NH_4)^+,$$

but the Brønsted concept is not much help in analyzing the analogous reaction of ammonia with silver ion to form the silver-ammonia complex:

$$2 :NH_3 + Ag^+ \rightleftharpoons Ag(NH_3)_2^+.$$

In the second reaction, $Ag^+$ is called a Lewis acid and the ammonia is the Lewis base. The bond which forms between the acid and base is called a coordinate bond and the reaction is given by the general equation

$$\underset{\text{Acid \quad Base}}{A + :B} \rightleftharpoons \underset{\text{Coordination Compound}}{A:B.} \qquad (4.65)$$

The acid is also called an *acceptor* and the base a *donor*. Donors which share electron pairs with cations are called ligands. Lewis theory permits us to classify as bases such substances as $H_2O$, $CO$, $Cl^-$, $SO_4^{2-}$, etc. The only qualification necessary is that the molecule or ion possess an unshared electron pair. Lewis acids include metal ions, $BF_3$, $AlCl_3$, $SiF_4$, and any other electron pair acceptor.

The unique aspect of Eq. 4.65 lies in the formation of A:B. A compound has been formed from two reactants, each capable of independent existence in solution, by virtue of the *sharing* of the electrons of the donor. Thus we distinguish this process from ordinary oxidation-reduction in which electrons are transferred between donor and acceptor and the resulting several products have an independent existence.

The equilibrium constant for the general reaction shown in Eq. 4.65 is written in the usual way:

$$K = \frac{[A:B]}{[A][:B]}. \qquad (4.66)$$

In the vocabulary of coordination chemistry such an equilibrium constant is called a *stability constant* or *formation constant*.

If, as with the silver-ammonia complex, two ligands can add to the acceptor, the second step is written

$$A:B + :B \rightleftharpoons B:A:B, \qquad (4.67)$$

and the formation constant is

$$K = \frac{[B:A:B]}{[A:B][:B]}. \qquad (4.68)$$

The constants are designated $K_1$ and $K_2$ and referred to as *stepwise* formation constants.

If we add together Eqs. 4.65 and 4.67 to obtain the overall reaction, the result is

$$A + 2 :B \rightleftarrows B:A:B. \tag{4.69}$$

The overall formation constant is

$$K = \frac{[B:A:B]}{[A][:B]^2}. \tag{4.70}$$

We can see on inspection that this is equal to $K_1 K_2$. When formation constants are expressed as overall constants the symbol $\beta$ is used to distinguish them from the stepwise constants. In the above sequence, $\beta_1 = K_1$, since Eq. 4.66 is for the overall reaction up to that point. However, $\beta_n = K_1 K_2 \ldots K_n$.

If a ligand can add to a series of acceptors, as is certainly true with ammonia and a number of metal ions, the formation constant is largest for the reaction leading to the greatest amount of coordination compound. In other words, the greater the stability of the compound, the greater the formation constant.

The Lewis acids of principal interest in biological science are the metal ions. The biochemical preeminence of coordination compounds such as hemoglobin, chlorophyll, the vitamin $B_{12}$ group, the cytochromes, and other metalloenzymes testifies to this fact. Coordination, however, is not restricted to the polyvalent ions. Even alkali metal ions show some tendency to coordinate, an observation which may be of significance in explaining the effect of $K^+$ and $Na^+$ on the activity of certain enzymes. The number of ligands with which a metal ion will coordinate depends on several factors, but for each metal ion in a specific oxidation state, that number will usually be constant and is called its *coordination number* (abbreviated as C.N.). The common coordination numbers of some metal ions are given in Table 4.8. Note that these numbers are either 2, 4, or 6. Odd coordination numbers (3, 5, 7) are known but are less common.

The most common ligand (donor) atoms are: C, N, O, F, Cl, Br, I, P, S, and As. The list of ligands in molecules of biological importance may be shortened to N, O, S, P, and C, which is less common than the first four. Ligands are either negatively charged or constitute the negative part of a dipole.

In our forthcoming brief treatment of coordination theory we will for the most part limit the discussion to the interaction of ligand and metal ion. The student should be aware, however, that in water solution, the metal ions are always aquated (that is, coordinated with water) to some

**TABLE 4.8** COORDINATION NUMBERS
OF CERTAIN METAL IONS

| $M^+$ | $M^{2+}$ | $M^{3+}$ |
|---|---|---|
| Li 4 | Ca 6 | Al 4,° 6 |
| Na 4 | Fe 6 | Fe 6 |
| Ag 2 | Co 4,° 6 | Co 6 |
| Au 2,4 | Ni 4,° 6 | Au 4 |
| Cu 2, 4 | Cu 4, 6° | Cr 6 |
| | Zn 4 | |
| | Mg 6 | |

°Less common.

extent—perhaps even fully. A chemical reaction often involves the displacement of a certain number of water molecules by ligands for which the metal ion has greater affinity. With equal concentrations of two or more ligands (or of two or more metal ions with a given concentration of ligand), the coordination compound with the highest stability constant will be the one which is present in greatest concentration at equilibrium.

Not only may ligands compete for a Lewis acid, but metal ions will compete for a specific ligand by displacing one another in accordance with the formation constants of their coordination compounds. At the risk of appearing facetious, one might truly describe this situation as the "law of fang and claw" since ligands are referred to as "monodentate" (one-tooth), "bidentate," etc., according to the number of binding sites they possess, and coordinated multidentate ligands are called "chelates" from the Greek *chele*, "claw." For an example, let us look at the formula for a fully chelated copper-glycine complex (coordination compound). The glycine is bidentate and forms stable five-membered rings with $Cu^{2+}$:

A full exposition of the theories of bonding in coordination compounds is beyond the scope of this presentation. Therefore, we will limit ourselves to a descriptive treatment of those presently in greatest

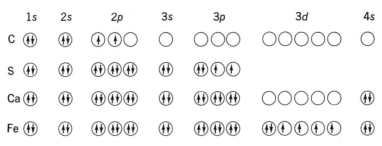

**Figure 4.15**  Electronic structures of C, S, Ca, and Fe.

use: the valence-bond theory and the ligand (crystal) field theory. More complete discussions are found in references listed in *Suggested Additional Reading.*

A satisfactory theory must permit the calculation of bond energies (and hence stabilities) and predict the properties of the coordination compounds, *e.g.,* color, geometry, etc. The basic question to be answered is, "What is the nature of the coordinate bond and how may it be described mathematically?" Is it covalent, ionic, or both? Valence bond theory assumes that the bond is covalent; ligand field theory assumes that the bond is principally electrostatic. The truth probably lies somewhere between these two concepts.

As a basis for a discussion of these theories, let us briefly review the electronic structure of the atom. Electrons are assigned to various energy levels, the first four such levels containing 2, 8, 18, and 32 electrons, respectively. Within the seven principal energy levels are sublevels of increasing energy which are designated *s, p, d,* and *f.* Within a sublevel there exist additional subdivisions of energy, called orbitals. The sublevels fill up according to the rule that an electron goes in the unfilled sublevel of lowest energy. An overlapping of sublevels occurs between principal energy levels, so that the $3d$ sublevel is of higher energy than the $4s$, the $4d$ is of higher energy than the $5s$, etc. The complete sequence is $1s, 2s, 2p, 3s, 3p, 4s, 3d, 4p, 5s, 4d, 5p, 6s, 4f, 5d, 6p, 7s, 5f,$ and $6d.$ The intervals between sublevels are not energetically equal, but that fact need not bother us in these elementary discussions.

We are interested principally in those elements which are in the first four periods and which, consequently, have relatively simple structures. Let us compare the electronic structures of some elements of biological significance: C, S, Ca, and Fe. In Fig. 4.15 we have represented each electron by a small arrow so as to show its direction of spin. We observe that the sublevels accept unpaired electrons of identical spin direction. This tendency appears to result from the mutual replusion of like charges and is one of *Hund's Rules.* Once the available sublevels are occupied by unpaired electrons, additional electrons will pair, but with opposing spin direction. Not that the $4s$ sublevel is filled before any electrons occupy the $3d$, in accordance with the energy level sequence given. However, the biologically important forms of Ca and Fe are the ions, in which case the $4s$ sublevel will be empty and we can direct our attention to unfilled sublevels.

As the electrons move in the $s$, $p$, and $d$ sublevels, they sweep out volumes of different geometries. The region in which one is most likely to find a particular electron is its orbital. In Fig. 4.16 are shown the shapes of the $s$, $p$, and $d$ orbitals. The $s$ orbital is centered at the intersection of the three coordinate axes and is spherical. The $p$ orbitals are dumbbell-shaped and are symmetrically skewered on the axis to which they correspond; that is, the $p_x$ lies on the $x$ axis, etc. Unlike the $p$ orbitals, the $d$ orbitals are not geometrically identical, but are of three different types. Three of them ($d_{xy}$, $d_{xz}$, and $d_{yz}$) have the shape of two intersecting dumbbells with lobes extending into the quadrants formed by the two designate axes. For example, $d_{xy}$ has lobes extending into the quadrants formed by the intersection of the $x$ and $y$ axes. The plane of the orbital ($xy$) is perpendicular to the third axis ($z$). Of the remaining two $d$ orbitals, one, $d_{(x^2-y^2)}$, is a double dumbbell skewered on the $x$ and $y$ axes, and the other, $d_{z^2}$, has the appearance of a dumbbell with a doughnut around its center and is skewered on the $z$ axis.

In the absence of ligands or any external field, electrons occupying the $d$ orbitals are assumed to be equal in energy (despite the variations in shape), a condition referred to as *degeneracy*.

The first successful qualitative theory of coordination compounds was proposed by Werner in 1893. Quantitative treatment was made possible by Pauling's adaptation (1930–1950) of the valence bond theory of Heitler and London. Only valence bond theory is helpful in accounting for the coordination compounds of the lighter elements, since ligand field theory cannot be applied to elements devoid of $3d$ electrons. The biological importance of calcium, magnesium, sodium, and potassium dictates that we essay an explanation of their coordination compounds.

The valence bond theory views the coordinate linkage as a covalent electron pair formed between a metal ion (the Lewis acid) and a ligand (the Lewis base). The orbital of the unshared electron pair of the ligand is presumed to overlap the hybridized empty orbitals of the metal ion. The concept of orbital hybridization was also developed by Pauling and is one that should be familiar to the student. Returning to Fig. 4.15, we note that the four outer electrons of carbon occur in two energy sublevels, $2s$ and $2p$. However, as every student of organic chemistry knows, these electrons become energetically equivalent by orbital hybridization, and give rise to a tetrahedral geometry. The shapes of certain hybridized orbitals are shown in Fig. 4.17.

Only a small number of hydridized orbitals are important in metal coordination compounds. These are $sp$, $sd$, and $dp$ (linear); $sp^3$ (tetra-

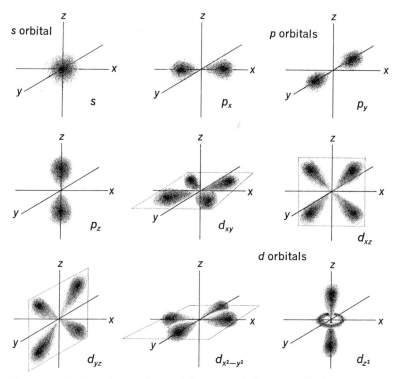

**Figure 4.16** Boundary surfaces of the $s$, $p$, and $d$ atomic orbitals.

hedral); $dsp^2$ (planar); and $d^2sp^3$ (octahedral). When the electron pair of the ligand enters the orbitals of the metal ion, the electronic configuration of the ligand is assumed to be unchanged. We center our attention on the orbitals of the metal ion.

Valence bond theory accounts for coordination compounds of a light ion such as $Al^{3+}$ in the same way as for those of a heavier ion such as $Fe^{2+}$. The electronic structures of Al and $Al^{3+}$ are

|  | 1s | 2s | 2p | 3s | 3p |
|---|---|---|---|---|---|
| Al | ⊕ | ⊕ | ⊕ ⊕ ⊕ | ⊕ | ⊕ ◯ ◯ |
| $Al^{3+}$ | ⊕ | ⊕ | ⊕ ⊕ ⊕ | ◯ | ◯ ◯ ◯ |

The ligand contributes electron pairs to the unfilled 3s and 3p orbitals. In the $AlCl_4^-$ complex the structure is regarded to be:

|  | 1s | 2s | 2p | 3s | 3p |
|---|---|---|---|---|---|
| $AlCl_4^-$ | ⊕ | ⊕ | ⊕ ⊕ ⊕ | ⊕ | ⊕ ⊕ ⊕ |
|  |  |  |  | 4Cl⁻ | |

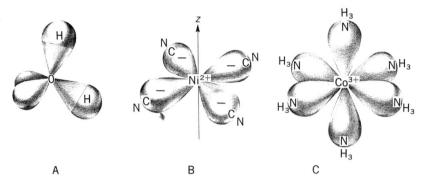

**Figure 4.17**  Boundary surfaces of certain hybrid orbitals.
A: Hybrid orbital representation of the water molecule. The $sp^3$ hybridization
gives rise to a tetrahedral geometry. A similar hybridization is displayed by
carbon compounds.
B: Hybridization of the $sp^2d$ type in $Ni^{2+}(CH^-)_4$ gives rise to a square-planar
geometry.
C: The $Co^{3+}(NH_3)_6$ complex is an example of $sp^3d^2$ hybridization which results
in octahedral geometry. [From *Physical Principles of Chemistry* by Cole, R. H.,
and J. S. Coles. W. H. Freeman and Company, Copyright © 1964.]

The hybridization is of the $sp^3$ type, resulting in a tetrahedral geometry.
For Fe and its ions, we need not reproduce the structure of the first
five sublevels ($1s$ through $3p$) since these are filled and remain so (see
Fig. 4.15). The sublevels of interest are $3d$, $4s$, $4p$ and $4d$, as shown below:

|  | $3d$ | $4s$ | $4p$ | $4d$ |
|---|---|---|---|---|
| Fe | ⊕⊕⊕⊕⊕ | ⊕ | ○○○ | ○○○○○ |
| $Fe^{2+}$ | ⊕⊕⊕⊕⊕ | ○ | ○○○ | ○○○○○ |
| $Fe^{3+}$ | ⊕⊕⊕⊕⊕ | ○ | ○○○ | ○○○○○ |

Either the $3d$ or $4d$ orbitals can be used for hybridization by $Fe^{2+}$. The
$3d$ orbitals are used in the $Fe(CN)_6^{4-}$ complex whereas it is presumed
that the $4d$ orbitals are used in $Fe(NH_3)_6^{2+}$. The former is called an "inner"
complex and the latter, an "outer" complex. Both types are of particular
biological interest because of the poisoning of hemoglobin and cytochromes by $CN^-$ and the mode of coordination of iron by N-bearing
ligands in the heme proteins. The structures are

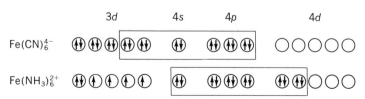

The electronic structure in these complexes is assigned principally on the basis of their magnetic properties. The inner complex of $Fe(CN)_6^{4-}$ has no unpaired electrons and is diamagnetic, whereas the outer complex of $Fe(NH_3)_6^{2+}$ is paramagnetic and therefore must possess unpaired electrons. Inner and outer complexes differ in the ease with which ligands are displaced, the inner complexes being relatively inert toward competing ligands and the outer complexes being more labile. The importance of $Fe^{2+}$ in enzyme catalysis suggests that many biologically active iron complexes are of the outer type, otherwise the bond formed between metal ion and substrate (ligand) would be too difficult to break. In addition, when the $Fe^{2+}$ of hemoglobin carries an $O_2$ molecule to tissues, it must be able to release the oxygen readily. The bond to $O_2$ is displaced by a stronger ligand, water, at the lower pH of the tissues.

The coordination compounds of the lighter alkali metals are regarded as principally electrostatic in nature, since these elements possess no partially filled $p$ or $d$ orbitals. Of the alkaline earth metals, we are particularly interested in calcium and magnesium since both are known to be enzyme activators and, in some cases, antagonists. Calcium ion, $Ca^{2+}$, can fill the $3d$ and $4s$ orbitals to attain a coordination number of 6 and form an octahedral complex. With $Mg^{2+}$, ligands will have to contribute electrons to the $3s$, $3p$, and $3d$ orbitals to form an octahedral $sp^3d^2$ complex. A specific example of $Ca^{2+}$ and $Mg^{2+}$ antagonism is seen in β-methylaspartase.[*]

Ligand field theory is applied to those metals of the 4th Period which possess partially filled $3d$ orbitals. One of these is iron, which we have already considered in the light of valence bond theory. Rather than begin with iron, however, we will choose titanium. Its deficiencies with regard to biological significance are fully compensated by its electronic simplicity. The ion, $Ti^{3+}$, possesses one $3d$ electron. According to ligand field theory, the $3d$ electron of the isolated metal ion can move in orbitals of equal energy known as degenerate orbitals. This is a way of saying that there is an equal probability of finding the electron in any of the five $3d$ orbitals. When ligands approach, two things happen: the energy of the five $3d$ orbitals is raised because of electrostatic repulsion between the metal ion and the negative ligands; and the energy of the orbitals redistributes itself unequally, resulting in a splitting into two or more energy levels. This splitting is usually referred to as crystal field (CF) splitting. The splitting is the result of the "electron pressure" of the ligand as it moves in toward the metal ion and distorts the geometry of the $d$ orbitals. In Fig. 4.18 we see an energy diagram representing (1) the Coulombic energy increase, and (2) the CF splitting which occurs when the $Ti^{3+}$ is approached by ligands. It is known that $Ti^{3+}$ will form an octahedral complex with six ligands (Fig. 4.19). If we compare the geometry of this complex with the shapes of the five $d$ orbitals (Fig. 4.16) we

---

[*]Williams, V. and J. Selbin, *J. Biol. Chem.*, 239:1635 (1964).

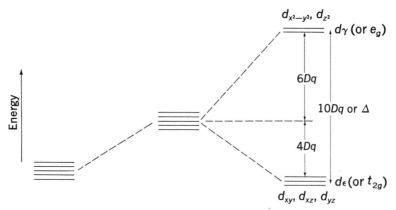

**Figure 4.18** Energy level diagram showing the crystal field splitting of the five degenerate $d$ orbitals in an octahedral ligand field. The higher energy orbitals are called the $d\gamma$ or $e_g$ orbitals; the lower energy orbitals are called the $d\epsilon$ or $t_{2g}$ orbitals. The energy difference, $\Delta$, is equal to 10 $D_q$.

can see that the greatest electron repulsion will occur between the electrons of the ligands and any electrons that may occupy the $d_{(x^2-y^2)}$ and $d_{z^2}$ orbitals of the metal ion. This repulsion results in a lower energy state for the three remaining $d$ orbitals: $d_{xy}$, $d_{xz}$, $d_{yz}$. The single $3d$ electron of $Ti^{3+}$ will prefer to occupy the orbitals of lower repulsion energy and the complex thereby achieves stabilization energy. Stabilization energy is represented by the symbol $\Delta_0$ and is dependent on the magnitude of the CF splitting, as may be seen in Fig. 4.18. Each electron placed in one of the three lower energy orbitals represents an energy stabilization of $+0.4\Delta_0$ energy units, and each electron placed in one of the two higher energy orbitals is equal to a stabilization energy loss of $-0.6\Delta_0$ energy units. If an electron is placed in each orbital, as might be the case of $Fe^{3+}$, the stabilization energy would be equal to zero:

$$3(0.4\Delta_0) + 2(-0.6\Delta_0) = 0.0\Delta_0.$$

A helpful pictorial representation is obtained by comparing a simple electrostatic model of the metal ion with the model provided by ligand field theory. In the simple electrostatic model the metal ion consists of a positive nucleus surrounded by a spherical electron cloud. Outside this electron cloud lie the ligands in their most stable geometry. The ligand field theory model depicts the electron cloud of the metal ion as being distorted so that the $3d$ electrons may avoid the positions occupied by the electrons of the ligands.

The CF splitting will be different for complexes of different geometry. In Fig. 4.20 we see the splitting of the $d$ orbitals for complexes having tetrahedral, octahedral, tetragonal, and square-planar geometries.

To apply ligand field theory to $Fe^{3+}$ with five $3d$ electrons and also to consider $Fe^{2+}$ with six, we must understand that the final distribution

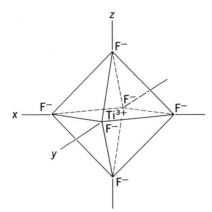

**Figure 4.19** Octahedral geometry of the $(TiF_6)^{3-}$ complex.

of the electrons will be in accordance with the Second Law, that is, the arrangement will be the one of greatest stability and lowest energy. The extent of the CF splitting depends on several factors, among which is the nature of the ligand. Large negative charge and small size of the ligand will both be conducive to a large splitting. Another factor is the extent of polarization of the ligand. A single electron pair will produce a greater splitting than two or more free pairs because it can be "focused" on an orbital. Thus, an $NH_3$ molecule produces a higher splitting than does a water molecule. Unsaturated molecules or groups with pi orbitals, *e.g.*, CO, $CN^-$, $CH_2\!\!=\!\!CH_2$, bring about the largest splitting of all, as a result of forming pi bonds with the metal ion. The oxidation state of the metal itself is important in determining the extent of splitting: $Co^{3+}$ will produce higher splitting than $Co^{2+}$.

When all of these factors have exerted their influence and the extent of CF splitting has been fixed, the distribution of the electrons will be the resultant of two factors: (1) the tendency of the electrons to distribute themselves singly in orbitals because of mutual repulsion (Hund's Rule) and (2) their tendency to occupy the lower energy levels and thereby to achieve stabilization. It may be impossible for the electrons to do both of these. Electron pairing requires the expenditure of energy; occupying the lower energy orbitals results in energy gain. If the CF splitting energy is large, the electrons will pair in order to achieve the lower energy orbitals; if the CF splitting energy is small, they will distribute themselves according to Hund's Rule. Complexes possessing the same number of unpaired electrons as in the free ion are called "high-spin" type; those with fewer unpaired electrons than in the free ion are called "low-spin" type. The two types are distinguished experimentally by measuring the extent of paramagnetism of the complexes.

The abilities of some common ligands to produce CF splitting are $I^- < Br^- < Cl^- < F^- < OH^- < H_2O < -NCS^- < NH_3 <$ ethylenediamine $<$ dipyridyl $< NO_2^- < CN^-$. Reversals do occur, particularly when the bonding is not predominantly electrostatic.

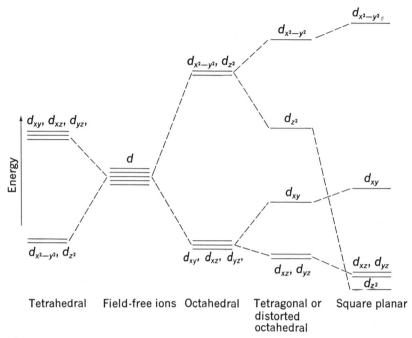

**Figure 4.20**   Crystal field splittings of the $d$ orbitals of a central ion into complexes having different geometries.

Just as various ligands differ in their ability to produce splitting and the subsequent energy stabilization, so do the metal ions differ in the stability of the complexes formed with a particular ligand. For many ligands, the stability order of the divalent ions of the 4th Period is Ca < V < Cr > Mn < Fe < Co < Ni < Cu > Zn. This is the order obtained by ranking the heats of hydration of the ions. Of these ions, Ca, Mn, and Zn lack stabilization energy because they possess 0, 5, and 10 $3d$ electrons, respectively. If the stabilization energy due to CF splitting is calculated and subtracted from the heats of hydration of the remaining ions, a regular sequence in order of atomic number from Ca to Zn is obtained.

These anion and cation sequences bring us back to the Hofmeister series, mentioned in connection with the salting-out of proteins. The cation series are $Li^+ < Na^+ < K^+ < Rb^+ < Cs^+$ and $Mg^{2+} < Ca^{2+} < Sr^{2+} < Ba^{2+}$. Although this sequence cannot be compared with that of the 4th Period metal ions just discussed, it is worth commenting that both the Hofmeister series and the 4th Period ion sequence represent the order of hydration of the ions, a clear-cut coordination phenomenon. Correlations of the Hofmeister series for the alkali metals with other properties of salt solutions also exist. For example, the series $Li^+$ $Na^+$ $K^+$ $Rb^+$ $Cs^+$ applies to:

(1) the activity of water in 3 molal solutions of the chlorides, bromides, or iodides at 25°,

(2) the partial unitary molal entropy of the ions in water at 25°,

(3) the order of $-B$, where B is a viscosity coefficient,

(4) the Walden product, *i.e.,* the product of the limiting equivalent ionic conductance and the solution viscosity,

(5) water proton chemical shift for aqueous ionic solutions at 25°, as determined by NMR relaxation measurements, and

(6) the ionic self-diffusion concentration parameter for water (quantity not determined for $Rb^+$).[*]

Since salts affect so many properties of water, it has been suggested that their ions alter the structure of the water close to them, with the extent of the alteration dependent upon which ions are involved. The relative effects of a series of salts on protein solubility resemble their relative effects on many—but not all—of the physical properties of salt solutions. Consequently the ability of salts to alter protein solubility may result from their alteration of the properties of the solvent. The emphasis here is not upon any direct interaction of the salt with the protein, but rather of salt with water. This proposal has recently been reviewed at considerable length.[†]

An alternative explanation for the results of salts upon proteins places the emphasis on a direct interaction between the salt and groups on the protein. The Hofmeister series has been shown to apply to the solubility of small compounds, such as N-acetyltetraglycine ethyl ester, a model for the polypeptide backbone.[‡] In this example the data were interpreted in terms of a direct interaction between the salt and the peptide bond. Crystals can be obtained by mixing aqueous solutions of lithium bromide with N-methylacetamide, another model for the peptide bond.[§] Determination of the structure of the crystals by x-ray diffraction shows that the ions are apt to interact with the peptide model as an ion-dipole interaction. The effects of salts upon the stability of proteins can be accounted for by polymer solution theory and the assumption that a direct interaction occurs between the protein and salt.[**]

The preceding paragraphs show that the effects of salts on proteins are not yet completely understood. The problem is currently under active investigation in several laboratories. At present the best view

[*]Von Hippel, P. H. and T. Schleich in *Structure and Stability of Biological Macromolecules,* Timasheff, S. N. and Fasman, G. D., Eds., Marcel Dekker, New York, 1969, p. 417.

[†]Von Hippel, P. H. and T. Schleich in *Structure and Stability of Biological Macromolecules,* Timasheff, S. N., and Fasman, G. D., Eds., Marcel Dekker, New York, 1969, p. 417.

[‡]Robinson, D. R. and W. P. Jencks, *J. Amer. Chem. Soc.,* **87**, 2470 (1965).

[§]Bello, J., D. Haas and H. R. Bello, *Biochemistry,* **5**, 2539 (1966).

[**]Mandelkern, L. and W. E. Stewart, *Ibem.,* **3**, 1135 (1964).

seems to be that the effectiveness of ligands in salting-out proteins relates to the extent to which they may be hydrated and the extent to which they may directly interact with the macromolecule. The properties of the ligands as Lewis bases are important in both cases. The general ideas of ligand field theory can be extended to the heavy metals, although the calculations are restricted to those elements with unfilled $d$ orbitals.

Of particular biochemical interest are the stabilities of complexes formed between metal ions and ligand groups on amino acids. The formula for the copper-glycine complex was given earlier (page 224). This complex is most likely planar. With other divalent ions of the transition metal group the 2:1 bidentate amino acid complexes may be tetragonal with remaining positions occupied by water molecules. The structures of a large number of complexes of metals with amino acids and small peptides are given in a review by Freeman.[*] A number of enzymes are active only when they are complexed with a metal ion. The best understood metalloenzyme is carboxypeptidase, the structure of which has been determined by x-ray study of its crystals.[†] The physiological form of the enzyme contains one atom of $Zn^{2+}$, although an active enzyme can also be obtained if $Co^{2+}$ is substituted for $Zn^{2+}$. The $Zn^{2+}$ is tetrahedrally coordinated, with three of the ligands being provided by the protein (two histidine side chains and one glutamic acid side chain) and the fourth ligand being a water molecule. When exposed to a substrate, glycyl-L-tyrosine, the carbonyl oxygen atom of the glycine residue replaces the water as the fourth ligand. The hemoproteins constitute another group of proteins which contain metals. The structures of myoglobin,[‡] hemoglobin,[‡] and cytochrome $c$[§] have been determined by x-ray crystallography. The iron atoms in these proteins have four coordination positions provided by the heme. The fifth coordination for the ionic atom is provided by the imidazole side chain of a histidine residue in the protein. The sixth coordination is provided by the sulfur atom of a methionine side chain in cytochrome $c$, while in myoglobin and hemoglobin, the sixth position may be vacant or occupied by a small molecule, such as water, oxygen, or carbon monoxide, depending upon the conditions under which the crystal was prepared.

Eigen and Hammes have measured the rate constants for water substitution in metal complex formation in the attempt to arrive at generalizations which would be useful in the area of metal activation of enzymes.[**]

---

[*] Freeman, H. C., *Advan. Protein Chem.*, **22**, 258 (1967).

[†] Quiocho, F. A. and W. N. Lipscomb, *ibid*, **25**, 1 (1971), and J. A. Hartsuck and W. N. Lipscomb in *The Proteins*, Boyer, P. D., ed., 3rd Ed., Academic Press, New York, 1971, Vol. 3, p. 1.

[‡] Dickerson, R. E. and I. Geis, *The Structure and Action of Proteins*, Harper and Row, New York, 1969, Chapt. 3.

[§] Dickerson, R. E., *Scientific American*, April, 1972, p. 58. Available as Scientific American Offprint No. 1245 from W. H. Freeman and Company, "The Structure and History of an Ancient Protein."

[**] Eigen, M. and G. G. Hammes, *Adv. Enzymol.*, **25**, 1 (1963).

The order of increasing substitution rate was reported by Eigen and Hammes to be $Cu = Ca < Zn < Mn < Fe < Co < Mg < Ni$. This sequence is different in several respects from the order of complex stability, but it must be emphasized that we are now comparing a rate pheonomenon with an equilibrium process. Other factors bearing on metal specificity are discussed by these authors. In addition the general topic of metalloproteins has been the subject of a recent book.[*]

Finally, to state the obvious, the area of coordination compounds of biological importance is an extremely large and significant one. Among the various biochemical processes involving the binding of small ions are (1) the activation of enzymes by anions and cations, (2) the polymerization of protein subunits, (3) the selective action of membranes, (4) the binding of RNA to ribosomes, (5) ion exchange in soils, bones, and other semi-solid systems, (6) phosphoryl group transfer and phosphate binding by divalent Mg, Mn, and Ca, and (7) the many reactions governed by the tetrapyrrole pigments. Further excursions into this territory lie beyond the general objectives of this presentation. The books by Bailar and by Dwyer and Mellor listed in the Suggested Additional Reading at the end of the chapter are excellent guidebooks for deeper explorations.

### 4.7   Cellulose Ion Exchangers For Protein Purification

Various ion exchange resins have been widely utilized in the purification of biological materials. The ion exchange resins consist of a macromolecular support, such as cellulose, or other polysaccharides, to which ionizable groups are attached by covalent bonds. Among the most widely used cellulose ion exchangers are diethylaminoethyl cellulose (DEAE-cellulose) and carboxymethyl cellulose (CM-cellulose). Partial structures are given below as these resins would exist at neutral pH.

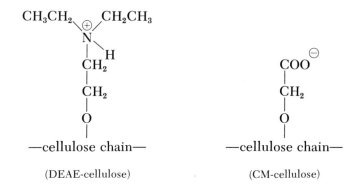

(DEAE-cellulose)                    (CM-cellulose)

---

[*]Vallee, B. L. and W. E. C. Wacker, *The Proteins,* H. Neurath, Ed., 2nd Ed., Vol. 5, Academic Press, New York, 1970.

Ion exchange resins can be made from many other combinations of polymer backbone and ionizable groups, *e.g.*, by the introduction of sulfonic acid groups into a copolymer of styrene and divinylbenzene.

The most obvious mode of interaction of proteins with ion exchange celluloses is electrostatic. A protein that bears a net negative charge at pH 7 would be bound by DEAE-cellulose, but not by CM-cellulose, if it were passed through an ion-exchange cellulose column using a buffer of pH 7 as the eluting agent. If a mixture of various proteins with different net charges were passed through an ion-exchange cellulose column, the protein with the net charge of the same sign as that of the ion-exchange cellulose would be eluted first. The proteins with a net charge of opposite sign to that of the ion-exchange cellulose column would be eluted later, with those proteins with the largest net charge being eluted last. In practice, the separation scheme can be made quite sophisticated by varying the pH of the buffer used for column elution, and thereby varying the net charge on the proteins, or by varying the ionic strength of the eluent, with increasing ionic strength reducing the electrostatic interaction between proteins and the ion exchange cellulose. Other types of interactions, those including hydrogen-bonding as well as other types of non-covalent bonds, are possible between proteins and ion exchange resins and can occasionally lead to results that could not be explained on the basis of electrostatic interactions alone.

There are two particular advantages to the use of ion-exchange celluloses in the purification of proteins. First, the polymer chains in the ion exchange celluloses are, on the average, relatively far apart, allowing the large protein molecules to diffuse into the matrix and interact with the charged groups. Second, the density of charged groups in the ion exchange celluloses is relatively low, so that a single protein molecule can interact with only one or a few charged groups on the resin at any time—an advantage that allows removal of the protein from the ion-exchanger under comparatively mild conditions.

### Suggested Additional Reading

#### General: Ionic Equilibria

Albert, A., and E. P. Sergeant, *Ionization Constants of Acids and Bases.* John Wiley and Sons, New York, 1962.

Clark, W. M., *Topics in Physical Chemistry*, 2nd Ed. Williams and Wilkins, Baltimore, 1952.

Freiser, H., and Q. Fernando, *Ionic Equilibria in Analytical Chemistry.* John Wiley and Sons, New York, 1963.

VanderWerf, C. A., *Acids, Bases, and the Chemistry of the Covalent Bond.* Reinhold, New York, 1961.

**Biochemical Applications**

Dixon, M., and E. C. Webb, *Enzymes,* 2nd Ed., Academic Press, New York.

Gurd, F. R. N. in *Physical Principles and Techniques of Protein Chemistry,* Part B, Leach. S. J., Ed., Academic Press, New York, 1970, p. 365.

Steinhardt, J. and J. A. Reynolds, *Multiple Equilibria in Proteins.* Academic Press, New York, 1969.

Tanford, C., *Adv. Protein Chem.,* **17,** 69 (1962).

Von Hippel, P. H. and T. Schleich in *Structure and Stability of Biological Macromolecules,* S. N. Timasheff and G. D. Fasman, Ed., Marcel Dekker, New York, 1969, p. 417.

**Coordination Chemistry**

Bailar, J. C., *Chemistry of the Coordination Compounds.* Reinhold, New York, 1956, (pp. 698–742).

Basolo, F., and R. Johnson, *Coordination Chemistry,* W. A. Benjamin, New York, 1964.

Day, M. C., and J. Selbin, *Theoretical Inorganic Chemistry.* Reinhold, New York, 1962.

Dwyer, F. P., and D. P. Mellor, editors, *Chelating Agents and Metal Chelates.* Academic Press, New York, 1964.

Freeman, H. C., *Advan. Protein Chem.,* **22,** 258 (1967).

Murmann, R. K., *Inorganic Complex Compounds.* Reinhold, New York, 1964.

**Problems**

**4.1** (a) A 0.1 $M$ solution of potassium sulfate ($K_2SO_4$) freezes at $-0.558°C$. What is the degree of dissociation?

(b) A 0.1 $M$ solution of dichloroacetic acid freezes at $-0.278°C$. What is the degree of dissociation?

*Answer.*
(a) 1.00.
(b) 0.495.

**4.2** At 25°C a 0.1 $m$ solution of acetic acid is 1.35 percent dissociated. Calculate the freezing point and osmotic pressure of the solution assuming that the percent dissociation is the same at the freezing point as at 25°C. What would the freezing point and osmotic pressure be if there were no dissociation?

*Answer.* $-0.189°C$, 2.48 atm; $-0.186°C$, 2.45 atm.

**4.3** Calculate the ionic strength of 0.1 $M$ solutions of
(a) sodium acetate,
(b) $Na_2SO_4$,
(c) $MgSO_4$,
(d) $K_2HPO_4$.

**4.4** Calculate the activity of $Mg^{2+}$ in a solution which is 0.001 $M$ in $Na_2SO_4$ and 0.005 $M$ in $MgCl_2$. The solvent is water and the temperature is 25°C.

**4.5** Calculate the activity of sodium ion in a solution which is 0.01 $M$ with respect to NaCl and 0.02 $M$ with respect to $Na_2HPO_4$. The solvent is water and the temperature is 25°C.

**4.6** Calculate the pH of each of the following solutions:
(a) 0.1 $M$ sodium formate,
(b) 0.2 $M$ monopotassium phthalate,
(c) 0.01 $M$ aniline hydrochloride,
(d) 0.05 $M$ monosodium malonate,
(e) 0.02 $M$ methylamine,
(f) 0.03 $M$ propionic acid.

*Answer.* (a) 8.37, (b) 4.14, (c) 3.29, (d) 4.27, (e) 11.5, (f) 3.20.

**4.7** Exactly 500 ml of 0.1 $N$ ammonium hydroxide is combined with 100 ml of 0.5 $N$ HCl. What is the final pH of the solution?

*Answer.* 5.2.

**4.8** A buffer is found to have a pH of 7.20 at 25°C. The buffer is then moved into a cold room (about 5°C) for use in an experiment. Will the pH still be 7.20? Explain the thermodynamic basis for your answer.

**4.9** What is the pH of a 0.05 $M$ solution of sodium borate at 25°C? What would be the pH of the same solution at 5°C? The $pK_a$ of boric acid is 9.44 at 5°, and $pK_W = 14.73$ at 5°C.

*Answer.*
25°C: 10.96.
 5°C: 11.44.

**4.10** Table 4.6 shows that ATP, ADP, and $P_i$ each exist predominantly in two forms at physiological pH. Write a balanced equation, taking account of all significant ionic forms, for the hydrolysis of one mole of ATP to yield one mole each of ADP and $P_i$ at pH 7.20.

*Answer.*
.25 $ATP^{3-}$ + .75 $ATP^{2-}$ + $H_2O$ =
       .5 $ADP^{2-}$ + .5 $ADP^-$ + .5 $H_2PO_4^-$ + .5 $HPO_4^{2-}$ + .75 $H^+$.

**4.11** The enzyme hexokinase is added to a solution of ATP and D-glucose which was originally at pH 7.40. In which direction will the pH change? Explain how this phenomenon could be utilized to measure the rate of the conversion of ATP and D-glucose into ADP and D-glucose-6-phosphate.

*Answer.* pH drops during reaction.

**4.12** Calculate the concentration of carbonate ion in a solution of 0.01 $M$ sodium bicarbonate at pH 8.0. $K_a$ for $HCO_3^-$ is $5.6 \times 10^{-11}$.

*Answer.* $5.56 \times 10^{-5}$ $M$.

**4.13** Five hundred ml of 0.2 $M$ propionic acid is mixed with 500 ml of 0.2 $M$ KOH. What is the pH of the solution?

*Answer.* 8.93.

**4.14** Given a liter of 0.1 $M$ acetic acid solution at 25°C, how many grams of sodium acetate must be added to raise the pH to 4.5? Assume no volume change.

*Answer.* 5.67 g of anhydrous sodium acetate based on $pK' = 4.66$.

**4.15** What are the pH and pOH of a 0.1 $N$ $NH_4Cl$ solution at 25°C?

**4.16** How many g of $NH_3$ are needed to make one L of solution with a pOH of 4.0?

**4.17** Sketch the titration curve for succinic acid. What is the pH when 1/4, 1/2 and 3/4 of the total succinic acid has been titrated? Are the ionizing groups in succinic acid sufficiently far apart so that they do not influence each other?

*Answer.* 4.2, 4.92, 5.63, no.

**4.18** Table 3.5 gives $\Delta G°$ for glucose-1-phosphate$^{2-}$ $\rightarrow$ glucose-6-phosphate$^{2-}$. Calculate $\Delta G°$ for the process glucose-1-phosphate$^{1-}$ $\rightarrow$ glucose-6-phosphate$^{1-}$ at 25°C.

*Answer.* 1.17 Kcal/mole.

**4.19** Calculate the $PO_4^{3-}$ concentration in a 0.1 $M$ $H_3PO_4$ solution at 25°C.

*Answer.* $1.05 \times 10^{-18}$ $M$.

**4.20** What must be the initial concentrations of $KH_2PO_4$ and KOH in the same flask to give a buffer of pH 7.5 and ionic strength 0.2? The $pK'$ for the primary-secondary phosphate anion equilibrium is 6.86.

**4.21** Calculate the ionic strength of a buffer resulting from mixing 100 ml of 0.1 $M$ $KH_2PO_4$ with 50 ml of 0.1 $M$ NaOH and diluting to 1 L. If the $pK_a'$ for the pertinent dissociation is 6.86, what will be the pH of this solution?

*Answer.* $I = 0.02$; pH $= 6.86$.

**4.22** The dissociation of an indicator can be expressed

$$\text{pH} = pK_I + \log\frac{\text{yellow conjugate base}}{\text{red conjugate acid}}$$

for its color change. The $pK_I$ is 4.5. Certain solutions must be adjusted to the following pH values: 6.0, 5.5, 5.0, 4.5, 4.0, 3.5, and 3.0. For which of these would you use this indicator? What would be the concentration ratio of yellow to red colored species in the corresponding standard solutions?

*Answer.* For all but 6.0 and 3.0; for these the eye is not sufficiently discriminating. The concentration ratios for the series 6.0 to 3.0 will be 31.6, 10.0, 3.16, 1.00, 0.316, 0.100, 0.0316.

**4.23** Describe how you would construct a thermometer in which the temperature was given by the color of a solution.

**4.24** An indicator has a $pK_I = 5.2$ and changes color from red to blue as the pH rises. No pH standards or pH meter was available in the laboratory, but it was found that the color of the indicator in the solution of unknown pH could be matched by looking through two tubes (held in tandem). The first tube contained 10 ml of 0.1 N HCl and 12 drops of the indicator, and the second, 10 ml of 0.1 N NaOH and 4 drops of indicator. Sixteen drops of indicator were added to 10 ml of the unknown solution to give approximately the same intensity of hue for comparison. What was the pH of the unknown solution?

**4.25** Pyridoxamine is 2-methyl-3-hydroxy-4-aminomethyl-5-hydroxy-methylpyridine. V. R. Williams and J. B. Neilands determined the apparent $pK_a$ values for the three dissociable groups: 3.54 for the phenolic hydroxyl, 8.21 for the pyridinium group, and 10.63 for the primary amino group (*Arch. Biochem. Biophys.*, 53:56, 1954).
(a) Write the structural formulas for the principal species present at pH 2.0, 6.0, 9.5, and 11.5.
(b) At what pH is the compound isoelectric?
(c) Draw the titration curve, labeling the axes clearly.

**4.26** Tyrosine has 3 dissociation constants corresponding to the following $pK_a'$ values:

$$-COOH \xrightarrow{OH^-} -COO^- \qquad 2.4;$$

$$-NH_3^+ \xrightarrow{OH^-} -NH_2 \qquad 9.6;$$

$$-C_6H_4-OH \xrightarrow{OH^-} -C_6H_4-O^- \qquad 10.1.$$

(a) What will be the pH of the solution of the isoelectric species?
(b) In what region of the pH scale is tyrosine a cation?
(c) Sketch the general appearance of the titration curve.
(d) Would tyrosine be a more effective buffer at pH 6.0, 9.4, or 9.8?

*Answer.* (d) 9.8.

**4.27** The compound 2-pyridone has the following formula:

The fully protonated structure is

The compound has two $pK_a'$ values: 0.75 and 11.6. These have been assigned on the basis of spectrophotometric studies:

$pK_1'$ is for the phenolic proton,
$pK_2'$ is for the pyridinium proton.

(a) Write the formulas corresponding to the structure of the compound at the following pH values: 0.0, 7.0, 14.0.
(b) What is the isoelectric point of the compound?
(c) Write formulas for the resonance isomers that will be in equilibrium at the isoelectric point.
(d) Sketch the titration curve of the compound, labeling all axes.
(e) If 190 mg of 2-pyridone is dissolved in 25 ml of water, how many ml of 1.0 N NaOH must be added to bring the solution to a pH equal to the $pK_2'$? The molecular weight of 2-pyridone is 95.

*Answer.* (b) 6.18, (e) one ml.

**4.28** Benesch and Benesch (*J. Am. Chem. Soc.*, **77**:5877, 1955) determined the $pK_a'$ for the thiol group of thioglycollic acid by a spectrophotometric titration. At 24.5°C, the $pK_a'$ was 10.32; at 38°C, it was 10.10. From these data calculate $\Delta H$ for the ionization. (See Eq. 4.51)

*Answer.* 6.9 kcal.

**4.29** A protein sample was tested for purity by the method of solubility curves. The resulting plot consisted of four sloping segments having the following connecting points:

| S(mg of protein dissolved) | W(mg of protein added) |
|:---:|:---:|
| 2.00 | 2.00 |
| 4.25 | 5.00 |
| 5.13 | 7.50 |
| 5.75 | 10.00 |

Consider that the first solid that separates is protein A, the second is protein B, etc. Calculate the composition of the mixture and the solubility of each individual component.

**4.30** Prepare a table of $Y$ values (Eq. 4.59) corresponding to pH 6.0, 6.5, 7.0, 7.5, 7.9, and 8.0. Check your answers by inspection of Table V, p. 214, Edsall, J. T. and J. Wyman, *Biophysical Chemistry*, Academic Press, N.Y., 1958.

**4.31** From the $Y$ values calculated in the preceding problem, calculate $K'_{equi}$ for the hydrolysis of ATP at pH 6 and pH 8 at 37°C.

# ELECTROCHEMICAL CELLS

Prometheus, they say, brought fire to the service of mankind: electricity we owe to Faraday.

*Sir William Bragg*

For a living cell to survive it must obtain energy from its environment and transform that energy into a biochemical fuel acceptable to the cellular enzymes whose business it is to synthesize proteins, carbohydrates, lipids, and accessory substances. All of the major synthetic processes depend principally on the chemical energy of adenosine triphosphate (ATP) to drive endergonic reactions. (The formula for ATP may be found in Chapter 3, Section 3.8.)

Green plants convert the radiant energy of sunlight into chemical energy by reactions whose initial events are still somewhat obscure. When chloroplasts absorb sunlight three products are formed: reduced ferredoxin, ATP, and $O_2$. The reduced ferredoxin transfers electrons to one of the pyridine nucleotides, nicotinamide-adenine-dinucleotide phosphate (NADP), which then joins forces with ATP to bring about the reduction of $CO_2$ and the subsequent synthesis of higher carbohydrates. The light-induced formation of reduced pyridine nucleotides and ATP is called *photophosphorylation*.

Animal cells and nonphotosynthetic plants couple ATP formation to the oxidation of cell nutrients in two ways: (1) A cellular nutrient or metabolite is phosphorylated with inorganic phosphate and the resulting phosphate compound is transformed by oxidation or dehydration into a high-energy phosphate; the high-energy group is then directly transferred without energy loss to adenosine diphosphate (ADP) to form ATP. (2) Cellular nutrients or metabolites are oxidized to carbon dioxide and water through a series of graduated-energy steps, three of which lead to ATP formation in a complex way yet to be fully understood. Method (1) is called *substrate-level phosphorylation*; method

(2) is called *oxidative phosphorylation.* The first is considered a more primitive pathway (in the evolutionary sense) because it is not oxygen-dependent, and it is limited to a few known compounds. Muscle and brain cells use substrate-level phosphorylation to generate ATP on demand. Oxidative phosphorylation is the "high-yield" energy machinery of living cells and can be utilized with any metabolite from which hydrogen atoms can be removed enzymatically. In many organisms the two pathways are directly linked, in that the end products of substrate-level phosphorylation may be channeled into the oxidative phosphorylation system.

These fundamental processes are related to electrochemical cells in the following way. In living systems undergoing oxidative phosphorylation, we find a series of compounds which can be reversibly reduced and oxidized, but which undergo reduction at different *electrical potentials* — completely analogous to the relative ease of reduction of simple metal ions and nonmetals. The *reduction potential* of a metal ion, nonmetal, or compound is a measure of the tendency of the substance to accept an electron from an electron donor. The higher the potential, as expressed in volts, the greater the affinity of the acceptor for the electron. Biochemical electron acceptors can be arranged in a series ranging from the least easily reduced, as measured by its reduction potential, to the most readily reduced (and, thereby, the least readily oxidized). Physical separation of these substances within the mitochondria prevents interaction between the extremes of the series and results in electrons from hydrogen atoms passing in orderly fashion from acceptor to acceptor with the energy of oxidation tapped off at three points to yield three molecules of ATP per hydrogen pair, as a maximum. Loss of chemical energy as heat is minimized by the relatively small differences between the reduction potentials of the successive members of the series. The details of the process, as far as it is presently understood, will be presented in the last section of the chapter. Fundamental to an understanding of biological oxidation is a firm grasp of the principles of oxidation-reduction in simple chemical systems. We will begin, therefore, by studying the principles of the galvanic cell as it relates electrical energy to chemical work.

## 5.1 Electrical and Chemical Work

In Chapter 3 the useful work that a chemical system can do isothermally was calculated from the concentrations of reactants and products and the equilibrium constant of the reaction: $\Delta G = -2.303\ RT \log K + 2.303\ RT \log Q$. The value of $\Delta G$ thus obtained is for the chemical reaction to which $K$ and $Q$ apply; it may be for one mole of reactant (or product, as the case may be) or for more than one mole depending upon the manner in which the balanced equation is written. Remember that at equilibrium $\Delta G = 0$ and that, since $K$ is a constant, the factor that really controls the amount of energy which is isothermally available from a chemical system is the extent by which $Q$ differs from $K$. For chemical systems in which the only kind of work that can be done isothermally is electrical work, a useful relationship exists between useful work ($\Delta G$) and electrical energy. This recapitulation of principles stated earlier is to emphasize that the relative concentrations or pressures of reactants and products determine the chemical work, available in a system. The equation is

$$\Delta G = -n\mathscr{E}\mathscr{F}, \tag{5.1}$$

where $n$ is the number of electrons transferred in the oxidation-reduction reaction, $\mathscr{E}$ is the electrical potential of the system in volts, and $\mathscr{F}$ is a proportionality constant called the Faraday, equal to 23,060 calories per volt. This equation holds under conditions of constant temperature and pressure, reversibility, and restriction to electrical work. The voltage can be determined directly if the chemical reaction is separated into a pair of electrode systems called half-cells, and if the electrical potential is measured under conditions wherein, ideally, no current is drawn from the cell.

The expressions for the free energy change in terms of chemical work and electrical work can be related.

$$\Delta G = \Delta G^\circ + 2.303 \ RT \log Q \qquad \text{(chemical work)},$$

and

$$\Delta G = -n\mathscr{E}\mathscr{F} \qquad \text{(electrical work)},$$

and therefore,

$$-n\mathscr{E}\mathscr{F} = \Delta G^\circ + 2.303 \ RT \log Q.$$

Since all terms are constant except $\mathscr{E}$ and $\log Q$, measuring these will enable us to calculate the standard free energy change, which is the chemical potential per mole. Substituting for $\Delta G^\circ$ from the relationship

$$\Delta G^\circ = -2.303 \ RT \log K \qquad \text{(Eq. 3.43)}$$

and solving for $\mathscr{E}$, we have

$$\mathscr{E} = \frac{2.303 \ RT}{n\mathscr{F}} \log K - \frac{2.303 \ RT}{n\mathscr{F}} \log Q. \tag{5.2}$$

When products and reactants are in their standard states (unit activity), $Q = 1$ and

$$\mathscr{E} = \frac{2.303 \ RT}{n\mathscr{F}} \log K = \mathscr{E}^\circ. \tag{5.3}$$

The $\mathscr{E}°$ is called the standard potential. Both $\mathscr{E}°$ and $\Delta G°$ are customarily measured at 25°C.

Substituting from Eq. 3.43 into Eq. 5.3 yields

$$\Delta G° = -n\mathscr{E}°\mathscr{F}. \tag{5.4}$$

Substituting from Eq. 5.3 into Eq. 5.2 gives still another form of the equation,

$$\mathscr{E} = \mathscr{E}° - \frac{2.303\ RT}{n\mathscr{F}}\log Q. \tag{5.5}$$

Equation 5.5 is called the Nernst Equation in recognition of the scientist who first derived it. It is the fundamental equation of electrochemistry and basic to the calculations presented in this chapter. Consideration will be limited to those cells which may be used as sources of electrical energy and hence are given the distinguishing name, *galvanic cells*. Cells in which physical or chemical changes are produced by applying an external source of electrical energy are called *electrolytic cells*.

### 5.2   The Galvanic Cell and Cell Conventions

The production of electricity from chemical reactions has been known since 1800, the year in which Volta described his electric pile. If a metal such as zinc is placed in a solution of its salt, zinc sulfate, we have what is known as a *couple* or half-cell: the oxidized ($Zn^{2+}$) and the reduced (Zn) form of a substance in contact with each other. A similar couple can be made of copper and copper sulfate solution. If the two solutions are in contact across a porous plate, salt bridge, or some arrangement which permits current to be carried without excessive mixing of the solutions, we have a simple galvanic cell. Connecting the two metal electrodes with a wire will close the circuit and permit the chemical reaction to proceed in its spontaneous direction. Of zinc and copper, zinc is the more readily oxidized and therefore it will go into solution as $Zn^{2+}$, electrons will move through the wire to the copper electrode where they will be picked up by $Cu^{2+}$ in contact with it. Copper will plate out as a result of the reduction. The oxidation-reduction will continue until equilibrium concentrations of ions are present in the half-cells. The process is illustrated in Fig. 5.1. With regard to the external circuit the zinc electrode is negative since electrons are released on its surface; the copper electrode is correspondingly positive. Several cells arranged in series are called a *battery*; the voltage of the battery is the sum of the voltages of the individual cells. Faraday's Law applies to the operation of all such cells: for every gram equivalent of substance oxidized, $6.02 \times 10^{23}$ electrons, the charge of which is 96,500 coulombs of

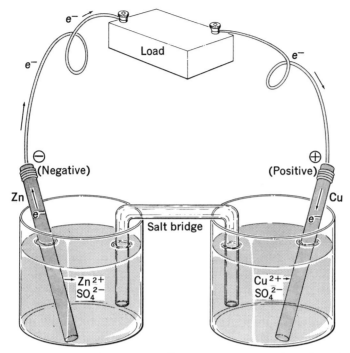

**Figure 5.1** Simple galvanic cell composed of a zinc electrode and a copper electrode, each immersed in a solution of its own ions and connected by a salt bridge.

electricity, travel along the wire to the electron acceptor and bring about the formation of one gram equivalent of the reduction product of the reaction. If the two half-cells have sufficiently different reduction potentials, the equilibrium constant will be so large that the reaction will go to completion for all practical purposes.

The rules for setting up electrical cells and making the various calculations are called *electrochemical conventions*:

1. The chemical equation describing the cell reaction is split into two electrode reactions. The oxidation reaction is always assigned to the left electrode; the reduction reaction is assigned to the right electrode. These assignments are based entirely on the way the equation is written and are without regard to the direction of spontaneity of the reaction. Suppose that the equation is written:

   $$Zn^{2+} + Cu \rightarrow Zn + Cu^{2+}$$

   (The anions do not enter into the oxidation-reduction and are, therefore, omitted in the interest of simplicity.)

The electrode reactions are

$$\text{Oxidation:} \quad Cu \rightarrow Cu^{2+} + 2e^-$$
$$\text{Reduction:} \quad 2e^- + Zn^{2+} \rightarrow Zn$$

Note that the sum of the electrode reactions must give the cell reaction. The electrode assignments are

$$Cu, Cu^{2+} \parallel Zn^{2+}, Zn.$$

In this notation, which depicts the cell, the metal electrodes are customarily written at the outside. The $\parallel$ represents the salt bridge connecting the two half-cells. Salt bridges will be discussed in the next section (5.3).

2. All standard half-cells or couples are ranked in order of their reduction potentials as determined with reference to the standard hydrogen half-cell, which is arbitrarily assigned a potential of 0.0000 volt. In the standard half-cells all solutions of ions are at unit activity and all gases at one atmosphere pressure. Pure solids and liquids, whether metals or nonmetals, are in their standard states and therefore at unit activity. Standard potentials are usually designated as being for 25°C.

3. In Table 5.1 are found the standard reduction potentials for a number of common half-cells. The most negative reduction potential is assigned to the Li, Li$^+$ couple, meaning that it has the least tendency to exist in the reduced state of any of the members of the series. Midway in the table is located the hydrogen couple and following it, a number of combinations in which the tendency to lose electrons is less than that for the standard hydrogen half-cell. By comparison with the hydrogen electrode, these half-cells have positive reduction potentials. Inspection of the table should call to mind the electromotive force series studied in general chemistry. This is precisely the same arrangement—with the data for the reduction potentials added. In Table 5.2 are given the standard reduction potentials for some biochemical reactions at pH 7.0. These half-cell potentials have been calculated from the known $\Delta G'$ values in accordance with Eq. 5.4, rather than measured experimentally, because they do not occur in the absence of a catalyst (usually an enzyme).

4. The symbol $\mathscr{E}°$ is used for the standard potential of each of the half-cells as well as for that of the cell: for the left electrode $\mathscr{E}°$ is the *oxidation potential*; for the right, it is the *reduction potential*. If reduction potentials are given, as in Table 5.1, oxidation potentials are obtained simply by reversing the algebraic sign. The standard potential of the cell is obtained by adding together the half-cell $\mathscr{E}°$ values. For example,

| Half-cell: | $Cu \rightarrow Cu^{2+}_{(a=1)} + 2e^-$ | $\mathscr{E}^\circ_{ox} = -0.337$ volt |
|---|---|---|
| Half-cell: | $Zn^{2+}_{(a=1)} + 2e^- \rightarrow Zn$ | $\mathscr{E}^\circ_{red} = -0.763$ volt |
| Cell: | $Zn^{2+}_{(a=1)} + Cu \rightarrow Cu^{2+}_{(a=1)} + Zn$ | $\mathscr{E}^\circ_{cell} = -1.100$ volt. |

Inspection of Table 5.1 will show that the oxidation potential for the copper half-cell was obtained by changing the sign of the reduction potential from plus to minus. Another widely used convention is to calculate half-cell potentials for biological reductions at pH 7.0. This is indicated by using $\mathscr{E}'$ or $\mathscr{E}'_0$ as the symbol instead of $\mathscr{E}^\circ$ (see Table 5.2). When using data from various literature sources, the student should be alert to the choice of conventions.

5. Many physical chemists prefer to use oxidation potentials in tabulating half-cell potentials. Biochemists commonly use reduction potentials because attention is traditionally focused on the path of hydrogen or electrons from donor to acceptor.

6. If the $\mathscr{E}$ of a cell is positive, the cell reaction is spontaneous as written. When $\mathscr{E}$ is positive, $\Delta G$ must be negative, since $\Delta G = -n\mathscr{E}\mathscr{F}$. Thus in the cell potential we have a second useful criterion of spontaneity. If the $\mathscr{E}$ of the cell is negative, the spontaneous reaction is the reverse of the chemical reaction written. If we desire to bring about a reaction whose $\mathscr{E}$ is negative, the cell must be driven by an external source of electrical potential slightly higher than the calculated cell voltage.

## 5.3 Types of Half-cells

The various electrode couples shown in Table 5.1 can be grouped into seven general classes. If the couple does not contain a metal that can serve as a conductor, an inert metal electrode, usually platinum, is immersed in the solution to provide a conducting path for electrons. Examples of the various types of electrodes are shown in Table 5.3.

Although any two half-cells can be combined theoretically to give a cell, experimental difficulties arise with certain actual combinations. If the solutions surrounding the electrodes come into contact with each other, precipitation or chemical action may occur, and a small potential, called the junction potential arises.* These difficulties are essentially eliminated by connecting the two half-cells through a salt bridge of potassium chloride. The salt bridge is made by preparing an aqueous solution of potassium chloride containing 2 percent agar, pouring it into a glass U-tube, and allowing it to set to a gel.

---

*For further consideration of junction potentials the student is referred to pages 437–441 in *Biophysical Chemistry* by Edsall and Wyman, listed in *Suggested Additional Reading*.

**TABLE 5.1** STANDARD ELECTRODE REDUCTION POTENTIALS AT 25°C

| Electrode | Electrode reaction | $\mathscr{E}°$ (volts) |
|---|---|---|
| Li, Li$^+$ | Li$^+$ + $e^-$ → Li | −3.045 |
| K, K$^+$ | K$^+$ + $e^-$ → K | −2.925 |
| Cs, Cs$^+$ | Cs$^+$ + $e^-$ → Cs | −2.923 |
| Ca, Ca$^{2+}$ | Ca$^{2+}$ + 2$e^-$ → Ca | −2.87 |
| Na, Na$^+$ | Na$^+$ + $e^-$ → Na | −2.714 |
| Al, Al$^{3+}$ | Al$^{3+}$ + 3$e^-$ → Al | −1.66 |
| Zn, Zn$^{2+}$ | Zn$^{2+}$ + 2$e^-$ → Zn | −0.763 |
| Fe, Fe$^{2+}$ | Fe$^{2+}$ + 2$e^-$ → Fe | −0.440 |
| Cd, Cd$^{2+}$ | Cd$^{2+}$ + 2$e^-$ → Cd | −0.403 |
| Sn, Sn$^{2+}$ | Sn$^{2+}$ + 2$e^-$ → Sn | −0.136 |
| Pb, Pb$^{2+}$ | Pb$^{2+}$ + 2$e^-$ → Pb | −0.126 |
| Fe, Fe$^{3+}$ | Fe$^{3+}$ + 3$e^-$ → Fe | −0.036 |
| Pt, D$_2$, D$^+$ | 2D$^+$ + 2$e^-$ → D$_2$ | −0.0034 |
| Pt, H$_2$, H$^+$ | 2H$^+$ + 2$e^-$ → H$_2$ | 0.0000 |
| Pt, Ti$^{3+}$, Ti$^{4+}$ | Ti$^{4+}$ + $e^-$ → Ti$^{3+}$ | +0.040 |
| Pt, Sn$^{2+}$, Sn$^{4+}$ | Sn$^{4+}$ + 2$e^-$ → Sn$^{2+}$ | +0.15 |
| Pt, Cu$^+$, Cu$^{2+}$ | Cu$^{2+}$ + $e^-$ → Cu$^+$ | +0.153 |
| Ag, AgCl, Cl$^-$ | AgCl + $e^-$ → Ag + Cl$^-$ | +0.222 |
| Hg, Hg$_2$Cl$_2$, Cl$^-$ | Hg$_2$Cl$_2$ + 2$e^-$ → 2Hg + 2Cl$^-$ | +0.2800 |
| Cu, Cu$^{2+}$ | Cu$^{2+}$ + 2$e^-$ → Cu | +0.337 |
| Pt, OH$^-$, O$_2$ | O$_2$ + 2H$_2$O + 4$e^-$ → 4 OH$^-$ | +0.401 |
| Pt, I$^-$, I$_2$ | I$_2$ + 2$e^-$ → 2 I$^-$ | +0.5355 |
| Pt, C$_6$H$_4$(OH)$_2$, C$_6$H$_4$O$_2$ | C$_6$H$_4$O$_2$ + 2H$^+$ + 2$e^-$ → C$_6$H$_4$(OH)$_2$ | +0.7000 |
| Pt, Fe$^{2+}$, Fe$^{3+}$ | Fe$^{3+}$ + $e^-$ → Fe$^{2+}$ | +0.771 |
| Ag, Ag$^+$ | Ag$^+$ + $e^-$ → Ag | +0.7991 |
| Hg, Hg$^{2+}$ | Hg$^{2+}$ + 2$e^-$ → Hg | +0.854 |
| Pt, Hg$_2^{2+}$, Hg$^{2+}$ | 2Hg$^{2+}$ + 2$e^-$ → Hg$_2^{2+}$ | +0.92 |
| Pt, Br$^-$, Br$_{2(l)}$ | Br$_{2(l)}$ + 2$e^-$ → 2Br$^-$ | +1.0652 |
| Pt, Cl$^-$, Cl$_2$ | Cl$_2$ + 2$e^-$ → 2Cl$^-$ | +1.3595 |
| Pt, Ce$^{3+}$, Ce$^{4+}$ | Ce$^{4+}$ + $e^-$ → Ce$^{3+}$ | +1.61 |

*Source:* Data are reproduced by permission from *Oxidation Potentials* by Latimer, W. M., Prentice-Hall, Englewood Cliffs, N.J., 1952.

### 5.4 Cells and Electrodes Meriting Special Consideration

**The reference cell or standard cell** Certain types of cells are very stable and give the same voltage for years. The Weston or cadmium cell shown in Fig. 5.2 is the accepted standard for measuring voltages. The half-cell reactions are

Oxidation: $\qquad$ Cd → Cd$^{2+}$ + 2$e^-$.

Reduction: $\quad$ Hg$_2$SO$_4$ + 2$e^-$ → 2Hg + SO$_4^{2-}$.

**TABLE 5.2** STANDARD HALF-CELL POTENTIALS FOR SOME BIOLOGICAL REDUCTIONS (AQUEOUS SOLUTION, 25°C, pH 7.0)

| Reaction | $\mathscr{E}_0'$ (volts) |
|---|---|
| Ferredoxin to reduced ferredoxin (chloroplast) | −0.43 |
| Xanthine to hypoxanthine + $H_2O$ | −0.37 |
| $TPN^+$ to TPNH + $H^+$ | −0.32 |
| $DPN^+$ to DPNH + $H^+$ | −0.32 |
| Acetone to 2-propanol | −0.30 |
| 1,3-diphosphoglyceric acid$^{4-}$ to glyceraldehyde-3-phosphate$^{2-}$ and $HPO_4^{2-}$ | −0.29 |
| Acetoacetate$^-$ to L-$\beta$-hydroxybutyrate$^-$ | −0.29* |
| Oxidized lipoate$^-$ to reduced lipoate$^-$ | −0.29 |
| Oxidized glutathione to reduced glutathione | −0.23* |
| Cystine to cysteine | −0.22* |
| Acetaldehyde to ethanol | −0.20 |
| Dihydroxyacetone phosphate$^{2-}$ to L-glycerol-1-phosphate$^{2-}$ | −0.19 |
| Pyruvate$^-$ to L-lactate$^-$ | −0.19 |
| Oxalacetate$^{2-}$ to L-malate$^{2-}$ | −0.17 |
| Coenzyme Q to reduced coenzyme Q | 0.00 |
| Cytochrome b to reduced cytochrome b | 0.00 |
| Fumarate$^{2-}$ to succinate$^{2-}$ | +0.03 |
| Dehydroascorbate$^-$ to ascorbate$^-$ | +0.06* |
| Cytochrome $c_1$ to reduced cytochrome $c_1$ | +0.22 |
| Cytochrome c to reduced cytochrome c | +0.26 |
| Cytochrome a to reduced cytochrome a | +0.29 |
| Cytochrome $a_3$ to reduced cytochrome $a_3$ (cytochrome oxidase) | +0.53 |
| $\frac{1}{2}O_2$ to $H_2O$ | +0.82 |

*Source:* Data are quoted from Burton, K., in Krebs, H. A., and H. L. Kornberg *Ergeb. Physiol. biol. Chem. u. exptl. Pharmakol.*, 49: 212, 1957; Montgomery. R. and C. A. Swenson, *Quantitative Problems in the Biochemical Sciences*, W. H. Freeman and Co., San Francisco, 1969, p. 148; and Green, D. E., and H. Baum, *Energy and the Mitochondrion*, Academic Press, New York, 1970, p. 166.
*30°C

The cell reaction is the sum of these:

$$Cd + Hg_2SO_4 \rightarrow Cd^{2+} + 2Hg + SO_4^{2-}.$$

If the cell is made up according to specifications established by the Bureau of Standards, its voltage will be 1.01463 volts at 25°C. When the cell is saturated with cadmium sulfate and mercurous sulfate, the concentration remains constant despite the electrode reactions. The cell is reversible: cadmium can be deposited and mercury ionized if the cell is connected to an external source of larger voltage.

**TABLE 5.3**  TYPES OF ELECTRODES AND EXAMPLES

| | |
|---|---|
| Metal, metal ion | $Cu, Cu^{2+}$ |
| Inert electrode, nonmetal in solution, ion | $Pt, I_{2(s)}, I^-$ |
| Inert electrode, ions in different states of oxidation | $Pt, Fe^{3+}, Fe^{2+}$ |
| Inert electrode, neutral solutes in different states of oxidation | $Pt,$ Hydroquinone, Quinone |
| Inert electrode, gas, ion | $Pt, Cl_2, Cl^-$ |
| Amalgam electrode, ion | $(Cd + Hg), Cd^{2+}$ |
| Electrode, insoluble salt or oxide, ion | $Hg, Hg_2Cl_2, Cl^-$ |

*Note:* If one of the components of a half-cell can exist in various physical states under ordinary conditions, the appropriate state is indicated by a subscript.

**Reference electrodes**  In many chemical operations, such as potentiometric titrations, a single reference electrode is needed, rather than a reference cell. For example, one of the commonly employed methods for measuring the amount of linear starch (amylose) in a mixture of linear and branched starch (amylopectin) utilizes an iodometric titration. The linear starch complexes the free iodine until the starch becomes saturated with iodine. At this point free iodine will appear in the solution and a break will appear in the plot of voltage versus volume of standard iodine solution added. The presence of free iodine in the solution, *i.e.*, the end point of the titration, can be determined potentiometrically by coupling the iodine-iodide electrode with a suitable reference electrode.

In the measurement of the standard voltages shown in Table 5.1, reference electrodes were used.

*The hydrogen electrode*  Although the hydrogen electrode has been selected as the zero reference point for relative electrode potentials, it is used less and less frequently in experimental measurements of other standard half-cells. Preparing a standard hydrogen electrode according to specifications is no simple matter. The hydrogen gas must be absolutely pure. The pressure should be one atmosphere; if it is not, the existing pressure must be carefully determined and properly handled in the calculations. The activity of the solution of hydrogen ion must be unity, and the platinum electrode must be scrupulously clean and free of impurities in the metal. The electrode must not be permitted to dry out for more than a few minutes or else the platinum becomes catalytically ineffective. A diagram of the hydrogen electrode is shown in Fig. 5.3. The solution must be free of all interfering oxidants or traces of substances which might poison the platinum catalyst. Additional inconveniences are the heavy hydrogen cylinder and the explosion hazard.

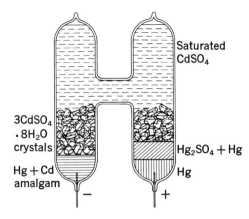

3CdSO$_4$
•8H$_2$O
crystals

Hg + Cd
amalgam

Saturated
CdSO$_4$

Hg$_2$SO$_4$ + Hg

Hg

−   +

**Figure 5.2**   Weston or cadmium cell.

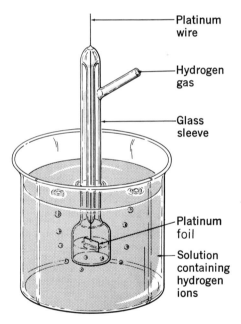

Platinum
wire

Hydrogen
gas

Glass
sleeve

Platinum
foil

Solution
containing
hydrogen
ions

**Figure 5.3**   Hydrogen electrode.

*The calomel electrode*   Probably the most commonly used reference electrode is the calomel electrode shown in Fig. 5.4. The half-cell is prepared by floating a paste of mercury and calomel ($Hg_2Cl_2$) on top of pure mercury, inserting a platinum wire into the mercury, solid phase, and immersing this part of the electrode in a potassium chloride solution. Contact with the other electrode of the cell is achieved by means

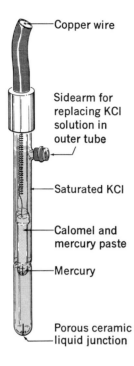

Copper wire

Sidearm for
replacing KCl
solution in
outer tube

Saturated KCl

Calomel and
mercury paste

Mercury

Porous ceramic
liquid junction

**Figure 5.4**  Saturated calomel electrode.

of a salt bridge or by a pinhole in the bottom of the glass sleeve as shown
in Fig. 5.4. The half-cell reaction, written in the reduction direction, is

$$Hg_2Cl_{2(s)} + 2e^- \rightarrow 2Hg_{(l)} + 2Cl^-.$$

Oxidized mercury is in the mercurous state, but does not appear in the
stoichiometry as an ion because of the insolubility of mercurous chloride.
Since the mercury and mercurous chloride are in their standard states
(pure liquid and pure solid) their activities are unity and the cell potential depends only on the concentration of chloride ion. The effect of
concentration may be seen from the reduction potentials at 25°C for
various concentrations of potassium chloride (note that these are concentrations, not activities; the standard calomel electrode is defined as that
in which the KCl solution is one normal, and, therefore, is an exception
to the unit activity rule):

|  |  |
|---|---|
| Saturated KCl | 0.2444 volt, |
| 1 N KCl | 0.2800 volt, |
| 0.1 N KCl | 0.3356 volt. |

Although the electrode containing saturated KCl solution is more
sensitive to temperature change than the others, laboratory personnel
usually prefer it because it is easy to maintain. Evaporation of solvent
does not result in undetermined concentration changes, as it does with

the 1 N or 0.1 N KCl electrodes. Saturated KCl is added to the electrode to replace losses resulting from evaporation; a few crystals of very pure KCl may be added to insure saturation.

*The silver-silver chloride electrode* Another commonly employed reference electrode is the silver-silver chloride electrode, for which the electrode reaction is analogous to that of the calomel electrode. It is usually prepared by electroplating silver onto a platinum wire and then depositing silver chloride onto the silver by a second electrolysis. The presence of a solid rather than liquid metal and a firmly deposited film of insoluble salt rather than a floating paste give this electrode obvious convenience advantages. An additional advantage is that the electrode can be made up in a very small size for use with samples of limited volume. As for the calomel half-cell, the potential of the silver-silver chloride electrode depends on the chloride ion concentration.

*The glass electrode* Since the invention of the glass electrode the measurement of pH has become a matter of simple laboratory routine. No longer is it necessary to cope with the hydrogen electrode or the quinhydrone electrode, both of which possess serious disadvantages for many biochemical systems. (Further consideration is given to these electrodes in the discussion of pH measurement, page 260.)

If a thin glass membrane separates two solutions of different hydrogen ion concentration, a very small potential arises across the membrane. Exactly how the potential originates is not well understood, although it is undoubtedly a consequence of interaction between two adjacent surfaces of adsorbed hydrogen ions. The potential is linearly proportional to the difference in pH and can be measured if suitably amplified. With the development of the modern vacuum-tube voltmeter, the use of this principle of pH measurement became a reality. A glass electrode is shown in Fig. 5.5. A reference electrode, usually of the silver-silver chloride type, is placed inside the tube which is filled with a solution of fixed pH, usually 0.1 N HCl. The glass is extremely thin either at the bottom of the electrode or in a narrow collar near the bottom. The thinness of the glass renders the electrode susceptible to damage if carelessly handled. Hitting the thin membrane hard against the bottom of a beaker or with the glass paddle of a stirring motor has literally been the final blow for many a glass electrode. The ordinary glass electrode cannot be used above pH 9. Ordinary glass, because of its chemical nature, is able to pass alkali metal ions at high pH and permanently alter the concentration of the internal reference solution. In recent years special glasses have been perfected for use in the high pH range.

The glass electrode is used with a reference calomel electrode in an instrument called a pH meter, probably a familiar device to most students. The meter is calibrated with standard buffer solutions daily. Both battery and line-operated pH meters with vacuum tube or solid state

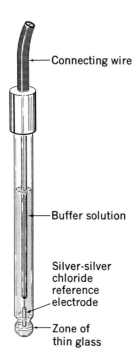

Connecting wire

Buffer solution

Silver-silver chloride reference electrode

Zone of thin glass

**Figure 5.5** Glass electrode.

electronics are available from instrument supply houses. A single probe combination glass-calomel electrode is now available with many pH meters. The narrow diameter of the probe permits its use with volumes of only a few milliliters.

**The concentration cell** A cell composed of two half-cells which are identical save for the concentrations of solutes is called a concentration cell. An example of this type of cell is

$$\text{Fe, Fe}^{2+}_{(a=x)} \parallel \text{Fe}^{2+}_{(a=y)}, \text{Fe.}$$

The standard oxidation potential is exactly equal and opposite in sign to the standard reduction potential, since the half-cells are identical, and therefore, the $\mathscr{E}°$ of the cell is equal to zero. The cell voltage, $\mathscr{E}$, then, depends only on the concentration ratio in the $Q$ term (Eq. 5.5). The $K_{equil}$ for a concentration cell is always unity. An example of a calculation involving a concentration cell is shown in Section 5.6.

**Fuel cells** Anyone within range of news media during the last few years is aware that our space vehicles are powered with electricity-generating devices called fuel cells.[*] The malfunctioning of these cells

---

[*] For the student who has done no previous reading on this subject, an excellent introductory article is Chopey, N. P., "What You Should Know About Fuel Cells," *Chem. Eng.*, **71**:(11), 125, 1964.

during one of the Gemini flights produced nationwide suspense and concern. The fuel cells presently used in the space program are of the hydrogen-oxygen type, although intensive research is being conducted on cells utilizing light hydrocarbons. Within a decade we may see significant commercial applications of the latter type.

Theoretically, any fuel which can be burned to yield energy stands as a potential direct source of electricity if it can be made to yield electrons to an electrode under controlled conditions. Even biological waste materials such as sewage could serve as electrical energy sources if the enzymes and electrodes that would provide a sufficiently rapid rate of electron release could be found. Research is actively pursued in this specific area in the hope of discovering new ways to utilize or dispose of human waste, both in space vehicles and in urban communities.

The principle of the hydrogen-oxygen fuel cell is the reverse of the electrolysis of water. Hydrogen and oxygen gases enter the cell, which contains a KOH electrolyte and specially coated carbon electrodes. The two half-cell reactions and the equation for water dissociation yield

$$
\begin{aligned}
H_{2(g)} &\rightarrow 2H^+ + 2e^- & \mathscr{E}^\circ_{ox} &= 0.000 \text{ volt} \\
2e^- + \tfrac{1}{2}O_{2(g)} + H_2O_{(l)} &\rightarrow 2OH^- & \mathscr{E}^\circ_{red} &= 0.401 \text{ volt} \\
\underline{2H^+ + 2OH^- \rightarrow 2H_2O_{(l)}} & & & \\
H_{2(g)} + \tfrac{1}{2}O_{2(g)} &\rightarrow H_2O_{(l)} & \mathscr{E}^\circ_{cell} &= 0.401 \text{ volt.}
\end{aligned}
$$

The cells actually used in the Gemini and Apollo capsules are not standard half-cells, since their KOH concentration may range from 6 to 14 $N$. The efficiency of the cell rises as the operating temperature goes up.

**The oxygen electrode**  The measurement of oxygen uptake by enzymes systems and tissue preparations can be followed by the Warburg technique (see Section 1.7). The Warburg respirometer cannot be used, however, for the direct measurement of oxygen tensions *in vivo*, in sea water, sewage, or in many other experiments in which oxygen consumption must be monitored.[*] The oxygen electrode, first used by Davies and Brink[†] with animal tissues, and those electrodes that have developed since lend themselves to both routine and novel experiments because of their small size and functional simplicity.

The modern oxygen probe (the term probe is widely used and is probably better than electrode because two electrodes constitute the basic elements of the device) works on a principle that is either voltametric (potential-current) or galvanic and is closely enough related to

---

[*]An excellent review of problems encountered and of appropriate analytical methods in the determination of dissolved oxygen can be found in *Analysis of Dissolved Oxygen in Natural and Waste Waters*, (1966), by Khalil H. Mancy and Theodore Jaffe, Public Health Service Publication No. 999-WP-37, U.S. Department of Health, Education and Welfare.

[†]Davies, P. W. and J. Brink, Jr., *Rev. Sci. Instr.* **13**, 524, 1942.

the discussion of special cells and electrodes to merit consideration here. In both the voltametric and galvanic applications the tip of the probe usually is coated with electrolyte and then is covered with a membrane such as polyethylene or teflon sheeting to hold the electrolyte in contact with the sensing element or the cathode. The membrane defines a diffusion layer with a thickness that is presumed to be independent of the hydrodynamic properties of the test solution and prevents direct contamination of the elements of the probe by the chemical constituents of the medium. Dissolved oxygen diffuses through the membrane, therefore time must be allowed for the system to come to equilibrium before data are taken. The importance of this membrane has resulted in the use of the term *membrane* electrodes.

In the voltammetric membrane-probe an external potential is applied across a large nonpolarizable reference electrode (Calomel or $Ag$-$Ag_2O$) and a stationary platinum microelectrode. The second electrode, wherein oxygen is reduced, is made several tenths of a volt more negative than the first electrode. The galvanic membrane-probe differs from the voltametric one in that the galvanic electrode system provides its own potential difference so that no external voltage need be applied. Specifically, oxygen, the species of interest, is a reactant at one of the two electrodes and thereby the galvanic system becomes a source of power for the probe. Since oxygen in both types is reduced at the cathode surface, a concentration gradient is created that leads to the diffusion of dissolved oxygen through the membrane in the direction of the cathode. At a given voltage a diffusion current develops that is ultimately limited by the rate of diffusion of oxygen. At this point the current is directly proportional to the activity of oxygen in solution. The current may be converted to a voltage by a variable resistance inserted in series with the cathode and the voltage then amplified and registered on a recorder, or the current may be read directly on a low-impedence microammeter or galvanometer. Temperature compensation in the apparatus design increases the accuracy and reproducibility of measurements.

The geometry of one particular type of galvanic probe is shown in Fig. 5.6. Oxygen is reduced at the silver cathode according to the following reaction,

$$O_{2(g)} + 2H_2O + 4e^- \rightarrow 4OH^-$$

and lead is consumed at the anode surface to form what is probably a hydrated oxy-anion of lead,

$$Pb + 4OH^- \rightarrow PbO_2^= + 2H_2O + 2e^-$$

for an overall balanced cell reaction of

$$O_2 + 2Pb + 4OH^- \rightarrow 2PbO_2^= + 2H_2O$$

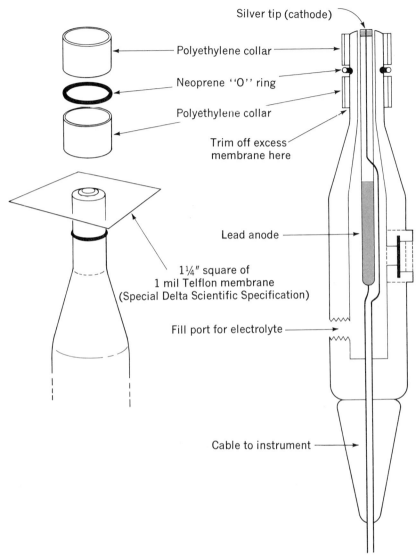

**Figure 5.6**  Schematic view of an "active" type oxygen probe, used in the Model 85 oxygen meter of Delta Scientific Corporation, Lindenhurst, N.Y., U.S.A. Reproduced by permission of the company. A temperature measuring thermistor is also included in the probe and temperature compensation is done automatically. The electrolyte is 1N potassium hydroxide.

Note that with continued use of the probe there occurs a decrease in concentration of hydroxyl ions around the anode and an increase around the cathode. This seems to offer no problem over a rather extended time since the electrolyte used in the probe of Fig. 5.6 is 1N KOH solution. When there is a steady-state flow of oxygen through the membrane, its rate will determine the diffusion current which is directly proportional to the activity of oxygen in the medium being investigated. At times care may be necessary in using this *activity* as the concentration or in converting it to concentration.

Several membrane electrode systems have been developed and are available for the determination of dissolved oxygen.* It is interesting that the membrane oxygen electrode system lends itself to the determination of oxygen in gaseous, nonaqueous, and aqueous samples.

The oxygen electrode possesses some advantages over the Warburg technique in that a series of additions to the reaction medium may be made when using the electrode assembly, whereas usually only two additions can be made to the Warburg reaction vessel. On the other hand, the Warburg respirometer has the advantage of a large oxygen reservoir for the uptake system, whereas the electrode technique limits the amount of oxygen present in the medium. Unless adequate precautions are taken with the oxygen electrode, considerable error can be incurred by slow diffusion of oxygen from the atmosphere into the reaction medium. The most obvious advantage of the oxygen electrode is its versatility, as it can be inserted into an artery or taken on an ocean-going research vessel for oceanographic studies.

## 5.5 The Determination of pH

The biochemist or biologist who studies the homeostasis of living systems or their pH-sensitive components finds the measurement and adjustment of pH to be part of the daily research routine. Before the advent of the pH meter, the choice lay between an indicator with a $pK_I$ suitable to the problem and a potentiometric arrangement involving a hydrogen or quinhydrone electrode versus standard calomel. Nearly always it was necessary to sacrifice part of the sample because the addition of indicator, quinhydrone, or hydrogen produced undesirable changes or contaminations. With the perfection of the glass electrode these problems disappeared. Both the glass and the reference calomel electrodes can be immersed in the sample without its being contaminated. Small electrodes have been constructed for use with samples of limited volume.

Theoretically, any half-cell reaction which consumes or produces hydrogen ions can be used for the measurement of pH. However, if the

---

* For a comparison between results obtained with the membrane electrode and the well known Winkler method, the reader should refer to the article by J. J. McKeown, L. C. Brown, and G. W. Gove in the *Journal of Water Pollution Control Federation*, **39**, 8, pp. 1323–1336, August 1967.

cell voltage is to depend only on [H$^+$], the half-cell reaction must be one in which it is possible to cancel out or equate to unity all other factors appearing in the $Q$ term (Eq. 5.5). The idea will be illustrated with the hydrogen and calomel electrodes, as one combination, and the quinhydrone and calomel electrodes, as a second combination.

**Normal calomel electrode and hydrogen electrode**  The cell is set up as follows:

$$\text{Pt, } H_{2(p\,=\,1\,\text{atm})},\ H^+_{(a\,=\,x)} \ \|\ \text{Normal calomel electrode.}$$

The half-cell reactions are

Oxidation: $\qquad\qquad \frac{1}{2}H_{2(p\,=\,1\,\text{atm})} \longrightarrow H^+_{(a\,=\,x)} + e^-$

Reduction: $\qquad\qquad \underline{\frac{1}{2}Hg_2Cl_{2(s)} + e^- \longrightarrow Hg_{(l)} + Cl^-_{(a\,=\,1)}}$

Cell reaction: $\frac{1}{2}Hg_2Cl_{2(s)} + \frac{1}{2}H_{2(p\,=\,1\,\text{atm})} \longrightarrow H^+_{(a\,=\,x)} + Hg_{(l)} + Cl^-_{(a\,=\,1)}.$

(The cell reaction has been written for a one-electron exchange, but no problem is presented by writing it for a two-electron exchange. The cell voltage is independent of $n$, the number of electrons transferred, for any individual reaction.)

From the Nernst Equation (Eq. 5.5),

$$\mathscr{E} = \mathscr{E}^\circ - \frac{2.303\ RT}{n\mathscr{F}}\log\frac{[H^+_{(a\,=\,x)}][Hg_{(l)}][Cl^-_{(a\,=\,1)}]}{[Hg_2Cl_{2(s)}]^{1/2}[H_{2(p\,=\,1\,\text{atm})}]^{1/2}}.$$

Except for [H$^+$], all factors in the $Q$ term are unity because all other components are in their standard states. The equation now simplifies to

$$\mathscr{E} = \mathscr{E}^\circ - \frac{2.303\ RT}{\mathscr{F}}\log\,[H^+_{(a\,=\,x)}].$$

Substituting pH for $-\log$ [H$^+$] and making the calculation for 25°C yields

$$\mathscr{E} = \mathscr{E}^\circ + 0.0592\ \text{pH}.$$

$\mathscr{E}^\circ$ is found to be

$$\mathscr{E}^\circ = \mathscr{E}^\circ_{\text{ox } H_2,\,H^+} + \mathscr{E}^\circ_{\text{red normal calomel}} = 0.0000 + 0.2800 = 0.2800\ \text{volt.}$$

Substituting this value for $\mathscr{E}^\circ$ in the Nernst Equation gives

$$\mathscr{E} = 0.2800 + 0.0592\ \text{pH}.$$

Even when the saturated or 0.1 $N$ calomel electrode is used, the $Q$ term remains uncomplicated because the [Cl$^-$] has been incorporated into the $\mathscr{E}$ listed for that concentration. The $\mathscr{E}$ values given for the various calomel electrodes have been experimentally determined, but the calculated values, based on activities, agree reasonably well. For example, $\mathscr{E}^{\circ}_{red}$ for the normal calomel electrode is 0.2800 volt and $\mathscr{E}$ for the 0.1 N calomel electrode is 0.3356 volt. The calculated difference would be 0.0592 log (0.1) subtracted from the standard $\mathscr{E}^{\circ}$. This is 0.2800 − (−0.0592) = 0.3392 volt, and in reasonable agreement with the experimental value.

**Quinhydrone electrode and normal calomel electrode** Although the quinhydrone electrode is not used in modern laboratory practice, it nevertheless is a convenient illustration of a half-cell reaction suitable for the measurement of pH.

Quinhydrone is a slightly soluble "molecular compound" of hydroquinone and quinone in a ratio of 1:1. When quinhydrone is dissolved in acid solution of low ionic strength, approximately equal concentrations of quinone and hydroquinone are present in solution. The equilibrium is

$2e^- + \quad + 2H^+ \rightleftharpoons \qquad\qquad \mathscr{E}^{\circ}_{red} = +0.7000 \text{ volt.}$

Let us combine this half-cell reaction with the normal calomel electrode to give a cell corresponding to

$$\text{Normal calomel electrode} \parallel H^+_{(a\,=\,r)}, \text{ Quinhydrone, Pt.}$$

The complete cell reaction for one equivalent of charge is

$$Hg_{(l)} + Cl^-_{(a\,=\,1)} + \tfrac{1}{2}C_6H_4O_2 + H^+_{(a\,=\,x)} \rightarrow \tfrac{1}{2}C_6H_6O_2 + \tfrac{1}{2}Hg_2Cl_{2(s)}.$$

The $Q$ term is

$$Q = \frac{[C_6H_6O_2]^{1/2}[Hg_2Cl_{2(s)}]^{1/2}}{[C_6H_4O_2]^{1/2}[Hg_{(l)}][Cl^-_{(a\,=\,1)}][H^+_{(a\,=\,x)}]}$$

This formidable combination reduces to $Q = 1/[H^+_{(a\,=\,x)}]$, because Hg, Cl$^-$ and Hg$_2$Cl$_2$ are in their standard states, and the ratio of hydroquinone to quinone is always unity. The Nernst Equation for the cell is

$$\mathscr{E} = \mathscr{E}^{\circ}_{\text{ox calomel}} + \mathscr{E}^{\circ}_{\text{red quinhydrone}} - 0.0592 \log \frac{1}{[H^+_{(a\,=\,x)}]},$$

$$\mathscr{E} = -0.2800 + 0.7000 - 0.0592 \text{ pH} = 0.4200 - 0.0592 \text{ pH.}$$

The cell reaction can be written without fractional coefficients.

$$2Hg + 2Cl^- + C_6H_4O_2 + 2H^+ \longrightarrow C_6H_6O_2 + Hg_2Cl_2$$

The Nernst Equation becomes

$$\mathscr{E} = \mathscr{E}^\circ - \frac{0.0592}{2} \log \frac{1}{[H^+_{(a=x)}]^2}.$$

Since $\frac{1}{2} \log 1/[H^+]^2 = \log 1/[H^+]$, the stoichiometry and the $n$ term are self-compensating and the cell voltage is independent of how the equation is written. The practical significance of this relationship lies in the corollary that the voltage of an electrode is theoretically independent of its size. Attention should be given, however, to the fact that the useful work which can be obtained from an electric cell is *dependent* on its size, as borne out in the equation for electrical work: $\Delta G = -n\mathscr{E}\mathscr{F}$. Unless otherwise stipulated, $\Delta G$ values are usually stated per mole. As mentioned in an earlier chapter, some authors use a special notation to represent molar thermodynamic quantities and thereby avoid any possible misunderstanding, *e.g.*, $\bar{G}$ or $\Delta\bar{G}$. Since that practice has not been followed generally in this text, the student must be alert when interpreting a problem. The free energy change for a reaction involving a two-electron change will be twice that for the same reaction written for a one-electron change.

### 5.6  Sample Calculations

### Cells in which reactants are at unit activity

*Example 5.1*  Given the cell at 25°C,

$$Pb, Pb^{2+}_{(a=1)} \parallel Sn^{2+}_{(a=1)}, Sn^{4+}_{(a=1)}, Pt,$$

(a) calculate the voltage; (b) write the cell reaction; (c) calculate the free energy change; and (d) calculate the equilibrium constant for the reaction.

(a) $\mathscr{E}^\circ = \mathscr{E}^\circ_{ox, Pb, Pb^{2+}} + \mathscr{E}^\circ_{red, Sn^{2+}, Sn^{4+}} = 0.126 + 0.15 = 0.276$ volt.

(b)  Oxidation: $\qquad\qquad Pb \longrightarrow Pb^{2+} + 2e^-$
Reduction: $\underline{Sn^{4+} + 2e^- \longrightarrow Sn^{2+}}$
$\qquad\qquad Sn^{4+} + Pb \longrightarrow Pb^{2+} + Sn^{2+}.$

(c) $\Delta G^\circ = -n\mathscr{E}^\circ\mathscr{F} = -(2)(0.276)(23,060) = -12,730$ calories.

(d) $\qquad \Delta G^\circ = -2.303\ RT \log K,$

$\qquad -12,730 = -2.303(1.987)(298)(\log K),$

$\qquad\quad \log K = 9.333,$

$\qquad\qquad K = 2.15 \times 10^9.$

## Cells in which one or more reactants are not at unit activity

*Example 5.2*  Given the cell at 25°C,

$$\text{Cu, Cu}^{2+}_{(a=0.01)} \parallel \text{H}^+_{(a=0.10)}, \text{H}_{2(0.9\ atm)}, \text{Pt}$$

(a) write the cell reaction; (b) calculate the cell potential; and (c) state whether the cell will operate spontaneously.

(a)  Oxidation:  $\text{Cu} \rightarrow \text{Cu}^{2+}_{(a=0.01)} + 2e^-$

Reduction:  $\underline{2e^- + 2\text{H}^+_{(a=0.1)} \rightarrow \text{H}_{2(0.9\ atm)}}$

$\text{Cu} + 2\text{H}^+_{(a=0.1)} \rightarrow \text{Cu}^{2+}_{(a=0.01)} + \text{H}_{2(0.9\ atm)}.$

(b)  $\mathscr{E} = \mathscr{E}^\circ - \dfrac{2.303\ RT}{n\mathscr{F}} \log Q$

$\quad = \mathscr{E}^\circ - \dfrac{0.0592}{2} \log \dfrac{[\text{Cu}^{2+}][\text{H}_2]}{[\text{Cu}][\text{H}^+]^2}$

$\quad = (-0.337 + 0.000) - \dfrac{0.0592}{2} \log \dfrac{(0.01)(0.9)}{(1)(0.1)^2}$

$\quad = -0.337 + 0.0014 = -0.3356.$

(c) The cell voltage is negative and therefore the cell will not operate spontaneously in the direction for which the equation was written.

*Example 5.3*  Calculate the activity of cerous ions in the following cell for which the cell potential is 1.406 volts.

$$\text{Pt, Ti}^{4+}_{(a=0.5)}, \text{Ti}^{3+}_{(a=0.3)} \parallel \text{Ce}^{4+}_{(a=0.002)}, \text{Ce}^{3+}_{(a=x)}, \text{Pt.}$$

First, write the cell reaction:

Oxidation:  $\text{Ti}^{3+}_{(a=0.3)} \rightarrow \text{Ti}^{4+}_{(a=0.5)} + e^-$

Reduction:  $\underline{\text{Ce}^{4+}_{(a=0.002)} + e^- \rightarrow \text{Ce}^{3+}_{(a=x)}}$

$\text{Ce}^{4+}_{(a=0.002)} + \text{Ti}^{3+}_{(a=0.3)} \rightarrow \text{Ti}^{4+}_{(a=0.5)} + \text{Ce}^{3+}_{(a=x)}.$

Next calculate $\mathscr{E}^\circ$ for the reaction:

$$\mathscr{E}^\circ = \mathscr{E}^\circ_{ox,\ Ti^{3+},\ Ti^{4+}} + \mathscr{E}^\circ_{red,\ Ce^{4+},\ Ce^{3+}}$$

$$= -0.040 + 1.610 = 1.570\ \text{volts.}$$

Then, write the Nernst Equation

$$\mathscr{E} = \mathscr{E}^\circ - \frac{2.303\ RT}{n\mathscr{F}} \log \frac{[\text{Ti}^{4+}][\text{Ce}^{3+}]}{[\text{Ce}^{4+}][\text{Ti}^{3+}]}.$$

Substituting the known values of the various concentrations and the cell voltage gives

$$1.406 = 1.570 - 0.0592 \log \frac{(0.5)(x)}{(0.002)(0.3)},$$

$$x = 0.7\ \text{moles per liter.}$$

*Example 5.4* Develop a formula for calculating the voltage of a cell consisting of two hydrogen electrodes at different pressures immersed in a solution of hydrogen ions. Calculate the voltage for the conditions: pressure of hydrogen at left electrode is 1 atm, pressure of hydrogen at right electrode is 0.25 atm, temperature is 25°C. First, set up the cell:

$$\text{Pt, } H_{2(p=x)}; \; H^+; \; H_{2(p=y)}, \text{ Pt.}$$

The cell reaction is

Oxidation: $\quad H_{2(p=x)} \rightarrow 2H^+ + 2e^-$

Reduction: $\quad 2H^+ + 2e^- \rightarrow H_{2(p=y)}$

$$\overline{2H^+ + H_{2(p=x)} \rightarrow 2H^+ + H_{2(p=y)}.}$$

The Nernst Equation is

$$\mathscr{E} = \mathscr{E}° - \frac{0.0592}{2} \log \frac{[H_{2(p=y)}]}{[H_{2(p=x)}]} = 0 - 0.0296 \log \frac{y}{x} = -0.0296 \log \frac{y}{x}.$$

This is the general equation asked for. Now substitute the given pressures:

$$\mathscr{E} = -0.0296 \log \frac{(0.25)}{(1)} = 0.0178 \text{ volt.}$$

## Concentration cells

*Example 5.5* Calculate the voltage for the cell,

$$\text{Pt, } Fe^{3+}_{(a=0.1)}, \; Fe^{2+}_{(a=0.5)} \; \| \; Fe^{3+}_{(a=0.4)}, \; Fe^{2+}_{(a=1)}, \text{ Pt.}$$

First write the cell reaction:

Oxidation: $\quad Fe^{2+}_{(a=0.5)} \rightarrow Fe^{3+}_{(a=0.1)} + e^-$

Reduction: $\quad e^- + Fe^{3+}_{(a=0.4)} \rightarrow Fe^{2+}_{(a=1)}$

$$\overline{Fe^{2+}_{(a=0.5)} + Fe^{3+}_{(a=0.4)} \rightarrow Fe^{3+}_{(a=0.1)} + Fe^{2+}_{(a=1)}.}$$

For the concentration cell, $\mathscr{E}° = 0$ because $\mathscr{E}°_{ox} = -\mathscr{E}°_{red}$. The last steps are to write the Nernst Equation and substitute values:

$$\mathscr{E} = \mathscr{E}° - \frac{2.303 \; RT}{n\mathscr{F}} \log Q = 0 - 0.0592 \log \frac{[Fe^{3+}][Fe^{2+}]}{[Fe^{2+}][Fe^{3+}]}$$

$$= -0.0592 \log \frac{(0.1)(1)}{(0.5)(0.4)} = 0.0178 \text{ volt.}$$

## 5.7 Oxidation-Reduction in Biological Systems

The relationship of biological oxidation to ATP synthesis was stressed in the opening paragraphs of the chapter. This introductory material should be read again by the student whose familiarity with biochemical processes is sketchy.

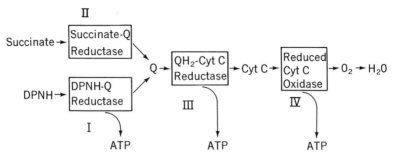

**Figure 5.7** Pathway of hydrogen from substrate to oxygen in mitochondria. The components of the complexes are:

I. Reduced nicotinamide-adenine-dinucleotide—Coenzyme Q Reductase.

II. Succinate—Coenzyme Q Reductase (contains flavin-adenine-dinucleotide).

III. Reduced Coenzyme Q—Cytochrome C Reductase.

IV. Reduced Cytochrome C—Cytochrome Oxidase.

[From D. E. Green and co-workers at the Institute for Enzyme Research at the University of Wisconsin (*Federation Proc., 22*:1460, 1963).]

Our present objective is to look quantitatively at the path of hydrogen and its electrons, that is, the electron-transport sequence. Can we calculate the energy differences that exist between individual members of the sequence and correlate these findings with energy requirements for ATP synthesis? Attention is directed only to those reactions which couple the energy of oxidation to ATP synthesis in mitochondria.

Mitochondria are subcellular particles which are popularly called the powerhouses or ATP-machines of living cells. The components of the electron-transport sequence, found within the mitochondria, constitute a kind of biochemical battery which produces electrical energy for the synthesis of ATP. These components are believed to be grouped into four enzyme complexes as shown in Fig. 5.7. Within the mitochondrion the complexes are linked together by lipoprotein structures, possibly of a membranous nature. ATP molecules are synthesized at the sites indicated in Fig. 5.7, three ATP molecules being obtained per pair of hydrogens if NADH is the starting point (Complex I), and two ATP molecules if oxidation begins with succinate (Complex II). The diagram represents the present opinion of the Madison group and is subject to some contention by other research workers. It has been selected as representative of current theories of biological oxidation.

In the simplest sense, each complex can be thought of as an electrochemical cell. Two components are shown for each complex; these constitute the half-cells. (Probably each complex contains additional components, which means that it is an approximation to treat a complex as a single electrochemical cell.) Figure 5.7 shows two alternate pathways for hydrogen: one starts with the removal of 2H from DPNH + H$^+$

and proceeds via Complexes II, III, and IV; the other starts with the removal of 2H from succinate and proceeds via Complexes II, III, and IV. Each pathway contains three complexes and is, therefore, a battery of three cells. Succinate and DPNH, the substances with the lowest reduction potentials, are separated by the complexes from oxygen, the substance with the highest reduction potential. As the hydrogens or electrons from hydrogen move from one complex to the next, successive potential differences occur. The potential differences accompanying the chemical reactions in Complexes I, III, and IV have to be sufficiently large in each case to furnish energy for the synthesis of an ATP molecule from ADP and inorganic phosphate.

The $\Delta G'$ of hydrolysis of the terminal high-energy phosphate of ATP is presently agreed to be of the order of $-7,000$ calories at $25°C$. Let us assume that approximately $-8,000$ calories are needed as a minimum. Is the known potential difference for each of the complexes compatible with the minimum energy requirement?

Since two electrons are transported per hydrogen pair, the potential difference required to yield $-8,000$ calories is

$$\mathscr{E}_0' = -\frac{\Delta G}{n\mathscr{F}} = \frac{8,000}{2(23,060)} = 0.17 \text{ volt.}$$

An $\mathscr{E}_0'$ equal to 0.17 volt is the minimum voltage for converting ADP to ATP, with all products and reactants (save $H^+$) in their standard states. If, on the other hand, the concentrations of ADP and inorganic phosphate are much higher than the concentration of ATP at the phosphorylation site, the necessary potential ($\mathscr{E}'$) will be considerably lowered.

All of the half-cells which are found in the biochemical "battery" of the mitochondrion are not positively identified as yet. Those which have been identified and their standard reduction potentials at pH 7.0 are

| | Half-cell | $\mathscr{E}_0'$ |
|---|---|---|
| (1) | DPN + $H^+$/DDN$^+$ | $-0.32$ |
| (2) | Reduced Flavoprotein, Flavoprotein | $-0.22$ |
| (3) | Reduced Coenzyme Q, Coenzyme Q | $0.00$ |
| (4) | Reduced Cytochrome C, Cytochrome C ($2Fe^{2+}$, $2Fe^{3+}$) | $+0.26$ |
| (5) | Reduced Cytochrome Oxidase, Cytochrome Oxidase ($2Fe^{2+}$, $2Fe^{3+}$) | $+0.53$ |
| (6) | $H_2O$, $\frac{1}{2}O_2$ | $+0.82$ |

The half-cell combinations that make up Complexes I, III, and IV of Fig. 5.7 are (1) and (3), (3) and (4), and (4) and (5), respectively.

Can each of these half-cell combinations yield the voltage required for the production of 8,000 calories, *i.e.*, 0.17 volt? As in calculating the voltage for simple chemical cells, we add the appropriate oxidation and reduction potentials for (1) and (3), (3) and (4), and (4) and (5):

|  | Complex I | Complex III | Complex IV |
|---|---|---|---|
| Oxidation: | (1) +0.32 | (3)  0.00 | (4) −0.26 |
| Reduction: | (3)  0.00 | (4) +0.26 | (5) +0.53 |
|  | $\mathscr{E}' = 0.32$ | +0.26 | +0.27 |

Despite the uncertainties and limitations of the calculation, the data and theory fit reasonably well: more than 0.17 volt is produced by each complex and the overall potential difference from (1) to (5) is 0.85 volt.

It is important for the student to remember that $\mathscr{E}'$ and $\Delta G'$ values do not correspond to the $\mathscr{E}$ and $\Delta G$ values for actual cellular processes. Until the concentrations of reactants and products at the active sites of enzymes can be measured, however, standard values are our best available estimates.

This area of biochemical knowledge is in an active state of research, and modifications of Fig. 5.7 may be anticipated. Additional details of the composition of the complexes and their mode of action for incorporating inorganic phosphate into ATP will be forthcoming. The principles by which biochemical oxidation-reduction operates, however, are the same as those for simple electrochemical cells. The complex nature of the biochemical half-cells should not obscure this basic fact.

### 5.8  Electrochemical Potentials and Ion Movement Through Membranes

Living membranes transport ions and neutral solutes in a ceaseless shuttling which involves simple diffusion, carrier-aided diffusion, and pumping machinery fueled by ATP. If a micro salt bridge is inserted into the aqueous solution on each side of a cell membrane and these salt bridges are connected to reference electrodes, a small electrical potential called the membrane potential can be detected. Ignoring the contributions from tiny liquid junction potentials and from those potentials that might occur in the unstirred layers of solution close to the membrane faces, let us examine the principal phenomenon itself, the potential that arises because of differences in ion distribution and in activities of solutes.

Solutes may be transported actively or passively. Active transport is defined as energy-linked uphill movement of solutes and the student wishing to pursue this subject may refer to Stein listed in the supplemental readings. Passive transport depends on diffusion, and we will limit ourselves to this topic.

If one of the properties of a system is changing over a finite distance, $x$, its rate of change with distance, $d(\text{property})/dx$, is called the gradient.

Thus, two examples are $dC/dx$ — a concentration gradient — and $d\mathscr{E}/dx$ — the voltage gradient (also called the electrical field strength). If we have a cell containing glucose at activity $a_i$ (inner) bathed in an outer solution containing glucose at an activity $a_o$, we already know from Eq. 3.41 that the free energy difference owing to the distribution of glucose is

$$\mu_{outer} - \mu_{inner} = \Delta\bar{G} = RT \ln\frac{a_o}{a_i} \cong RT \ln\frac{C_o}{C_i}$$

where $C_o$ and $C_i$ are outer and inner solute concentrations respectively. The gradient in this case is a *chemical potential gradient, $d\mu/dx$*, where $\mu$ is the chemical potential (see Section 3.1). If the glucose can eventually distribute itself uniformly across the membrane, $d\mu/dx$ and $\Delta G$ will fall to zero. This treatment and the following one assume that no solute is dragged along by migrating water molecules. A rigorous derivation must include this phenomenon also.

On the other hand, if the solute is an ion, *e.g.*, $Na^+$, unequally distributed across a membrane, two potential gradients are involved: the *chemical potential gradient, $d\mu/dx$*, dependent on the concentration distribution and the *electrical potential gradient, $n\mathscr{F}d\mathscr{E}/dx$*, dependent on the charge distribution. These forces acting on the sodium ion create the combined effect that is called the *electrochemical potential gradient, $dU/dt$*, where $U$ is the electrochemical potential of the ion.[*] Thus, the total force acting on the ion is:

$$\frac{dU}{dx} = \frac{d\mu}{dx} + \frac{n\mathscr{F}d\mathscr{E}}{dx}.$$ (5.6)

When the sodium ion attains passive equilibrium across the membrane (no active transport involved),

$$U_{Na^+\,outer} = U_{Na^+\,inner}.$$ (5.7)

Substituting from Eq. 5.6,

$$\mu_o + n\mathscr{F}\mathscr{E}_o = \mu_i + n\mathscr{F}\mathscr{E}_i.$$ (5.8)

From Eq. 3.41,

$$RT \ln a_o + n\mathscr{F}\mathscr{E}_o = RT \ln a_i + n\mathscr{F}\mathscr{E}_i.$$ (5.9)

To yield,

$$\mathscr{E}_i - \mathscr{E}_o = \mathscr{E} = \frac{RT}{n\mathscr{F}} \ln\frac{a_o}{a_i}$$ (5.10)

---

[*]The symbol $\bar{\mu}$ or $\bar{a}$ are commonly used for the electrochemical potential, but are avoided here because of the general convention of using the superior bar to denote partial molal quantities.

where $\mathscr{E}_i - \mathscr{E}_o$ is the electrical potential difference across the membrane at equilibrium. Equation 5.10 is recognizable as the Nernst Equation for the sodium ion, now in passive equilibrium across the membrane. In the dilute solutions present in living cells it is a safe approximation to substitute concentrations for activities in Eq. 5.10.

The type of equilibrium involved here, like any chemical equilibrium, is a dynamic or flux equilibrium in which the reaction is preceeding in both directions simultaneously without changes in the net concentrations of the participating species. Unlike the concentration cell, however, which reaches equilibrium only when the concentrations of solutes are the same in each half-cell, the passive equilibrium state just described may involve solutes at different concentrations across the membrane, as long as the difference in chemical activity is counterbalanced by a difference in electrical potential. According to Adrian,[*] $K^+$ and $Cl^-$ are in a state of passive equilibrium in frog sartorious muscle fibres. If the ion accumulated by the cell can be shown to be at a higher electrochemical potential than its counterpart in the outer fluid, it is assumed that work is being done on the ion and that it is being actively transported. Passive equilibria of this nature have been observed in living cells.

For the ion to be in a state of passive equilibrium, one well may wish to know the cause of the asymmetric distribution that leads to an equilibrium different from that in the concentration cell. One possibility is the presence of nondiffusible ions inside the cell, bringing about a Donnan equilibrium distribution (see page 129), and giving rise to a "Donnan potential" which can be calculated from the Nernst Equation. For the more interesting cases of passive equilibrium observed in animal cells, however, Donnan phenomena probably play a minor role. Ion pumps probably are the major cause of membrane potentials in animal cells. In many of the cells studied, sodium ion is pumped out of the cell by an energy-linked active transport system and is passively transported inward as a result of the gradient thus established.

A criterion for distinguishing passive transport from other types was developed by Ussing a number of years ago.[†] His essential ideas are discussed in the paragraphs that follow.

The flow of a penetrating species across a membrane is called a flux and is denoted by the symbol $J$. After a steady state has been established, the flux of the ion in question across an area perpendicular to the direction of diffusion is

$$J = -\frac{A}{f} \cdot C \cdot dU/dx \qquad (5.11)$$

---

[*]Adrian, R. H., *J. Physiol.* (London) **151**:154, 1960.
[†]Ussing, H. H., *Acta Physiol. Scand.*, **19**: 43, 1949.

where $A$ is the area, $f$ is the molar frictional coefficient for the ion, $C$ is the concentration and $dU/dx$ is the electrochemical potential gradient for the ion, or the total force acting on a mole of ions. In parallel with Eq. 3.41 the electrochemical potential, $U$, is related to the electrochemical activity of the ion, $a_{ec}$,

$$U = RT \ln a_{ec} \tag{5.12}$$

so that we may write

$$J = -\frac{ART}{f} \cdot C \cdot d \ln a_{ec}/dx. \tag{5.13}$$

We would now like to eliminate the $C$ term from Eq. 5.13 and can do so by defining the electrochemical activity, $a_{ec}$, in a way that follows directly from Eqs. 5.12 and 5.7–5.9:

$$RT \ln a_{ec} = RT \ln a + n\mathscr{F}\mathscr{E}. \tag{5.14}$$

Since the chemical activity of an ion is related to its concentration by the familiar equation (Eq. 4.4)

$$a = \gamma C$$

it follows that

$$RT \ln a_{ec} = RT \ln C + RT \ln \gamma + n\mathscr{F}\mathscr{E} \tag{5.15}$$

From 5.15 we obtain

$$C = \frac{a_{ec}}{\gamma} \cdot \exp\left[\frac{-n\mathscr{F}}{RT} \cdot \mathscr{E}\right]. \tag{5.16}$$

Combining 5.13 and 5.16 yields

$$J = -\frac{ART}{f} \cdot \frac{a_{ec}}{\gamma \exp\left[\dfrac{n\mathscr{F}}{RT} \cdot \mathscr{E}\right]} \cdot \frac{d \ln a_{ec}}{dx} \tag{5.17}$$

or

$$J = -\frac{ART}{f\gamma \exp\left[\dfrac{n\mathscr{F}}{RT} \cdot \mathscr{E}\right]} \cdot \frac{da_{ec}}{dx}. \tag{5.18}$$

Equation 5.18 contains too many unknowns to permit us to solve it. In addition, many of the terms may vary with $x$. However, since for identical ions, $A$, $f$, $\gamma$, and $\mathscr{E}$ are the same for a given value of $x$, we can use Eq. 5.18

to obtain a meaningful relationship about flux ratios. If $J_{i \to o}$ is the flux from inner to outer solution and $J_{o \to i}$ is the flux in the opposite direction, then

$$\frac{J_{i \to o}}{J_{o \to i}} = \frac{\dfrac{da_{ec(i)}}{dx}}{\dfrac{da_{ec(o)}}{dx}} \tag{5.19}$$

where $a_{ec(i)}$ and $a_{ec(o)}$ are the electrochemical activities of the inner and outer solutions respectively. The fluxes can be measured experimentally by tracer experiments. The terms on the right-hand side can be evaluated if we first integrate. If we assume that all the ions originating from the outer solution are isotopically distinguishable from the ions in the inner solution, we may impose the following boundary conditions:

for $x = 0$: $\quad a'_{ec(i)} = a_{ec(i)} \quad$ and $\quad a_{ec(o)} = 0$

for $x = x$: $\quad a_{ec(i)} = 0 \quad$ and $\quad a_{ec(o)} = a'_{ec(o)}$

where the primed terms denote a constant value of a variable in one of the solutions in contact with the membrane. Integrating numerator and denominator on the right-hand side of Eq. 5.19 between the limits imposed by the boundary conditions yields

$$\frac{J_{i \to o}}{J_{o \to i}} = \frac{a'_{ec(i)}}{a'_{ec(o)}}. \tag{5.20}$$

Substituting from 5.16,

$$\frac{J_{i \to o}}{J_{o \to i}} = \frac{\gamma'_i}{\gamma'_o} \cdot \frac{C'_i}{C'_o} \exp \frac{n\mathscr{F}}{RT} \cdot (\mathscr{E}_i - \mathscr{E}_o). \tag{5.21}$$

All of these quantities can be measured so that both sides of the equation can be evaluated numerically and thus provide a fundamental test for passive transport. When $J_{i \to o} = J_{o \to i}$, there is flux equilibrium and Eq. 5.21 reduces to the Nernst Equation. Deviations from Eq. 5.21 are not necessarily evidence for active transport, since other phenomena may be involved. Such deviations and other related topics have been presented in an excellent view by Dainty.* Special emphasis has been given to plant cells in a recent book by Nobel.† Ussing (loc. cit.) has also derived equations for the case involving solvent drag of solute and further points out that a full set of similar expressions are readily obtained for nonelectrolytes by omitting the term for the electrical potential, $\exp [n\mathscr{F}\Delta\mathscr{E}/RT]$.

---

*Dainty, J., *Ann. Rev. Plant Physiol.* 13: 379, 1959.
†Nobel, Park S., *Plant Cell Physiology: A Physiochemical Approach*, Chapt. 3, Freeman, 1970.

**Suggested Additional Reading**

### General: Electrochemical Cells

Daniels, F., and R. A. Alberty, *Physical Chemistry*, 3rd Ed. John Wiley and Sons, New York, 1966.

Moore, W. J., *Physical Chemistry*, 4th Ed. Prentice-Hall, Englewood Cliffs, N.J., 1972.

### Biochemical Applications

Green, D. E., and H. Baum, *Energy and the Mitochondrion*, Academic Press, New York, 1970.

Dowben, R. M., ed., *Biological Membranes*, Little, Brown and Company, Boston, 1969.

Edsall, J. T., and J. Wyman, *Biophysical Chemistry*, Vol. I. Academic Press, New York, 1958.

Lehninger, A. L., *Bioenergetics*, W. A. Benjamin, New York, 1965.

Neilands, J. B., and P. K. Stumpf, *Outlines of Enzyme Chemistry*, 2nd Ed. John Wiley and Sons, New York, 1958.

Stein, W. D., *The Movement of Molecules Across Cell Membranes*, Academic Press, New York, 1967.

### pH Measurement

Bates, R. G., *Determination of pH*. John Wiley, New York, 1964.

**Problems**

**5.1** (a) Explain why the $\mathscr{E}^\circ$ of a concentration cell is equal to zero.

(b) Explain why the potential of the calomel half-cell is dependent upon the concentration of chloride ion alone:

$$2Hg + 2Cl^- = Hg_2Cl_2 + 2e^-.$$

(c) Briefly describe how pH is determined today: the choice of electrodes used, and the reasons for choosing them.

**5.2** Calculate $\mathscr{E}^\circ$ at 25°C for the electrode Fe, $Fe^{3+}$ from the standard electrode potentials for the electrodes Fe, $Fe^{2+}$ and Pt, $Fe^{2+}$, $Fe^{3+}$.

**5.3** The potential of the cell

$$\text{Pt, buffer, quinhydrone} \parallel \text{normal calomel electrode}$$

is 0.0042 volt at 25°C. What is the pH of the buffer? The first buffer is replaced by a second buffer. The polarity of the cell is thereby reversed and a potential of 0.2175 volt is observed. What is the pH of the second buffer?

*Answer.* 7.17; 3.42.

**5.4** If a cell operates spontaneously,
(a) on which side does oxidation occur?
(b) which is the negative electrode (outside the cell)?
(c) is the sign of the potential positive or negative?
(d) is the sign of the free-energy change positive or negative?

**5.5** In biological oxidations the following reaction occurs:

$$DPNH + H^+ + FP \rightarrow FPH_2 + DPN^+.$$

The reduction potentials corrected to pH 7 are $-0.32$ volt for the $DPNH/DPN^+$ couple and $-0.060$ for the $FPH_2/FP$ couple.
(a) What is the equilibrium constant for the above reaction at 25°C?
(b) What is $\Delta G'$ for the reaction and what use is made of this available energy?

*Answer.*
(a) $6.1 \times 10^{15}$;
(b) $-12$ kcal; in mitochondria it is used for ATP synthesis.

**5.6** What is the activity of ferrous ion in a solution at 25°C that originally had an activity of 0.01 of ferric ion, after enough cuprous ion has been added so that the activity of cuprous ion in the solution is equal to 0.0005?

**5.7** Table 3.5 gives $\Delta G'$ for the reaction

$$Pyruvate^- + NADH + H^+ \rightarrow Lactate^- + NAD^+$$

Confirm this result using the data in Table 5.2.

**5.8** Given the galvanic cell at 25°C

$$Zn, Zn^{2+}_{(1.0 M)} \parallel Pb^{2+}_{(1.0 M)}, Pb,$$

calculate the voltage, free energy change, and equilibrium constant.
*Answer.* 0.637 volt; $-2.93 \times 10^4$ cal; $3.3 \times 10^{21}$.

**5.9** Consider the cell Zn, $Zn^{2+} \parallel I_2, I^-$, Pt at 25°C
(a) Write the reaction in the direction in which it occurs under standard conditions.
(b) What is $\mathscr{E}°$ for the reaction in (a)?
(c) What is the $\Delta G°$ for the reaction in (a)?
(d) What is the equilibrium constant for the reaction in (a)?
(e) Describe two methods by which you could make the reaction occur in the opposite direction to your answer to part (a).

**5.10** Derive the Nernst Equation using thermodynamic functions and the following relationship between electrical work and potential:

$$w_{net} = -n\mathscr{E}\mathscr{F}$$

**5.11** If a hydrogen electrode is used to measure the pH of a solution, and the hydrogen gas pressure fluctuates 0.2 atm, what percentage error could occur in the calculated pH values, if the $H_2$ pressure is assumed to be constant at 1 atm?

**5.12** Given the cell

$$Ag, Ag^+_{(0.01\,M)} \parallel Ag^+_{(0.1\,M)}, Ag$$

at 25°C, calculate the voltage and free energy change if one mole of electrons is transferred.

*Answer.* +0.059 volt; −1.36 kcal.

**5.13** Calculate the voltage and write the chemical reaction for the cell

$$Pt, H_{2(1\,atm)}, H^+_{(0.1\,M)} \parallel KCl_{(satd)}, Hg_2Cl_2, Hg$$

at 25°C.

*Answer.* 0.3036 volt; $Hg_2Cl_{2(s)} + H_{2(g)} \rightleftarrows 2Hg_{(l)} + 2Cl^- + 2H^+$.

**5.14** For the cell in Problem 5.10 find an expression that will relate the pH to the voltage. Then calculate the voltage for a solution of pH 3.0.

**5.15** For a solution at 25°C containing $Fe^{2+}$, $Fe^{3+}$, $Sn^{2+}$, and $Sn^{4+}$, calculate $K$ relating these ionic species:

$$K = \frac{[Sn^{4+}][Fe^{2+}]^2}{[Sn^{2+}][Fe^{3+}]^2}.$$

*Answer.* $10^{21}$.

**5.16** Calculate the voltage, free energy change for $n = 2$, and write the cell reaction for the cell

$$Cu, Cu^{2+}_{(0.1\,M)} \parallel Cl^-_{(0.2\,M)}, Cl_{2(0.5\,atm)}, Pt$$

at 25°C.

**5.17** Table 5.2 gives $\mathscr{E}'$ values for the following reactions:

$$CH_3CHO + [2H] \xrightarrow{\text{alcohol dehydrogenase}} CH_3CH_2OH;$$

$$CH_3COCOO^- + [2H] \xrightarrow{\text{lactate dehydrogenase}} CH_3CHOHCOO^-.$$

If these reactions were coupled in an electrochemical cell to give a positive voltage,
(a) what would be the chemical equation for the spontaneous reaction?

(b) how many joules could be produced from 25 moles of each reactant at 25°C?

(c) how should initial concentrations be changed (from standard states) to give a larger voltage?

*Note*: In the above equations the brackets are used to denote hydrogen from a donor molecule rather than molecular hydrogen or hydrogen ions.

**5.18** Is an aqueous solution of NADH at 25° and pH 7.0 thermodynamically stable if it is exposed to the atmosphere?

**5.19** The oxidation potential of the standard mercury-mercurous ion couple is $-0.789$ volt. Using this value and the standard potential of the normal calomel electrode, calculate the solubility product of $Hg_2Cl_2$ at 25°C.

*Answer.* $6.3 \times 10^{-18}$

**5.20** Given the standard half-cell potential

$$H_2S_{(1\,M)} \rightleftharpoons S_{(S)} + 2H^+ + 2e^- \qquad \mathscr{E}° = -0.141 \text{ volt},$$

calculate the free energy of formation of hydrogen sulfide corresponding to the equation

$$H_{2(g)} + S_{(s)} \rightarrow H_2S_{(1\,M)}.$$

*Answer.* $-6.5$ kcal.

# 6 | KINETICS

The key to a knowledge of enzymes is the study of reaction velocities, not of equilibria.

*J. B. S. Haldane*

In the previous chapters attention has been centered on equilibrium processes, and especially on the use of $K_{equil}$ as it relates to acid-base equilibria, electrolytic dissociation, various thermodynamic parameters (particularly $\Delta G$ and $\Delta H$), and the reduction potential of cells. Now we are ready to make an excursion into a different aspect of the chemical reaction: the special province of reaction rates and mechanisms known as *kinetics.*

The foundation of the theory of kinetics is the fundamental pronouncement of Guldberg and Waage that the *rate* of a chemical reaction is proportional to the active masses of the reactants. Application of this principle leads to general rate expressions in which each active mass or activity (and for our purposes, concentration) is raised to the power of its coefficient in the corresponding stoichiometric equation. For the simplest example,

$$A \rightarrow \text{Products.}$$

The rate of decomposition of $A \propto [A]$, and when

$$nA + mB \rightarrow \text{Products,}$$

$$\text{Rate} \propto [A]^n[B]^m.$$

The use of a single arrow in these equations shows that we are considering the rate in that direction only, otherwise we would have to specify "forward rate" or "reverse rate," as the case might be. If calculations are to be made (the inevitable outcome of these discussions), the proportionality sign must be replaced by the equals sign, an operation which is permitted by introducing a proportionality constant, $k$. We now write for the equation of the single reactant

$$\text{Rate}_{\text{forward}} = k[A],$$

and for the two reactants,

$$\text{Rate}_{\text{forward}} = k[A]^n[B]^m,$$

the two $k$'s not being identical. These proportionality constants are called *reaction rate constants* or *velocity constants* and are usually assigned an identifying subscript. For example, if we are considering two consecutive reactions and need to identify the rate constants for all forward and reverse steps, we label the constants and write them over the appropriate arrows:

$$A + B \underset{k_{-1}}{\overset{k_1}{\rightleftarrows}} C; \qquad C \underset{k_{-2}}{\overset{k_2}{\rightleftarrows}} D.$$

Some kineticists prefer to number the rate constants with consecutive positive numbers and would use $k_1$, $k_2$, $k_3$, and $k_4$ in the place of $k_1$, $k_{-1}$, $k_2$, and $k_{-2}$. Another system is to use $k_{1-2}$, $k_{2-1}$, $k_{3-4}$, and $k_{4-3}$. The system illustrated in the equations, however, has many advantages and has been selected out of personal preference.

### 6.1 Writing the Rate Equation

Before becoming involved in a consideration of the various reaction types and their appropriate and most useful rate expressions, the student should acquire some basic technique in translating the information stated by the *chemical* equation (or equations) into a *rate* equation. The process is simple. Consider the reaction

$$A \overset{k_1}{\longrightarrow} B.$$

for which, as we have seen, the rate of the reaction is equal to $k_1[A]$. Since the reaction velocity is the time rate-of-change of (increase in) the concentration of $B$, $d[B]/dt$, we write

$$\text{Rate} = d[B]/dt = k_1[A].$$

The rate of formation of $B$ is the rate of disappearance of $A$, or

$$d[B]/dt = -d[A]/dt$$

where the minus sign expresses the fact that $[A]$ is decreasing with time and that it is the rate of disappearance of $A$ which is being described.

If we wish to allow for the possibility that $B$ may react to give $A$, we can write the reaction as

$$A \underset{k_{-1}}{\overset{k_1}{\rightleftarrows}} B.$$

The rate of the forward reaction, *i.e.*, of the reaction of $A$ to yield $B$, is still $K_1[A]$. The rate of the reverse reaction is $k_{-1}[B]$. However, the rate of change of the concentration of $B$ with time is no longer $k_1[A]$, as it was in the preceding case. The reason is that $k_1[A]$ only accounts for the rate at which $B$ is formed from $A$. In order to know how the concentration of $B$ varies with time, we need to know not only how fast it is being formed from $A$ but also how rapidly it is being used up to generate $A$. Assuming that $A$ goes directly to $B$ without the formation of intermediates, we can take account of both effects by writing

$$\frac{d[B]}{dt} = \text{rate of formation of } B - \text{rate of utilization of } B$$

$$= k_1[A] - k_{-1}[B].$$

Similarly, we find for the rate of change in the concentration of $A$:

$$\frac{d[A]}{dt} = \text{rate of formation of } A - \text{rate of utilization of } A$$

$$= k_{-1}[B] - k_1[A]$$

The relationship

$$\frac{d[B]}{dt} = -\frac{d[A]}{dt}$$

is still valid. At equilibrium the forward and reverse rates are equal, and therefore for the equilibrium situation

Rate of disappearance of $A$ = Rate of formation of $A$,

$$k_1[A] = k_{-1}[B],$$

$$\frac{k_1}{k_{-1}} = \frac{[B]}{[A]} = K_{\text{equil}}.$$

(We seem unable to avoid $K_{\text{equil}}$ altogether.) Proceeding with the development of this notation, we can also write for the equilibrium case

(Rate of disappearance of $A$) − (Rate of formation of $A$) = 0,

which is usually expressed mathematically as

$$\frac{d[A]}{dt} = 0.$$

For the condition of equilibrium, we translate the differential equation as the change in the concentration of $A$ with time is zero.

The ideas of formation and disappearance can be applied to several reactions proceeding sequentially. Suppose that the reaction

$$A \underset{k_{-1}}{\overset{k_1}{\rightleftarrows}} B,$$

is immediately followed by a second reaction, $B \xrightarrow{k_2} C$, which proceeds at such a rate that the concentration of $B$ does not change with time, *i.e.*, its rate of formation is equal to its rate of disappearance. This situation is not the result of an equilibrium because neither of the reactions is at equilibrium. The situation is called a "steady state." As a molecule of $B$ is transformed into $C$, another molecule of $B$ takes its place, thanks to the rate of decomposition of $A$, and the concentration of $B$ remains "steady" or constant. For such a steady state, we write

Rate of disappearance of $B$ = Rate of formation of $B$,

and

$$d[B]/dt = 0.$$

Supplying the rate terms which apply to the formation and disappearance of $B$ gives

Rate of formation of $B = k_1[A]$;

Rate of disappearance of $B = k_{-1}[B] + k_2[B]$.

The steady state equation for $B$ is

$$d[B]/dt = k_1[A] - (k_{-1} + k_2)[B] = 0.$$

If the reader is willing to work through one additional and more complicated reaction, the technique of writing rate expressions should be within his grasp. In the following equations, several products and reactants appear:

$$A + B \underset{k_{-1}}{\overset{k_1}{\rightleftarrows}} C; \quad C + D \underset{k_{-2}}{\overset{k_2}{\rightleftarrows}} E; \quad E \xrightarrow{k_3} C + F.$$

We describe the sequence in the following way: $A$ and $B$ react to produce an intermediate compound, $C$; $C$ reacts with $D$ to form a new intermediate, $E$; finally, $E$ decomposes to regenerate $C$ and to form the final product, $F$. The last reaction is assumed to be irreversible, so only one arrow is shown. Since $C$ appears in all three equations let us write a steady state equation for $C$ and assume that, once the reaction reaches steady state, the concentration of $C$ will remain constant.

$$d[C]/dt = 0 = k_1[A][B] + (k_{-2} + k_3)[E] - k_{-1}[C] - k_2[C][D].$$

$$\underbrace{\phantom{k_1[A][B] + (k_{-2} + k_3)[E]}}_{\text{formation of } C} \qquad \underbrace{\phantom{k_{-1}[C] - k_2[C][D]}}_{\text{disappearance of } C}$$

The basic technique is to find all the arrows pointing toward $C$ and to combine the rate constant and *reactant* (*s*) for each arrow into a rate term, so that there is a term for each arrow. These will be the *formation* terms. The same process is applied to the arrows leading away from $C$ to obtain the *disappearance* terms. Then, the sum of the disappearance terms is subtracted from the sum of the formation terms and equated to zero for the steady state condition. Sections 6.6 and 6.7 will show how the steady state equation is used to yield valuable kinetic information.

## 6.2  Reaction Order and Molecularity: Definitions

In kinetics, reactions are described as being of first-, second-, third-, and zero-order, or even of fractional-order. *The order of a reaction is a number equal to the sum of the exponents of the concentration terms in the rate equation.* To illustrate: If

$$\text{Rate} = k[A],$$

the exponent of $[A]$ is unity, and therefore the reaction is first-order. If

$$\text{Rate} = k[A]^2,$$

the exponent of $[A] = 2$, and therefore the reaction is second-order. If

$$\text{Rate} = k[A][B],$$

the sum of the exponents $= 2$, and the reaction is second-order. Finally, if

$$\text{Rate} = k[A]^2[B][C]^3,$$

the reaction, which is an improbable one, is sixth-order. The last equation can also be described as being first-order in $B$, second-order in $A$, and third-order in $C$. If no concentration term appears in a rate equation, *i.e.*, rate $= k$, the order is zero. Zero-order reactions proceed at constant velocity and are independent of the concentration of the reactant. For example, in a catalyzed reaction where the concentration of reactant so greatly exceeds the concentration of catalyst that the catalyst is at all times fully "saturated" with the reactant, the reaction will be proceeding at maximum velocity and will not be influenced by increasing the concentration of the reactant.

The molecularity of a reaction refers to the reaction mechanism and describes the number of molecules reacting in an elementary reaction process. If the reaction $A \rightarrow B$ means that one molecule of $A$ decomposes in a simple process, it would be described as unimolecular. If two molecules must react for the product to be formed, as for the reaction, $H_2 + I_2 \rightarrow 2HI$, the reaction is properly described as bimolecular. In these two examples the orders of the reaction would parallel the molecularity, *i.e.*, the unimolecular reaction would be first-order and the bimolecular reaction would be second-order. However, in many reactions the kinetic order, as determined experimentally, does not parallel the molecularity. In such instances we speak of the *apparent* order of the reaction.

A careful kinetic study must be made before conclusions are drawn. With complex reactions (Section 6.7), a reaction of a certain order may consist of several steps of different molecularity. Although the reaction

$$2NO + O_2 \rightarrow 2NO_2$$

proves to be third-order kinetically, the reaction is not termolecular. That is, the reaction mechanism does not consist of the simultaneous collision of two NO molecules with an $O_2$ molecule. Statistically, such collisions would be highly improbable. The mechanism is believed to consist of two bimolecular steps:

$$2NO \underset{k_{-1}}{\overset{k_1}{\rightleftarrows}} N_2O_2;$$

$$N_2O_2 + O_2 \overset{k_2}{\rightarrow} 2NO_2.$$

### 6.3 Several Forms of the First-order Rate Equation

All of the rate equations written thus far are of the *differential* type because the expression $d[A]$ is known mathematically as a differential. (The idea expressed by the differential is that $d[A]$ represents a change in $[A]$, just as $dt$ represents a change in time.) These changes are too small to be measured experimentally, and therefore we will incur appreciable error if we try to use the differential form to calculate a rate constant.

The first-order rate equation,

$$\frac{-d[A]}{dt} = k[A],$$

may be rearranged to give

$$\frac{-d[A]}{[A]} = k\ dt. \tag{6.1}$$

Integration of Eq. 6.1 gives

$$-\ln [A] = kt + \text{constant.} \tag{6.2}$$

Useful application of this equation imposes limits, boundary conditions, on the values of the concentration of $A$ at time, $t$, and at time, $t = 0$, e.g.,

$$-\ln [A]\big|_{A_0}^{A} = kt\big|_{0}^{t} + \text{constant.}$$

We thus obtain the most useful form of the first-order rate equation:

$$\ln \frac{A_0}{A} = kt. \tag{6.3}$$

Converting to $\log_{10}$ gives

$$2.303 \log \frac{[A_0]}{[A]} = kt. \tag{6.4}$$

Often, kinetic data are obtained for the formation of a product rather than the disappearance of a reactant. If the reaction mechanism is known, and it is certain that the reaction is not complex, the rate constant may be conveniently calculated from such measurements. For example, in the thermal decarboxylation of an organic acid such as

$$HOOC\!-\!CH(NH_2)\!-\!CH(CH_3)\!-\!COOH \rightleftharpoons$$

$$HOOC\!-\!CH(NH_2)\!-\!CH_2\!-\!CH_3 + CO_2$$

the reaction can be followed by measuring the rate of $CO_2$ evolution. If we let the original concentration of dicarboxylic acid be $a$ and the amount of $CO_2$ produced at time $t$ be $x$, the dicarboxylic acid remaining at time $t$ would be equal to $(a - x)$. We can then write the integrated rate expression

$$2.303 \log \frac{a}{a - x} = kt. \tag{6.5}$$

This particular notation may be somewhat more convenient for certain reactions and the student may exercise his own preference in the choice of notation.

*Example 6.1* Benzene diazonium chloride undergoes first-order thermal decomposition in water at 50°C with a rate constant $(k)$ of 0.071 $min^{-1}$. If the initial concentration of the compound is 0.01 $M$, how long must the solution be heated at 50° to reduce the concentration to 0.001 $M$?

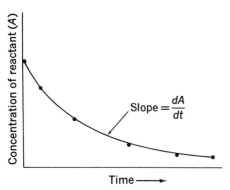

**Figure 6.1** Concentration versus time plot for a first-order reaction.

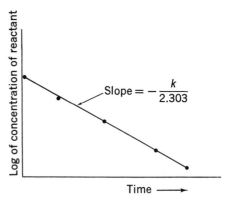

**Figure 6.2** Log concentration versus time plot for a first-order reaction.

The initial and final concentrations and the rate constant are inserted into Eq. 6.4:

$$2.303 \log \frac{A_0}{A} = kt;$$

$$2.303 \log \frac{0.01}{0.001} = 0.071(t);$$

$$t = 32.4 \text{ min.}$$

Note that the first-order rate equation contains the *ratio* of initial to final concentrations of the reactant. This means that percentages or any other measure of relative amounts of $A$ can be used as conveniently as actual concentrations. If this problem had asked how long it would take for the reaction to be 90 percent complete, the answer would be the same:

$$2.303 \log \frac{0.01}{0.001} = 0.071(t);$$

$$t = 32.4 \text{ min.}$$

This result shows that the time required for a first-order reaction to proceed to a certain stage, that is, 50 percent complete, 75 percent complete, etc., is independent of the initial concentration of the reactant.

The graphical relationships between concentration, log concentration, and time are shown in Figs. 6.1 and 6.2. In Fig. 6.1, the slope of the line is $dA/dt$. A plot of rate versus concentration should yield a straight line.

In order to characterize the speed of a reaction, it is convenient to define a quantity $t_{1/2}$ called the half-life. This quantity is also useful in estimating reaction order since there is a unique relationship between $t_{1/2}$ and the reaction rate constant. For first-, second-, and third-order (or higher) reactions, the time necessary for a reaction to be half-completed shows a characteristic dependency that differs from order to order, and therefore may be used to identify the order of a reaction. For the first-order rate expression

$$2.303 \log \frac{[A_0]}{[A]} = kt \qquad \text{(Eq. 6.4)}.$$

Let $t_{1/2}$ represent the time necessary for the reaction to be half-completed, at which time, $[A] = \frac{1}{2}[A_0]$. It follows that

$$2.303 \log \frac{[A_0]}{\frac{1}{2}[A_0]} = kt_{1/2}. \qquad (6.6)$$

The left-hand side of the equation reduces to 2.303 log 2 or 0.693, so

$$t_{1/2} = \frac{0.693}{k}, \qquad (6.7)$$

showing that the half-life of a first-order reaction depends only on the reaction rate constant and is independent of the initial concentration of the reactant. This will not be true of second- and third-order reactions.

*Example 6.2* Calculate the half-life of benzene diazonium chloride in aqueous solution at 50°C from the data given in Example 6.1.
From Eq. 6.7,

$$t_{1/2} = \frac{0.693}{k} = \frac{0.693}{0.071} = 9.76 \text{ min.}$$

The answer to Example 6.1 may be estimated by successive approximations. How many half-lives will be required to decompose 90 percent of the benzene diazonium chloride? The data may be tabulated for convenience:

| Percent of total Amount decomposed | Percent remaining | Total time required (min) |
|---|---|---|
| 50 | 50 | 9.76 |
| 75 | 25 | 2(9.76) = 19.5 |
| $87\frac{1}{2}$ | $12\frac{1}{2}$ | 3(9.76) = 29.3 |
| $93\frac{3}{4}$ | $6\frac{1}{4}$ | 4(9.76) = 39.0 |

The time required for 90 percent to decompose will be about half-way between 29.3 and 39.0 minutes, or about 34 minutes. The exact answer, as given in Example 6.1, is 32.4. The method of successive approximations is often useful in detecting calculation errors.

### 6.4 The Second-order Rate Equation

Second-order reactions are of two types: two molecules of the same substance react or one molecule each of two different substances react together. Examples are

$$2I_{(g)} \longrightarrow I_{2(g)}$$

$$CH_3COOC_2H_5 + Na^+OH^- \longrightarrow CH_3COO^-Na^+ + C_2H_5OH.$$

The first example yields a rate expression of simple form because there can be only one concentration of $I_{(g)}$, despite the fact that the reaction is second-order. The general form of the differential rate expression is obtained in the following way:

$$2A \longrightarrow Products.$$

$$Rate_{forward} = \frac{-d[A]}{dt} = k[A]^2.$$

This is rearranged to give

$$\frac{-d[A]}{[A]^2} = k\,dt, \tag{6.8}$$

which integrates to

$$\frac{1}{[A]} = kt + constant. \tag{6.9}$$

Integration of Eq. 6.8 between limits for initial values $(A_0, 0)$ and values at any time $(A, t)$ as for the first-order reaction, gives

$$\frac{1}{[A]} - \frac{1}{[A_0]} = kt. \tag{6.10}$$

In the notation using $a$ and $x$, Eq. 6.10 takes the form

$$\frac{1}{a-x} - \frac{1}{a} = \frac{x}{a(a-x)} = kt. \tag{6.11}$$

Equations 6.10 and 6.11 are the alternate forms from which $k$ may be calculated.

> *Example 6.3* The most familiar examples of second-order reactions are saponifications or esterifications. Hinshelwood and Hutchinson (*Proc. Roy. Soc.* 111A: 380, 1926), however, discovered that the thermal decomposition of acetaldehyde in the gas phase followed second-order kinetic law:
>
> $$2CH_3CHO \longrightarrow 2CH_4 + 2CO \quad (518°C).$$

A small sampling of their data is given below. The initial pressure of the system, which is proportional to the initial concentration of acetaldehyde, is 363 mm. The pressure increase due to the formation of CO (also the pressure increase due to the formation of $CH_4$) is proportional to the concentration of product formed.

| $t$ (sec) | Pressure increase due to formation of CO (mm) | $P_{acet}$ (mm) | $1/P_{acet}$ $(mm^{-1})$ |
|---|---|---|---|
| 0 | 0 | 363 | $2.76 \times 10^{-3}$ |
| 42 | 34 | 329 | $3.04 \times 10^{-3}$ |
| 105 | 74 | 289 | $3.46 \times 10^{-3}$ |
| 300 | 152 | 211 | $4.74 \times 10^{-3}$ |
| 450 | 189 | 174 | $5.75 \times 10^{-3}$ |
| 650 | 223 | 140 | $7.14 \times 10^{-3}$ |
| 840 | 244 | 119 | $8.40 \times 10^{-3}$ |

Prove that the reaction follows second-order kinetic law in two ways: (a) analytically, by calculating $k$, the reaction rate constant, and determining whether it is invariant with time and (b) graphically, by plotting $1/P_{acetone}$ vs. time to see whether the data give a linear relationship. Using Eq. 6.11, $a = 363$ and selected values in the second column above correspond to $x$.

$$\frac{x}{a(a-x)} = kt$$

$$\frac{34}{363(363-34)} = k(42); \qquad k = 6.79 \times 10^{-6} \text{ sec}^{-1} \text{ moles}^{-1} \text{ liter}$$

$$\frac{74}{363(363-74)} = k(105); \qquad k = 6.71 \times 10^{-6} \text{ sec}^{-1} \text{ moles}^{-1} \text{ liter}$$

$$\frac{244}{363(363-244)} = k(840); \qquad k = 6.72 \times 10^{-6} \text{ sec}^{-1} \text{ moles}^{-1} \text{ liter}$$

The values of $k$ are all sufficiently close to one another to justify describing the reaction kinetics as second-order.

A plot of the data is shown in Fig. 6.3. On inspection it may be seen that a satisfactory straight line relationship is obtained between $1/P_{acetone}$ and time, indicating that the reaction follows second-order kinetics.

The half-life expression is readily obtained from Eq. 6.11. At $t_{1/2}$, $x = \frac{1}{2}(a)$. Substituting in 6.11 gives

$$t_{1/2} = \frac{1}{ka}. \tag{6.12}$$

For the second-order reaction, the half-life definitely depends on the initial concentration of reactant and is inversely proportional to it.

*Example 6.4* Equation 6.12 can be used for a second-order reaction of the type $2A \rightarrow$ products, or of the type $A + B \rightarrow$ products, *provided the*

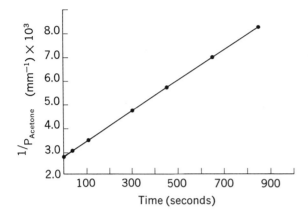

**Figure 6.3** Data of Example 6.3, a second-order reaction plotted as $\frac{1}{P}$ versus time.

concentrations of $A$ and $B$ are the same. Consider an example of the second type.

If the second-order constant for the saponification of ethyl acetate by sodium hydroxide at 30°C is 0.66 liters-mole$^{-1}$ min$^{-1}$, what will be the difference in the time required for the reaction to be 50 percent complete if the reaction is started with 0.5 $M$ ethyl acetate and 0.5 $M$ sodium hydroxide, or with 0.01 $M$ concentrations of each reactant?

Since $t_{1/2}$ is inversely proportional to the initial concentration of the reactants, it is immediately apparent that $t_{1/2}$ for the 0.01 $M$ system will be 50 times larger than $t_{1/2}$ for the 0.5 $M$ system.

The actual time required for the 0.5 $M$ reactants is

$$t_{1/2} = \frac{1}{0.66(0.5)} = 3 \text{ min.}$$

Now let us consider a second-order reaction in which two different molecules are reacting to form products:

$$A + B \longrightarrow \text{Products.}$$

$$\text{Rate} = \frac{-d[A]}{dt} = \frac{-d[B]}{dt} = k[A][B]. \tag{6.13}$$

If the reaction is carried out with equal concentrations of $A$ and $B$, the rate equation will simplify to the form shown in Eq. 6.9. If different concentrations of $A$ and $B$ are used, however, the integrated form of the second-order rate expression is more complex. Integration of 6.13 between limits gives

$$\frac{2.303}{[A_0] - [B_0]} \log \frac{[B_0][A]}{[A_0][B]} = kt, \tag{6.14}$$

where $[A_0]$ and $[B_0]$ represent the initial concentrations of $A$ and $B$ at time $= 0$, and $[A]$ and $[B]$ are concentrations at time $= t$. This form is

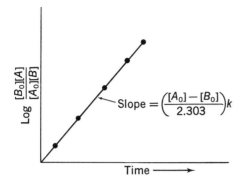

**Figure 6.4** A second-order reaction plotted as $\log \dfrac{[B_0][A]}{[A_0][B]}$ versus time.

not quite as cumbersome as it may first appear since the coefficient of the left-hand term is a constant and needs to be calculated only once. If $k$ is to be determined graphically, $\log ([B_0][A])/([A_0][B])$ is plotted against $t$. The slope is then equal to $k$ divided by $2.303/([A_0] - [B_0])$.

The form equivalent to Eq. 6.14 in the $x$ and $a$ notation is

$$\frac{2.303}{a - b} \log \frac{b(a - x)}{a(b - x)} = kt. \qquad (6.15)$$

## 6.5 The Third-order Rate Equation

Even though some chemical reactions do have rate equations of the third-order type, it should be stressed again that a truly termolecular reaction is believed to be unlikely. The equation

$$3A \rightarrow \text{Products},$$

is mechanistically improbable but is a convenient point from which to begin the derivation. For this equation,

$$\text{Rate}_{\text{forward}} = \frac{-d[A]}{dt} = k[A]^3. \qquad (6.16)$$

The alternate forms of the integrated expression corresponding to Eq. 6.16 are

$$\frac{1}{2}\left(\frac{1}{[A]^2} - \frac{1}{[A_0]^2}\right) = kt \qquad (6.17)$$

and

$$\frac{1}{2}\left(\frac{1}{(a - x)^2} - \frac{1}{a^2}\right) = kt. \qquad (6.18)$$

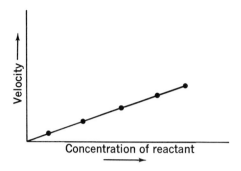

**Figure 6.5**   Velocity versus concentration plot for a first-order reaction.

If the reaction is of the type

$$2A + B \longrightarrow \text{Products},$$

where $[B]$ is not equal to $[A]$, the integration of the third-order rate expression is much more complicated. The student is referred to page 337 of the book by Moore that is listed in the *Suggested Additional Readings* on page 320.

### 6.6   The Zero-order Rate Equation

When no concentration terms appear in the rate equation, the sum of the exponents is necessarily zero and the rate is designated as zero-order:

$$\text{Rate}_{\text{forward}} = \frac{-d[A]}{dt} = k. \tag{6.19}$$

The zero-order reaction is easily recognized experimentally by measuring the reaction velocity at different concentrations of reactant. Whereas with a first-order reaction, velocity diminishes linearly with decreasing concentration of reactant, with the zero-order reaction, velocity remains constant. See Figs. 6.5 and 6.6. The zero-order reaction is of particular interest in catalyzed reactions where the amount of catalyst available may be rate-limiting. This situation is a familiar one in the realm of enzyme kinetics and will be discussed in Section 6.8.

> *Example 6.5*   A number of enzymatic reactions may be followed in the spectrophotometer because of the optical characteristics of the reaction components. In the deamination of β-methylaspartate by the enzyme β-methylaspartase, only the olefinic reaction product, mesaconate, absorbs light in the ultraviolet region of the spectrum. The reaction is therefore conveniently followed by measuring the increase in absorption at 240 mμ. At the end of the chapter the student will find several problems based on studies of this enzyme system conducted in the authors' laboratory.

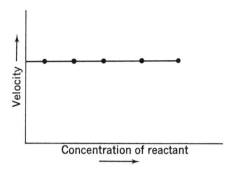

**Figure 6.6** Velocity versus concentration plot for a zero-order reaction.

During the standard experimental procedure with β-methylaspartase, the reaction becomes zero-order when the concentration of β-methylaspartate reaches 0.01 M. That is, increasing the concentration of reactant has no further effect on reaction rate. The data might look something like

| Concentration of L-β-methylaspartate (moles/L) | Reaction Rate ($\Delta A_{240}$/min) |
|---|---|
| 0.01 | 0.330 |
| 0.02 | 0.331 |
| 0.05 | 0.329 |
| 0.10 | 0.330 |

Plotting these data in the manner of Fig. 6.6 will give a horizontal line parallel to the abscissa, characteristic of a zero-order system.

## 6.7 Complex Reactions

Many chemical reactions, including all enzyme-catalyzed reactions, proceed through one or more intermediate steps, so that the reaction kinetics cannot be assumed to correspond to the overall stoichiometry as represented by the balanced equation giving initial reactants and final products. The experimenter becomes aware that he is dealing with a complex reaction when he fails to obtain a linear plot, or an unchanging value of the rate constant, with the rate expression which appears to be most appropriate to the reaction equation. This situation imposes upon the research chemist the necessity of working out the *mechanism* of the reaction, that is, the sequence of steps by which the reactants are transformed into the final products. The equations describing the mechanism will contain formulas for various transitory intermediates which are regarded as participating in the complex reaction. A well-known example is the hydrolysis of sucrose. The chemical equation is written

$$C_{12}H_{22}O_{11} + H_2O \xrightarrow{H^+} C_6H_{12}O_6 + C_6H_{12}O_6. \tag{6.20}$$

sucrose                    glucose        fructose

Hydrogen ion is required for the hydrolysis to occur; in its absence, sterile aqueous solutions of sucrose are stable for long periods of time. If the concentration of $H^+$ is followed throughout the time-course of any particular experiment, it is found to be unchanging and therefore its effect must be catalytic. However, the rate constant $k$ does change if the concentration of acid is varied from one experiment to the next. Moreover, if the rate of reaction is followed in a polarimeter or by other suitable means, the rate constant is found to be unchanging if log [sucrose] is plotted versus time, *i.e.*, the rate is first-order with respect to sucrose. To account for the role of hydrogen ion and the apparent first-order rate, the following mechanism has been proposed. First, the sucrose molecules are protonated by the acid catalyst:

$$C_{12}H_{22}O_{11} + H^+ \underset{k_{-1}}{\overset{k_1}{\rightleftarrows}} (C_{12}H_{22}O_{11})H^+. \tag{6.21}$$

This reaction is assumed to occur very rapidly, so rapidly in fact, that it is regarded as reaching equilibrium almost immediately. In truth, it is *not quite* an equilibrium because the next step in the sequence constantly removes the protonated sucrose molecules. This next step is the hydrolysis of the protonated sucrose molecules:

$$(C_{12}H_{22}O_{11})H^+ + H_2O \overset{k_2}{\rightarrow} C_6H_{12}O_6 + C_6H_{12}O_6 + H^+. \tag{6.22}$$

This final step occurs at a much slower rate than the protonation step and is written as being irreversible. The slowest step in any sequence is called the *rate-controlling* or *pace-maker* step, since it is the bottleneck in the mechanism that determines the experimentally measured rate. The reaction rate, then, will be equal to the rate of the second step:

$$\text{Rate}_{\text{forward}} = k_2[(C_{12}H_{22}O_{11})H^+][H_2O]. \tag{6.23}$$

Now we may proceed by either of two methods, the *equilibrium method* or the *steady state method*. The former has the advantage of simplicity, and the latter, the advantage of fewer limitations. The problem is to eliminate the term for the intermediate, $[(C_{12}H_{22}O_{11})H^+]$, from Eq. 6.23, since there is no way to measure its concentration.

Taking the equilibrium method first, we assume that the initial step (Eq. 6.21) comes to equilibrium.

$$K_{\text{equil}} = k_1/k_{-1} = \frac{[(C_{12}H_{22}O_{11})H^+]}{[C_{12}H_{22}O_{11}][H^+]}. \tag{6.24}$$

The concentration of the intermediate species is then

$$[(C_{12}H_{22}O_{11})H^+] = \frac{k_1}{k_{-1}}[C_{12}H_{22}O_{11}][H^+]. \tag{6.25}$$

All the terms on the right-hand side of Eq. 6.25 can be measured. All that remains to be done is to substitute Eq. 6.25 into Eq. 6.23, eliminating the $[(C_{12}H_{22}O_{11})H^+]$ term. This leaves

$$\text{Rate}_{\text{forward}} = \frac{k_2 k_1}{k_{-1}} [C_{12}H_{22}O_{11}][H^+][H_2O]. \qquad (6.26)$$

We appear to be confronted with a paradox at this point. The experimentally determined rate follows first-order kinetics but the derived rate expression is third-order. However, in any individual experiment $[H^+]$ is constant. Because it is present in such excess, $[H_2O]$ is constant within limits of measurement. Therefore,

$$\text{Rate}_{\text{forward}} = k'[C_{12}H_{22}O_{11}], \qquad (6.27)$$

where $k'$ incorporates all the rate constants and other constant terms. The value of Eq. 6.26 lies in its ability to explain the observed effect of $[H^+]$ on the experimentally determined rate constant, when $[H^+]$ is varied from one experiment to the other. Since $[H^+]$ is incorporated into the overall rate constant, changing $[H^+]$ will change $k'$. A rate constant such as $k'$, which combines a number of individual constants, is more appropriately called an *apparent rate constant*.

Turning to the steady state method, we stipulate that the first step in the mechanism (Eq. 6.21) does not reach true equilibrium. However, the first step produces the intermediate species, $(C_{12}H_{22}O_{11})H^+$, initially more rapidly than it can be removed by the second step. Although the rate constants for the production and removal of $(C_{12}H_{22}O_{11})H^+$ are different as concentrations change, a constant or steady concentration of the intermediate is reached within a few seconds after the reaction begins. Reviewing Section 6.1, the reader will recall that writing the rate equation for the steady state has already been discussed. Since the concentration of the intermediate is unchanging with time,

$$\frac{d[(C_{12}H_{22}O_{11})H^+]}{dt} = 0.$$

In the next step all the *disappearance* terms are subtracted from the *formation* terms, and this result is set equal to zero:

$$\frac{d[(C_{12}H_{22}O_{11})H^+]}{dt} = 0 = k_1[C_{12}H_{22}O_{11}][H^+] - k_{-1}[(C_{12}H_{22}O_{11})H^+]$$

$$- k_2[(C_{12}H_{22}O_{11})H^+][H_2O]. \qquad (6.28)$$

Collecting terms gives

$$0 = k_1[C_{12}H_{22}O_{11}][H^+] - [C_{12}H_{22}O_{11})H^+](k_{-1} + k_2[H_2O]),$$

and solving for the concentration of the intermediate species,

$$[(C_{12}H_{22}O_{11})H^+] = \frac{k_1[C_{12}H_{22}O_{11}][H^+]}{k_{-1} + k_2[H_2O]}. \tag{6.29}$$

Comparison of Eq. 6.29 with Eq. 6.25 shows that the steady-state method gives an expression for the reactive intermediate species that contains an extra term, $k_2[H_2O]$. The steady-state method always gives more complicated expressions for the intermediate species because it considers all formation and disappearance reactions in defining the concentration of the reactive intermediate, whereas the equilibrium method limits itself to the initial equilibrium step. To solve the rate equation in terms of measurable quantities, we substitute Eq. 6.29 into Eq. 6.23 and obtain

$$\text{Rate}_{\text{forward}} = \frac{k_1k_2[C_{12}H_{22}O_{11}][H^+][H_2O]}{k_{-1} + k_2[H_2O]}. \tag{6.30}$$

This equation is identical with Eq. 6.26 except for the $k_2[H_2O]$ term. If $[H^+]$ is constant and $[H_2O]$ does not change, *i.e.*, $k_{-1} + k_2 [H_2O]$ is also constant, Eq. 6.30 reduces to

$$\text{Rate}_{\text{forward}} = k'[C_{12}H_{22}O_{11}]. \tag{6.27}$$

The difference between the two methods is that the steady state method reveals that the kinetic order with respect to water is complex. Helpful approximations can frequently be made in rate expressions containing two terms in the denominator, by considering the consequences of one term being much larger than the other. In Eq. 6.30, since the concentration of water is large, $k_2[H_2O]$ may be larger than $k_{-1}$. If so, as a useful approximation we can drop $k_{-1}$ from the sum and examine what remains. The resulting expression is

$$\text{Rate}_{\text{forward}} = \frac{k_1k_2[C_{12}H_{22}O_{11}][H^+][H_2O]}{k_2[H_2O]}. \tag{6.31}$$

which reduces to

$$\text{Rate}_{\text{forward}} = k_1[C_{12}H_{22}O_{11}][H^+]. \tag{6.32}$$

This is just another way of arriving at the same conclusion, *i.e.*, when water is in excess, the rate of reaction is independent of the concentration of water. Admittedly we already knew this fact, but the steady state approach has proved its worth in the realm of enzyme kinetics and leads to conclusions that are not so readily obvious.

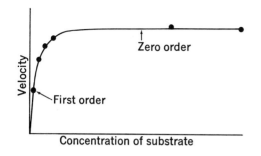

**Figure 6.7** Velocity versus concentration plot for an enzyme-catalyzed reaction. Curve shown is an actual data plot for the enzyme $\beta$-methylaspartase. [From Williams, V. R., and J. Selbin, *J. Biol. Chem.*, **239**:1636, 1964.]

## 6.8 Kinetics of Enzyme-catalyzed Reactions

**Rate equations**  The earliest studies on the effect of reactant concentration on the rate of enzymatic reactions revealed a fundamental characteristic of enzyme catalysis: the kinetics of such reactions is complex. At low concentrations of reactant (called *substrate* in the terminology of enzymology) the rate is first-order, but at high concentrations of substrate, the rate becomes independent of concentration and is, therefore, zero-order. Such biphasic kinetics yield velocity-concentration plots of the general appearance shown in Fig. 6.7.

One of the early investigators, Henri, proposed that the enzyme forms an intermediate compound or complex with the substrate. At high concentrations of substrate all the enzyme will be converted to the enzyme-substrate complex, and the reaction velocity will reach a maximum. In other words, it is the decomposition of the enzyme-substrate complex which is rate-controlling in the enzyme-catalyzed reaction. We may express this theory in the following way, allowing $E$ to represent the enzyme and $S$ to represent the reactant or substrate:

Formation of complex:

$$E + S \underset{k_{-1}}{\overset{k_1}{\rightleftharpoons}} ES \text{ (fast step);} \tag{6.33}$$

Decomposition of complex:

$$ES \overset{k_2}{\rightarrow} \text{Products} + E \text{ (slow step);} \tag{6.34}$$

$$\text{Rate}_{\text{forward}} = k_2[ES]. \tag{6.35}$$

In the light of the discussion of complex reactions (Section 6.7), the above equations should present a familiar pattern. We can pursue the solution of the rate equation either by the equilibrium method or by the steady state method. If there is experimental evidence to indicate that a near-equilibrium is established in the fast step, the equilibrium method is a valid solution. Historically, the equilibrium method is of special interest, because this was the original approach of Michaelis and Menten* to the problem of devising a mathematical treatment for the experimental data of the type indicated in Fig. 6.7. The problem is the same as encountered previously in complex reactions: we must find an expression for ES in terms of quantities which we can measure, so as to eliminate it from the reaction rate equation (6.35). The equilibrium constant for the first step is (6.33):

$$K = \frac{k_1}{k_{-1}} = \frac{[ES]}{[E][S]}. \tag{6.36}$$

Therefore,

$$[ES] = \frac{k_1}{k_{-1}}[E][S]. \tag{6.37}$$

The problem is not solved at this point because $[E]$ is the concentration of free enzyme, which we cannot measure experimentally. The following equation, however, describes the equilibrium between free and bound enzyme:

$$[E_{total}] = [E_{free}] + [ES]. \tag{6.38}$$

This equation is called the Enzyme Conservation Equation. Total enzyme can be measured. We can eliminate $[E]$, the variable which we cannot measure, from Eq. 6.37 by substituting $[E_t] - [ES]$ for it. The result is

$$[ES] = \frac{k_1}{k_{-1}}([E_t] - [ES])[S].$$

Solving for $[ES]$ we obtain

$$[ES] = \frac{[E_t][S]}{k_{-1}/k_1 + [S]}. \tag{6.39}$$

We can incorporate the two rate constants into a single $K$, which, in this case, is the reciprocal of the equilibrium constant for the reaction shown in Eq. 6.33. The new $K$ is actually the dissociation constant for $ES$, whereas the equilibrium constant which we wrote in Eq. 6.36 was the formation constant for $ES$. The new $K$, the quantity $k_{-1}/k_1$, is traditionally called the Michaelis constant and usually is written with an identifying subscript, *i.e.*, $K_m$. Introducing this notation into Eq. 6.39 gives

$$[ES] = \frac{[E_t][S]}{K_m + [S]}. \tag{6.40}$$

---

*Michaelis, L. and M. L. Menten, *Biochem. Z.*, **49**:333, 1913.

We now have an expression for $[ES]$ which we can substitute into the overall rate expression, Eq. 6.35. That is,

$$\text{Rate}_{\text{forward}} = k_2[ES] \qquad \text{(Eq. 6.35)};$$

$$\text{Rate}_{\text{forward}} = \frac{k_2[E_t][S]}{K_m + [S]} \tag{6.41}$$

One further simplification remains. When all of the enzyme is combined with substrate (at high levels of substrate), the reaction will be going at maximum velocity. Under these conditions, no free enzyme will remain and $[E_t] = [ES]$. The rate under these conditions is called the $V_{\text{max}}$ in the notation of enzymology. It is a constant, characteristic of the reaction under zero-order conditions. The mathematical relationship is

$$\text{Rate}_{\text{maximum}} = V_{\text{max}} = k_2[E_t]. \tag{6.42}$$

Inspection of Eq. 6.41 shows that the term $k_2[E_t]$ appears in the numerator. We now substitute $V_{\text{max}}$ in its place. Such a substitution is highly advantageous, because the maximum rate is easily measured experimentally for any fixed enzyme concentration, and the determination of $E_t$ and $k_2$ may be difficult. For certain enzymes it is now possible to use a pure preparation of known molecular weight and to calculate the molar concentration of $E_t$. However, for all impure preparations and for enzymes of unknown molecular weight, the use of the $V_{\text{max}}$ term in place of $k_2[E_t]$ is an essential simplification.

The final expression now is

$$\text{Rate}_{\text{forward}} = \frac{V_{\text{max}}[S]}{K_m + [S]} \qquad \text{where} \qquad K_m = \frac{k_{-1}}{k_1}. \tag{6.43}$$

Usually the notation for the left-hand term is $v_f$ or, simply, $v$:

$$v_f = \frac{V_{\text{max}}[S]}{K_m + [S]} \qquad \text{(Michaelis-Menten Equation)}.$$

If this equation is to serve as the mathematical analysis of the biphasic plot shown in Fig. 6.7, it must reduce to a first-order expression for low values of $S$ and to a zero-order expression for high values of $S$. Examining the equation reveals that $[S]$ can be eliminated from the denominator when $[S]$ is small in comparison to $K_m$. The equation reduces to

$$v_f = \frac{V_{\text{max}}[S]}{K_m} = k'[S] \qquad \text{(first-order)}. \tag{6.44}$$

At high levels of substrate, $S$ may be considered much larger than $K_m$. Eliminating $K_m$ from the denominator leaves

$$v_f = V_{\text{max}} = \text{constant} \qquad \text{(zero-order)}. \tag{6.45}$$

For many enzyme-catalyzed reactions the steady state approach is considered to be more realistic than the equilibrium method. Let us see what difference this will make in the final rate equation. The initial steps are the same as for the equilibrium method.

$$\text{(fast step)} \qquad E + S \underset{k_{-1}}{\overset{k_1}{\rightleftharpoons}} ES \qquad \text{(Eq. 6.33)};$$

$$\text{(slow step)} \qquad ES \overset{k_2}{\longrightarrow} \text{Products} + E \qquad \text{(Eq. 6.34)};$$

$$\text{Rate} = k_2[ES] \qquad \text{(Eq. 6.35)}.$$

When a steady state is reached, $[ES]$ will be constant and its rate of change will be zero:

$$d[ES]/dt = 0 = \underset{\text{(formation)}}{k_1[E][S]} - \underset{\text{(disappearance)}}{(k_{-1}[ES] + k_2[ES])},$$

$$[ES] = \frac{k_1[E][S]}{(k_{-1} + k_2)}. \qquad (6.46)$$

The difference between this expression and Eq. 6.37 lies in the constants: the equilibrium method gives a term $k_1/k_{-1}$ whereas the steady state method gives $k_1/(k_{-1} + k_2)$. The reciprocal of the latter will be a new $K_m$, a Michaelis constant for steady state conditions. All the remaining steps are the same as the simplifications and substitutions made in the first derivation (Eqs. 6.38 to 6.43). We write as the final form of the steady state rate equation:

$$\text{Rate}_{\text{forward}} = v_f = \frac{V_{\text{max}}[S]}{K_m + [S]} \qquad \text{where} \qquad K_m = \frac{k_{-1} + k_2}{k_1}. \qquad (6.47)$$

This rate equation is the one used by an enzyme kineticist, unless he has persuasive evidence that the initial fast reaction (Eq. 6.33) comes to a rapid equilibrium. It cannot come to *true* equilibrium, because the subsequent slow reaction is constantly removing $ES$.

Various forms of the Michaelis-Menten Equation have been devised because Eq. 6.47 is the equation of a hyperbola and the constant terms, $K_m$ and $V_{\text{max}}$, are not easily determined from the plot. The most successful of these alternate forms is the Lineweaver-Burk Equation (6.48), which permits the investigator to plot the data in a linear form, provided the system does not behave abnormally. If Eq. 6.47 is written

$$v = \frac{V_{\text{max}}[S]}{K_m + [S]},$$

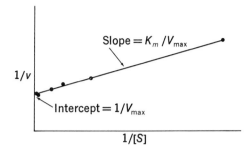

**Figure 6.8** Lineweaver-Burk plot for the data of Fig. 6.7. [From Williams, V. R., and J. Selbin, *J. Biol. Chem.*, **239**:1636, 1964.]

and then both sides of the equation are expressed as reciprocals,

$$\frac{1}{v} = \frac{K_m + [S]}{V_{max} \, [S]} \, .$$

The right-hand side can be separated into two terms:

$$\frac{1}{v} = \frac{K_m}{V_{max}} \cdot \frac{1}{[S]} + \frac{1}{V_{max}} . \tag{6.48}$$

The two variables, $v$ and $[S]$, have now been conveniently separated, and we have obtained the equation of a straight line in which the variables are $1/v$ and $1/[S]$, $K_m/V_{max}$ is the slope, and $1/V_{max}$ is the intercept. Many enzymatic reactions yield data which plot linearly in this form, commonly referred to as the *double reciprocal* form of the Lineweaver-Burk Equation. A typical set of data are shown in Fig. 6.8. With certain enzyme systems anomalies occur in the double reciprocal plots. For example, an enzyme may be inhibited by its substrate at high substrate concentrations, or, on the other hand, it may appear to be activated by excess substrate. Substrate saturation curves for enzymes displaying allosteric effects frequently are sigmoid in shape and often can be fitted to a form of Eq. 6.48 in which $[S]$ is replaced by $[S]^n$.

If the experimental anomalies occur at low substrate concentrations, these points will lie out to the right in the double reciprocal plot shown in Fig. 6.8, and may not interfere with the calculation of $K_m$ and $V_{max}$. If, however, departures from linearity are found at high substrate concentrations, these points will cluster near the intercept and pose problems in extrapolating the linear portions of the curve. An alternate form of the Lineweaver-Burk Equation may be used in such cases. If both sides of Eq. 6.48 are multiplies by $[S]$, we obtain

$$\frac{[S]}{v} = \frac{1}{V_{max}} \cdot [S] + \frac{K_m}{V_{max}} . \tag{6.49}$$

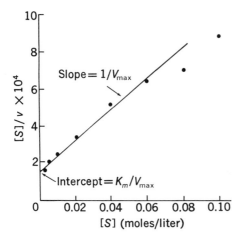

**Figure 6.9** $S/v$ versus $S$ plot for the aspartase reaction. $S$ is substrate concentration; $v$ is reaction velocity. [From Williams, V. R., and R. T. McIntyre, *J. Biol. Chem.*, **217**:467, 1955.]

Plotting $[S]/v$ versus $[S]$ gives a straight line of slope $1/V_{max}$ and intercept $K_m/V_{max}$. See Fig. 6.9. Any anomalies occurring at high values of $[S]$ will lie out to the right and may, at the discretion of the investigator, be ignored.

One additional alternate form will be mentioned, since it is gaining popularity with many investigators and has the advantage of magnifying departures from linearity which might be overlooked in the plots illustrated in Figs. 6.8 and 6.9. If Eq. 6.48 is multiplied by $V_{max}(v)$ and the equation is rearranged,

$$v = -K_m \left( \frac{v}{[S]} \right) + V_{max}. \tag{6.50}$$

An illustration of this plot may be seen in Fig. 6.10. Note that the intercept on the ordinate is $V_{max}$ and that the intercept on the abscissa is $V_{max}/K_m$. The slope of the line is $-K_m$.

Fitting the best straight line to a plot of experimentally determined points requires a large measure of empiricism and personal prejudice when it is done by eye. Many investigators, therefore, prefer to use the method of least squares in data handling. Bliss and James,[*] among others, have published an analysis of the statistical problems associated with fitting data to the Michaelis-Menten Equation. The calculations discussed by these authors may be carried out on a digital computer run on a Fortran IV program.[†]

**The significance of $K_m$ and $V_{max}$**   The Michaelis constant, $K_m$, is routinely determined in characterizing a particular enzyme because it is a reproducible constant, independent of enzyme concentration. It has a

---

[*]Bliss, C. R., and A. T. James, *Biometrics*, **22**: 573, 1966.
[†]Hanson, K. R., R. Ling and E. Havir, *Biochem. Biophys. Res. Comm.*, **29**, 194, 1967.

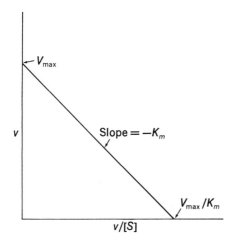

**Figure 6.10** Plot of the $v$ versus $v/S$ type. $S$ is substrate concentration; $v$ is reaction velocity.

special relationship to the $V_{max}$ in that *it is equal to the substrate concentration at which the enzyme activity is half-maximum.* A little algebraic manipulation of the Michaelis-Menten Equation proves the point. We can write the equation and then insert the condition that $v = 0.5\ (V_{max})$:

$$v = \frac{V_{max}\ [S]}{K_m + [S]}.$$

Substituting $v = 0.5\ (V_{max})$ yields

$$0.5\ V_{max} = \frac{V_{max}\ [S]}{K_m + [S]}.$$

Dividing both sides by $V_{max}$ gives

$$0.5 = \frac{[S]}{K_m + [S]};$$

$$K_m = [S]. \tag{6.51}$$

$K_m$ and $[S]$ are always expressed in the same units: moles per liter. In the past, considerable emphasis has been given to the relationship of $K_m$ to the affinity of the enzyme for its substrate or for any of several substrates, if there is more than one. It was often stated that the substrate of lowest $K_m$ was the "natural substrate" of the enzyme or that for which it had the greatest affinity. Now it is considered injudicious to designate any one of several substrates as being the natural or preferred substrate of the enzyme. Furthermore, to discuss $K_m$ values in the light of enzyme-substrate affinities is permissible only if it can be proved that $K_m$ is the dissociation constant of $ES$, *i.e.*, $k_{-1}/k_1$. If it should prove that $K_m$ is actually equal to $(k_{-1} + k_2)/k_1$ (the steady state constant), or even $k_2/k_1$

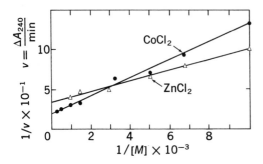

**Figure 6.11** Lineweaver-Burk plots for divalent metal ion activators in the β-methylaspartase system. [From Williams, V. R., and J. Selbin, *J. Biol. Chem.*, **239**:1636, 1964.]

(for the system in which $ES$ does not dissociate to form $E$ and $S$), or some other complex function, then there is no simple interpretation of the variation of $K_m$ from one substrate to another.

The use of the $K_m$ concept is not reserved for substrate alone. If an enzyme requires an activator, *e.g.*, a divalent metal ion, the substrate concentration can be fixed at some desired level and the activity of the enzyme studied as a function of the concentration of activator. If we represent the concentration of activator as $[A]$, $1/v$ plotted versus $1/[A]$ will yield $K_m$ and $V_{max}$ values characteristic of the activator. In this fashion, a variety of metal ion activators can be compared by a standard procedure. The interpretation of results remains the problem of the investigator. Lineweaver-Burk plots for metal activators in the β-methyl-aspartase system are shown in Fig. 6.11. In this figure, $1/[M]$ is used in place of $1/[A]$.

The fundamental significance of $V_{max}$ results from its relationship to $k_2$. In Eq. 6.41 we substituted $V_{max}$ for $k_2[E_t]$ for the special condition that all the enzyme was combined in the $ES$ form, making $[ES]$ at a maximum:

$$\text{Rate} = k_2[ES] = \text{maximum, } i.e., V_{max}.$$

If the enzyme concentration is known in moles per liter, and the rate of substrate disappearance is expressed in moles per liter per minute, $k_2$ is equal to the maximum *turnover number* of the enzyme:

$$k_2 = \text{Turnover number}$$

$$= \frac{\text{Moles of substrates transformed per min}}{\text{Moles of enzyme}}. \tag{6.52}$$

Turnover numbers are expressed for a specific temperature, since the reaction rate is necessarily temperature dependent. The disadvantage of using the turnover number is that the molecular weight of the enzyme must be known, as well as its molar concentration in the solution. Measurement of the turnover number may lead to ambiguous results because the enzyme is assumed to act upon only one substrate molecule at a time, whereas it actually may possess more than one catalytic center (called an *active site*). The use of the turnover number concept is not uniform; for example, the turnover number for amylase is calculated from the moles of product formed per minute rather than moles of substrate transformed.

**The Michaelis-Menten Equation and enzyme mechanisms** The simple mechanism presented in Eqs. 6.33 to 6.35 serves as an introduction to the topic of enzyme kinetics, but in reality possesses many defects. The most apparent of these deficiencies is the implication that there is only one intermediate complex, that is, *ES*. A more realistic view would allow at least one additional complex, the one which is first formed in the reverse reaction: $E + P \rightarrow EP$, where $P$ represents the product. Surely the initial combination of enzyme and product cannot be the same as the initial combination of enzyme and substrate. As an improvement on the original mechanism the equation should be written

$$E + S \underset{k_{-1}}{\overset{k_1}{\rightleftharpoons}} ES \underset{k_{-2}}{\overset{k_2}{\rightleftharpoons}} EP \rightleftharpoons E + P. \tag{6.53}$$

If a forward rate equation is derived for this mechanism, it proves to be identical in *form* with the Michaelis-Menten Equation, *i.e.*, both are equations for hyperbolas. Determinations of slope and intercept values for the Lineweaver-Burk form will yield the constants, $K_m$ and $V_{max}$. For the more complicated mechanism, however, $K_m$ contains additional reaction velocity constants and there is no simple-minded interpretation of it.

Actually, the hyperbolic form of the Michaelis-Menten Equation is given by many different mechanisms. One should be cautious, therefore, in espousing the simple mechanism shown in Eqs. 6.33 to 6.35 or in Eq. 6.53, just on the tenuous basis of linear Lineweaver-Burk plots.

**General velocity equation** An additional defect of the Michaelis-Menten Equation is that it describes only the initial forward velocity of the reaction, whereas we would frequently like to have a general equation for the velocity at any point in the reaction between initiation and equilibrium. Such an equation can be derived and will contain all the velocity constants for all the individual rates in addition to terms for the concentration of reactant and of product. The simplest form of this equation, written for a single substrate $S$ and a single product $P$ corresponding to the reaction shown in Eq. 6.53 is:

$$v = \frac{(k_1 k_2 [S] - k_{-1} k_{-2} [P]) [E_t]}{(k_{-1} + k_2) + k_1 [S] + k_{-2} [P]}. \tag{6.54}$$

This equation may be derived by assuming that $d[ES]/dt = 0$ and that $v = d[P]/dt$. If all terms in $P$ are eliminated and both numerator and denominator are divided by $k_1$, Eq. 6.54 reduces to Eq. 6.47, the steady state expression for the initial forward velocity. This and other complex equations are considered in detail in more advanced treatments and are mentioned here only for the purpose of opening a little wider the door to the fascinating subject of enzyme kinetics.

**Enzyme inhibition**  Enzymes may be inhibited by substances which produce chemical transformations that render the enzyme inactive or by substances which compete with the substrate for the active site of the enzyme and thereby slow down the rate of substrate transformation. Any process which removed enzyme from the scene of action will slow down the reaction rate. Chemical transformations may also affect the affinity of the enzyme for its substrate and thereby alter the reaction kinetics. Chemical transformations may be produced by metal ions, acylating agents, alkylating agents, etc., *i.e.*, any reagent which reacts with the functional groups of a protein. These changes are not reversed by adding more substrate to the enzyme solution. The inhibition may be reversible, but not through the action of the substrate. This type of inhibition is called *noncompetitive inhibition*. On the other hand, if the enzyme is inhibited by the attachment of a pseudosubstrate or substrate analog to its active site, such inhibition is reversed by increasing the concentration of the substrate in the solution. This type of inhibition is called *competitive inhibition*. A third possibility is that the inhibitor combines with some form of the enzyme other than $E$, the free enzyme; that is, the inhibitor might combine with $ES$ to form an enzyme-substrate-inhibitor complex. Inhibition of this type may be noncompetitive or *uncompetitive*, the latter being kinetically distinguishable from the other types. The characteristic appearance of the double reciprocal plots is shown in Figs. 6.12, 6.13, and 6.14.[*] The inhibition equations in the double reciprocal forms are

Noncompetitive Inhibition:

$$\frac{1}{v} = \frac{K_m}{V_{max}}\left(1 + \frac{[I]}{K_I}\right)\frac{1}{[S]} + \frac{1}{V_{max}}\left(1 + \frac{[I]}{K_I}\right); \tag{6.55}$$

Competitive Inhibition:

$$\frac{1}{v} = \frac{K_m}{V_{max}}\left(1 + \frac{[I]}{K_I}\right)\frac{1}{[S]} + \frac{1}{V_{max}}; \tag{6.56}$$

---

[*]The derivation of the kinetic equations for the various types of enzyme inhibition may be found in a standard reference such as the one by Dixon and Webb included in *Suggested Additional Reading*.

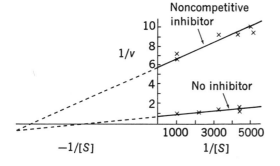

**Figure 6.12** Apparent noncompetitive inhibition of arylsulfatase by 0.001 *M* hydroxylamine. Substrate is *p*-nitrophenyl sulfate. [From the Ph.D. dissertation of R. C. Ellerbrook, Louisiana State University, August 1965.]

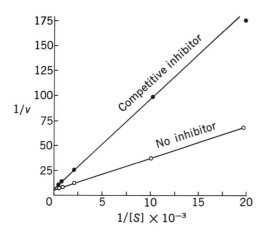

**Figure 6.13** Competitive inhibition of β-methylaspartase by 0.005 *M* β-*threo*-methyl-β-hydroxyaspartate. [Substrate is β-*threo*-methyl-aspartate. [From Williams, V. R., and W. Y. Libano, *Biochim. et Biophys. Acta*, **118**:124, 1966.]

Uncompetitive Inhibition:

$$\frac{1}{v} = \frac{K_m}{V_{\max}} \cdot \frac{1}{[S]} + \frac{1}{V_{\max}} \left(1 + \frac{[I]}{K_I}\right). \tag{6.57}$$

Let us consider the three equations together and see what generalizations can be made. Two new terms appear: [I], the molar concentration of the inhibitor, and $K_I$, the inhibitor constant. These new terms are combined in a coefficient: $(1 + [I]/K_I)$. Other algebraic arrangements of the equations are preferred by some kineticists, but the form shown in Eqs. 6.55 to 6.57 has the merit of being easily remembered. If we examine the plots which are characteristic of the equations, Figs. 6.12, 6.13, and 6.14, we note that all of the plots are straight lines, and that the differences between the control (no inhibitor) and the inhibitor plots are: their slopes in Fig. 6.13, their intercepts in Fig. 6.14, and both

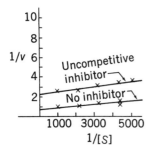

**Figure 6.14** Uncompetitive inhibition of arylsulfatase by 0.004 $M$ hydroxylamine. Substrate is $p$-nitrophenyl sulfate. [From the Ph.D. dissertation of R. C. Ellerbrook, Louisiana State University, August 1965.]

their slopes and their intercepts in Fig. 6.12. *Whenever the slope or the intercept of an inhibitor plot differs from the control, that term in the corresponding equation will contain the "inhibitor coefficient,"* $(1 + [I]/K_I)$. If we write the Lineweaver-Burk Equation for the uninhibited reaction and indicate at which points the inhibitor coefficient appears, we can summarize the three inhibitor equations:

$$\left(1 + \frac{[I]}{K_I}\right) \qquad \left(1 + \frac{[I]}{K_I}\right)$$

$$\frac{1}{v} = \frac{K_m}{V_{max}} \downarrow_{\uparrow} \cdot \frac{1}{[S]} + \downarrow_{\uparrow} \frac{1}{V_{max}}.$$

Competitive          Uncompetitive
Noncompetitive     Noncompetitive

One further detail should be noted by inspection of Figs. 6.12 to 6.14. Since the abscissa represents $1/[S]$, any points lying on the ordinate correspond to the system at infinite substrate concentration (if $1/[S] = 0$, then $[S] = \infty$). In competitive inhibition, the inhibition is reversed by excess substrate, and therefore $V_{max}$ must be the same for both systems at infinite substrate concentration, *i.e.*, the control system and the inhibited system must have a common intercept. Fig. 6.13 illustrates this. In noncompetitive inhibition, the plots for the control system and the inhibited system are usually depicted as intersecting on the $-1/[S]$ axis. To be exact, it must be added that this situation prevails only for a particular kind of noncompetitive inhibition, called "simple linear noncompetitive inhibition." If the inhibition is not simple, the plots may intersect above or below the $-1/[S]$ axis.* See Fig. 6.12.

---

*The student who is interested in learning more about enzyme kinetics can educate himself with the help of an excellent, straightforward exposition to be found in the book by Reiner that is included in the listing in *Suggested Additional Reading*. Reiner assumes no prior knowledge of kinetics on the part of the reader.

**Analysis of enzyme mechanisms by use of product inhibition patterns** In most enzyme-catalyzed reactions we find several substrates and several products that do not conform to the simple single substrate and single product model as treated in Eqs. 6.33–6.35. Since enzyme-catalyzed reactions necessitate binding of the chemical reaction constituents to the surface of the protein catalyst, several interesting possibilities present themselves in considering a reaction involving two substrates, $A$ and $B$, and two products, $P$ and $Q$. The general reaction expression will be

$$A + B \underset{}{\overset{\text{enzyme}}{\rightleftarrows}} P + Q,$$

although this equation tells us nothing of the order of binding of the substrates and of release of products. Some of the possibilities are: $A$ must bind to the enzyme before $B$ can bind, or vice versa; $A$ must bind and $P$ be released before $B$ can bind; neither $A$ nor $B$ must bind in preferential order, etc. A choice can be made between the alternative by studying the type of inhibition (competitive, noncompetitive, or uncompetitive) which results from adding the products one at a time to mixtures of $A$ and $B$ at specified relative concentrations. This method of analysis originated with W. W. Cleland[*] and will be presented in part as a simplified introduction to his approach. To begin, we must first familiarize ourselves with the special terminology of kinetic pattern analysis that follows.

   (a) *The number of kinetically important substrates or products in a mechanism will be designated by the syllables Uni, Bi, Ter, Quad.*
   (b) *The number of reactants involved in the reaction in one direction will be the reactancy in that direction.*

   *Example:*
      Bi Bi is bireactant in both directions,
$$A + B \rightleftarrows P + Q.$$
      Uni Bi is unireactant in the forward direction and bireactant in the reverse,
$$A \rightleftarrows P + Q.$$
   Others: Ter Ter or Ter Bi,
$$A + B + C \rightleftarrows P + Q + R, \quad A + B + C \rightleftarrows P + Q.$$
   Terreactant is not the same as termolecular.

[*]Cleland, W. W., *Biochim. et Biophys. Acta*, **67**: 104–196, 1963.

(c) *A "sequential" mechanism will be one where all substrates must be present on the enzyme before any products may leave.*
*A "sequential" mechanism will be called "ordered" if the substrates add in a definite obligatory order and the products leave in an obligatory order. If the order is not obligatory, the sequence will be called "random."*

(d) *If one or more products are released before all the substrates have added to the enzyme, the mechanism will be called "ping pong." If the mechanism is not obvious it will be further described by the use of Uni, Bi, Ter, to indicate the successive groups of substrate additions and product dissociations that occur.*

(e) *Substrates will be designated by the letters A, B, C, D, in the order in which they add to the enzyme.*

(f) *Products will be designated by P, Q, R, S, in the order in which they leave the enzyme.*

(g) *Stable enzyme forms (those incapable of unimolecular degradation with liberation of a substrate or product, or isomerization into such a form) will be designated by E, F, G, H, with E being the least complex or "free enzyme,"* if such a distinction is possible. Sequential mechanisms include only one stable form (free enzyme). Ping-pong mechanisms include two or more stable forms.

(h) *Transitory enzyme forms (those capable of unimolecular degradation with liberation of a substrate or product, or isomerization into such a form) will be designated by combinations of letters chosen to represent their composition, such as EA, EAB, FB, etc.*

(i) *Transitory forms that cannot participate in bimolecular reaction steps with substrates or products, but can only undergo unimolecular degradation with release of substrates or products, or isomerize into such a form, will be called central complexes. These will be distinguished from other transitory forms by being enclosed in parentheses. (EAB) or (EAB-EPQ).* The various isomeric forms of the central complex will be treated as one complex since steady state kinetics even when combined with other types of data can give no evidence about the isomerization of central complexes, as opposed to the isomerization of other transitory forms.

To illustrate Cleland's method we will restrict ourselves to Bi Bi mechanisms. Many other types are presented in the original reference cited above. The sequence of events in a Bi Bi mechanism can be presented graphically from left to right as will be shown below. A horizontal line represents the enzyme; vertical arrows are used to denote substrate additions and product dissociations. Rate constants may be shown, if desired, by writing those for the forward reaction to the left of the appropriate arrows and those for the reverse reaction to the right.

Three examples of Bi Bi mechanism patterns are shown below. The first two are sequential mechanisms: Ordered Bi Bi and Rapid Equilibrium Random Bi Bi; the third is Ping Pong Bi Bi.

Ordered Bi Bi (with one central complex)

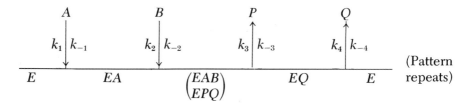

Rapid Equilibrium Random Bi Bi (with one central complex) (rate constants omitted)

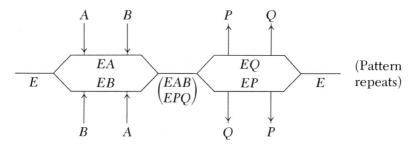

Ping Pong Bi Bi (rate constants omitted)

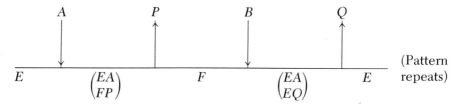

It is equally meaningful to write the Ping Pong Bi Bi pattern by beginning with $F$:

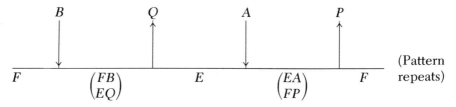

In the Ping Pong Bi Bi pattern the free enzyme exists in two forms ($E$ and $F$) as in the case of glutamate-oxalacetate transaminase. When the $B_6$ coenzyme (bound to the transaminase) is in the aldehydic form

(—CHO) the enzyme binds L-glutamate, but when the coenzyme is in the amino form (—CH$_2$NH$_2$) the enzyme can bind only the keto acid.

Other Bi Bi reaction mechanisms can be written. In addition, there are Uni Bi, Ter Bi, and Ter Ter mechanisms, but limitations of space require that we simply mention these in passing.

Rate equations for complicated steady-state mechanisms can be derived by a geometric method devised by King and Altman.[*] The method is straightforward and avoids the use of complicated algebra and large determinants. This approach has been used by Cleland to derive the steady-state equations for the various mechanisms which he presents. Anyone who likes mathematical games will enjoy learning to work out rate equations by the King and Altman method. With a little practice the student will discover that he has mastered a powerful technique that enables him to derive equations that would, by ordinary algebra, be entirely beyond his ability.

Our purpose now is to show how we may determine the correct mechanism for a Bi Bi reaction by experimentally determining whether the reaction products act as competitive, noncompetitive, or uncompetitive inhibitors of the various substrates.

Substances that inhibit enzymatic reactions may do so by combining with various enzyme forms, substrates, or activators. We will restrict ourselves to combinations between inhibitors and enzyme forms since this is the basis for "product inhibition." If an ordinary chemical reaction is proceeding in the forward direction at a measured rate and one of the reaction products is added to the system, the net forward rate of the reaction slows down because the rate of the reverse reaction has been increased. In a similar way, the addition of product to an enzyme catalyzed reaction slows down the forward reaction. However, the latter situation is more complicated because we must take into consideration the formation of a specific enzyme-product complex. Actually, it is this complication that permits us to use product inhibition as a mechanistic probe. Before we see how the method works, certain basic rules for product inhibition analysis must be explained.

(a) *Whenever a product and a substrate combine individually with the same enzyme form, the product will be a competitive inhibitor of the substrate.* For example, if $A$ combines reversibly with $E$ to form $EA$ and $QE$ dissociates reversibly to release $Q$ and $E$, then $Q$ will display competitive inhibition toward $A$. If an experiment is carried out and the data plotted according to the Lineweaver-Burk form, the results should look like Fig. 6.13. In the product inhibition case, the concentration of $A$ is varied at some fixed concentration of $Q$. If the reaction is Bi Bi, then

---

[*] King, E. L., and C. Altman, *J. Phys. Chem.*, **60**: 1375, 1956.

the second substrate $B$ should be present in all samples so that the reaction is not limited by insufficient $B$.

(b) *Whenever a product and a substrate combine individually with different forms of the enzyme, the product will be a noncompetitive inhibitor of the substrate if the path between the product and the substrate contains no irreversible steps.* There are two types of irreversible steps: the addition of a substrate at a saturating level and the release of a product at zero concentration. Let us consider the first type. In the partial sequence shown below,

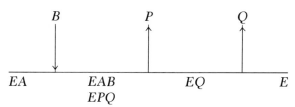

the path between $P$ and $A$ is reversible if $B$ is present in less than saturating concentrations, but is irreversible if $B$ is at a saturating level. For a pathway to be reversible, all the components of that pathway, including the various enzyme forms, must be present at some finite concentration. If the level of $B$ is saturating, then the concentration of $EA$ will be negligible because it will all be converted to $EAB$ as fast as it is formed. With $EA$ essentially removed from the reaction system the path between $A$ and $P$ is irreversible.

Similarly, if a second product — one lying between the substrate and product being considered — is absent from the system (present at zero concentration), then the reaction cannot be reversed along this path because one of the essential reaction components will be missing. Consider the partial sequence shown below:

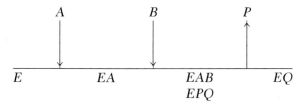

If the reaction system for which initial velocity measurements are being taken contains no added $P$, then the path between $B$ and $Q$ for the section of pattern shown is an irreversible path because the absence of $P$ blocks the sequence on the left side of $Q$. If $P$ is present at a finite concentration, then all steps between $B$ and $Q$ are reversible and $Q$ will behave as a noncompetitive inhibitor of $B$.

(c) *Whenever a product and a substrate combine with different enzyme forms along an irreversible pathway the product will be an uncompetitive inhibitor of that substrate, with one exception.* If the

**TABLE 6.1**  PRODUCT INHIBITION PATTERNS FOR BI BI MECHANISMS

| Mechanism | Inhibitory product | Variable substrate | | | |
|---|---|---|---|---|---|
| | | A | | B | |
| | | Unsaturated with B | Saturated with B | Unsaturated with A | Saturated with A |
| Ordered Bi Bi | P | NC° | UC | NC | NC |
| | Q | Comp | Comp | NC | — |
| Rapid Equilibrium | P | NC | — | Comp | Comp |
| Random Bi Bi, dead-end EAP complex | Q | Comp | — | Comp | — |
| Rapid Equilibrium | P | NC | — | Comp | Comp |
| Random Bi Bi, dead-end EAP, EBQ complexes | Q | Comp | Comp | NC | — |
| Ping Pong Bi Bi | P | NC | — | Comp | Comp |
| | Q | Comp | Comp | NC | — |

° The abbreviations used are: Comp, competitive; UC, uncompetitive; NC, noncompetitive; —, no inhibition. *Source:* W. W. Cleland, *Biochim. Biophys. Acta.* **67**, 123, 1963, and personal communication.

irreversibility results from the addition of a second substrate at saturation and the product in question is a competitive inhibitor of the second substrate, then there will be *no* inhibition. This exception will be illustrated below.

These simple rules cover only the basic ideas of product inhibition analysis. If the compound being studied combines with two or more enzyme forms, additional rules must be considered.

Restricting ourselves still to the Bi Bi mechanisms, we perform the experiments by varying the concentration one substrate (called the variable substrate) while holding constant the concentration of the second substrate (called the changing fixed substrate). The second substrate is called the changing fixed substrate because it may be included at saturation in one series of initial velocity measurements while included at a lower level in another series. For any individual series of measurements, however, it will be constant.

For the three Bi Bi mechanisms presented earlier the product inhibition patterns are those shown in Table 6.1. The student should work through these analyses, applying the rules stated above. It is vitally important to remember that the patterns repeat. Although a path is irreversible in one direction, it may be reversible in another direction. In the sequence shown below, the path between *B* and *Q* is irreversible on the left of *Q* when *P* is absent, but is reversible on the right of *Q* when *A* is not at saturating levels.

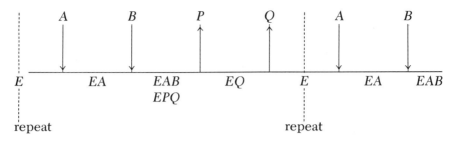

Note that for the Ordered Bi Bi mechanism (Table 6.1) the addition of A at saturating levels removes the inhibitory effect of Q on B. This is the exception described under Rule (c). Saturating levels of A remove all the free E from the system so that none is left to combine with Q; therefore, Q has no effect and no inhibition is observed.

In the Ping Pong mechanism, P combines with F and is therefore a noncompetitive inhibitor of A which combines with E. It is a competitive inhibitor of B, regardless of the concentration of A. However, the addition of B as the fixed substrate at saturation, removes the inhibition of P for A as the variable substrate, because the concentration of F has been reduced to a negligible level (it is all converted to FB).

If the student has found this material interesting, he should consult the original reference for additional Bi Bi mechanisms and their product inhibition patterns.

### 6.9 The Arrhenius Equation: The Effect of Temperature on Reaction Rate

The rate of any simple chemical reaction is accelerated by an increase in temperature, provided the higher temperature does not produce secondary alterations in the reactants or catalyst. A case in point is the decrease in the rate of an enzyme-catalyzed reaction at temperatures that denature the enzyme. If the overall velocity of a reaction appears not to increase with temperature, complicating factors exist that are obscuring the basic rate processes. The general effect of temperature on reaction rates may be analyzed as the dependence of rate on the kinetic energy of the reactants. As stated in the kinetic theory, kinetic energy depends only on the absolute temperature.

A frequently quoted approximation, known as the van't Hoff rule, states that the reaction rate doubles for a ten-degree temperature rise. The ratio of the rate constants for two temperatures differing by ten degrees is called the $Q_{10}$ of a reaction. The generalization $Q_{10} = 2$ holds for a number of reactions because they have nearly the same *activation energy*, a fundamental term appearing in the Arrhenius Equation, various forms of which are given below.

The efforts of van't Hoff to establish a mathematical relationship between reaction temperatures and equilibrium constants led to the equation bearing his name:

$$\ln\frac{K_2}{K_1} = \frac{\Delta H°(T_2 - T_1)}{R(T_2 T_1)} \qquad \text{(Eq. 3.47)}.$$

This integrated form arises from a differential form, which expresses the rate at which $\ln K$ changes with temperature:

$$\frac{d(\ln K)}{dT} = \frac{\Delta H°}{R} \cdot \frac{1}{T^2}.$$

Since the equilibrium constant of a reaction is the ratio of the forward and reverse rate constants, Arrhenius suggested that a similar mathematical form might properly describe the effect of temperature on the reaction rate constant:

$$\frac{d(\ln k)}{dT} = a \cdot \frac{1}{T^2}. \qquad (6.58)$$

The constant term $a$ is assumed to be similar to $\Delta H/R$, with the quantity $E_a$, the activation energy, replacing $\Delta H$. On integration Eq. 6.58 becomes

$$\ln k = -\frac{E_a}{RT} + \ln A. \qquad (6.59)$$

The exponential form of the equation is

$$k = A(e^{-E_a/RT}), \qquad (6.60)$$

and the form similar to Eq. 3.47 is

$$\ln\frac{k_{T_2}}{k_{T_1}} = \frac{E_a(T_2 - T_1)}{RT_2 T_1}. \qquad (6.61)$$

The subscripts $T_2$ and $T_1$ rather than simply 2 and 1 are used to distinguish the two rate constants in the hope of avoiding confusion. Earlier we used $k_2$ and $k_1$ to denote rate constants for *different* reactions. In Eq. 6.61, the two rate constants are for the *same* reaction at two *different* temperatures. Eqs. 6.59, 6.60, and 6.61 are all known as the Arrhenius Equation.

Equation 6.59 is a convenient form for plotting data. The slope of the line, when $\ln k$ is plotted versus $1/T$, is $-E_a/R$, from which the activation energy can be readily obtained. In Fig. 6.15 are plotted data from a manometric study of catecholase activity. In this graph, as is frequently done, the logarithm of the reaction rate rather than its rate constant has been plotted versus $1/T$. If initial concentrations of all components of a reaction mixture are held constant, and only temperature is varied, the

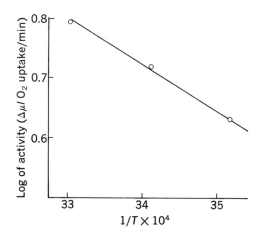

**Figure 6.15**  Arrhenius plot of crustacean catecholase activity measured by the manometric method. [From the Ph.D. dissertation of M. E. Bailey, Louisiana State University, June, 1958.]

reaction rate itself will obey the Arrhenius Equation. In the catecholase experiments, for a specific temperature the reaction rate depends on the concentrations of enzyme and substrate:

$$\text{Rate} = k[E][S].$$

If concentrations of the components are held invariant,

$$\frac{\text{Rate at } T_2}{\text{Rate at } T_1} = \frac{k_{T_2}}{k_{T_1}}.$$

The data shown in Fig. 6.15 are unusual in that $E_a$ is found to be approximately 3,000 cal/mole. This is low for an activation energy, but appears to be characteristic of phenol oxidases.

The activation energy is always a positive quantity (unlike $\Delta H$), and is equal to the molar increase in energy required by the reactants before products can be formed. If we diagram an exothermic reaction along its energy coordinate we obtain the potential energy map shown in Fig. 6.16. The reaction is exothermic because the products lie at a lower energy level than the reactants. This energy difference is $\Delta E$ for a constant volume process and $\Delta H$ for a constant pressure process. At the top of the energy barrier the reactants are considered to be in an intermediate state called the *activated complex* or *transition state*. The energy difference between the activated complex and the initial energy of the reactants is equal to $E_a$. In the reverse reaction, products and reactants change roles and $E_a$ is the energy barrier marked $E_{a(\text{reverse})}$. The difference between the two activation energies is the reaction enthalpy:

$$E_{a(\text{forward})} - E_{a(\text{reverse})} = \Delta H_{\text{reaction}}. \tag{6.62}$$

The sign of $\Delta H$ depends on the relative magnitude of these two activation energies.

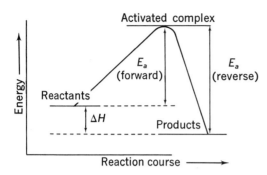

**Figure 6.16** Energy diagram for an exothermic reaction (not catalyzed).

Although the Arrhenius Equation was empirical, it proved remarkably successful in predicting the temperature dependence of $k$, the reaction rate constant. In an attempt to relate the equation to molecular events, Arrhenius proposed that $E_a$ was the critical activation energy and that $e^{-E_a/RT}$ was the fraction of molecules possessing this energy. The term $e^{-E_a/RT}$ had earlier been derived by Boltzmann in his theoretical treatment of energy distribution in molecular systems. In addition to the exponential term, the reaction rate was regarded to depend on two other factors: the molecular collision frequency, and the probability that the collisions would occur with the molecules in a favorable orientation. The term $A$ in Eq. 6.60 incorporates these two quantities and is called the *frequency factor*. The probability factor is often regarded as being unity. These ideas form the basis of the *collision theory* of reaction rates. The activated state is described in terms of two colliding molecules and thus suffers the limitation of being most clearly applicable to bimolecular processes.

A modern kinetic analysis of the degrees of freedom of colliding molecules led to a theoretical derivation of the Arrhenius Equation in the early 1900's. The few known second-order gas reactions were examined in the light of the new derivation with considerable success. The frequency factor, $A$, was discovered to be somewhat temperature dependent, a relationship not apparent from plots of experimental data. Furthermore, on a theoretical basis there is no reason to assume a temperature-independence for $E_a$. On occasion, a small temperature dependence has been observed.

As already stated, the principal defect in the collision theory is its awkwardness in describing unimolecular reactions. We have to postulate that a molecular collision leads to the formation of an activated species which undergoes first-order decomposition to products. Schematically, this is

$$A + A \rightleftarrows A\dagger + A^*$$

$$A^* \longrightarrow \text{products}$$

where $A^*$ is the activated species and $A\dagger$ is a molecule of $A$ in an energy state lower than the initial state. The kinetic development is handled by the now-familiar steady-state method. It is assumed that $d(A^*)/dt = 0$, and a rate expression is derived in terms of the concentration of $A$. This approach to a collision theory for unimolecular reactions was devised by Lindemann in 1922. Further details will not be considered here. Suffice it to say that it has proved mathematically satisfactory in a number of instances. Some of these, however, have subsequently been discovered to follow more complex reaction courses involving free radicals.

A fresh approach to reaction rate theory was devised by Eyring in 1935 and the following years. Eyring's theory is called the *transition state theory* since it selects as the fundamental process the breakdown of the activated complex at the top of the energy barrier. It can be applied to one or to several molecules and describes the activation process as a distortion of the reactant molecule(s) that favors conversion to products. It regards the activated state as a sort of halfway house in which loose complexes and unusual bond orientations are the prevailing but transient order of things. This concept is particularly helpful in the analysis of enzyme-catalyzed reactions. The substrate combines with the enzyme in a loose complex (otherwise it could never be dislodged) and the substrate is partially altered by enzyme binding so that it achieves an intermediate form possessing high instability. The reaction is completed when the transition state disintegrates and products are released. Leaving the catalyzed reaction for the moment, let us consider a simple bimolecular reaction, for example,

$$H_{2(g)} + I_{2(g)} \rightleftarrows 2HI_{(g)}.$$

We assume that the reactants form an activated complex $(H_2 - I_2)$ which is in equilibrium with them:

$$H_2 + I_2 \rightleftarrows (H_2 - I_2).$$

The concentration of the complex can be calculated by the equilibrium method:

$$[(H_2 - I_2)] = K^{\ddagger}[H_2][I_2], \tag{6.63}$$

where a special notation, $K^{\ddagger}$, is used for this type of equilibrium constant.

The reaction rate will be equal to the concentration of the activated complex multiplied by the rate at which the complex crosses the energy barrier, *i.e.*, the rate at which the activated complex dissociates to form products. The complex breaks down when its energy of bond vibration becomes equal to the bond potential energy. These two relationships are

$$E_{\text{vibration}} = h\upsilon, \tag{6.64}$$

where $h$ is Planck's constant and $v$ is the vibrational frequency; and

$$E_{\text{potential}} = k_B T, \tag{6.65}$$

where $k_B$ is the Boltzmann constant (the gas constant per molecule). Equation 6.63 may also be written

$$E_{\text{potential}} = \frac{RT}{\mathfrak{N}},$$

where $R$ is the molar gas constant and $\mathfrak{N}$ is Avogadro's number. When $E_{\text{vibration}} = E_{\text{potential}}$, $h\nu = k_B T$. Rearranging gives the frequency of complex dissociation (to products)

$$\nu = \frac{k_B T}{h}. \tag{6.66}$$

If the reaction rate is, as we have stated above, equal to the product of $[(H_2 - I_2)]$ times frequency of complex dissociation, then combining Eq. 6.63 and Eq. 6.66 gives

$$\text{Rate} = K^{\ddagger}[H_2][I_2]\frac{k_B T}{h}. \tag{6.67}$$

The rate must also be

$$\text{Rate} = k[H_2][I_2], \tag{6.68}$$

where $k$ is the second-order velocity constant. We can see from a comparison of Eq. 6.67 and Eq. 6.68 that the reaction rate constant, $k$, is related both to the fundamental constants and also to the equilibrium constant for complex formation in the following way:

$$k = \frac{k_B T}{h} K^{\ddagger}. \tag{6.69}$$

If $K$ can be determined, the reaction rate constant can be calculated from Eq. 6.69. This derivation assumes that the complex will dissociate into products rather than go back to reactants. Equation 6.69 is not precise: The probability of such a dissociation is actually between 50 and 100 percent and must be included in the equation for precise calculations.

We now return to the thermodynamic relationship between $K_{\text{equil}}$ and the free energy change to complete our brief exposition of Eyring theory. The standard free energy change for a chemical reaction is:

$$\Delta G° = -RT \ln K \qquad (\text{Eq. 3.43}).$$

For the activation process the corresponding expression is:

$$\Delta G^{\ddagger} = -RT \ln K^{\ddagger}. \tag{6.70}$$

The relationship between changes in free energy, enthalpy, and entropy for the activation reaction at constant temperature is also a familiar equation with special notation:

$$\Delta G^{\circ\ddagger} = \Delta H^{\circ\ddagger} - T\,\Delta S^{\circ\ddagger}. \tag{6.71}$$

These functions, $\Delta G^{\circ\ddagger}$, $\Delta H^{\circ\ddagger}$, and $\Delta S^{\circ\ddagger}$ are called the free energy of activation, enthalpy of activation, and entropy of activation, respectively. The enthalpy of activation, $\Delta H^{\circ\ddagger}$, is related to $E_a$ in the following way:

$$E_a = \Delta H^{\circ\ddagger} + RT.$$

Combining Eqs. 6.69 and 6.70 yields

$$k = \frac{k_B T}{h}\, e^{-\Delta G^{\circ\ddagger}/RT}. \tag{6.72}$$

Substituting Eq. 6.71 into Eq. 6.72 gives

$$k = \frac{k_B T}{h}\, e^{\Delta S^{\circ\ddagger}/R}\, e^{-\Delta H^{\circ\ddagger}/RT}. \tag{6.73}$$

In Eq. 6.73 we see something new that lends significance to the concept of the probability factor postulated in collision theory. The exponential term now appears as the product of two exponentials: one is a clearly recognizable probability and "orientation" term $e^{\Delta S^{\circ\ddagger}/R}$; the second exponential, $e^{-\Delta H^{\circ\ddagger}/RT}$ contains the activation energy. If the reaction velocity constant, $k$, and $E_a$ are determined experimentally, substitution into Eq. 6.73 permits us to calculate the entropy of activation. This quantity is of great interest in exploring the reaction mechanism.

For the student of biological science, the enzyme-catalyzed reaction is of paramount interest. Catalyzed reactions differ from simple chemical ones in that the energy level of reactants and products is altered by adsorption to the catalyst, and especially important, in that the energy barrier is lowered. These important distinctions permit enzyme-catalyzed reactions to occur at moderate temperatures in neutral dilute solutions. An energy contour map is shown in Fig. 6.17 for the system:

$$E + S \underset{k_{-1}}{\overset{k_1}{\rightleftarrows}} ES \text{ complex} \underset{k_{-2}}{\overset{k_2}{\rightleftarrows}} EZ \text{ transition state} \underset{k_{-3}}{\overset{k_3}{\rightleftarrows}} EP \text{ complex} \underset{k_{-4}}{\overset{k_4}{\rightleftarrows}} E + P.$$

The reaction enthalpy, $\Delta H$, has the same definition as previously and is calculated from the van't Hoff Equation or measured by microcalorimetry. It cannot, however, be calculated from Eq. 6.62. Fig. 6.17 is helpful in relating certain concepts of enzyme kinetics to general kinetic theory. If $K_m$, the Michaelis constant (that is, $K_m = k_{-1}/k_1$), is assumed to be the dissociation constant of $ES$, as it sometimes is, the energy difference between $E + S$ and $ES$ may be calculated. For $ES$ formation,

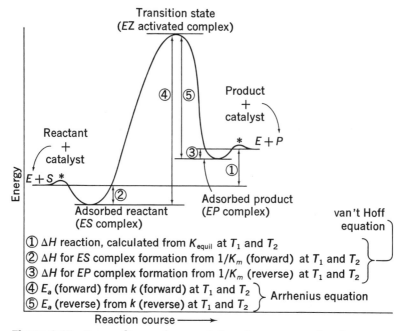

**Figure 6.17** Energy diagram for an endothermic enzyme-catalyzed reaction. Symbols are $E$ for enzyme; $S$, substrate; $P$, product; $EZ$, transition state of activated complex; *, small energy barriers between $E + S$ and $ES$, and between $E + P$ and $EP$.

$1/K_m = K_{equil}$. We substitute $1/K_m$ at $T_1$ and $T_2$ into the van't Hoff Equation and solve for $\Delta H_{ES}$. The same approach may be used with the reverse reaction to determine the energy difference between $E + P$ and $EP$. A full exposition of the use of velocity constants in determining various thermodynamic quantities for enzyme reactions is given by Dixon and Webb.* Especially valuable is their specific application of theory to fumarase, for which they have prepared a complete energy contour map and thermodynamic balance sheet.

## Suggested Additional Reading

### General: Kinetics

Daniels, F., and R. A. Alberty, *Physical Chemistry*, 3rd Ed. John Wiley and Sons, New York, 1966.

Moore, W. J., *Physical Chemistry*, 4th Ed. Prentice-Hall, Englewood Cliffs, N.J., 1972.

---

*Pages 145–166 of the reference listed in *Suggested Additional Reading*.

**Biochemical Applications**

Bray, H. G., and K. White, *Kinetics and Thermodynamics in Biochemistry,* Academic Press, New York, 1957.

Dixon, M., and E. C. Webb, *Enzymes,* 2nd Ed. Academic Press, New York, 1964.

Jencks, W. P., *Catalysis in Chemistry and Enzymology.* McGraw-Hill, New York, 1969, Chapter 11.

Plowman, K. M., *Enzyme Kinetics.* McGraw Hill, New York, 1972.

Reiner, J. M., *Behavior of Enzyme Systems.* 2nd Ed. Reinhold, New York, 1968.

**Problems**

**6.1** (a) If 0.01 mole reacts in 1 second in a zero-order reaction, how much reacts in 10 seconds?

(b) If we start with 1000 molecules and 500 are destroyed in 1 second in a first-order reaction, how many would be destroyed in 2 seconds?

**6.2** Describe what data would be necessary, and how you would handle the data, in order to distinguish between the following two possible reaction mechanisms

$$A \longrightarrow \text{products}$$

$$A + A \longrightarrow \text{products}$$

**6.3** The overall reaction for the oxidation of NO is

$$2NO + O_2 \rightleftharpoons 2NO_2.$$

The reaction mechanism is believed to consist of the two steps shown at the end of Section 6.2. Assume the second step to be rate determining and derive the rate expression for the overall reaction by the equilibrium method.

*Answer.* Rate $= (k_1 k_2 / k_{-1}) [NO]^2 [O_2]$

**6.4** Find an expression for $k_{-3}$ in terms of other rate constants for the mechanism

$$A \underset{k_{-1}}{\overset{k_1}{\rightleftharpoons}} B$$

$$A \underset{k_{-2}}{\overset{k_2}{\rightleftharpoons}} C$$

$$B \underset{k_{-3}}{\overset{k_3}{\rightleftharpoons}} C$$

**6.5** Given the reaction course

$$A + B \underset{k_{-1}}{\overset{k_1}{\rightleftarrows}} C \underset{k_{-2}}{\overset{k_2}{\rightleftarrows}} A + D,$$

find expressions for
(a) rate of decrease of $A$;
(b) rate of formation of $C$.

**6.6** Consider the following series of irreversible first-order reactions

$$A \overset{k_1}{\longrightarrow} B \overset{k_2}{\longrightarrow} C$$

The initial concentrations are $[A] = 0.1$ m, $[B] = [C] = 0$. Make a sketch of the concentrations of $A$, $B$, and $C$ vs. time when
(a) $k_1 = 0.693$ min$^{-1}$ and $k_2 = 693$ min$^{-1}$
(b) $k_1 = 693$ min$^{-1}$ and $k_2 = 0.693$ min$^{-1}$

**6.7** The following reaction sequence was found to represent the correct mechanism for a reaction:

$$A + B \underset{k_{-1}}{\overset{k_1}{\rightleftarrows}} X + C;$$

$$X \overset{k_2}{\longrightarrow} D.$$

Derive the reaction rate expression (a) by the equilibrium method and (b) by the steady state method. Compare the two.

**6.8** Calculate the half-life period for the saponification of methyl acetate in sodium hydroxide solution when the starting concentrations are 0.03 $M$ for each reactant. The reaction takes place at 25°C, and the forward velocity constant at this temperature is 11.5 L/mole-min.

*Answer.* 2.9 min.

**6.9** A reaction has the overall stoichiometry $2A + B \longrightarrow$ Products. The mechanism has been investigated by following the disappearance of $A$ and $B$ with three different initial concentrations (concentrations are in moles/liter). Suggest a mechanism for the reaction, identify the rate-determining step, and evaluate the rate constant for the rate-determining step.

| | Exp. 1 | | Exp. 2 | | Exp. 3 | |
|---|---|---|---|---|---|---|
| Time, min. | A | B | A | B | A | B |
| 0 | 0.100 | 0.100 | 0.100 | 0.200 | 0.050 | 0.100 |
| 1 | 0.050 | 0.075 | 0.050 | 0.175 | 0.033 | 0.092 |
| 3 | 0.025 | 0.0625 | 0.025 | 0.1625 | 0.020 | 0.085 |
| 9 | 0.010 | 0.055 | 0.010 | 0.155 | 0.009 | 0.080 |

*Answer.* Reaction mechanism,

(1) $A + A \xrightarrow{k_1}$ [Intermediate]
     slow

(2) [Intermediate] $+ B \rightarrow$ Products
        fast

    (1) is rate determining

$$k_1 = 10 \text{ liters mole}^{-1} \text{ min}^{-1}$$

**6.10** When compound $A$ is exposed to light, it is converted to compound $B$. The initial rate of product production was measured under three experimental conditions, with the following results

| Experiment | $[A]_0$, m/l | Light intensity (arbitrary units) | Initial $d[B]/dt$, m/l min |
|---|---|---|---|
| 1 | 0.0001 | 1 | 0.000001 |
| 2 | 0.0001 | 10 | 0.000010 |
| 3 | 0.0010 | 1 | 0.000010 |

Suggest a possible mechanism for the reaction.

**6.11** The following data were obtained in the photooxidation of $\beta$-methylaspartase with methylene blue. Since the methylene blue was present in excess, the reaction should be pseudo first order in enzyme concentration. Prove that the following data are consistent with a description of the mechanism as pseudo first order.

| Enzyme sample | Time (min) | $2.3 \times \log 1/[Enzyme]$ |
|---|---|---|
| A | 0 | 0.995 |
| | 5 | 1.233 |
| | 10 | 1.373 |
| | 15 | 1.663 |
| | 25 | 1.996 |
| B (a 1:3 dilution of A) | 0 | 1.787 |
| | 5 | 2.006 |
| | 10 | 2.231 |
| | 15 | 2.410 |
| | 25 | 2.967 |

*Answer.* $t_{1/2}$ is independent of $[E]$; $k$ for sample $A$ is 0.042 min$^{-1}$, $k$ for sample $B$ is 0.048 min$^{-1}$.

**6.12** Cysteine is photooxidized with methylene blue, and samples are withdrawn and titrated spectrophotometrically with $N$-ethylmaleimide. Calculate the half-life for the photooxidation of cysteine, initially 0.01 $M$, from the following data:

| Time (min) | *Absorbancy at 300 mμ* (*proportional to concentration*) |
|---|---|
| 0 | 0.145 |
| 1 | 0.120 |
| 2 | 0.091 |
| 4 | 0.052 |
| 5 | 0.033 |

**6.13** In studies on the photooxidation of $\beta$-methylaspartase with methylene blue, the addition of substrate greatly affected the half-life of the enzyme. From the following data, determine what effect the addition of substrate had on the stability of the enzyme to oxidation:

| Sample | Time (min) | ΔA/min (proportional to enzyme concentration) |
|---|---|---|
| Unprotected enzyme | 0 | 0.100 |
|  | 2.5 | 0.072 |
|  | 5.0 | 0.056 |
|  | 10.0 | 0.033 |
| Enzyme protected by substrate | 0 | 0.104 |
|  | 5.0 | 0.092 |
|  | 10.0 | 0.082 |
|  | 30.0 | 0.044 |

*Answer.* $t_{1/2}$ for unprotected enzyme is 5.95 min; $t_{1/2}$ for protected enzyme is 26.6 min. Substrate greatly stabilizes the enzyme to oxidation.

**6.14** A certain chemical reaction is of the type

$$A + B \rightleftarrows C + D.$$

With the initial concentrations of $A$ and $B$ always equal, the following data were obtained for the half-life of the reaction:

| Initial concentration (molarity) | $t_{1/2}$ (min) |
|---|---|
| 0.005 | 304 |
| 0.010 | 154 |
| 0.050 | 30.6 |
| 0.100 | 15.2 |

Prove that the reaction follows second-order kinetics.

**6.15** The following data are cited by Daniels and Alberty for the hydrolysis of ethyl acetate at 25°C (Daniels, F., and R. A. Alberty, *Physical Chemistry*, 3rd Ed. John Wiley and Sons, New York, 1966). Plot the data according to the integrated form of the second-order rate law to test for kinetic order.

| Time (sec) | NaOH (molarity) | Ethyl Acetate (molarity) |
|---|---|---|
| 0 | 0.00980 | 0.00486 |
| 178 | 0.00892 | 0.00398 |
| 273 | 0.00864 | 0.00370 |
| 531 | 0.00792 | 0.00297 |
| 866 | 0.00724 | 0.00230 |
| 1510 | 0.00645 | 0.00151 |
| 2401 | 0.00574 | 0.00080 |

**6.16** (a) Write the Michaelis-Menten Equation and show that it reduces to a first-order rate expression at low concentrations of S and to a zero-order expression at high concentrations of S.

(b) Prove that $K_m = S$ when $v = \frac{1}{2}V_{max}$.

**6.17** The following data illustrate the advantage possessed by one particular form of the Lineweaver-Burk Equation over another. Plot the data according to Eqs. 6.48, 6.49, and 6.50. Calculate $K_m$ from each plot. From which plot is $K_m$ most easily evaluated? These data are for the enzyme aspartase.

| [S] (molarity) | v (ΔA per min) |
|---|---|
| 0.002 | 0.045 |
| 0.005 | 0.115 |
| 0.020 | 0.285 |
| 0.040 | 0.380 |
| 0.060 | 0.460 |
| 0.080 | 0.475 |
| 0.100 | 0.505 |

*Answer.* $K_m$ calculated from Eq. 6.49 is 0.022 M.

**6.18** The following data were obtained in a study of the competitive inhibition exhibited by L-β-hydroxy-β-(*threo*)-methylaspartate in the β-methyl-aspartase system. The inhibitor was studied in a system containing the best substrate for the enzyme, L-*threo*-β-methylaspartate. The concentration of inhibitor was $5 \times 10^{-4}$ M.

| [$S$] (molarity) | $v-$ no inhibitor ($\Delta A_{240}$ per min) | $v-$ plus inhibitor ($\Delta A_{240}$ per min) |
|---|---|---|
| $5 \times 10^{-5}$ | 0.014 | 0.006 |
| $1 \times 10^{-4}$ | 0.026 | 0.010 |
| $5 \times 10^{-4}$ | 0.092 | 0.040 |
| $1.5 \times 10^{-3}$ | 0.136 | 0.086 |
| $2.5 \times 10^{-3}$ | 0.150 | 0.120 |
| $5.0 \times 10^{-3}$ | 0.165 | 0.142 |

Calculate apparent $K_m$ for substrate and $K_I$ for inhibitor.

*Answer* $K_m = 6.5 \times 10^{-4}\ M$; $K_I = 2.7 \times 10^{-4}\ M$.

**6.19** Will the rate of a chemical reaction be more sensitive to changes in temperature if $E_a$ is large? Assuming that the catalyst is not affected by changes in temperature, would a catalyzed or a non-catalyzed reaction be more sensitive to changes in temperature?

**6.20** There is a similarity in the appearance of Equations 3.47 and 6.61. If we are considering reactions in which there are no thermal alterations of reactants or catalysts, explain why $\Delta H$ in Equation 3.47 may be positive, negative, or zero, while $E_a$ in Equation 6.61 must be positive.

**6.21** Write the kinetic mechanism patterns for the six Ter Ter mechanisms. For answer consult W. W. Cleland, *Biochim. Biophys. Acta,* **67**: 110–111, 1963.

**6.22** The glutamic dehydrogenase reaction is reported to be Ordered Ter Bi (C. Frieden, *J. Biol. Chem.*, 234: 2981, 1959). Summarize the evidence for this mechanism.

**6.23** The following Bi Bi mechanism is called Theorell-Chance and was originally proposed for alcohol dehydrogenase (H. Theorell and B. Chance, *Acta Chem. Scand.*, **5**: 1127, 1951). Work out the product inhibition patterns for $P$ and $Q$ as inhibitors (as is done for other mechanisms in Table 6.1).

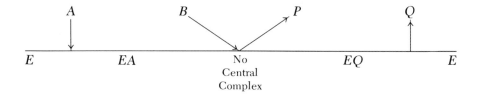

Consult W. W. Cleland, *Biochim. Biophys. Acta* **67**: 123, 1963, for the correct answer.

**6.24** A chemical reaction is studied at two temperatures, 27°C and 37°C. The forward reaction rate constants for the two temperatures are 3.4 min$^{-1}$ and 8.5 min$^{-1}$, respectively. Calculate the energy of activation for the reaction. The equilibrium constants are also known for these two temperatures: $K_{27°} = 2.3 \times 10^{-6}$ and $K_{37°} = 5.1 \times 10^{-6}$. Calculate the heat of reaction from these data (see Eq. 3.47). Finally, draw an energy contour map for the reaction.

*Answer.* $E_a = 1.7 \times 10^4$ cal; $\Delta H = 1.46 \times 10^4$ cal.

**6.25** A certain reaction was found to have an activation energy of 15,500 calories and a specific reaction rate constant of 1.2 min$^{-1}$ at 20°C. Calculate the specific reaction rate constant at 0°C.

**6.26** Prepare an energy scale diagram for the fumarase reaction at 25°C from the following thermodynamic data:

| | *calories* | |
|---|---|---|
| *Individual reaction* | $\Delta H$ | $E_a$ |
| Fumarate + $H_2O$ → Malate | −3,600 | |
| Fumarate + fumarase → Fumarate-fumarase complex | +4,200 | |
| Fumarate-fumarase complex → Transition state | | +6,100 |
| Malate + fumarase → Malate-fumarase complex | −1,200 | |
| Malate-fumarase complex → Transition state | | +15,100 |

(These data are from Massey, V., *Biochem. J.*, **53**:72, 1953.)

# 7 | SYSTEMS OF LARGE PARTICLES: MACROMOLECULAR SOLUTIONS AND COLLOIDAL DISPERSIONS

It has become evident that in order to understand the properties and reactions of protein solutions, the theories accepted for the state of true solutions must be applied.

*S. P. L. Sørensen*

The molecular world of the biologist and the biochemist, in contrast to that of the chemist, is populated with natural high polymers called macromolecules. The vital business of the living cell, such as the transmission of coded genetic information and subsequent protein synthesis, depends on giant molecules of precisely controlled composition and architecture. Many of these possess molecular weights of $10^5$ or more.

All of the principles discussed in the foregoing chapters apply to these macromolecules. Their basic thermodynamic and kinetic properties, their behavior as solutes, their oxidation-reduction characteristics, etc., are in accord with the laws laid down for smaller molecules. One property not shared with the smaller molecules, however, imparts to a macromolecule special characteristics: the extensive surface and subsequent surface energy. This surface gives rise to kinetic, electrokinetic, optical, and adsorption characteristics not found in solutions of small molecules.

The study of macromolecules constitutes a specialized area of chemistry which is not confined to true molecules, but includes polymers, aggregates of smaller molecules, and particles lacking structural organization. The criterion for inclusion in this area is *size*: at least one dimension of the molecule or particle should be between a micron and a millimicron, as an approximation. Larger particles will settle out of solution and smaller ones will fail to display a significant number of the macromolecular characteristics. The term "particle" is a general one and can be applied to any species in or near this size range. Some oil droplets, soap particles, and metal oxide particles fall within these size limitations and give rise to systems traditionally referred to as *colloidal dispersions*. The particles in these systems are frequently called *micelles*, as opposed to the term macromolecules, which is reserved for individual large molecules or a very low-order polymer of giant molecules. For systems of micelles we use the terms *particle weight* or *micellar weight* rather than molecular weight. A problem in molecular weight designation arises with certain of the high polymers, such as amylose and some proteins. A starch solution may contain a number of species of different molecular weights, and a protein solution may be composed of subunits in several states of aggregation. Fortunately, it is usually possible to determine the degree of heterogeneity of the solute molecules by special separation techniques.

While devoting our major interest to the natural high polymers we should not forget that the industrial domain of gels, foams, aerosols, and emulsions is one of long-standing economic importance. In addition, accolades are due the chemists who have filled the commercial markets and our homes with a variety of synthetic macromolecules: polyethylene, polystyrene, polyacrylate, silicones, nylon, dacron, and on and on.* The presentation which follows falls roughly into three major topics: surface chemistry (Sections 7.1, 7.2, and 7.3), transport properties of macromolecular systems (Section 7.5), and optical properties of macromolecular systems (Section 7.6).

## 7.1 The Surface Energy

That a surface could be capable of performing useful work may seem strange at the outset, but we can prove the fact experimentally. If a film is stretched out in a suitable supporting frame, it will, on release, spontaneously contract and lift a small weight in so doing. The lifting of the weight is linear work and comes from the surface energy of the film. The free energy of the surface is defined as

$$G_{\text{surface}} = \gamma A, \tag{7.1}$$

---

*A recent interesting article is O. A. Battista, "Colloidal Macromolecular Phenomena." *Amer. Sci.*, **53**:151, 1965.

where $\gamma$ is the surface tension and $A$ is the surface area. If either surface tension or area changes, the resulting gain or loss in energy will be $\Delta G_{surface}$. The surface tension may be thought of as the work necessary to increase a surface by one square centimeter. It has the units of dynes per cm.

The spontaneous contraction of the surface just described is a consequence of the Second Law of Thermodynamics, that is, the natural tendency of a system to reduce its free energy. The Belgian physicist Plateau has shown that surfaces tend to assume a curvature such that, if $r_1$ and $r_2$ are the principal radii of curvature at a point,

$$\frac{1}{r_1} + \frac{1}{r_2} = \text{constant.} \tag{7.2}$$

Surfaces for which this relationship holds are surfaces of minimum area. A molecular explanation of the tendency of surfaces to contract is based on the strong attractive forces between the interior molecules and the surface molecules. With a liquid surface, the transition between the liquid and the air is sharp, and consists of a zone only one or two molecules thick. A globular protein molecule is more difficult to describe. The surface is uneven, perhaps even containing holes or crevices, regardless of how compact the tertiary folding may be. The surface cannot be of homogeneous character, and is especially complicated for an enzyme possessing only one or two active sites. We shall have to design a "model" molecule of uniform surface and describe its behavior, trusting that the globular proteins will behave in a manner that is reasonably similar. The sphere is the model usually chosen. For unfolded helical structures a long thin rod could be used as a model, but the model structure usually preferred is the prolate ellipsoid. Writing appropriate mathematical equations for the sphere and the ellipsoid is easier than for other possible models.

If a colloidal particle cannot reduce its surface area by contracting, it may reduce its energy by adsorbing substances onto its surface, provided the substances will lower the surface tension, or by preferentially concentrating itself in the surface of the liquid. Oil films which spread out on water reduce the interfacial tension of the supporting liquid and thereby achieve a more stable state. The ancient practice of "pouring oil on troubled waters" (rough seas) has a firm foundation in the Second Law. Surface adsorption is observed for lipids that are carried along by the blood as protein-coated droplets, thus attaining stability and water-solubility at the same time. The passage of substances through living membranes by the transport machinery of the cell very likely also involves adsorption phenomena.

## 7.2 Adsorption Isotherms

The general thermodynamic treatment of adsorption was derived by Gibbs in 1876. Ostwald developed a simplified derivation; however, both treatments lead to the equation:

$$a = -\frac{C}{RT}\frac{(d\gamma)}{dC},\qquad(7.3)$$

where $a$ is the molar excess of solute per square centimeter of the surface area, $C$ is the bulk concentration, and $\gamma$ is the surface tension. The "surface excess," $a$, has been defined in various ways. Gibbs' choice was to consider a portion of the liquid containing a surface of unit area. A portion of the solution in the interior of the liquid containing exactly the same number of molecules of solvent was taken for comparison. The number of moles of the solute present in the first portion in *excess* of the number of moles of the same solute in the second portion (interior) was taken to be the molar excess of solute. The equation gives a mathematical basis for a useful generalization, that is, if a substance tends to reduce surface tension it will collect in the surface layer of the liquid, and conversely. In this context, "in the surface" means in that volume of liquid bound by the interface and extending into the liquid phase to a depth of only a few molecules. Inspection of the equation shows that, if the rate of change of the surface tension with concentration is negative (surface tension decreases with increasing concentration of adsorbed material), $a$ will be positive, meaning that solute is present in the surface in excess of its concentration in the body of the liquid. If $d\gamma/dC$ is positive (surface tension increases with increasing concentration of solute), $a$ will be negative, meaning that the solute concentration in the surface is less than the concentration in the body of the liquid.

Equation 7.3 contains simplifications based on the Perfect Gas Law, and, therefore, is subject to the limitations imposed by invoking ideality. Although measuring concentrations of solutes in surface layers only a few molecules thick is very difficult, soaps, detergents and other "surface-active" substances have been shown experimentally to be present in the surface layers in concentrations exceeding those in the interiors of solutions. All such surface-active substances lower the surface tension of aqueous solutions. With most inorganic solutes, increasing the concentration leads to a small increase in surface tension and a surface excess of solute does not occur.

The Gibbs Equation was not derived from an analysis of the mechanism of the adsorption process. Langmuir, however, derived his equation in just such a manner. He proposed for consideration a crystal surface on which gas molecules were colliding and rebounding. He reasoned that the rate at which molecules would be adsorbed would depend on: (1) the number striking the surface per second, (2) the fraction of incident molecules that adhered, and (3) the area not already covered by the adsorbed molecules. The temperature is considered to remain constant throughout the process. Recognizing that a surface of a crystal is irregular, he proposed that adsorption would take place only at certain sites, designated as active centers. If the number of active centers is $a'$, and the number occupied is $a$, then the number of free sites is $(a' - a)$. The rate of adsorption of molecules will be proportional to $(a' - a)$ and to the pressure, $p$. The rate of adsorption, $r_1$, is then defined as

$$r_1 = \beta(a' - a)p,$$

where $\beta$ is a proportionality constant. The rate of desorption, $r_2$, will be proportional to the number of molecules already bound, and therefore,

$$r_2 = \alpha a,$$

where $\alpha$ is a proportionality constant for the desorption process. At equilibrium, $r_1 = r_2$ and

$$\beta(a' - a)p = \alpha a.$$

Hence,

$$\frac{a}{a'} = \frac{\beta p}{\alpha + \beta p} = \frac{1}{1 + \dfrac{\alpha}{\beta p}}.$$

Replacing $\beta/\alpha$ by $b$ we obtain

$$a = \frac{a'bp}{1 + bp}. \tag{7.4}$$

If the molecules being adsorbed onto the surface are solute molecules in a solvent, pressure is replaced by concentration and we write

$$a = \frac{a'bC}{1 + bC}. \tag{7.5}$$

At low pressures (or concentrations) $1 > bp$ and therefore,

$$a = a'bp.$$

The coverage of the surface will be directly proportional to the pressure. At high pressures, $bp > 1$ and $a = a'$, meaning that all active centers are occupied. A plot of $a$ versus $p$ gives a curve of the form shown in Fig. 7.1. The plot in Fig. 7.1 is an isotherm, *i.e.*, temperature is not a variable in the system described. The parallel between the Langmuir model and an

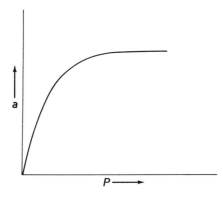

**Figure 7.1** General shape of the Langmuir adsorption isotherm.

enzyme adsorbing substrate onto its active site is most striking, as is the parallel between the Langmuir plot and the biphasic velocity-concentration plot for an enzymatic reaction (Fig. 6.7).

If adsorption is multilayer, that is, if molecules are being adsorbed in $n$ successive layers, the relationships are described by the Brunauer, Emmett, and Teller Equation, commonly known as the BET isotherm. It is

$$a = \frac{a'hP[1 - (n+1)P^n + nP^{n+1}]}{(1-P)[1 + (h-1)P - hP^{n+1}]}. \tag{7.6}$$

The terms $a$ and $a'$ are defined as in the Langmuir equation; $h$ is a constant term. $P = p/p_0$, where $p_0$ is the saturated vapor pressure and $p$ is the observed vapor pressure of the substance being adsorbed. As an approximation,

$$h = e^{-(E_1 - E_n)/RT},$$

where $(E_1 - E_n)$ is the difference in heats of adsorption of the first layer $(E_1)$ and the last layer $(E_n)$. It is assumed that $E_n$ is approximately equal to the ordinary heat of condensation of the gas. Multilayer adsorption gives rise to plots of the general shape shown in Fig. 7.2. If only one layer exists, $n = 1$, and the BET Equation reduces to

$$a = \frac{a'hP}{1 + hP}, \tag{7.7}$$

which is nearly identical with the Langmuir equation. If we assume that the heats of adsorption of all layers except the first are nearly equal,

$$\frac{p}{a(p_0 - p)} = \frac{1}{a'h} + \frac{(h-1)p}{a'hp_0}. \tag{7.8}$$

Plotting $p/a(p_0 - p)$ versus $p/p_0$ should give a straight line.

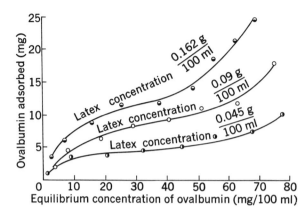

**Figure 7.2** Adsorption of ovalbumin on latex at different latex concentrations. The pH of the system was 6.98. [From Williams, H. B., and A. R. Choppin, *J. Gen. Physiol.*, 34:183, 1950. By permission of the Rockefeller Institute Press.]

Historically, the Freundlich adsorption isotherm is the oldest, and has probably been the most widely used. Unlike those just given, it is empirical. Freundlich proposed that the amount of gas or solid adsorbed per gram of adsorbing material would be proportional to the equilibrium pressure (for a gas) or the equilibrium concentration (for a solid), raised to some fractional exponent. In other words, the higher the pressure or the greater the concentration of solute, the more substance would be adsorbed onto the surface, but the proportionality would not be simple, rather, it would be exponential. The equation may be written

$$x = kp^{1/n} \quad \text{or} \quad x = kC^{1/n}, \tag{7.9}$$

where $x$ is the grams of substance adsorbed per gram of adsorbing material, $p$ is the equilibrium pressure or $C$ is the equilibrium concentration, and $1/n$ is the empirical exponent necessary to linearize the data. If we write the equation in the logarithmic form, then

$$\log x = \frac{1}{n}(\log C) + \log k. \tag{7.10}$$

Plotting $\log x$ versus $\log C$ should give a straight line of intercept $\log k$ and slope $1/n$. Although the Freundlich Equation has no theoretical basis, it has been useful as a means of presenting data. In Table 7.1 are some classical experimental data for the adsorption of solutes on charcoal at 25°C. The resulting plots are shown in Fig. 7.3.

None of the mathematical treatments of adsorption presented in the preceding paragraphs is of significant use in describing or analyzing such biochemical phenomena as the adsorption of substrates to the active sites of enzymes or the transport of solutes through living membranes. Our knowledge of these processes — which involve a variety of forces — is still too limited to propose suitable models. For the adsorption

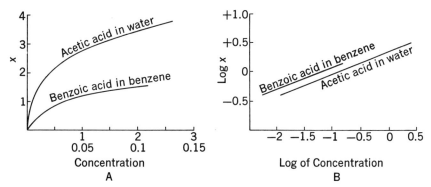

**Figure 7.3** Freundlich plots prepared from the data of Table 7.1. A is a plot according to Eq. 7.9; B is a plot according to Eq. 7.10. [From Getman, F. H., and F. Daniels, *Theoretical Chemistry*, 6th Ed. John Wiley and Sons, New York, 1937.]

of substrates, we usually assume that Coulombic forces associated with ionic bonds are of great importance as are van der Waals forces arising from the formation of apolar bonds between hydrocarbon groups and from the formation of polar bonds between hydrophilic groups. With the increasing vigor of research on mechanisms of enzyme action, active site mapping, and all topics related to these areas, adsorption phenomena will eventually receive their share of scrutiny. A major difficulty is that the theory is not sufficiently powerful in its present state of development and is too much colored by empiricism.

**TABLE 7.1** ADSORPTION OF SOLUTES BY CHARCOAL AT 25°C

| Acetic acid in water | | Benzoic acid in benzene | |
|---|---|---|---|
| C | x | C | x |
| 0.018 | 0.467 | 0.006 | 0.44 |
| 0.031 | 0.624 | 0.025 | 0.78 |
| 0.062 | 0.801 | 0.053 | 1.04 |
| 0.126 | 1.11 | 0.118 | 1.44 |
| 0.268 | 1.55 | | |
| 0.471 | 2.04 | | |
| 0.882 | 2.48 | | |
| 2.79 | 3.76 | | |

*Source:* Reproduced from Getman, F. H., and F. Daniels, *Theoretical Chemistry*, 6th Ed. John Wiley and Sons, New York, 1937.

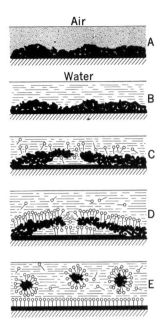

**Figure 7.4** A schematic representation of the mechanism of detergent action in removing particles of soil from a surface.

A: Surface is covered with greasy particles of soil.

B: Water does not remove dirt because it fails to wet the hydrophobic surface.

C: A detergent is added to the water, as represented by the wand-shaped particles. The head of the wand is the hydrophilic end of the molecule and the handle represents the long hydrocarbon tail (hydrophobic end).

D: The detergent molecules orient themselves so that the hydrocarbon tails are adsorbed to the greasy particles of soil. The dirt may now be solubilized when the surface is agitated.

E: Particles of dirt have been solubilized by the adsorption of detergent molecules. Note the adsorption of detergent to the surface. [From "Synthetic Detergents" by Kushner, L. M., and J. I. Hoffman. Copyright © October 1951 by Scientific American, Inc. All rights reserved.]

## 7.3 Adsorption on Surfaces

The preceding discussion of adsorption of gases and solids on surfaces may appear only remotely related to the macromolecular systems present in living cells. A closer look, however, reveals much that is germane. The macromolecules of protein, polysaccharide, or nucleic acid composition may not adsorb gases or foreign solids on their surfaces, but they do bind ions and immovable water films only a few molecules thick. The acquisition of both a solvent shield and a charge greatly increases the stability of macromolecules in solution. Since all biological systems are aqueous ones, lipid particles can avoid precipitation by acquiring an external protein film which is then solvated and charged. Large particles or macromolecules that have polar surfaces and are readily solvated by water are called *hydrophilic* (water loving) and those particles with nonpolar surfaces are called *hydrophobic* (water hating). The tendency of hydrophilic substances to coat the surfaces of hydrophobic particles in an aqueous system is called *protective colloid action.*

A graphic molecular picture of protective colloid action is seen in the detergent properties of soaps. The soap molecule, corresponding to the general formula $RCOO^-Na^+$, has a highly polar head attached to a long hydrocarbon tail. As shown in Fig. 7.4, the hydrocarbon tails tend to dissolve in or bind to the surface of an oil droplet, leaving the polar heads oriented toward the water phase. The particle then becomes highly ionic in character and is soluble in the solvent. Particles of soil are thus removed from the fibers of clothing and other surfaces. Agitation helps to loosen the dirt and speed the process of emulsification.

Films of oriented soap layers can be built up by repeatedly dipping a glass plate into water which has been previously covered with a film of barium stearate. The barium stearate orients itself on the water surface with the polar carboxyl group in contact with the water and the hydrocarbon chains extended above the surface. When the glass plate contacts the film, the carboxylate groups bind to the glass surface and the hydrocarbon chains orient outward. On the next passage into the film, the second layer orients itself with apolar groups touching apolar groups, leaving the carboxylate groups on the outside. The third passage reverses the direction of the soap film again, and so on. If we can build up stable lipid layers of regular structure this easily, it is understandable how living membranes achieve stability through alternating protein and lipid layers as was shown earlier in Fig. 3.4.

Protective colloid action was studied by Zsigmondy, who devised a quantitative measurement called the gold number. *Gold sol* (a *sol* is a colloidal dispersion of a solid in a liquid) contains negatively charged hydrophobic colloidal particles. In the presence of various hydrophilic substances, stability of varying degrees is achieved by protective colloid action. The gold number is the weight (mg) of protective colloid that when added to 10 ml of a Zsigmondy gold sol, just fails to prevent the color change from red to blue when 1 ml of 10 percent NaCl is added. With the unprotected gold sol, the addition of sodium chloride brings about charge neutralization, aggregation of the sol particles, and finally, precipitation. The various steps in the precipitation of gold sol are shown in Fig. 7.5.

The concept of protective colloid action described above, however, is a deceptively simple one. In a detailed study of gold sol stabilization, Williams and Chang found that low concentrations of protective substances, such as gelatin and gum arabic, resulted in sensitization rather than stabilization—that is, the system with the "protective" substance was more readily precipitated by electrolytes than the gold sol alone.[*] Apparently, when the gold sol particles greatly outnumber the hydrophilic ones, aggregates of gold sol particles form around hydrophilic nuclei, producing a system that is readily precipitated. When the concentration of hydrophilic particles exceeds a certain critical value, the nature of the aggregates is reversed, and the hydrophilic particles form clusters with gold sol nuclei. Proteins, detergents, and polysaccharides differ in their effectiveness as protective colloids.

The gold number of spinal fluid can be used by the clinical chemist in the diagnosis of certain diseases in which the proteins of the spinal fluid are altered.

---

[*]Williams, H. B., and L. T. Chang, *J. Phys. and Colloid Chem.*, **55**:719, 1951.

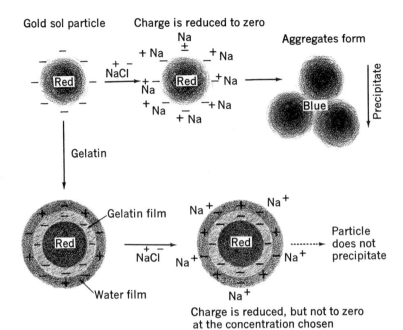

**Figure 7.5**  Protective colloid action of gelatin on gold sol. The gold sol particle seen in the upper left is charged by virtue of the adsorption of anions. These could be citrate or formate ions, depending on the method of sol preparation. In the absence of protective colloid, the sol particle does not possess a protective solvent sheath and is, therefore, easily precipitated by the addition of sodium chloride to the solution, as shown in the upper right. The charge on the gold sol is neutralized by the sodium ions, the particles no longer repel one another, and aggregation results. If gelatin is added to the sol, the protein is attracted to the sol particle and forms a protective film around it. Additional charges, both positive and negative, are contributed by the ionic surface of the gelatin. The protein surface adsorbs water molecules, and thereby affords additional protection to the gold sol particle. The addition of sodium chloride to the solution will reduce the charge on the sol, but not as effectively as in the absence of protective colloid. The degree of protection given by the gelatin is quantitatively expressed in the gold number, as defined in the text.

Apropos of the terminology introduced thus far, *i.e.*, gold sol, hydrophilic, hydrophobic, etc., the student should be aware that this realm of interfacial phenomena is so replete with terms of specialized technical meaning that it approximates a language separate from the rest of chemistry. It has seemed the better part of wisdom to omit most of them.[*]

Of more interest to the biochemist than the gold number and related phenomena, however, is the matter of charging the surface of the

---

[*] See page 74 of the book by Jirgensons and Straumanis that is included in the *Suggested Additional Reading.*

macromolecule. How does the charge originate? (1) The surface can produce its own charge by virtue of anionic or cationic groups on the molecule. Proteins contain carboxylate anions, which contribute negative charges, and protonated basic groups, which contribute positive charges. Nucleic acids are negatively charged because of the dissociation of the phosphate groups that alternate with pentose molecules in the backbone structure. (2) If no ionizing groups are present in the molecular structure, as for amylose or cellulose, the surface will bind ions from the solvent or solution because of the polarization around hydroxyl groups or ring oxygen atoms. Even an apparently inert surface such as that of a droplet of paraffin oil will nevertheless be nonuniform in its electron distribution and therefore susceptible to induced polarization leading to binding of ions from the solvent. This extended surface of charge is one of the properties of large particles not encountered in ordinary solutes. The existence of the charge was deduced from early observations on the movement of particles in an electric field. Tiny particles of glass, oil, and rust, or bacteria move across the field of the microscope when an electric potential is placed across the solution, and the direction of movement is reversed when the polarity of the field is reversed. This phenomenon is called *electrophoresis* and is the first of the transport properties of macromolecules discussed in Section 7.5.

## 7.4 Types of Molecular Weights

The apparent molecular weight of a heterogeneous sample depends on the method by which it is determined. Any method of molecular weight determination which involves molecule-counting (freezing point depression, osmotic pressure measurements, titration methods, etc.) gives a type of molecular weight known as a *number-average molecular weight*, $M_n$. Other methods of molecular weight determination that involve terms related to the size and shape of the molecule (methods which will be discussed in Sections 7.5 and 7.6 — diffusion, sedimentation, flow birefringence, and light scattering) give a molecular weight known as the *weight-average molecular weight*, $M_w$. *Viscometric molecular weights*, $M_v$, constitute a third type. In addition, there are types of averages called z average, (z + 1) average, etc., that give molecular weights larger than $M_n$, $M_w$, and $M_v$.

The number-average and weight-average molecular weights are the most frequently used. The relationship between the five types mentioned for polydisperse samples is

$$M_n < M_v \le M_w < M_z < M_{(z+1)}.$$

Number-average and weight-average molecular weights are statistically weighted averages based on the number fraction and the weight fraction, respectively. If $N_i$ is the number of molecules of the $i$th species present in the mixture, then $W_i$ is the weight of the $i$th species, and $M_i$ is the corresponding molecular weight. The difference between a number-average and a weight-average molecular weight is shown in the following definitions:

$$M_n = \sum_i n_i M_i \quad \text{where} \quad n_i = \frac{N_i}{\sum_i N_i} = \text{number fraction.} \quad (7.11)$$

$$M_w = \sum_i w_i M_i \quad \text{where} \quad w_i = \frac{W_i}{\sum_i W_i} = \text{weight fraction.} \quad (7.12)$$

Since $N_i = W_i/M_i$, we may define $M_n$ and $M_w$, as well as $M_z$ and higher orders of averages, in a uniform notation, as shown below.

$$M_n = \frac{\sum_i N_i M_i}{\sum_i N_i} \tag{7.13}$$

$$M_w = \frac{\sum_i N_i M_i^2}{\sum_i N_i M_i} \tag{7.14}$$

$$M_z = \frac{\sum_i N_i M_i^3}{\sum_i N_i M_i^2} \tag{7.15}$$

$$M_{(z+1)} = \frac{\sum_i N_i M_i^4}{\sum_i N_i M_i^3} \tag{7.16}$$

The summation in each equation must be extended over all possible species, up to and including the $i$th species. If a sample is truly homogeneous, then all equations will give the same molecular weight.

For the number-average molecular weight, it may be more convenient to express the number of molecules of each type as a percentage or as a mole fraction. For these cases we write

$$M_n = \frac{\sum_i M_i(\%_i)}{100} \tag{7.17}$$

and

$$M_n = \sum_i X_i M_i. \tag{7.18}$$

The difference between the number-average and weight-average molecular weight may be illustrated by a simple example. Suppose that we know the distribution of the different molecular weights making up a polydisperse sample. Let us say that the sample contains

| Percent | Molecular weight |
|---------|------------------|
| 20 | 10,000 |
| 75 | 40,000 |
| 5 | 80,000 |

According to Eq. 7.17 (a special form of Eq. 7.13),

$$M_n = \frac{(20)\,(10{,}000) + (75)\,(40{,}000) + (5)\,(80{,}000)}{100} = 36{,}000.$$

We may calculate $M_w$ from Eq. 7.14, which is also expressed in a percentage form comparable to Eq. 7.17.

$$M_w = \frac{(20)\,(10{,}000)^2 + (75)\,(40{,}000)^2 + (5)\,(80{,}000)^2}{(20)\,(10{,}000) + (75)\,(40{,}000) + (5)\,(80{,}000)} = 42{,}800.$$

### 7.5  The Transport Properties of Macromolecular Systems

The movement of matter, heat, or any other entity under the influence of an appropriate gradient is called a *transport* phenomenon. Examples are: matter moving across a concentration, pressure, or potential energy gradient; charged particles driven by a difference in electrical potential; heat flowing across a temperature gradient; and, for viscosity, momentum being transported across a velocity gradient.

In general, the transport properties of a macromolecule are studied to obtain information about its size and shape or its charge. Five phenomena relating to the transport properties of macromolecules will be discussed in this section: electrophoresis, diffusion, sedimentation, viscosity, and flow birefringence. The osmotic pressure of macromolecules was discussed in Chapter 3.

**Electrophoresis**  The movement of a macromolecule (or large particle) in an electric field is called *electrophoresis*. Helmholtz undertook a mathematical analysis of the electrical potential of large particles, both in an ion-free solvent and in the presence of ions of opposite charge. The Helmholtz derivation is as follows.

If we take a sphere as a suitable model of the macromolecular particle and assign to it a charge $Q$ and a radius $r$, the potential at a point on the surface of the sphere is given by

$$\text{Potential} = \frac{Q}{Dr}, \tag{7.19}$$

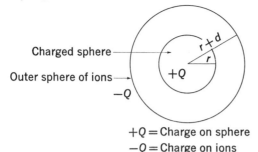

Charged sphere

Outer sphere of ions

$+Q$

$-Q$

$r+d$

$r$

$+Q =$ Charge on sphere
$-Q =$ Charge on ions

**Figure 7.6** The Helmholtz model of the "double layer."

where $D$ is the dielectric constant of the medium. This is the potential in the absence of any significant concentration of ions. This relationship is readily derived from the expression for the potential energy of attraction of ions, discussed in the first part of Section 1.1. On page 3, we showed that

$$\text{Energy of attraction} = \frac{Q_1 Q_2}{Dr}.$$

Since electrical energy can be expressed as the product of potential times charge (Table 1.3), we can write

$$\text{Energy} = \text{Potential} \times \text{charge } (Q) = \frac{Q_1 Q_2}{Dr}.$$

Dividing this equation by $Q$, the charge, yields Eq. 7.19. The potential may also be defined as the work necessary to bring a unit charge from infinity to a point $r$ centimeters from the center of the sphere, *i.e.*, the work per unit charge.

When small ions are present in the solvent, those of opposite sign are attracted to the charged sphere and tend to orient themselves around it. Helmholtz likened this situation to the two charged plates of a condenser, actually a condenser made of two concentric spheres, as shown in Fig. 7.6. The two plates of this condenser were referred to as the "rigid double layer." The outer layer of ions reduces the potential on the sphere by an amount equal to $- Q/D(r + d)$, where $d$ is the thickness of the double layer (see Fig. 7.6). The resultant potential was called the *zeta potential*.

$$\zeta = \frac{+Q}{Dr} - \frac{Q}{D(r + d)} = \frac{Q}{Dr} \cdot \frac{d}{(r + d)}. \qquad (7.20)$$

When $d \ll r$

$$\zeta = \frac{Qd}{Dr^2}. \qquad (7.21)$$

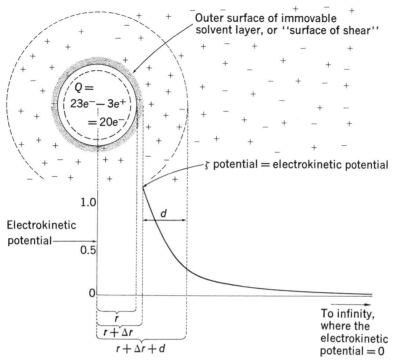

**Figure 7.7** A large spherical particle of negative charge with a firmly bound solvent layer and a diffuse double layer of gegenions. This figure should be compared with Fig. 7.6. [After Abramson, H. S., L. S. Moyer, and M. H. Gorin, *Electrophoresis of Proteins*. Reinhold, New York, 1942, p. 32.]

The Helmholtz model of the rigid concentric spheres is an intuitive concept that has been modified and extended by Gouy, Debye, and Hückel to yield a more realistic model which is mathematically more complex. The double layer is now conceived to be composed of the charged surface of the particle (still spherical for the sake of simplicity) and a diffuse, enveloping "ion atmosphere" in which the ions of charge opposite to that of the surface (called *gegen*-ions, from the German *gegen*, against) are most concentrated near the surface of the sphere and thin out as distance from the surface increases. Figure 7.7 presents this concept in a diagrammatic way. Immediately next to the surface of the particle is an immovable water layer. The distance $d$ is measured from the outer boundary of the immovable water film, sometimes referred to as the surface of shear. You will note in Fig. 7.7 that $d$ is depicted as that distance which will define an area containing some of the *gegen*-ions, rather than all of them as in the Helmholtz model. If $d$ is so defined as to sweep out an area containing all of the *gegen*-ions, an unrealistically large area would be involved because of the diffuse nature of the double layer. The theoretical value of the zeta potential would not agree with that determined experimentally.

Despite the simplicity of the model shown in Fig. 7.7, it serves as a basis for calculating the electrostatic interactions that arise as a consequence of the protein surface being a polyionic fabric. These calculations provide a quantitative explanation for the experimental observation that certain $pK'_a$ values within a large protein may be shifted as much as 1.5 units away from the $pK'_a$ of that same group in an amino acid (review Table 4.7). A high net positive charge on the protein will repel hydrogen ions and tend thereby to lower the $pK'_a$ of groups which become protonated at low pH, just as a high net negative charge will repel hydroxyl ions and similarly displace the $pK'_a$ of groups which dissociate at high pH. The equations for calculating electrostatic interactions have not been included in this presentation.*

Returning to Eq. 7.20, we can rearrange it to give the form

$$\zeta = \frac{Q}{Dr} \cdot \frac{1}{(1 + r/d)}. \tag{7.22}$$

The Debye-Hückel parameter $\kappa$ has the dimensions of reciprocal centimeters

$$\kappa = \sqrt{\frac{8\pi \mathfrak{N} e^2}{1000\, DkT}} \cdot \sqrt{I},$$

where $I$ = the ionic strength of the solution,
   $k$ = the Boltzmann constant,
   $D$ = the dielectric constant of the medium,
   $e$ = the charge on the electron,
   $\mathfrak{N}$ = Avogadro's number,
   $T$ = the absolute temperature.

Substituting $\kappa$ for $1/d$ in Eq. 7.22 gives

$$\zeta = \frac{Q}{Dr} \cdot \frac{1}{(1 + \kappa r)} \tag{7.23}$$

for a sphere whose radius is less than the thickness of the double layer. The following equation is the most general expression for the zeta potential of a sphere of any size relative to the thickness of the double layer:

$$\zeta = \frac{Q}{Dr} \cdot \frac{f(\kappa r)(1 + \kappa r_i)}{(1 + \kappa r + \kappa r_i)}, \tag{7.24}$$

---

*Further discussion of electrostatic interactions is given by Edsall and Wyman in the reference work listed in *Suggested Additional Reading*.

where $r_i$ is the average radius of the ions in solution. The term $f(\kappa r)$ stands for a function of $\kappa r$ and must be evaluated separately.

Now that we have a satisfactory expression for the potential—in a sense the "net potential"—of the charged macromolecule, we are ready to analyze the forces involved in electrophoresis.

The derivation consists of three steps: analyzing the opposing forces which control the phenomenon, defining these forces mathematically, and expressing them in an equation which may be rearranged to suit our needs. The same technique will be used in deriving the equation of sedimentation.

The charged solute particle, *e.g.*, a macromolecule, moves toward the electrode of opposite charge under the influence of an electrical force. This movement is opposed by a frictional force resulting from the solvent molecules dragging against and obstructing the particle. When the particle moves with constant velocity, these two forces exactly balance and

$$\overrightarrow{\text{Electrical Force}} = -\overleftarrow{\text{Frictional Force}}$$
$$\text{(driving force)} \qquad \text{(retarding force)}$$

We need to define a number of terms at this point:

1. $\overrightarrow{\text{Electrical force}} = QX$

2. $\overleftarrow{\text{Frictional force}} = -fv$ or $-f\dfrac{dx}{dt}$

$Q =$ charge on the particle
$X =$ electrical field strength
$f =$ frictional coefficient
$v =$ velocity of the particle in cm/sec
$\dfrac{dx}{dt} =$ velocity of the particle in cm/sec
$\eta =$ viscosity of the solvent
$r =$ radius of the particle in cm

3. Molecular frictional coefficient for a spherical particle $= f_0$, where $f_0 = 6\pi\eta r$ (Stoke's Law)

4. Molecular frictional coefficient for a particle of any size and shape $= f$, where $f = kT/\mathcal{D}$

$k =$ Boltzmann constant
$T =$ absolute temperature
$\mathcal{D} =$ the diffusion constant of the particle

5. Molar frictional coefficient for spherical particles $= f_0$, where
$$f_0 = 6\pi\eta\mathfrak{N}\left(\frac{3M}{4\pi\mathfrak{N}\sigma}\right)^{1/3}$$

6. Molar frictional coefficient for particles of any size and shape $= f$, where $f = RT/\mathcal{D}$

$M =$ the molecular weight of the particles
$\sigma =$ the density of the particles
$\mathfrak{N} =$ Avogadro's number

To proceed with the derivation, we equate the two forces:

$$QX = -(-fv) = fv. \tag{7.25}$$

Electrical force $= -$Frictional force

If the electrophoretic *mobility*, $u$, is defined as the velocity per unit field of force, it follows that

$$u = \frac{v}{X} = \frac{Q}{f} \tag{7.26}$$

The mobility has the units of cm$^2$ volt$^{-1}$ sec$^{-1}$. We may write two equations, each for an isolated particle

$$u = \frac{Q}{6\pi\eta r} \qquad\qquad u = \frac{Q\mathcal{D}}{kT} \tag{7.27}$$

(spherical particle)       (particle of any shape)

These are the fundamental equations of electrophoresis. Note that the electrophoretic mobility is independent of the applied electrical field. The electrophoretic mobility can be measured experimentally as can the diffusion constant. The radius of a spherical particle can be readily calculated from its molecular weight and density. We would particularly like to calculate the charge on the particle, but we need a more complete equation than 7.27, because we must take the ion atmosphere into consideration. The simple derivation just presented applies only to a charged particle in pure solvent. Let us retrieve our original definition of the potential on the surface of a sphere in the absence of ions:

$$\zeta = \frac{Q}{Dr} \qquad \text{(Eq. 7.19)}.$$

Inspection of Eq. 7.27 for spherical particles

$$u = \frac{Q}{6\pi\eta r}$$

shows that $Q$ and $r$ are common to both expressions. Substituting $\zeta D$ for $Q/r$ gives

$$u = \frac{\zeta D}{6\pi\eta}. \tag{7.28}$$

We now have an electrophoresis equation in terms of zeta potential rather than charge. As was shown in Eq. 7.23, the effect of the ion atmosphere on the zeta potential is to reduce it by $1/(1 + \kappa r)$. Let us make

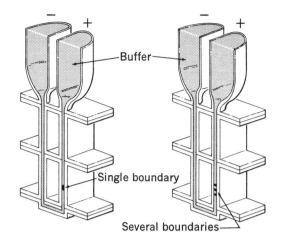

**Buffer**

**Single boundary**

**Several boundaries**

**Figure 7.8** Cross sections of Tiselius cells showing the appearance of a single boundary on the left and several boundaries on the right. These boundaries are not visible to the eye unless the protein is colored (*e.g.*, cytochrome *c*, etc.).

that same modification in Eq. 7.28. For $\zeta D$ we will substitute $Q/r \cdot 1/(1 + \kappa r)$. This gives an electrophoresis equation of the form

$$u = \frac{Q}{6\pi\eta r} \cdot \frac{1}{(1 + \kappa r)} \qquad (7.29)$$

for a spherical particle in a solution containing ions. The complete electrophoretic treatment employs the same correction factor as shown in Eq. 7.24:

$$u = \frac{Q}{6\pi\eta r} \cdot \frac{f(\kappa r)(1 + \kappa r_i)}{(1 + \kappa r + \kappa r_i)}. \qquad (7.30)$$

An application of this equation is shown in Problem 7.7.*

The experimentally measured quantity is the velocity of the macro-molecules in the electric field. Since the individual molecules cannot be seen, we measure the velocity of the protein front as it moves through the solvent. This zone can be detected by optical methods and is called a *boundary*. It will be defined more precisely later. A description of the experimental technique of electrophoretic measurements is inappropriate to the purpose of this presentation. Detailed accounts can be found in many reference works.† In Fig. 7.8 is shown diagrammatically the boundary of a pure protein and the boundaries from a protein mixture in an electrophoresis cell. Tiselius designed the first electrophoresis apparatus with optical attachments for the accurate measurement of

---

*Data were taken from *Electrophoresis of Proteins* by Abramson, Moyer and Gorin, Reinhold, New York, 1942. Reprinted by Hafner Publishing Co., New York, 1964.

†Greenberg, D. M., Ed., *Amino Acids and Proteins*, Charles Thomas, Springfield, Ill., 1951.

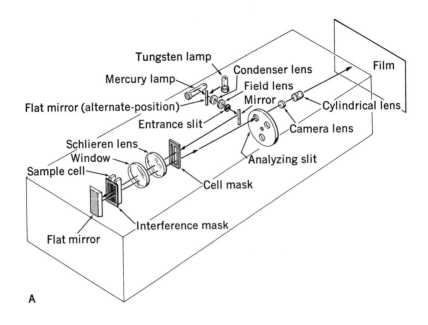

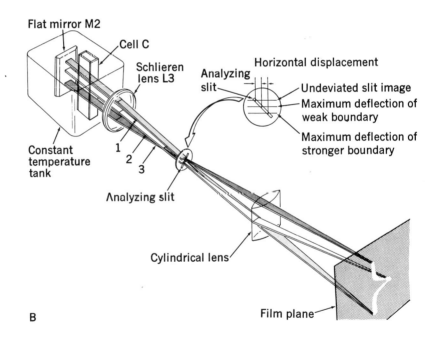

**Figure 7.9** Optical schematic of the Perkin-Elmer Model 238 electrophoresis apparatus.
A: Layout of optical system. The two available systems are shown: modified Philpot-Svennson cylindrical lens system with diagonal slit and tungsten source, and alternate Rayleigh interference fringe system (with or without diagonal phase plate), using mercury lamp source. The system is folded at the cell by a mirror to double-pass the

*(cont'd.)*

velocities. For this accomplishment he received the Nobel Prize. The essential features of the cell and optical system are shown in Fig. 7.9. The technique has many applications, including use as a means of separating mixtures of macromolecules, as a criteria of purity of proteins, and as a means of determining the isoelectric point of a protein in a particular buffer. When a solution of a mixture of proteins is placed in an electrophoresis cell and an electrical potential is applied, each protein will move with a velocity that depends on its *charge density*, which is the net charge of the particle divided by the surface area. In the ideal situation, each individual protein will display a velocity different from the others, although it is not unusual for two proteins to have the same velocity. If, however, no protein has the same velocity as another, if

cell. The schlieren lens cell, $3\frac{1}{2}$ inches in diameter, is filled with dry nitrogen to prevent lens fogging. In the schlieren system, light from the tungsten lamp passes through the entrance slit, which lies at the focal plane of the schlieren lens. Light emerging from the entrance slit is therefore collimated by the lens and directed onto the flat mirror (extreme left) after passing through the Tiselius cell. The mirror reflects the light through the cell, and it is focused by the schlieren lens onto the analyzing slit. Light from the slit passes through the camera lens and cylindrical lens onto the film (extreme right). A detailed representation of this part of the system is shown in part B of the figure. If the Rayleigh interference system is used, the mercury source is selected. A mechanical linkage connected to the source selector switch causes the grid and appropriate slit to move into the optical system (at the point labeled "entrance slit") when the switch is actuated. An interference mask is placed on the back of the Tiselius cell before the cell is immersed in the constant-temperature bath. The diagonal slit is replaced by an open diaphragm. (Certain other combinations are possible.) The grid converts the single horizontal slit into a large number of point sources, which are spaced to match the interference effect produced by the cell mask. The beams from the point sources, after being collimated by the schlieren lens, pass through the interference mask and the cell and cell flange as in a standard Rayleigh interferometer. If a uniform sample is present in the cell, the light passes through both parts of the cell and produces interference maxima and minima at the focal plane of the schlieren lens. The cylindrical lens magnifies the line pattern in a horizontal direction so that it covers the entire area occupied by the film and separates the lines— making them readily visible. When a boundary is present within the cell, the difference in velocity of light above and below the boundary produces a horizontal displacement of any fringe as it is traced across the boundary. The fringes are counted and the concentration of solute at the boundary is calculated accordingly.
B: Detail of optical system. The flat mirror on the left reflects light through the cell as previously described above. Three rays are shown: rays 1 and 3 pass through solvent only; ray 2 passes through a boundary. The schlieren lens focuses the light at the plane of the analyzing slit. The greater the refractive index at the boundary, the more the light is bent downward. As may be seen in the diagram, the outer rays converge at a point in the plane of the slit, whereas the middle ray is displaced downward. The camera lens, which is located directly behind the analyzing slit, focuses the image of the cell on the film. When a boundary in the cell diverts a portion of the beam, the deflection caused by the analyzing slit is magnified in the horizontal direction by the cylindrical lens, and the corresponding portion of the cell image is displaced to one side. [Schematic diagrams and explanatory notes supplied by Perkin-Elmer Corporation, Norwalk, Connecticut.]

the cell is of suitable size, and if sufficient time is allowed, a mixture of proteins can be separated into individual components. Sections of the cell can be closed so as to trap the individual proteins in separate compartments, and in this way the electrophoretic technique can be used to prepare and purify proteins.

One of the criteria for purity of a protein is that it exhibit only one boundary during electrophoresis. If a protein, hitherto thought to be a pure preparation, separates into two boundaries in the electrophoresis cell, it is obviously not homogeneous. It may, of course, be a mixture of subunits in various degrees of polymerization. This is more the exception than the rule. A mixture which does not separate under specific electrophoretic conditions may do so under other conditions, such as electrophoresis at a different pH, or sedimentation in a centrifugal field of force. In establishing the purity of a protein, therefore, it is necessary to show that the preparation is homogeneous by several criteria.

The *isoelectric point* is the pH at which the protein does not move in an electric field. Usually it is defined for a particular buffer and ionic strength, since it may vary with experimental conditions. Conceptually, it is taken as the pH at which the particle possesses a net charge of zero: all the anions on the particle are exactly matched by the cations. As has already been stressed, these ionic groups arise from both the protein and the ions in the buffer. Protein free of all foreign ions is charged solely through surface ionization. At the pH of zero net charge each protonated group will be exactly matched by an anion from which a proton has dissociated. This pH is called the *isoionic point* of the protein. Whether this state is ever completely achieved is not known. It can be approached by exhaustively dialyzing protein against distilled water or by passing the protein solution through a column of mixed bed resin. Under these conditions, of course, many proteins precipitate. Because of these uncertainties and difficulties, attention is directed to the more readily determined parameter, the isoelectric point.

As we have previously commented, reducing the net charge on the particle to zero collapses the double layer and greatly increases the instability of the particle. Hydrophobic particles, such as gold sol, aggregate and precipitate under these circumstances, and proteins become highly sensitive to precipitation. The addition of any component that competes with the hydrophilic macromolecular particle for its water sheath will usually initiate precipitation. Organic solvents, for example ethanol or acetone, and neutral salts are the reagents usually chosen. Isoelectric precipitation is a standard technique of protein purification.

The isoelectric point may be determined experimentally by measuring the electrophoretic mobility of a protein over a range of pH in which it reverses charge. Proteins are positively charged below their isoelectric points and negatively charged above them. The point at which the plot intersects the zero mobility line is the pH of the isoelectric point; the method is illustrated in Fig. 7.10.

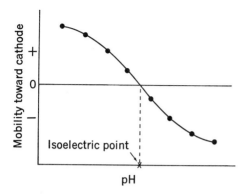

**Figure 7.10** Schematic experimental plot illustrating the method for determining the isoelectric point of a protein. The pH of zero net charge is the isoelectric point.

For an enzyme sample too small for study in electrophoresis equipment employing boundary methods (or too impure to be examined by such a technique), the electrophoretic mobility as a function of pH may be followed on a solid support such as granular starch, polyacrylamide gel, or paper. The enzyme is located by sectioning the supporting medium and testing the individual fractions for enzyme activity. If the enzyme is denatured above or below its isolectric point, the mobility is measured over a pH range in which the enzyme is stable and the isoelectric point is obtained by extrapolation. This method has been used successfully for aspartase (Fig. 7.11).

**Diffusion**   All solutes tend to diffuse through solutions until the composition is homogeneous throughout. Small molecules and ions move with sufficient velocity to distribute themselves throughout the solvent rapidly. Macromolecules have a large mass and correspondingly low

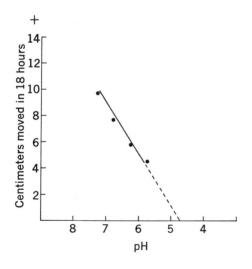

**Figure 7.11**   Determination of the isoelectric point of aspartase by plotting its rate of migration on starch plate as a function of pH. [From Wilkinson, J. S., and V. R. Williams, *Arch. Biochem. Biophys.*, **93**:81, 1961.]

velocity (since at any fixed temperature all particles in the solution have the same average kinetic energy). Their diffusion depends on their own sluggish motion plus Brownian motion resulting from bombardment by solvent and other solute molecules. The size and shape of the molecule is paramount in determining its diffusion characteristics.

The rate at which a substance diffuses across a uniform cross-sectional area (customarily, one square centimeter), depends not only on the molecular size and shape but also on the concentration gradient of that substance. In the absence of any other influencing factors, matter moves spontaneously from a region of high concentration toward one of lower concentration. We express this idea—the more the concentration is *decreasing,* the more the diffusion rate *increases*—by writing the gradient terms as $-dC/dx$, where $C$ is the concentration and $x$ is the distance. If we let $dm/dt$ represent the rate at which $m$ grams of solute cross the barrier or reference plane, Fick's First Law of Diffusion (1855) states that

$$\frac{dm}{dt} = -\mathcal{D}(A)\frac{dC}{dx}, \tag{7.31}$$

where $\mathcal{D}$ is the diffusion constant of the solute and $A$ is the cross-sectional area across which the solute is diffusing. The mass in grams which crosses one square centimeter in unit time will be

$$\frac{dm/dt}{A} = -\mathcal{D}\frac{dC}{dx}$$

The increment of solute, $dm$, which crosses a unit area per unit time is the "flow" and is represented by $J$, so that we may also write Fick's First Law in the form

$$J = -\mathcal{D}\frac{dC}{dx} \qquad \text{(an alternate form of Eq. 7.31)}.$$

The equation expresses mathematically the fact that the rate of diffusion of a neutral solute across a unit area is determined by the diffusion constant of the solute, and the concentration gradient. The equation assumes that temperature and viscosity of the solvent remain constant. The diffusion constant (or coefficient), $\mathcal{D}$, is of the order of $10^{-7}$ cm$^2$ per sec for enzymes.

The accurate measurement of diffusion constants is not an easy matter. Two methods are used: the porous disc method and the free-diffusion method. In the former, Eq. 7.31 is applied directly through the use of the simple apparatus shown in Fig. 7.12. The solute is allowed to diffuse through a porous disc made of alundum or sintered glass. The rate of transfer through the disc is measured for a solute of known diffusion constant. Then, the rate of transfer of a second solute is measured and its diffusion constant is calculated on the basis of the calibration. Relative rather than absolute values are obtained for $\mathcal{D}$.

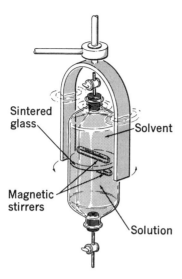

Sintered glass

Solvent

Magnetic stirrers

Solution

**Figure 7.12**   Closed porous disc diffusion cell. [After Stokes, R. H., *J. Am. Chem. Soc.*, **72**:763, 1950.]

In the free diffusion method, one may use the electrophoresis apparatus, the analytical ultracentrifuge, or a special diffusion apparatus. A sharp line of separation is formed between a solution containing macromolecules and a zone of solvent, either by glass partitions or by sedimentation. The partition is then removed or the centrifuging is continued at low speed and the solute is permitted to diffuse across the reference plane at constant temperature in a vibration-free system. The calculation of $\mathfrak{D}$ is based on Fick's Second Law

$$\frac{dC}{dt} = \mathfrak{D}\frac{d^2C}{dx^2}. \tag{7.32}$$

In this equation, $dC/dt$, the rate at which concentration is changing with time, is related to the first derivative of the concentration gradient, $dC/dx$, with respect to distance. That is, $d^2C/dx^2$ is the rate at which the *gradient* is changing with distance. Optical devices which measure the refractive index over the distance of the cell are very useful in the application of Eq. 7.32. The refractive index gradient $(dn/dx)$ will be the greatest where the concentration gradient is the steepest, *i.e.*, where concentration is changing most rapidly with distance. For this reason, refractive index measurements over the cell give directly curves of the type shown in part C of Fig. 7.13. The boundary is defined as the point at which $dC/dx$ is a maximum. (This is the exact definition promised earlier.) In Fig. 7.13 the boundary lies on the dashed horizontal line at $x = x_m$. Usually the change in concentration of solute is followed with respect to both time and distance. Various integrated equations are also used in diffusion calculations. By the free diffusion method absolute values for $\mathfrak{D}$ are obtained.

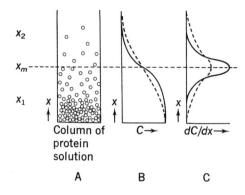

$x_2$

$x_m$ - - - - - - - - - - - - - - - - - - - - - - - -

$x_1$

$x$ ↑

Column of protein solution

$x$ ↑

$C \rightarrow$

$x$ ↑

$dC/dx \rightarrow$

A          B          C

**Figure 7.13**   Distribution of a dissolved substance at a boundary.
A: Column of protein solution.
B: Variation of concentration with distance.
C: Variation of concentration gradient with distance. [From Greenberg, D. M., editor, *Amino Acids and Proteins.* Charles C. Thomas, Springfield, Ill., 1953, p. 333.]

For a complete mathematical exposition of the analysis of the experimentally obtained diffusion gradient curves, the reader is referred to a classical paper by Neurath.[*] Detailed descriptions of procedures and calculations may be found in a number of reference works, an exceptionally lucid presentation being that of Lundgren and Ward.[†] An equally readable discussion which includes recent technical improvements has been prepared by Schachman.[‡]

From diffusion constants and sedimentation constants, one may determine the size of monodisperse proteins, the degree of homogeneity of protein solutions, and a combination of degree of hydration and deviation from spherical shape. The sedimentation constant will be discussed in the next section. The ratio of the frictional coefficient for nonspherical particles, $f$, to the frictional coefficient for spherical particles, $f_0$, is known as the *dissymmetry constant*. (These frictional coefficients were defined in the preceding discussion on electrophoresis.) The dissymmetry constant, $f/f_0$, can be determined from any measure of the apparent molecular shape that may be available, for example, from viscosity, flow birefringence, or dielectric dispersion. The diffusion constant, molecular weight, and dissymmetry constant are related as expressed by the equation

$$M = K \frac{1}{\mathcal{D}^3 \bar{V}} \left(\frac{f_0}{f}\right)^3, \tag{7.33}$$

where

$$K = \frac{R^3 T^3}{162 \pi^2 \eta^3 \mathfrak{N}^2},$$

[*]Neurath, H., "The Investigation of Proteins by Diffusion Measurements." *Chem. Rev.*, **30**:357, 1942.

[†]Greenberg, D. M., Ed., *Amino Acids and Proteins.* Charles Thomas, Springfield, Ill., 1951, p. 312–342.

[‡]Schachman, H. K., in *Methods in Enzymology*, Vol. IV, edited by Colowick, S. P., and N. Kaplan, Academic Press, New York, 1957, p. 32.

and where $\bar{V}$ is the partial specific volume of the particles, and all other terms are as previously defined. The partial specific volume is the reciprocal of the density. Equation 7.33 may be used for the calculation of any one of the three parameters (diffusion constant, molecular weight, or dissymmetry constant) if the other two are known. The dissymmetry constant (or number) is also called the molar frictional ratio. Although it is not equal to the axial ratio of the particles, it does give pertinent information regarding molecular shape. For nonspherical molecules, $f/f_0$, will always be greater than unity. For proteins, it ordinarily lies in the range 1 to 2. For rod-shaped molecules it will be larger, *e.g.*, tobacco mosaic virus has a molar frictional ratio of 3.12, although the true axial ratio is of the order of 35–40, judging from observations in the electron microscope. The molar frictional ratio is an index of deviation from spherical shape rather than an absolute statistic.

**Sedimentation** Macromolecules have the property of being nearly insensitive to gravitational settling, although over long periods of time concentration gradients will develop in undisturbed solutions. This phenomenon is even displayed by the earth's atmosphere, which becomes increasingly less dense as altitude increases. Gravitational sedimentation, however, has no widespread application in the study of macromolecules. Efforts to purify and characterize macromolecules by centrifuging their solutions were unspectacular until 1923, at which time Svedberg and Nichols announced the development of a high-speed centrifuge, ingeniously designed to permit the transmission of light through the centrifuge cell and record the movement of boundaries of macromolecules. A high-speed centrifuge of the Svedberg type is usually referred to as an analytical ultracentrifuge to distinguish it from other high-velocity instruments lacking optical attachments. The essential features of a modern analytical ultracentrifuge are diagrammed in Fig. 7.14. Since the solution cannot be directly sampled, concentrations are determined by optical procedures. If ultraviolet light is passed through the cell, the concentration of macromolecular solute will be proportional to the darkening of a photographic film positioned above the cell. If refractometric methods are used, the position of the boundary will be given by a peak, as previously explained in connection with Fig. 7.13. A third technique for following the boundary employs an interferometer to determine the position and concentration of the solute. Interference fringes are counted and the concentration can be determined with an accuracy of the order of 0.1 percent.

Further information on analytical procedures is given in the reference by Schachman, included in the *Suggested Additional Reading* on page 392.

Molecular weights are determined in ultracentrifuges by three techniques: the sedimentation velocity method, the sedimentation equilibrium method, and density gradient centrifugation.

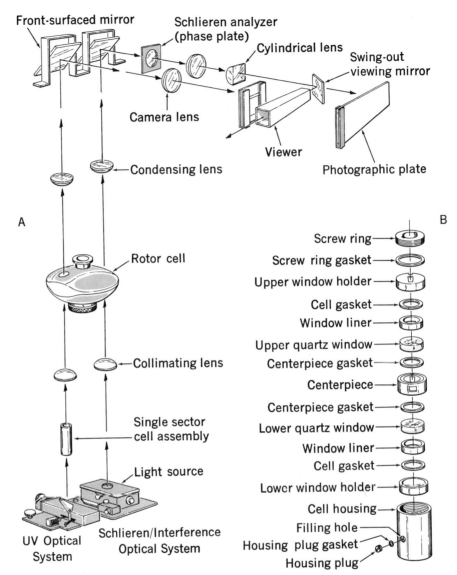

**Figure 7.14** Optical systems and cell assembly in Spinco Model E ultracentrifuge. A: The schlieren and interference optical systems, shown on the right, share many common components. The light source for both systems is a 1000-watt, water-cooled, mercury-arc lamp, which is situated at the focal plane of a collimating lens. The collimated light passes through the cell and into the condensing lens. In an interference run, two narrow slits are mounted with the cell; light rays passing through these slits cause interference fringes to form at the focal plane of the condensing lens. The light, after passing through the cell, is then focused by the condensing lens on the plane of the phaseplate; it then travels to the camera and cylindrical lenses, which focus it on the photographic plate. In a schlieren run, the condensing lens first focuses the light on the phaseplate (the schlieren analyzer element), which is located between the

*(cont'd.)*

*Sedimentation velocity method* Of the three sedimentation methods, only the sedimentation velocity method measures the speed of the moving boundary. In a mixture of macromolecules being subjected to high-speed centrifugation, molecules of different molecular weight move at different velocities because of variations in their sizes and shapes. The fundamental equation of sedimentation permits us to calculate the molecular weight from the velocity of sedimentation and certain fundamental constants.

Electrical forces of the type considered in the preceding section have some influence on the sedimentation velocity, but these effects can in general be minimized by using isoelectric protein or adding a neutral salt like 0.1 *M* KCl. The pH or ionic strength is adjusted so that the macromolecules are electrically neutral with respect to the medium, otherwise double-layer phenomena will produce an ion cloud to impede the movement of the large particles. In the derivation that follows, electrical forces are not considered.

The particles move because of centrifugal force and that movement is resisted by a frictional or viscous force corresponding to the one just considered in the discussion of electrophoresis. As with the equation of electrophoresis, the derivation consists of analyzing the opposing forces, defining the forces mathematically, and expressing the relationship in an equation which may be put in a convenient form.

$$\overrightarrow{\text{Centrifugal Force}} = \overleftarrow{-\text{Frictional Force}}$$
$$\text{(driving force)} \qquad \text{(retarding force)}$$

Centrifugal force is defined as

$$F_c = (M\omega^2 x)(1 - \bar{V}\rho), \tag{7.34}$$

where M is the molecular weight of the macromolecular solute, $\omega$ is the angular velocity of the centrifuge rotor in radians per second, $x$ is the distance in centimeters of the macromolecular boundary from the center

---

condensing and camera lenses. This phaseplate is used to translate the refracted light to a line showing rate of change in refractive index. From the camera lens, the light first travels to the cylindrical lens, which magnifies perpendicularly to the direction of sedimentation, and then to the photographic plate.

B: The ultraviolet adsorption system, shown on the left, is completely separate, sharing none of the schlieren/interference components. The adsorption light source is a medium-pressure mercury-arc lamp with an ultraviolet-transmitting glass envelope. Light is transmitted from the light source to the collimating lens, which renders the light parallel during its passage through the cell; the condensing lens focuses the light at some position along the optical track near the camera lens. The camera lens then projects the image of the illuminated cell onto the photographic film. As the molecules in the cell sediment, a transparent solvent region is created and the image of this portion of the cell becomes darker. [Schematic diagrams and explanatory notes supplied by Beckman Instruments, Inc., Palo Alto, California.]

of the rotor, and $(1 - \bar{V}\rho)$ is a buoyancy correction factor in which $\rho$ is the density of the solvent and $\bar{V}$ is the partial specific volume of the macromolecular solute, as explained in the discussion on diffusion.

The frictional force equation is the same as before:

$$F_f = -fv \quad \left(\text{or } F_f = -f\frac{dx}{dt}\right).$$

Since the molecular weight appears in the centrifugal force equation (Eq. 7.34) we are concerned with the force acting on a mole of solute and will require the molar frictional coefficient. Because it is unlikely that the sedimenting macromolecules are spherical, we will use the general expression for the molar frictional coefficient: $f = RT/\mathfrak{D}$, where $\mathfrak{D}$ is the diffusion constant as previously defined.

Combining the two force equations gives

$$(M\omega^2 x)(1 - \bar{V}\rho) = fv = \left(\frac{RT}{\mathfrak{D}}\right)v. \tag{7.35}$$

Centrifugal Force $= -$Frictional Force.

We now wish to define a new quantity, the sedimentation coefficient, $s$, which will be equal to the velocity of the particle per unit centrifugal field of force. (Recall that a similar step was performed in deriving the equation of electrophoresis: we defined the mobility, $u$, as the velocity of the particle per unit electrical field of force.)

The definition of $s$ is

$$s = \frac{v}{\omega^2 x}.$$

Substituting the above equation into Eq. 7.35 and rearranging terms so as to solve for M, we obtain

$$M = \frac{sRT}{\mathfrak{D}(1 - \bar{V}\rho)}. \tag{7.36}$$

This is the Svedberg Equation. The experimentally measured quantity is $s$, the sedimentation constant (or coefficient). It has the general order of magnitude of $10^{-13}$ seconds, which has been called a Svedberg unit, abbreviated S. Molecular weights obtained from Eq. 7.36 are independent of the shape of the solute particle.

The sedimentation coefficient is calculated from the slope of a plot of log $x$ versus time in a sedimentation velocity run ($x$ is the distance of the boundary to the axis of rotation). Returning to the definition of the sedimentation coefficient (see above), we write

$$s = \frac{v}{\omega^2 x} = \frac{dx/dt}{\omega^2 x}.$$

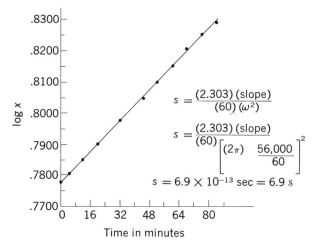

**Figure 7.15** Determination of sedimentation coefficient of aspartase from a plot of log $x$ ($x$ is the distance of boundary to axis of rotation) versus $t$ ($t$ is the time in minutes). (V. R. Williams, unpublished data.)

This rearranges to

$$\frac{dx}{x} = s\omega^2 dt.$$

Integrating both sides yields

$$2.303 \log x = s\omega^2 t + \text{integration constant.} \tag{7.37}$$

An example of a plot of this type is shown in Fig. 7.15.

If we assume the molecules to have a spherical shape (which is rather improbable), M can be estimated without evaluating the diffusion constant. The value of M so obtained is a minimum molecular weight and is treated as an informed estimate. The equation for calculating M from $s$ and $f_0$, rather than $f$, is obtained by defining the radius of the particle, $r$, in terms of the molecular weight, M, and partial specific volume of the solute, $\bar{V}$. As we have previously defined, the frictional coefficient for a spherical particle is given by Stoke's Law: $f_0 = 6\pi\eta r$. The radius of the particle is related to its volume by the relationship: $v = \frac{4}{3}\pi r^3$. The mass of a single molecule is equal to $M/\mathfrak{N}$. Multiplying the partial specific volume, $\bar{V}$, by $M/\mathfrak{N}$ gives $v$, the volume of a single particle. Therefore it must follow that $M\bar{V}/\mathfrak{N} = \frac{4}{3}\pi r^3$. Solving for $r$, we find that $r = (3M\bar{V}/4\mathfrak{N}\pi)^{1/3}$. We can now readily see the origin of the complicated term for the molar frictional coefficient (previously defined in the discussion on electrophoresis)

$$f_0 = 6\pi\eta\mathfrak{N}\left(\frac{3M\bar{V}}{4\pi\mathfrak{N}}\right)^{1/3}.$$

We substitute the right-hand term for $RT/\mathfrak{D}$ in Eq. 7.37 and obtain, after expanding and collecting terms,

$$M = \left[\frac{s^3}{(1-\bar{V}\rho)^3} \cdot 162\pi^2\eta^3\mathfrak{N}^2\bar{V}\right]^{1/2}. \tag{7.38}$$

The preceding derivations all assume that the particles are unhydrated spheres. For most proteins, $\bar{V}$ is in the range 0.70–0.75 ml per gram.

Rather than calculate a molecular weight from Eq. 7.38, it is often more convenient to characterize a macromolecule in terms of its sedimentation coefficient. In the research literature proteins, nucleic acid fractions, ribosomal preparations, and antibodies are often identified as $5\,\mathrm{S}$, $30\,\mathrm{S}$, $50\,\mathrm{S}$, etc. particles. The number is the Svedberg constant (or coefficient, as it is also called) converted into Svedberg units.

*Sedimentation equilibrium method* The molecular weights of particles can be determined in the ultracentrifuge by a method which dispenses with diffusion and sedimentation velocity measurements. This technique is known as the sedimentation equilibrium method. A protein solution, for example, is centrifuged at a speed which exactly balances the tendency of the protein to disperse itself by diffusion. Such speeds are relatively low. By the use of short columns and over-speeding techniques, equilibrium may be attained within a matter of hours. The concentration distribution is then determined by optical methods. With recent improvements in sedimentation equilibrium techniques, the sedimentation velocity method has diminished in importance. In addition, weight- and $z$-average molecular weights can both be obtained from equilibrium experiments. The derivation of the equation of sedimentation equilibrium follows.

According to Fick's First Law (Eq. 7.31) the flow of solute through a unit area of the boundary, as a result of diffusion, is given by the expression

$$\overleftarrow{J} = -\mathfrak{D}\left(\frac{dC}{dx}\right) \qquad \text{(Eq. 7.31)}.$$

An equivalent form is obtained by substituting the $\mathfrak{D}$ in terms of $f$, the frictional coefficient:

$$\overleftarrow{J} = -\frac{RT}{f}\left(\frac{dC}{dx}\right). \tag{7.39}$$

The flow of solute in the opposite direction as a consequence of centrifugal force is given by the equation

$$\overrightarrow{J} = \frac{C\omega^2 x M(1-\bar{V}\rho)}{f}, \tag{7.40}$$

where all terms are as previously defined, $C$ being the concentration of the solute at the specific point in the cell under consideration. When equilibrium is reached, the centrifugal driving force is exactly balanced by the diffusional retarding force and we may say: $\overrightarrow{J} = -\overleftarrow{J}$. Equating 7.39 and 7.40 in this fashion and solving for M, we obtain

$$M = \frac{RT}{\omega^2(1 - \bar{V}\rho)} \frac{dC/dx}{xC},\tag{7.41}$$

where $C$ and $dC/dx$ are the concentration and concentration gradient at the position $x$. If these three terms can be determined, M can be calculated. Refractive index methods are extensively used in the application of Eq. 7.41.

Integration of Eq. 7.41 yields

$$M = \frac{2RT(\ln C_2 - \ln C_1)}{\omega^2(1 - \bar{V}\rho)(x_2^2 - x_1^2)}.\tag{7.42}$$

Absorption optics are used to calculate values of $C$ corresponding to values of $x$. The molecular weights obtained from the above equations are weight-average molecular weights for unhydrated molecules. Equations 7.41 and 7.42 are ordinarily used with homogeneous preparations, although data for heterogeneous systems can be handled by other mathematical treatments.

The basic equation of sedimentation equilibrium (7.41) will hold *at any time* (nonequilibrium) for any parts of the cell in which there is no flow of particles. There are two such places, the meniscus of the solution and the bottom of the cell. If the term $(dC/dx \cdot 1/xC)$ is determined for various values of $x$ and a plot is made of the data, the lines may be extrapolated to $x =$ the distance of the meniscus from the center of the rotor, and $x =$ the distance of the bottom of the cell from the center of the rotor. The general shape of such a plot is shown in Fig. 7.16. The intercept value for the corresponding value of $x$ is then inserted into Eq. 7.41 and M is calculated. This particular application of sedimentation equilibrium is known as the Archibald Method.[*] In Fig. 7.16, which is based on actual experimental results, the two intercept values are not the same and will not lead to the same value of M when substituted into Eq. 7.41 unless corrections are made for the difference in activity coefficient of the solute at the meniscus and at the bottom of the cell. Several runs should be made at different initial concentrations of protein and the true molecular weight calculated by extrapolation of the results to $C = 0$.[†] For a complete discussion of the various mathematical and experimental

---

[*] Archibald, W. J., *J. Phys. and Colloid Chem.*, **51**:1204, 1947.
[†] See pages 284–290 in the reference by Tanford that is listed in *Suggested Additional Reading.*

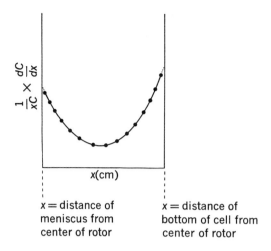

x = distance of meniscus from center of rotor

x = distance of bottom of cell from center of rotor

**Figure 7.16** The Archibald method for determination of molecular weight.

treatments of sedimentation phenomena, including the experimental determination of the partial specific volume, $\bar{V}$, the student should consult the book by Schachman that is listed on page 392.

*Density gradient centrifugation* A new technique in sedimentation, density gradient centrifugation, is achieving wide popularity. It has two advantages: a preparative ultracentrifuge is used, and the fractionated material can be separated into individual components by a simple technique. A rotor known as a swinging-bucket type is used, and the separation is carried out in a plastic centrifuge tube.

There are two kinds of density gradient centrifugation techniques, velocity (sucrose) and equilibrium (cesium chloride). Preformed gradients are used in the velocity technique; in the equilibrium technique, the centrifugal field forms the gradient during the experiment.

For *velocity sedimentation,* the sucrose gradient can be prepared with the aid of an automatic gradient-maker. The solution of macromolecular particles is carefully layered on top of the density gradient column. Although the addition of this layer produces a negative gradient at the top of the liquid column, the layer of particles is supported by the steep positive density gradient of the low molecular weight solute, thereby preventing the sedimentation of droplets of liquid and insuring that the large molecules will migrate at a steady rate. As centrifugation takes place, the particles separate into zones as they migrate through the liquid column under the influence of centrifugal force. Each zone consists of one type of molecule, moving through the gradient at its characteristic sedimentation velocity. After a predetermined length of time, the centrifuge is stopped, the tubes are removed, and fractions are collected by one of several techniques. Since the solution cannot be

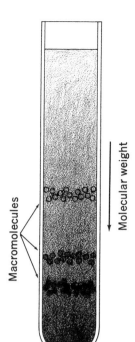

**Figure 7.17** Schematic illustration of density gradient centrifugation. Particles of different size sediment in the sucrose or salt gradient separating according to their hydrodynamic characteristics and molecular weights.

viewed and the position of the various boundaries observed, one is forced to "fly blind" in an experiment of this type. Several preliminary runs may be necessary to determine the optimum length of time to achieve the desired separation.

In the *equilibrium technique* the low molecular weight solute and the macromolecules are uniformly distributed throughout the solution before the start of the centrifugation. Inorganic salts such as CsCl, RbCl, and NaBr are used to give solutions of high density and low viscosity. Interest in salt gradients has been stimulated by the excellent work of Meselson, Stahl, and Vinograd* on DNA. Since the gradient is formed during centrifugation, much longer times are required than with preformed gradients, but overspeeding techniques can be used to shorten the time required for the attainment of equilibrium. The resulting gradient is calculable in terms of thermodynamic equations and the properties of the molecules forming the gradient. With preformed gradients, the exact value of the gradient at the conclusion of the experiment cannot be known with certainty.

A diagrammatic representation of the tube and its contents is shown in Fig. 7.17. The entire tube and contents may be frozen and then sliced

*Meselson, M., F. W. Stahl, and J. Vinograd, *Proc. Nat. Acad. Sci. (U.S.)*, **43**:581, 1957, and Meselson, M., and F. W. Stahl, *Proc. Nat. Acad. Sci. (U.S.)*, **44**:671, 1958.

into thin sections, or the bottom of the tube may be pierced with a needle and the solution permitted to drip out slowly. Since the individual boundaries cannot be seen, the investigator detects the various components by collecting small individual fractions, *e.g.*, 20 drops each, and testing them separately. Streaming along the walls of the tube or development of small areas of turbulence may plague the inexperienced investigator. These problems have led to the design of an electrical apparatus for siphoning off the layers, and even of a centrifuge which will spin down the material and siphon it off. Density gradient velocity centrifugation is both an analytical and a preparative technique, and is most frequently used for the determination of approximate molecular weights. In velocity sedimentation a "marker" of suitable molecular weight is added to the solution layered on top of the density gradient. Catalase, $M = 244,000$, and peroxidase, $M = 40,000$, are frequently used in determinations on proteins. The position of the marker is located in the fractions by appropriate measurements, and the relative molecular weight of the major solute is determined by the method of Martin and Ames.[*] The molecular weights are related to the distances moved through the gradient as follows:

$$\frac{M_1}{M_2} = \left(\frac{\text{Distance moved by Component 1}}{\text{Distance moved by Component 2}}\right)^{3/2}$$

A density gradient molecular weight is of the same order of accuracy as one determined from $s$ and $f_0$. It is a minimum molecular weight based on assumptions of spherical shape. The Martin and Ames relationship may be derived from Eq. 7.38, since the distance moved by a component (macromolecule) is directly proportional to $s$. The two components are assumed to have the same partial specific volume ($\bar{V}$).

**Viscosity** The viscosity characteristics of macromolecular solutions are regarded by some experimenters as easier to measure and interpret than sedimentation and diffusion constants. The apparatus required is much simpler, direct numerical values are obtained for flow times and densities, and the viscosity is measured on a homogeneous solution free of concentration gradients. Viscosity measurements are especially valuable in determining the shape characteristics of macromolecules. On the other hand, some disadvantages should be noted: ordinary apparatus requires larger samples of material than those required for sedimentation studies, the viscosity method does not yield an unequivocal value for the molecular weight, and it is not convenient to use at low temperatures.

Many types of apparatus have been designed for viscosity determinations. Three important ones are based on: flow through capillaries, the rotation of a cylinder immersed in the solution, and the rate of fall of a

[*] Martin, R. G., and B. N. Ames, *J. Biol. Chem.*, **236**:1372, 1961.

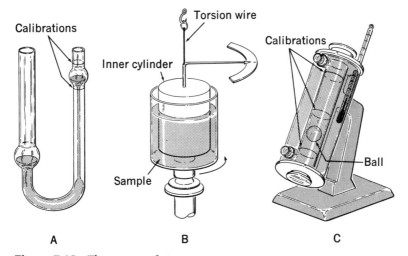

**Figure 7.18** Three types of viscometers.
A: A capillary viscometer of the Ostwald type. The rate at which the liquid
drains down between the two calibrations is measured.
B: The Couette rotating-cylinder viscometer. The liquid rotating in the outer
cylinder causes a torque to be applied to the torsion wire attached to the
inner cylinder. This torque is measured.
C: The falling ball viscometer. The cylinder is filled with the liquid to be
measured. The time required for the ball to fall the distance marked by the
calibrations is measured. From the various measurements noted above, the
viscosity can be calculated by appropriate equations.

ball through the solution. These three types are shown diagrammatically
in Fig. 7.18.

As an illustration of the method, the capillary viscometer will be
discussed. Most modern capillary viscometers are modifications of the
original apparatus of Ostwald. The determination of viscosity by measuring the flow time is based on Poiseuille's Law, which states that the
volume $V$ of a liquid that flows through a capillary tube is directly proportional to the flow time $t$, the pressure $p$ under which the liquid flows,
and the fourth power of the capillary radius, $r$. Also, the volume of liquid
flow is inversely proportional to the length of the capillary, $l$ and to the
viscosity, $\eta$.

$$V = \frac{\pi t r^4 p}{8l\eta}. \qquad (7.43)$$

The *absolute* viscosity is

$$\eta = \frac{\pi r^4 p t}{8Vl} \qquad \text{(Eq. 7.43 rearranged)}.$$

The viscosity unit is the *poise* and has the units of g sec$^{-1}$ cm$^{-1}$.

If only the *relative* viscosity is desired, that is, the viscosity of one solution or liquid relative to that of a second one, $V$, $l$, and $r$ will be constant for any particular viscometer. These terms and $\pi/8$ can be incorporated into a single constant, $k$. The viscosity is then

$$\eta = kpt, \tag{7.44}$$

which states that the viscosity is proportional to the flow time if the pressure driving the liquid through the viscometer is a constant. The driving pressure is the pressure difference at two points at opposite ends of the liquid column (the calibrations in Fig. 7.18), and it depends on the density of the solution. If a solution of density $d_1$ flows through the viscometer in $t_1$ seconds and the pure solvent, or reference solution, of density $d_0$ flows through in $t_0$ seconds, the relative viscosity will be

$$\eta_{rel} = \frac{t_1 d_1}{t_0 d_0}. \tag{7.45}$$

If the densities of the two solutions, the solution and the reference solvent, are nearly identical, we may further simplify the expression to

$$\eta_{rel} \cong \frac{t_1}{t_0}. \tag{7.46}$$

The *specific viscosity* is defined as

$$\eta_{sp} = \eta_{rel} - 1 \cong \frac{t_1 - t_0}{t_0}. \tag{7.47}$$

The specific viscosity should be proportional to the concentration, $C$, of the macromolecular solute, expressed in grams per ml or per 100 ml. The value of $\eta_{sp}/C$ extrapolated to infinite dilution is called the *intrinsic viscosity*, and is indicated by the special notation $[\eta]$:

$$\lim \, (\eta_{sp}/C)_{C \to 0} = [\eta]. \tag{7.48}$$

The specific viscosity of a particular solute is measured in solutions of different known concentration and the intrinsic viscosity is calculated by extrapolation to infinite dilution as shown in Fig. 7.19. The quantity, $\eta_{sp}/C$, is sometimes called the *reduced viscosity* and may be strongly concentration dependent. This concentration dependency is shown by the Huggins Equation

$$\frac{\eta_{sp}}{C} = [\eta] + k[\eta]^2 C, \tag{7.49}$$

where $k$ is the Huggins constant. The value of $k$ ranges from 2.0 for solid uncharged spheres to 0.35 for flexible coils.

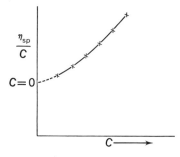

**Figure 7.19** Plot for the determination of intrinsic viscosity. Data for the reduced viscosity obtained at a series of concentrations are extrapolated to the abscissa (where $C = 0$).

With certain polymers the Wagner Equation is applicable. Fewer experimental data are needed for this equation, which is

$$[\eta] = \frac{\ln \eta_{rel}}{C} + k \left( \frac{\ln \eta_{rel}}{C} \right)^2. \tag{7.50}$$

Theoretically, one may calculate $[\eta]$ from one relative viscosity determination at one concentration. For studies on amylose a concentration of 0.2 percent was selected and $k$ was taken to be 0.032. Excellent agreement was obtained between intrinsic viscosities calculated in this way and those found by extrapolation.[*]

Frequently intrinsic viscosities are determined for a series of polymer fractions known to differ in molecular weight, and no further calculations are made. The intrinsic viscosities are used, in this case, as an index of molecular weight. Staudinger established that the intrinsic viscosity is quantitatively related to the molecular weight. The Staudinger Equation contains empirical constants which must be evaluated for each particular solute and solvent system. Fortunately, one can often find in the research literature previously determined Staudinger constants suitable to the polymer one wishes to study. The general form of the modified Staudinger Equation is

$$[\eta] = K'M^a, \tag{7.51}$$

where $K'$ and $a$ are empirical constants and M is the molecular weight. The constant $a$ is related to the shape of the molecule, being 0.5 to 1.0 for flexible random coils and 1.8 for rods. Everett and Foster found $K'$ to be $0.85 \times 10^{-1}$ deciliters per gram and $a$ to be 0.76 for amylose solutions in 1 $N$ potassium hydroxide.[†] The constants were used by Phillips and Williams to calculate the molecular weight of crystalline amylose preparations isolated from rice starch. Data for amylose and other typical macromolecules are given in Table 7.2. Molecular weights for amylose

[*]Wolff, I. A., L. J. Gundrum, and C. E. Rist, *J. Am. Chem. Soc.*, **72**:5188, 1950.
[†]Everett, W. W., and J. F. Foster, *J. Am. Chem. Soc.*, **81**:3464, 1959.

**TABLE 7.2** MOLECULAR WEIGHTS AND INTRINSIC VISCOSITIES OF VARIOUS MACROMELECULES

| Substance | $M_v$ | $[\eta]$ in $cc/g$ | Solvent |
|---|---|---|---|
| Rice Amyloses | | | |
|   Rexoro variety | 325,000 | 209 | 1N KOH |
|   Century Patna | | | |
|   variety | 100,000 | 84 | 1N KOH |
| Ribonuclease | 13,683 | 3.3 | aq. salt solution |
| Serum albumin | 65,000 | 3.7 | aq. salt solution |
| Tropomyosin | 93,000 | 52 | aq. salt solution |
| Collagen | 345,000 | 1150 | aq. salt solution |
| DNA | 6,000,000 | 5000 | aq. salt solution |

*Sources:* Data for rice amyloses are from Phillips, A. T., and V. R. Williams, *J. Food Sci.*, **26**:573, 1961; for ribonuclease from Buzzell, J. G., and C. Tanford, *J. Phys. Chem.*, **60**:1204, 1956; for serum albumin from Tanford, C., and J. G. Buzzell, *J. Phys. Chem.*, **60**:225, 1956; for tropomyosin from Tsao, T-C., K. Bailey, and G. S. Adair, *Biochem. J.*, **49**:27, 1951; for collagen from Boedtker, H., and P. Doty, *J. Am. Chem. Soc.*, **78**:4267, 1956; and for DNA from Reichmann, M. E., S. A. Rice, C. A. Thomas, and P. Doty, *J. Am. Chem. Soc.*, **76**:3047, 1954.

fractions are usually averages because of the inherent heterogeneity of the preparation. With purified proteins such as tropomyosin, however, the sample should be homogeneous as to species and yield a representative molecular weight. Many proteins are globular and have intrinsic viscosities of about 3 cc/g. Molecular weights obtained by viscometry are called $M_v$ molecular weights and differ from $M_n$ and $M_w$ as discussed in Section 7.4. For most molecular weight distributions $M_v$ agrees fairly well with $M_w$.

The intrinsic viscosity may be combined with data from sedimentation or diffusion to obtain molecular weights, by treatments which assume that the macromolecules possess some degree of flexibility and hydration. The appropriate equations for these calculations have been given by Schachman[*]

$$M = \frac{4690\,(S_{20})^{3/2}[\eta]^{1/2}}{(1 - \bar{V}\rho)^{3/2}}, \tag{7.52}$$

and

$$M = \frac{6.58 \times 10^{-16}}{(\mathcal{D}_{20})^3[\eta]}. \tag{7.53}$$

The sedimentation coefficient, $S_{20}$, is in Svedbergs. The solvent is assumed to be water and the temperature, 20°C. The intrinsic viscosity

---

[*]Schachman, H. K., in *Methods in Enzymology*, Vol. IV, edited by Colowick, S. P., and N. Kaplan, Academic Press, 1957, p. 103.

is in reciprocal concentration units, $(g/100 \text{ ml})^{-1}$. Equation 7.52 is a form of the Scheraga-Mandelkern Equation,* and should be compared with Eq. 7.38.

Of the various hydrodynamic properties of macromolecules, none appears to be so intimately related to their shape as is viscosity. Solutions of spherical particles have relatively low viscosities that are not greatly dependent on particle size, but the viscosities of linear polymers are high and increase markedly with particle size. When we survey the entire array of high polymers, both natural and synthetic, it becomes apparent that variety in shape is the order of the day. Every type, from rigid cylindrical rods to flat discs, is represented. Protein molecules are usually treated mathematically as either prolate ellipsoids or random coils. The possible variations of ellipsoidal shape are shown in Fig. 7.20. Since the viscosity is obviously related to shape in a qualitative way, many attempts have been made to derive mathematical equations which would quantitatively relate viscosity to the axial ratio of the particles. One of these is the Simha Equation

$$\text{As } \phi \rightarrow 0, \frac{\eta_{\text{sp}}}{\phi} = \frac{\left(\dfrac{L}{D}\right)^2}{15\left(\ln\dfrac{2L}{D} - \dfrac{3}{2}\right)} + \frac{\left(\dfrac{L}{D}\right)^2}{5\left(\ln\dfrac{2L}{D} - \dfrac{1}{2}\right)} + \frac{14}{15}, \qquad (7.54)$$

where $\phi$ is the volume fraction occupied by the particles and $L/D$ is the ratio of length to diameter, *i.e.*, the axial ratio. Equation 7.54 is for prolate (rod-shaped) ellipsoids. For oblate (disc-shaped) ellipsoids, the appropriate equation is

$$\frac{\eta_{\text{sp}}}{\phi} = \frac{16}{15} \frac{\left(\dfrac{L}{D}\right)}{\tan^{-1}\left(\dfrac{L}{D}\right)}. \qquad (7.55)$$

These equations apply only to molecules with large axial ratios. The complete equations applicable for all axial ratios will be found in the research papers of Simha and co-workers.†

Viscometry is a technique of wide application, as one may observe by inspecting various research papers on synthetic polymers, purified polysaccharides, nucleic acids, proteins, and polypeptides. It is generally preferred for molecular weight determinations on polymers of helix or random-coil conformation. Viscometry has a decided advantage over sedimentation in its simplicity and low cost.

A special problem arises in the viscometry of high molecular weight systems containing either stiff coils of the type found in DNA, or very

---

*Scheraga, H. A., and L. Mandelkern, *J. Am. Chem. Soc.*, **75**:179, 1953.

†Simha, R., *J. Phys. Chem.*, **44**:25, 1940; Mehl, J. W., J. L. Oncley, and R. Simha, *Science*, **92**:132, 1940.

Polar axis/equatorial diameter in percent

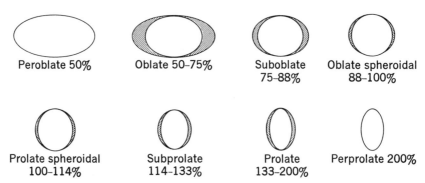

Peroblate 50%    Oblate 50–75%    Suboblate 75–88%    Oblate spheroidal 88–100%

Prolate spheroidal 100–114%    Subprolate 114–133%    Prolate 133–200%    Perprolate 200%

**Figure 7.20**  Ellipsoids of various axial relationships. [Used by permission of G. Morales.]

asymmetric molecules, or extremely flexible molecules. In these cases the viscosity is dependent on the flow rate, rather than being independent of it, as defined in Eq. 7.43. As a consequence, the molecules change their shape, or their orientation distribution, or both, in response to the shear stress in the viscometer, and the viscosity decreases. Ordinarily, the thermal motion of the molecules would counteract any effects due to flow itself and the liquid or solution would display *Newtonian flow.* Liquids or solutions in which the viscosity is affected by flow rate are said to display *non-Newtonian flow.*

The Couette viscometer shown in Fig. 7.18 tends to minimize the effects of non-Newtonian flow, whereas the usual capillary viscometer produces a very high shear stress and is unsatisfactory for DNA solutions. Conventional Couette viscometers have a number of practical drawbacks that have been largely overcome by recent modifications.*

**Flow birefringence**  Information about the shape of nonspherical macromolecules may be obtained by studying their birefringence under conditions of hydrodynamic shear or fluid flow. A crystalline substance which is *birefringent* has two indices of refraction, each along a different axis, and displays certain recognizable optical characteristics when studied in the polarizing light microscope. Even macromolecular "crystals" such as starch granules will give rise to recognizable patterns (dark crosses and other shapes) when examined by this method. The birefringence arises from the anisotropy of the molecular arrangement, which causes light to travel along one axis of the crystal with a different

*Zimm, B. H., "A New Rotating-Cylinder Viscometer and Its Use on DNA Solutions." *Fractions*, No. 3, Beckman Instruments, Inc., Palo Alto, California, 1965.

A                 Particles at rest

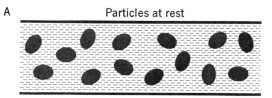

B   Particles moving along stream lines at low speed

C   Particles moving along stream lines at high speed

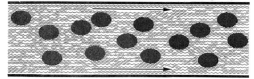

**Figure 7.21** The tendency of particles to align themselves so as to offer the least resistance to the moving liquid.

velocity from that at which it travels along the second axis. Birefringence is also called *double refraction*. When *solutions* of anisotropic substances, rather than crystals, are examined under the polarizing light microscope, however, no birefringence is observed because the molecules are randomly oriented. If the molecules can be caused to align themselves in a uniform way, birefringence will be observed. The alignment can be achieved either by imposing an electrical field or by subjecting the molecules to hydrodynamic shear. The first method is called *electrical birefringence*, the second is called *flow birefringence, double refraction of flow* or *streaming birefringence*. The orientation of the particles along stream lines (lines of flow) is shown in Fig. 7.21.

Unusual optical properties resulting from the alignment of the macromolecules can be observed with the naked eye and were recognized for many years prior to the development of special equipment for measuring birefringence. If an aged colloidal dispersion of vanadium pentoxide is slowly stirred, the path of the stirring rod lights up because of the orientation of the particles along the stream lines. This effect is due to a difference in the amount of light reflected by the symmetrically aligned particles and that reflected by the randomly oriented ones. In a similar way, the field of the polarizing light microscope will light up when a solution of nonspherical macromolecules is caused to flow, the brightness being proportional to the birefringence.

The intensity of the birefringence will depend on the flow rate and the shape of the particles, but will be independent of the birefringence

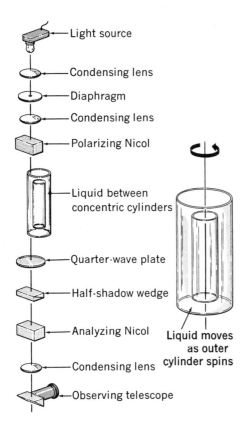

- Light source
- Condensing lens
- Diaphragm
- Condensing lens
- Polarizing Nicol
- Liquid between concentric cylinders
- Quarter-wave plate
- Half-shadow wedge
- Analyzing Nicol
- Condensing lens
- Observing telescope

Liquid moves as outer cylinder spins

**Figure 7.22** Diagram of the optical system of the flow birefringence apparatus. Light from the source is concentrated by the condensing lens on the opening in the diaphragm, which then serves as the effective light source. From the diaphragm the light passes through the second condensing lens and Nicol polarizer into the liquid between the concentric cylinders. The quarter-wave plate and half-shadow wedge are inserted only for the measurement of double refraction and are removed when extinction angles are to be determined. Light then passes through the analyzing Nicol prism. The final condensing lens focuses the light so that the optical patterns can be seen through the observing telescope. [After Edsall, J. T., *et al.*, *Review Sci. Instruments*, **15**:250, 1955.]

of the *individual* particles. In fact, the individual particles need not be *optically* anisotropic at all, provided they have an asymmetrical shape. A randomly coiled molecule of polyvinyl chloride possesses no birefringence of its own, but a solution of these asymmetric particles will exhibit flow birefringence because of the anisotropy of the system.

Quantitative measurements of flow birefringence are made in an apparatus which contains two concentric cylinders. A schematic drawing of the instrument is shown in Fig. 7.22. A macromolecular solution is placed between the cylinders, and the outer one is then caused to rotate. The solution begins flowing in the same direction as the rotating cylinder and the large asymmetric particles so align themselves as to offer minimum resistance to the fluid flow. Their position will be somewhat tangential to the direction of flow at lower velocities, but as the speed of the moving cylinder is increased and the solution flows correspondingly faster, the angle between the long axis of the particle and the stream lines decreases. The situation is rather like the behavior of a small stick floating down a meandering stream. Where the current is slow, the stick will twist and turn somewhat, although keeping itself generally aligned with the direction of flow. Where the current is swift, the stick hardly turns at all, but floats along with its long axis parallel

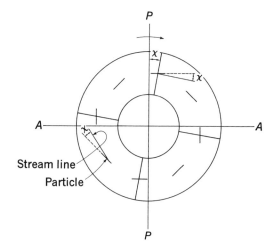

P

A ————————— A

Stream line

Particle

P

**Figure 7.23** The cross of isocline, which is observed during measurements of flow birefringence. The extinction angle $\chi$ is the angle made by the intersection of the long axis of the particle with the stream line. It is also given by the intersection of the polarizer axis (*P–P*) or the analyzer axis (*A–A*) with the cross of isocline; from this latter relationship it is measured. [After Cerf, R., and H. A. Scheraga, *Chem. Rev.*, **51**:189, 1952.]

to the stream line. Unlike the stick, however, macromolecules possess sufficient kinetic energy to turn themselves about, albeit slowly, a characteristic that is expressed by the *rotational diffusion constant* of the particle.

The optical effect that results from the birefringence of the oriented particles is called the *cross of isocline*, as shown in Fig. 7.23. The dark cross forms an angle with analyzer axis (marked A in the figure) and to the same extent with the polarizer axis (marked P in the figure) of the microscope. This angle is called the *extinction angle*, and is given the symbol $\chi$. As shown in the figure, the angle $\chi$ made by the cross of isocline with the optical axes of the microscope is the same as the angle made by the long axis of the particles with the stream line. As has already been stated, the angle decreases as the speed of the rotating cylinder increases, and the approximate dependence is given by the equation

$$\tan 2\chi = \frac{6\Theta}{G}, \tag{7.56}$$

where $\Theta$ is the rotational diffusion constant, and $G$ is the flow gradient. We now see that the tangent of the angle will depend on the rotational diffusion constant and will be inversely proportional to the flow gradient. For any given substance, $\Theta$ is a constant. It has the units of reciprocal seconds and can range from 370 for tobacco mosaic virus to 840,000 for serum albumin. When the rotating system attains a steady state, the particles will spend most of their time in a preferred orientation. This orientation is given by the parameter, $\alpha$, which is a function of the size and shape of the solute molecules. It is defined as

$$\alpha = \frac{G}{\Theta}. \tag{7.57}$$

The rotational diffusion constant is related to the rotary frictional coefficient in the same way that the translational diffusion constant, $\mathfrak{D}$, is related to the translational frictional coefficient, $f$:

$$\Theta = \frac{kT}{\zeta},\tag{7.58}$$

where $\zeta$ is the rotary frictional coefficient, $k$ is the Boltzmann constant and $T$ is the absolute temperature (recall at this point that $\mathfrak{D} = kT/f$). The quantity which is measured experimentally is $\Theta$, and $\zeta$ is calculated from it.

The rotary frictional coefficient is related to shape of the particle in terms of its axial lengths. Thus, for a long prolate ellipsoid with semiaxes $a$ and $b$, where $a > 5b$,

$$\zeta = \frac{16\pi\eta_0 a^3}{3\left[-1 + 2\ln\left(\frac{2a}{b}\right)\right]}.\tag{7.59}$$

As before, $\eta_0$ is the viscosity of the solvent. To use Eq. 7.59, we must determine independently the axial ratio, $a/b$, usually from intrinsic viscosity measurements (see Eq. 7.54).

For oblate ellipsoids, the appropriate equation is

$$\zeta = \frac{32\eta_0 b^3}{3}.\tag{7.60}$$

In this notation $b$ is the longer of the two axes.

The experimentally determined quantities are $\chi$ and $G$. From Eq. 7.56 we may calculate $\Theta$. Another procedure is to measure $\chi$ for various values of $G$. These data are plotted as shown in Fig. 7.24. For values of $\alpha$ less than 1.5, a more complicated equation reduces to

$$\chi = \frac{\pi}{4} - \frac{G}{12\Theta}.\tag{7.61}$$

Thus, the initial slope of the plot shown in Fig. 7.24 can be related to $G$ by

$$\frac{d\chi}{dG} = -\frac{1}{12\Theta}.\tag{7.62}$$

The experimental operation consists of placing a solution in the apparatus and causing the outer cylinder to rotate at a steady speed. The coupled crossed Nicol prisms of the polarizing microscope are rotated until the small visible portion of the cross of isocline is brought to the center of the field (see Fig. 7.23). The direction of rotation of the cylinder is reversed and the procedure repeated. The difference between the two

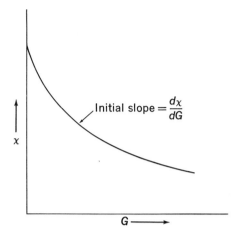

**Figure 7.24** A schematic plot for the determination of $\Theta$.

positions is equal to $2\chi$. Some representative data for macromolecules, which were collected by Tanford, are given in Table 7.3.

The birefringence, or double refraction, $\Delta n$, of systems of macromolecules is related to their weight-average molecular weight. The equation expressing this relationship

$$\frac{\Delta n}{C} = B_n M_i, \tag{7.63}$$

where $B_n$ is a constant characteristic of the substance, and $C$ is the concentration of the solution, is similar in form to the original Staudinger

**TABLE 7.3** THE LENGTH OF THE MAJOR AXIS, *a*, IN ANGSTROMS FROM FLOW BIREFRINGENCE MEASUREMENTS

| Macromolecule | Molecular Weight | $\Theta$ sec$^{-1}$ | Calculated from Eqs. 7.59 and 7.60 | Determined by other methods |
|---|---|---|---|---|
| Tobacco mosaic virus | $39 \times 10^6$ | 370 | 3,400 | 3,200 (light scattering) |
| Fibrinogen | $33 \times 10^4$ | $39.4 \times 10^3$ | 670 | 710 (intrinsic viscosity) |
| Serum albumin | $67 \times 10^4$ | $84 \times 10^4$ | 190 | 168 (frictional coefficient) |

*Source:* Data are quoted from Tanford, C., *Physical Chemistry of Macromolecules.* John Wiley and Sons, New York, 1961, p. 444. Original references are given by the author, and certain of the data are discussed.

Equation for relating the intrinsic viscosity of a solution to the molecular weight of the solute

$$\lim \ (\eta_{sp}/C)_{C \to 0} = K_n M_i \qquad \text{(see Eqs. 7.48 and 7.51).}$$

The extinction angle, $\chi$, is also related to the molecular weight:

$$\tan 2\chi = B_x \frac{\sum\limits_i C_i M_i}{\sum\limits_i C_i M_i^3}. \tag{7.64}$$

The birefringence may be measured by using a quarter-wave plate which functions as a Senarmont compensator in the polarizing light microscope. If $\phi$ is the angle of rotation of the analyzer, in degrees, when using the compensator, the birefringence is given by the equation

$$\Delta n = \frac{\lambda \phi}{180S}, \tag{7.65}$$

where $\lambda$ is the wavelength of the incident light in vacuo, and $S$ is the path length through the solution. Other mathematical relationships of interest are given in more extensive treatments of the subject.[*]

### 7.6 The Optical Properties of Macromolecular Systems

Light can be used in a variety of ways to unmask the size, shape, and structural characteristics of a macromolecule. The overall molecular dimensions can be ascertained by electron microscopy, light scattering, and X-ray diffraction. In addition, X-ray diffraction can under favorable circumstances permit us to calculate the positions of individual atoms in the macromolecular architecture. The high degree of refinement now obtained in the X-ray studies of myoglobin, hemoglobin, and a few other proteins have given us the first complete three-dimensional projections for giant molecules. This achievement has been one of the truly exciting advances in the biochemistry of this decade.

The absorption of light (spectroscopy) has been used to count certain functional groups in proteins, to study the contribution of various resonance forms to the peptide linkage, and to obtain information about the tertiary folding of proteins, to cite only a few applications. Polarization of fluorescence has been utilized for the determination of molecular weights of macromolecules. Information about the secondary structure of polymers can be obtained from measurements of optical rotatory dispersion.

---

[*] Cerf, R., and H. Scheraga, *Chem. Rev.*, **51**:185, 1952. See also pages 437–451 of the book by Tanford that is included in the listing in *Suggested Additional Reading.*

Of these various optical methods, two of wide application have been selected for presentation in this section: optical rotatory dispersion and light scattering. Spectroscopy and X-ray diffraction have been omitted because of the lengthy discussions required to do them justice. Electron microscopy is sufficiently similar to light microscopy in its overall approach and results to be familiar to most readers. Polarization of fluorescence is the least used of all these methods, at least at the present, and has been omitted for that reason. An interesting discussion of this method has been prepared by Weber.*

**Optical rotatory dispersion**  The dependence of the optical rotation of a substance on the wavelength of light is called its *optical rotatory dispersion*. This is so widely used that it is commonly abbreviated to ORD. Although proteins, polysaccharides, and nucleic acids are the most biochemically important macromolecules which display optical rotatory dispersion, extensive research has also been done on the synthetic polypeptides.

The student of organic chemistry has much contact with the subject of optical rotation, but acquires little fundamental knowledge about it. The basis for this state of affairs has been well stated by Schellman.†

Optical rotation is a field of study in which a powerful and general theory has made little contact with numerous precise experimental investigations. The origin of this situation is to be found in the quantum mechanical formula for optical rotation which shows that rotatory power is the sum of a large, but unknown, number of terms containing coefficients which in general are not susceptible to experimental measurement or to theoretical evaluation. The experimental measurement of these coefficients requires an investigation in the very deep ultraviolet region and this has so far presented insurmountable difficulties. On the other hand, a really adequate theoretical treatment would require a very detailed knowledge of the electronic structure of molecules, which has not yet been attained. . . . The utility of optical rotation lies in its unparallelled sensitivity to spatial configuration, to which other physical methods are generally insensitive.

A very brief introduction to the theory of optical rotation begins with the premise that all matter, whether structurally asymmetric or not, interacts with electromagnetic radiation. Light, which consists of perpendicular electric and magnetic vectors moving in phase, induces electric and magnetic moments in matter as it interacts with it. If molecules are not optically active, the effect of such interaction will be limited

---

*Weber, G., *Advan. Protein Chem.*, **8**:439, 1953.
†Schellman, J. A., *Comptes rendus des travaux du Laboratoire Carlsburg*, **30**:363, 1958.

to the induction of oscillating electrical dipoles in the irradiated matter. These dipoles will radiate continually and will set up secondary wavelets which travel in the same direction as the incident radiation but lag behind it 90°. The wave train is thus retarded, and the light is said to be refracted. Optical rotation does not occur, because the secondary wavelets are polarized in the same direction as the incident radiation. The induced magnetic moment in this case is so small as to be negligible.

On the other hand, if the molecules are optically active, the induced magnetic moment is appreciable and is related to the rate at which the electrical field strength is changing with time. Similarly, the expression for the induced electric moment contains a term which shows dependency on the rate at which the magnetic field strength is changing with time. These crossovers between electrical and magnetic effects lead to equations containing terms for secondary wavelets which are polarized *perpendicularly* to the original radiation. The superposition of the primary and secondary waves results in rotation in the direction of polarization. The student is referred to the article by Schellman for a comprehensive theoretical exposition and the pertinent equations.

Optical rotation is usually expressed as the *specific rotation,* which is a parameter analogous to the specific viscosity and other concentration-related terms. The definition of the specific rotation is

$$[\alpha]_\lambda^t = \frac{\alpha}{lC}, \tag{7.66}$$

where $\alpha$ is the observed rotation, $C$ is the concentration in grams per cc, $l$ is the length of the light path in decimeters, $t$ is the temperature of the solution, and $\lambda$ is the wavelength of light used in the polarimeter. The substitution of the subscript $D$ in place of a specific wavelength means that sodium light ($\lambda = 589$ m$\mu$) was used in the measurements.

The optical activity of proteins, polysaccharides, and nucleic acids arises from their optically active components — amino acids and sugars, and from their asymmetric secondary structure, which is in the form of helices of right- or left-handed screw. A denatured protein is a random coil and, therefore, will have a different optical rotation from its native helical counterpart. Similarly, amylopectin is largely nonhelical and the optical rotation of its solutions will differ from those of freshly prepared amylose (helical), these comparisons being based on the same weight of glucose. Changes in the secondary structure of macromolecules can be observed by measuring the specific rotation at a single wavelength. It has been known for a long time that the specific rotation of a protein solution becomes progressively more negative as denaturation proceeds. Completely denatured proteins and randomly coiled polypeptides have $[\alpha]_D$ values ranging from $-90°$ to $-125°$, whereas the specific rotation of native proteins is about 100° higher. Changes in the conformation of proteins in response to pH changes are also reflected in the specific rotation. All of these properties of protein solutions are known from

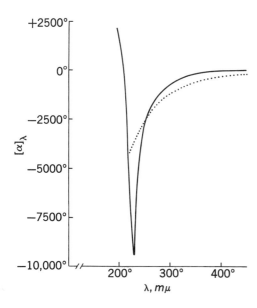

**Figure 7.25** Dependence of the specific rotation of proteins on the wavelength of light. The solid line depicts an idealized case of a native protein with a high $\alpha$-helix content. The dotted line represents the same protein following denaturation. With proteins there is no Cotton effect at the absorption maximum (270–280 m$\mu$). [From Jirgensons, B., *Proc. Seed Protein Conference*, January 21–23, 1963. U.S.D.A., New Orleans, Louisiana.]

observations of specific rotation at a single wavelength, usually the *D* line of sodium.

When the wavelength of light is systematically varied by attaching a monochromator to the polarimeter, the specific rotation is found to change also. As the wavelength approaches certain of the absorption bands of the substance investigated, for example, 225 m$\mu$ for proteins, a striking increase or decrease occurs, accompanied by a change in sign of the optical rotation as the absorption band is passed. This is known as the Cotton effect, and is illustrated in Fig. 7.25.

For wavelengths of light longer than the principal absorption bands of the substances investigated, the dependence of the specific rotation $[\alpha]$ on the wavelength, $\lambda$, follows a general relationship proposed by Drude in 1907:

$$[\alpha] = \frac{K}{\lambda^2 - \lambda_0^2}, \tag{7.67}$$

where $K$ and $\lambda_0$ are constants.

This is known as the one-term Drude dispersion equation. It was rearranged by Yang and Doty to give a linear expression

$$\lambda^2[\alpha] = \lambda_0^2[\alpha] + K. \tag{7.68}$$

Plots of $\lambda^2[\alpha]$ versus $[\alpha]$ gave straight lines for a number of native and denatured proteins in the wavelength range, 350–600 m$\mu$.[*] The dispersion constant $\lambda_0$ was evaluated from the slope of the straight line.

---

[*]Yang, J. T., and P. Doty, *J. Am. Chem. Soc.*, 79:761, 1957.

The optical rotatory dispersion of many proteins has been studied by Jirgensons.[*] Simple rotatory dispersion (obeying Eq. 7.68) was observed for a number of proteins, most of which are believed to exist in solution as random coils, as $\beta$-chains, or with a highly disordered secondary structure of partial helical character.

Variations in $K$ and $\lambda_0$ have produced an operational classification of proteins: Class I, proteins for which $\lambda_0$ is higher than the average for random polypeptide chains, presumably as a consequence of the presence of right-handed $\alpha$-helices; Class II, proteins for which $\lambda_0$ is about the same as the average for random polypeptide chains and therefore most probably are in a random conformation; and Class III, proteins for which $\lambda_0$ is lower than the average for random polypeptide chains, presumably as a result of the presence of $\beta$-structures or left-handed $\alpha$-helices or some other unknown structures.

A considerable number of proteins are known, however, which give anomalous plots when Eq. 7.68 is applied. Their dispersion is described as being *complex* and follows a two-term equation, first proposed by Moffitt

$$[\alpha] = \frac{A\lambda_0^2}{\lambda^2 - \lambda_0^2} + \frac{B\lambda_0^4}{(\lambda^2 - \lambda_0^2)^2}, \qquad (7.69)$$

where $A$, $B$, and $\lambda_0$ are constants.

The Moffitt Equation has been applied to proteins in an interesting way by Doty.[†] The equation is modified by expressing $[\alpha]$ as the *mean residue rotation*, $[m']$:

$$[m'] = \frac{M_0[\alpha]}{100} \cdot \frac{3}{(n^2 + 2)}, \qquad (7.70)$$

where $M_0$ is the average weight of an amino acid residue (approximately 125 for most proteins) and $n$ is the refractive index of the medium.

The first right-hand term, $M_0[\alpha]/100$, is the molar rotation of the average amino acid residue and the second right-hand term is the Lorentz correction for refractive index.

The general constants $A$ and $B$ are replaced by $a_0$ and $b_0$, where $a_0 = (a_0^R + a_0^H)$. The superscripts refer to the contribution made by the amino acid residue $(R)$ and the *alpha*-helix $(H)$. The constant $b_0$ is attributed to helical rotation entirely. The equation is then

$$[m'] = (a_0^R + a_0^H)\left(\frac{\lambda_0^2}{\lambda^2 - \lambda_0^2}\right) + b_0\left(\frac{\lambda_0^2}{\lambda^2 - \lambda_0^2}\right)^2. \qquad (7.71)$$

---

[*]Jirgensons, B., *Arch. Biochem. Biophys.*, **74**:57, 70, 1958; **78**:235, 1958; **85**:89, 1959; **93**:172, 1961.

[†]Doty, P., *Proc. IV International Congress Biochem.*, **IX**:8–41, 1958.

This equation is identical in form with Eq. 7.69. The dispersion constant, $\lambda_0$, is assumed to be the same in both right-hand terms and is selected by trial and error. For poly-$\gamma$-benzyl-L-glutamate and poly-L-glutamic acid, $\lambda_0 = 212$ m$\mu$.

If we divide both sides of Eq. 7.71 by $\lambda_0^2/(\lambda_2 - \lambda_0^2)$, we obtain

$$\frac{[m']}{\left(\dfrac{\lambda_0^2}{\lambda^2 - \lambda_0^2}\right)} = b_0\left(\frac{\lambda_0^2}{\lambda - \lambda_0^2}\right) + (a_0^R + a_0^H). \tag{7.72}$$

For a number of proteins, it was found that plotting the data in this form gave a straight line of slope $b_0$ and intercept $(a_0^R + a_0^H)$. Values for $b_0$ were found to cluster around $-630$ for a group of polypeptides presumed to be helical. The intercept term varied from one polypeptide to another, presumably because of the different contributions of $a_0^R$ and $a_0^H$, but ranged between $+250$ and $-30$.

Assuming that $a_0$ and $b_0$ are correctly defined, in any solvent in which the polypeptide is randomly coiled, $a_0^H$ will equal zero and $a_0^R$ will be equal to the intercept. Doty and Yang assumed this to be the case for poly-L-glutamic acid and poly-L-lysine in aqueous solution at pH 8 and ionic strength of 0.2 $M$. When comparisons are to be made between different solvent systems, it is mandatory that $\lambda_0$ be the same in both solvents.

These studies have established ranges for the specific rotation of polypeptides. Completely denatured proteins should have $[\alpha]$ values of the order of $-110°$, and those for the pure helix should be about $+5°$.

Some interesting possibilities have been considered by Doty and Yang. If a copolymer of $D$ and $L$ acids in a right-handed helix could be synthesized, the observed rotation should be due to $a_0^H$ and $b_0$ only, since the residue rotations should cancel. This situation has not been realized experimentally. A second interesting possibility would be to examine the rotation of an equal mixture of right- and left-handed helices. In this case, the rotation should be that of a random coil.

Yang and Doty have proposed that Eq. 7.71 be arranged in a general form which will permit the calculation of the percentage of the structure that is helical in a protein. This form is

$$[\alpha] = 1.39\left[(\Sigma a_0^R + f a_0^H)\left(\frac{\lambda_0^2}{\lambda^2 - \lambda_0^2}\right) + f b_0\left(\frac{\lambda_0^2}{\lambda^2 - \lambda_0^2}\right)^2\right], \tag{7.73}$$

where $f =$ fraction of residues in helical form and $\Sigma a_0^R =$ the sum of all $a_0^R$ values characteristic of each residue. Rotatory dispersion measurements are made on the protein in aqueous solution and in the denatured state. If the equation is now rearranged to give the form similar to Eq. 7.72, a plot of $([\alpha](\lambda^2 - \lambda_0^2)/1.39\lambda_0^2)$ versus $\lambda_0^2/(\lambda^2 - \lambda_0^2)$ will have a slope, $f b_0$, and an intercept, $(\Sigma a_0^R + f a_0^H)$. If we assume $b_0 = -630$ (as

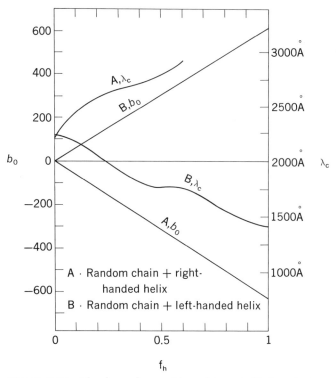

**Figure 7.26** The dependence of $b_0$ and $\lambda_c$ on helical content for mixtures of right-handed helices with a random coil and left-handed helices with a random coil; $f_h$ is fractional helical content. [From *The Proteins Composition Structure and Function*, 2nd Ed., Vol. 2, Hans Neurath, editor, 1964, p. 80, in the article by J. A. Shellman and C. Shellman. Academic Press Inc., New York. Reprinted by permission.]

established in the polypeptide studies), $f$ is determined from the slope. For a denatured protein, $f = 0$ and the intercept $= \Sigma a_0^R$. If it is known that $\lambda_0 = 212$ m$\mu$, only one wavelength need be used. From the denatured protein, $a_0^R$ is evaluated and $f a_0^H$ is obtained from the equation.

On the basis of this mathematical approach, the percentage helical content of a number of proteins has been calculated. Yang and Doty (*loc. cit.*) plotted dispersion curves that they obtained from various mixtures of $\alpha$-helices and random coils and thereby obtained calibration curves with which to compare the apparent helicity of a given protein molecule. If the dispersion constant, $\lambda_0$ for the helically coiled protein does not change when the protein unfolds, then $b_0$ can be used as an index of the percentage helicity. The approximations so obtained tend to be less accurate as the percentage helicity decreases. In Fig. 7.26 are shown plots of $b_0$ and $\lambda_0$ (the notation $\lambda_c$ specifically refers to the dispersion constant for the random coil) as a function of $f$, the fraction of resi-

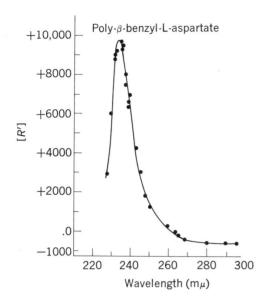

**Figure 7.27** Ultra-violet rotatory dispersion of poly-$\beta$-benzyl-L-aspartate in methylene dichloride solution. These data show the beginning of a *positive* Cotton effect. [*Aspects of Protein Structure*, G. N. Ramachandran, editor, 1963, p. 241, in the article by E. R. Blout. Academic Press Inc. (London) Limited. Reprinted by permission.]

dues in helical form. Note that $b_0$ is equal to $-630$ for the right-handed helix, but equals $+630$ for the left-handed helix. Here we have a powerful tool for detecting the helix sense of a protein.

Several other techniques can be used to investigate the screw sense of the helix, provided the polypeptide does not contain regions of opposite helicity. Whereas most proteins and synthetic polypeptides contain right-handed helices and display negative Cotton effects in the ultraviolet region of the spectrum (Fig. 7.25), Blout[*] has reported that the rotatory dispersion curve for poly-$\beta$-benzyl-*L*-aspartate reveals a positive Cotton effect (Fig. 7.27). Since this homopolymer also possesses a $b_0$ value equal to $+630$, one must conclude that a positive Cotton effect is evidence for the presence of left-handed helices. Similar results have been obtained by Blout[*] and coworkers for complexes of acridine orange and either poly-$\alpha$,*L*-glutamic acid or poly-$\alpha$,*D*-glutamic acid. The acridine orange:*D*-glutamic acid polymer exhibits a positive Cotton effect (left-handed helicity presumably) whereas the acridine orange:*L*-glutamic acid polymer displays a negative Cotton effect (right-handed helicity).

The relationship between optical rotatory dispersion and chain conformation, however, is not as simple as we might hope it to be. Many proteins which are believed to have helical conformations, based on X-ray diffraction data, obey the one-term Drude Equation. In addition, recent work by Hanlon and Klotz[†] casts serious doubt on the assertion

---

[*] Blout, E. R. in *Aspects of Protein Structure*, G. N. Ramanchandran, editor, Academic Press, New York, 1963, p. 241.

[†] Hanlon, S., and I. M. Klotz, *Federation Proc.*, **23**:215, 1964.

that rotatory dispersion changes accompanying a substitution of solvent are due only to conformational alterations in the polypeptide. A consideration pointed out by Tanford is that the unfolding of the coiled polypeptide chain in water is accompanied by transfer of side-chains from the hydrocarbon-like atmosphere of the interior of the helix to the polar environment of the solvent. This is, in itself, a solvent change which will be reflected in the specific rotation of the asymmetric unit.*

Despite the difficulties in arriving at an unambiguous interpretation of optical rotatory dispersion data, there is little doubt that the optical rotation is a highly sensitive probe with which to explore the structure of macromolecules. The work of Doty and Yang has been intensely stimulating to this general area of physical biochemistry. The protein molecule is, however, much more complex in composition than the synthetic polypeptides for which the constants, $b_0$ and $\lambda_0$, were evaluated. The problems inherent in extrapolating the polypeptide data to proteins have been clearly set forth by Tanford† and by Schellman and Schellman.‡ Continued research on the rotatory dispersion of macromolecules may enable us, in time, to make corrections for these complications. Recently, Shechter and Blout§ have proposed a modified two-term Drude Equation which appears to have definite advantages over other methods of analyzing optical rotatory dispersion data. The one-term Drude Equation and the Moffitt Equation are shown by these authors to be approximate forms of their modified two-term Drude Equation. The equation which they propose is fitted *only* by systems of α-helices, random coils, or mixtures of these two. Therefore, a criterion is afforded for detecting the presence of other conformations, such as proline helices and β-chains.

**Light scattering**  In 1871 Rayleigh stated that light scattered by particles small compared with the wavelength of light was essentially a diffraction phenomenon. His theory stated that the electrons of an isotropic particle of sufficient optical density, upon encountering a beam of light, were caused by the incident electromagnetic impulse to vibrate in unison with it—thus inducing an oscillating electric moment in the particles. These oscillating electrons became the sources of the scattered (or diffracted) light which for the most part had the same frequency as the exciting wave. Accordingly, scattered light from the particle weakened the intensity of the transmitted beam. These principles are

---

*Weber, R., and C. Tanford, *J. Am. Chem. Soc.*, **81**:3255, 4032, 1959; Siegel, S., and G. V. Smith, *J. Am. Chem. Soc.*, **82**:6082, 1960; Tanford, C., and K. De Paritosh, *J. Biol. Chem.*, **236**:1711, 1961.

†See pages 123–124 of the book by Tanford that is included in the listing in *Suggested Additional Reading.*

‡Schellman, J. A., and C. Schellman, in *The Proteins* (edited by H. Neurath), Vol. II, 2nd Ed. Academic Press, New York, 1964.

§Shechter, E., and E. R. Blout, *Proc. Nat. Acad. Sci. (U.S.)*, **51**:695, 794, 1964.

exactly the same ones cited in the preceding section on optical rotatory dispersion.

Since the induced electric moments in small isotropic particles are parallel to the electrical vector of the exciting beam, the light scattered at 90° will be plane polarized regardless of the state of polarization of the incident beam. When the incident beam is unpolarized, however, the light scattered at angles other than 90° will have both a vertical and horizontal component, and the scattering intensity will vary with the angle according to the factor $(1 + \cos^2 \theta)$, where $\theta$ is the angle of scattering.

Rayleigh formulated an equation which states mathematically the relationship between the scattered portion of the incident light, its wavelength, the particle size, and the optical properties of the scattering medium.* The equation takes into account the polarization of the particle due to its exposure to incident light, and is

$$\frac{i}{I} = \frac{8\pi^4 \nu \alpha^2}{\lambda'^4 r^2} (1 + \cos^2 \theta), \tag{7.74}$$

where

$i =$ the intensity of scattered light per unit volume (1 cc) of the scattering system,

$\nu =$ the number of scattering particles per unit volume,

$I =$ the intensity of the incident beam,

$r =$ the distance between the observer and the scattering system,

$\theta =$ the angle between the observer and the incident beam,

$\lambda' =$ the wavelength of the light falling on the particles in solution ($\lambda' = \lambda/n_0$ where $\lambda$ is the wavelength of light in a vacuum and $n_0$ is the refractive index of the medium),

$\alpha =$ the polarizability or induced moment per unit electrical field strength of the small isotropic particles.

In determining molecular weights for systems of small isotropic particles at high dilution it is more convenient to work in terms of the attenuation of the incident beam by the scattering solution. The term *turbidity* has been given to this extinction due to scattering and can be directly evaluated from the intensity of scattered light so that the need for consideration of $r$, as included in Eq. 7.74, is eliminated.

The phenomenon of light with initial intensity $I_0$ losing intensity through scattering as it passes through a medium can be represented by an expression similar to Lambert's Law:

$$\ln \frac{I_0}{I} = \tau x, \tag{7.75}$$

---

*A derivation of the Rayleigh Equation may be found on pages 10–12 of the book by Stacey that is included in the listing in *Suggested Additional Reading*.

where $\tau$ is the turbidity and $x$ is the distance traversed by the beam in the solution. Since the proportionality constant $\tau$ depends on the number and size of the scattering particles, it can be defined on a basis similar to the Rayleigh concept. Debye has shown on the basis of the theory of fluctuations, that for light scattered at 90° to the incident beam in ideal systems, the turbidity can be expressed by

$$\tau = \frac{32\pi^3}{3} \cdot \frac{n_0^2(n-n_0)^2}{\lambda^4} \cdot \frac{1}{a}, \tag{7.76}$$

where $n$ and $n_0$ are respectively the indices of refraction of the solution and solvent, $\lambda$ is the wavelength of the light in a vacuum, and $a$ is the number of solute particles per cc. Replacement of $a$ by its equivalent, $C\mathfrak{N}/M$, introduces the molecular weight into the expression, since $C$ is the concentration of solute in grams per cc and $\mathfrak{N}$ is Avogadro's number.

Equation 7.76 can be written in the form

$$\frac{\tau}{C} = H\mathrm{M}, \tag{7.77}$$

in which the proportionality constant $H$ $(\mathrm{cm}^2/\mathrm{g}^2)$ is found from refraction measurements and is defined by the relationship

$$H = \frac{32\pi^3}{3} \cdot \frac{n_0^2}{\mathfrak{N}\lambda^4} \cdot \left[\frac{(n-n_0)}{C}\right]^2. \tag{7.78}$$

The term $(n - n_0)/C$ is called the *refractive index increment* and is sometimes written $dn/dC$.

For nonideal systems Eq. 7.77 takes the form

$$\frac{HC}{\tau} = \frac{1}{M} + 2BC, \tag{7.79}$$

where the constant $B$ is termed the interaction constant and has been shown by osmotic pressure studies to be independent of M but dependent upon the solvent used. Equation 7.79 should give linear plots from which M can readily be evaluated. This is the equation used to calculate the molecular weights of macromolecules from light scattering data. Halwer, Nutting, and Brice[*] have plotted data for lysozyme, ovalbumin, and lactoglobulin according to Eq. 7.79. The plots obtained had slightly negative slopes, except for ovalbumin, which had a slope of zero. In Fig. 7.28 are presented data from a light scattering study of gum arabic. The $y$ intercept is 1/M. Values of M must be corrected for polarization effects, if high accuracy is desired. The polarization correction is obtained experimentally by interposing a Polaroid analyzer in

---

[*]Halwer, M., G. C. Nutting, and B. A. Brice, *J. Am. Chem. Soc.*, **73**:2786, 1951.

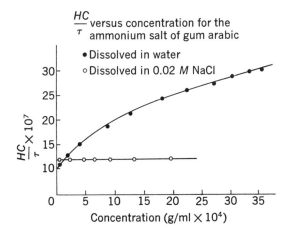

$\dfrac{HC}{\tau}$ versus concentration for the ammonium salt of gum arabic

• Dissolved in water
○ Dissolved in 0.02 *M* NaCl

**Figure 7.28** Experimental plot for the determination of the molecular weight of gum arabic by light scattering. Molecular weight is approximately $10^6$. [From unpublished data of G. D. Lee and H. B. Williams.]

the scattered beam. In most practical cases, the correction involved is of the order of only a few percent and can be safely neglected, since it is usually less than the experimental error. Molecular weights obtained by light scattering are weight-average molecular weights. The method is unusually sensitive to the presence of aggregates and frequently yields values that are considerably higher than those obtained by other methods.

When the particles in solution are greater in at least one linear dimension than 1/20 the wavelength of the incident light, the molecules cannot be considered as point sources and the light scattered from different parts of the same particle may produce interference. The effect of this interference differs in magnitude for different sizes and shapes of molecules, but for all molecules over 1/20 the wavelength of light the 90° scattering is no longer a true measure of the turbidity. The scattering is less in the direction backward to the incident beam than in the forward direction. Because of this asymmetric scattering pattern, measured by the scattering dissymmetry, there is a diminution of intensity at 90° and corrections to the observed turbidity are necessary.

Debye, in developing simple functions to represent the variation in angular scattering for particles of different size and shape, considered the solute molecule as divided into a series of independent dipole oscillators (or submolecules) that radiate according to the Rayleigh principle. The average sum of the contributions of these submolecules may be calculated from equations which Debye and others have developed. Assuming that the difference between the refractive index of the particle and of the solvent is very small and that there is no secondary, intermolecular interference in the solution, the following simplified equations are obtained for $I_\theta$, the intensity of the scattered light at an angle $\theta$.

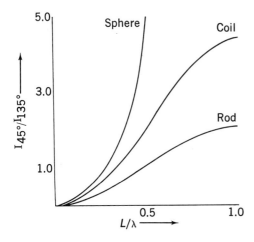

**Figure 7.29**   Theoretical curves for the dependence of the scattering dissymmetry on the particle shape. [After Jirgensons, B., and M. E. Straumanis, *A Short Textbook of Colloid Chemistry*, 2nd Rev. Ed. Macmillan, New York, 1962, p. 228.]

$$\text{Sphere: } I_\theta = \left[\frac{3}{x^3}(\sin x - x \cos x)\right]^2, \quad x = \frac{2\pi D}{\lambda}\sin\frac{1}{2}\theta. \qquad (7.80)$$

$$\text{Rod: } \quad I_\theta = \frac{1}{x}\int_0^{2x}\left(\frac{\sin x}{x}\right)dx - \left(\frac{\sin x}{x}\right)^2, \quad x = \frac{2\pi L}{\lambda}\sin\frac{1}{2}\theta. \qquad (7.81)$$

$$\text{Coil: } \quad I_\theta = \frac{2}{x^2}[e^{-x} - (1-x)], \; x = \frac{8}{3}\left(\frac{\pi L}{\lambda}\sin\frac{1}{2}\theta\right)^2. \qquad (7.82)$$

In these equations $D$ is the diameter of the spheres and $L$ is the length of rigid rods or the mean distance between the ends of a randomly kinked coil. From these equations one may calculate values for $I_\theta$ by assuming certain values of $L$ or $D$ for definite $\theta$ and $\lambda$. Experimental values are then compared with theoretical values. From these equations one may also calculate theoretical curves for the dependence of the scattering dissymmetry ($I_{45°}/I_{135°}$) on the shape of the particle. The general shape of these plots is shown in Fig. 7.29. The general shape of particles in solution may be determined by measuring the dissymmetry at various wavelengths and plotting the data. The shape of the resulting curve is compared with the theoretical curves (Fig. 7.29) to determine whether the particles are spheres, coils, or rods. If the shape of the particle is known, the particle length may be determined from the theoretical curves. The dissymmetry is measured over a suitable concentration range and the data are extrapolated to zero concentration. The intercept on the $I_{45°}/I_{135°}$ axis is called the *intrinsic dissymmetry*. Using the intrinsic dissymmetry and the appropriate theoretical curve, one may obtain $L/\lambda$ from the abscissa. The wavelength being known,

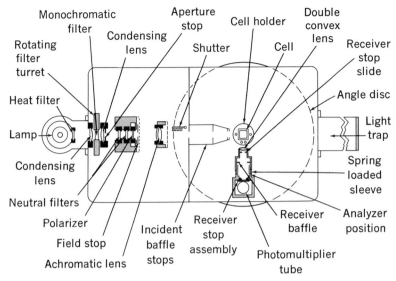

**Figure 7.30** Optical plan of the Aminco light scattering photometer. The incident optical system consists of a mercury vapor lamp, a rotating filter turret for positioning monochromatic filters in the beam, neutral filters for attenuation of beam intensity, a polarizer for vertical and horizontal polarization, a shutter, and a cell holder. The receiver optical system directs light scattered from the sample to the photomultiplier tube. The system includes a double convex lens; two precision apertures, which control the geometry of the scattered light received; means for positioning an analyzer, which is used for measuring depolarization factors; a photomultiplier tube, which converts light to electrical energy; and an angle disc to position the receiver around the sample. As shown, the instrument is set for measurements at a scattering angle of 90°, but other angles are readily measured by moving the turntable. [Diagram and explanatory notes supplied by American Instrument Company, Inc., Silver Spring, Maryland.]

$L$ is readily determined. A comprehensive review of light scattering in protein solutions has been prepared by Doty and Edsall.[*]

Light scattering is measured in an instrument called a light scattering photometer. A schematic illustration of such an apparatus is shown in Fig. 7.30. Various types of cells are available, so that scattering may be measured at any desired angle. The solutions to be examined by this technique must be prepared with great care to insure the absence of dust particles, which would lead to false scattering values. Filtration through fine-pore filters or high-speed centrifugation can be used to

---

[*] Doty, P., and J. T. Edsall, *Advan. Protein Chem.*, **VI**:35, 1951.

**TABLE 7.4** VARIOUS METHODS FOR DETERMINING THE MOLECULAR WEIGHT
OF MACROMOLECULES

| Method | Size range | Type of molecular weight | Equation | Remarks |
|---|---|---|---|---|
| Chemical analysis | $M < 2 \times 10^5$ | $M_n$ | —— | Depends on special groups or atoms in the molecule |
| Osmotic pressure | $M < 10^6$ | $M_n$ | 3.34 | Instrumentation in this area is much improved. See Chapter 3 |
| Sedimentation velocity alone | $M < 5 \times 10^7$ | $M_w$ | 7.38 | Gives minimal molecular weight based on spherical shape |
| Diffusion | $M < 10^6$ | $M_w$ | 7.33 | Gives minimal molecular weight based on spherical shape unless $f/f_0$ is known |
| Sedimentation velocity and diffusion | $M < 5 \times 10^7$ | $M_w$ | 7.36 | Is being replaced by the sedimentation equilibrium method. Is independent of shape |
| Sedimentation equilibrium | $M < 5 \times 10^6$ | $M_w, M_z$ | 7.41 7.42 | Vastly improved by new techniques that shorten the time required. Is independent of shape |
| Viscosity | unknown | $M_v$ | 7.51 | The Staudinger Equation is a limiting equation for loose structures |
| Viscosity and sedimentation velocity ⎫ Viscosity and diffusion ⎭ | unknown | $M_w$ | 7.52 7.53 | ⎧ Model is permitted some degree of flexibility and hydration ⎭ |
| Birefringence of flow | unknown | $M_w$ | 7.63 7.64 | Requires the evaluation of empirical constants |
| Light scattering | $M < 10^4$ | $M_w$ | 7.79 | Highly sensitive to the presence of aggregates |

remove dust from solutions. The cells and other glassware are rinsed by condensing acetone on their surfaces.

The refractive index term, $(n - n_0)$, is measured in a differential refractometer. The instrument is usually calibrated with sucrose or KCl solutions at a wavelength corresponding to the $D$ line of sodium.

### 7.7 Comparison of Methods for the Determination of Molecular Weight

So much space and emphasis has been given to the problem of molecular weight determinations that it seems fitting to conclude the chapter with a summary tabulation. Of the methods listed in Table 7.4, only *chemical* determinations have not been discussed in the preceding sections. In brief, chemical methods involve colorimetric, titrimetric, and other types of analyses that depend on the presence of a particular chemical group or entity in the macromolecular structure. Thus, if the number of such groups is known per molecule, a chemical analysis will permit a kind of molecule-counting that gives a number-average molecular weight. Unfortunately, chemical methods for proteins, carbohydrates, or other macromolecules are few in number. Some examples are the periodate oxidation method for linear starch, iron analysis methods for the iron-bearing proteins, and riboflavin analysis of flavoproteins containing a known number of prosthetic groups.

Most of the physical methods which have been discussed rely on assumptions regarding the molecular geometry and/or density, which involve an appreciable degree of uncertainty. X-ray diffraction and electron microscopy have, in many cases, shown the shapes of macromolecules to be much more irregular than the spheres, ellipsoids, and long cylinders which have served as such convenient models for our calculations. Protein molecules for which the tertiary structure has been determined with exactness are known to have holes and crevices in their surfaces. If solvent flows through these holes as the molecule moves through the solution, the hydrodynamic characteristics of the molecule will be different from those of an impenetrable solid particle. It remains to be seen how successfully the simple equations listed in Table 7.4 can be retailored to fit the shapes of real molecules instead of "model" molecules.

### Suggested Additional Reading

#### General: Macromolecular Chemistry

Edsall, J., and J. Wyman, *Biophysical Chemistry I*. Academic Press, New York, 1958.

Fasman, G. D., Ed., *Poly-α-Amino Acids*. Marcel Dekker, New York, 1967.

Flory, P. J., *Principles of Polymer Chemistry*. Cornell Univ. Press, Ithaca, New York, 1953.

Morawetz, H., *Macromolecules in Solution*. Wiley-Interscience, New York, 1965.

Tanford, C., *The Physical Chemistry of Macromolecules*. John Wiley and Sons, New York, 1961.

Timasheff, S. N. and G. D. Fasman, Eds., *Structure and Stability of Biological Macromolecules*. Marcel Dekker, New York, 1969.

Veis, A., *Biological Polyelectrolytes*. Marcel Dekker, New York, 1970.

### Surface Chemistry

Adamson, A. W., *Physical Chemistry of Surfaces*, 2nd Ed. Interscience, New York, 1967.

### Physical Methods

Beychok, S., *Science*, **154**, 1288 (1966).
Bowen, T. J., *An Introduction to Ultracentrifugation*. Wiley-Interscience, New York, 1970.
Cann, J. R., *Interacting Macromolecules*. Academic Press, New York, 1970.
Crabbé, P., *ORD and CD in Chemistry and Biochemistry*. Academic Press, New York, 1972.
Hjerten, S. in *Methods of Biochemical Analysis*, **18**, 55 (1970).
Leach, S. J., Ed., *Physical Principles and Techniques of Protein Chemistry*. Academic Press, New York, Part A, 1969, and Part B, 1970.
Schachman, H. K., *Ultracentrifugation in Biochemistry*. Academic Press, New York, 1959.
Tinoco, I. in *Methods of Biochemical Analysis*, **18**, 81 (1970).
Urry, D. W., Ed., *Spectroscopic Approaches to Biomolecular Conformation*. American Medical Association, 1970.

### Problems

**7.1** A soap film is stretched out on a small frame whose dimensions are 2.5 cm and 5.1 cm. The surface tension of the film is 8 dynes per cm. The frame settings are now adjusted so that the film shrinks to an area of 2.5 by 3.5 cm. Theoretically, how far could this change in surface energy lift a 1 mg weight?

**7.2** The following data are obtained for the average number of active sites per molecule that are saturated at various concentrations of substrate at 25°C. Plot the data according to the Langmuir adsorption isotherm and determine the integral number of active sites per molecule.

Suggestion: plot the equation in the double reciprocal form for simpification.

| Active Sites Saturated Per Molecule | Concentration of Substrate $\times 10^3$ (Molarity) |
|---|---|
| 1.50 | 0.05 |
| 2.45 | 0.10 |
| 3.12 | 0.20 |
| 3.50 | 0.30 |
| 3.70 | 0.40 |
| 3.80 | 0.60 |

*Answer.* 4 active sites per molecule

**7.3** Rearrange the Langmuir adsorption isotherm in a double reciprocal form like the Lineweaver-Burk form of the Michaelis-Menten Equation. What advantages would this form possess? What are the theoretical similarities between the two equations?

**7.4** The adsorption of a certain dye on activated charcoal may be represented by the equation

$$x = 1.903C^{0.7}$$

where $x$ is in g of dye per g of charcoal, and $C$ is in g per L. It is desired to remove, in one operation, 90 percent of the dye from 50 L of a solution originally containing 0.04 g of dye per L of solution. Calculate the weight of charcoal which must be used.

**7.5** The gold number of gelatin is determined by mixing 0.1 ml of a 1 percent gelatin solution with 9.9 ml of specially prepared gold sol and then diluting the initial solution serially 1:2 ten times. Ten percent sodium chloride is added and the sixth tube is the first one to be distinctly blue. What is the gold number?

*Answer.* 0.03

**7.6** A sample of a synthetic polymer contains the following distribution of molecular weights. Calculate $M_n$ and $M_w$.

| Percent | Molecular Weight |
|---------|------------------|
| 10 | 25,000 |
| 20 | 35,000 |
| 50 | 50,000 |
| 15 | 55,000 |
| 5 | 70,000 |

**7.7** From Eq. 7.30 and the information given below, calculate the charge on the ovalbumin molecule in valence units.
To convert charge $(Q)$ into valence units $(Z)$ substitute:

$$Q = Z \frac{(4.8 \times 10^{-10})}{300}$$

This converts coulombs to electrostatic units and then divides by the charge on the electron. In the same operation volts are converted to absolute volts.
Assume the radius of the albumin molecule to be $27.8 \times 10^{-8}$ cm. Assume the average radius of the ions in solution to be $2.5 \times 10^{-8}$ cm. The Debye-Hückel parameter, $\kappa$, is $1.033 \times 10^7$ cm$^{-1}$ under the conditions of the experiment.

$$6/f(\kappa r) = 5.47$$

The viscosity of water under the conditions of the experiment is 0.01792 abs. poise.

Assume a mobility ($u$) of 1 micron per sec per volt per cm.

*Answer.* +17.5

**7.8** What pH would you use to separate, by electrophoresis, two enzymes, $A$ and $B$, with isoelectric points at 5 and 8, respectively? What would be the sign of the net electrical charge of $A$ and $B$ at pH 4, 5, 6, 7, 8, and 9?

**7.9** Calculate the isoelectric point of serum albumin from the following pH-mobility relationships.

| pH | $u$ (microns per sec) |
|----|----------------------|
| 4.03 | 0.640 |
| 4.36 | 0.356 |
| 5.67 | −0.487 |
| 5.86 | −0.750 |
| 6.52 | −1.00 |
| 7.00 | −1.23 |
| 7.68 | −1.33 |

*Answer.* 4.76

**7.10** Calculate the molecular weight of a protein from the following data.

$$\mathscr{D}_{20} = 7.76 \times 10^{-7} \text{ cm}^2 \text{ sec}^{-1};$$

$$T = 20°C;$$

$$\eta = 1.1 \times 10^{-2} \text{ abs. poise};$$

$$f/f_0 = 1.17;$$

$$\bar{V} = 0.748 \text{ ml per g.}$$

*Answer.* Approximately 33,500

**7.11** A solution of an enzyme in a dilute buffer was subjected to centrifugation in an ultracentrifuge operated at 27° and 10,000 rpm. The solute had a density of 1.00 $g/cm^3$ and the enzyme had a partial specific volume of 0.73 $cm^3/g$. When equilibrium was attained, the concentration was 1.12 mg/ml at 7.2000 cm from the center of rotation and 5.30 mg/ml at 7.5000 cm from the center of rotation. What is the molecular weight of the enzyme?

*Answer.* 59,000

**7.12** An isolated enzyme has the following characteristics:

0.2 percent of a prosthetic group having $M = 200$;

$s_{20} = 8.0 \times 10^{-13}$ sec;

$\mathcal{D}_{20} = 4.0 \times 10^{-7}$ cm$^2$ sec$^{-1}$;

Density of protein $= 1.35$ g/cc.

How many moles of prosthetic group per mole of enzyme?

**7.13** Tanford, Kawahara, and Lapanje (*J. Amer. Chem. Soc.*, **89**, 729 (1967)), found that twelve denatured proteins obeyed the relationship

$$[\eta] = 0.716 \ n^{0.66}$$

in 6 M guanidine hydrochloride containing 0.1 M mercaptoethanol at 25°. The number of amino acid residues in the polypeptide chain is signified by n. Tanford *et al*, found that bovine serum albumin had an intrinsic viscosity of 52.2 cc/g in 6 M guanidine hydrochloride $+ 0.1$ M mercaptoethanol at 25°. What is the molecular weight of denatured bovine serum albumin? The average weight of an amino acid residue in bovine serum albumin is 110 g/mole.

**7.14** Calculate the molecular weight of aspartase from the following data obtained in a density gradient centrifugation experiment. The tube contents were separated into 31 fractions of equal volume. The fractions were numbered consecutively, beginning with the first fraction to drip out of the bottom of the tube. Catalase was in fraction No. 7 and aspartase in fraction No. 11. The molecular weight of catalase is 244,000.

*Answer.* 185,000

**7.15** A number-average molecular weight of 100,000 was determined for an enzyme in 0.1 M potassium phosphate buffer, pH 7.2, using an osmometer. The enzyme was then denatured, and a weight-average molecular weight of 33,000 was determined by sedimentation equilibrium. Explain why these experiments do *not* prove that denaturation splits the enzyme into three subunits. Suggest a possible combination of four subunits which would be consistent with the above data.

*Answer.* One possible combination would be that denaturation results in one subunit with a molecular weight of 50,000 and three subunits with a molecular weight of 16,700. Many other combinations are also possible.

**7.16** The following data were taken in a sedimentation velocity experiment on a bacterial enzyme. The rotor speed was 56,000 rpm. Complete the table of data and then calculate the sedimentation coefficient in Svedbergs. $R_n$ is the distance in centimeters from the center of the boundary to the reference line on the photographic plate when measured on the comparator. In column 3 $R_n$ is to be multiplied by 0.459 to compensate for the magnification factors involved. In column 4, the corrected value of $R_n$ from column 3 is added to 5.720 cm which is the distance from the reference hole (appears as a line on the photographic plate) to the center of the rotor. Column 4, thus, is x, the corrected radial distance of the boundary in centimeters. Column 5 is to be the common logarithm of the corresponding value in column 4.

| 1 | 2 | 3 | 4 | 5 |
|---|---|---|---|---|
| | | | x | |
| *Time* | $R_n$ | $R_n \times 0.459$ | *(Corrected* $R_n$ *+ 5.720)* | log x |
| *min* | *cm* | *cm* | *cm* | |
| 0 | 0.6204 | | | |
| 4 | 0.7003 | | | |
| 12 | 0.8205 | | | |
| 20 | 0.9783 | | | |
| 44 | 1.443 | | | |
| 52 | 1.644 | | | |
| 60 | 1.774 | | | |
| 68 | 1.963 | | | |
| 76 | 2.122 | | | |
| 84 | 2.241 | | | |

**7.17** A viscosity experiment was performed to determine the molecular weight of a polystyrene sample. Various amounts of polystyrene were dissolved in toluene at 20°C, and the viscosities were measured. The Staudinger constants for polystyrene have been determined to be $K = 3.6 \times 10^{-4}$, $a = 0.64$. Calculate the molecular weight from the following data.

Experimental results

| C (g/100 ml) | time (sec) |
|---|---|
| 1.5 | 77 |
| 1.0 | 74 |
| 0.5 | 67 |
| 0.25 | 62 |
| 0.125 | 59 |
| 0.000 | 55 |

*Answer.* Approximately 100,000

**7.18** Doty, Bradbury, and Holtzer (*J. Amer. Chem. Soc.*, **78**, 947 (1956))
report the following results for five samples of poly-γ-benzyl-L-
glutamate in dichloroacetic acid at 25°:

| Molecular weight | [η], cc/g |
|---|---|
| 21,400 | 16.2 |
| 66,500 | 45.0 |
| 130,000 | 77.5 |
| 208,000 | 119 |
| 262,000 | 141.5 |

Evaluate $K$ and $a$ for this system.

**7.19** The Staudinger constants for amylose in dimethyl sulfoxide are
$K = 3.06 \times 10^{-4}$ and $a = 0.64$. From the following data, calculate the
molecular weight.

| $\eta_{rel}$ | C (g/100 ml) |
|---|---|
| 1.09 | 0.15 |
| 1.12 | 0.20 |
| 1.19 | 0.30 |
| 1.26 | 0.40 |
| 1.00 | 0.00 |

**7.20** (a) Calculate the term $H$ for lysozyme from the following light
scattering data.

specific refractive increment 0.1955;
wavelength               436 mμ;
$n_o$                       1.337;
$T.$                        298°K.

(b) Calculate the molecular weight from the following data.

| $HC/\tau \times 10^5$ | C in g/100 ml |
|---|---|
| 6.6 | 0.4 |
| 6.3 | 0.8 |
| 6.0 | 1.2 |

*Answer.* See Halwer, M., G. C. Nutting, and B. A. Brice, *J. Am.
Chem. Soc.*, **73**; 2786, 1951.

# 8 | THE PRINCIPLES OF NUCLEAR CHEMISTRY AND SOME BIOLOGICAL APPLICATIONS

Life is not easy for any of us. But what of that? We must have perseverance and above all confidence in ourselves. We must believe that we are gifted for something, and that this thing, at whatever cost, must be attained.

*Marie S. Curie*

A knowledge of the basic chemistry of nuclear reactions has for some years been a part of the standard equipment of any physical scientist. Today, because of the grave social implications of "the bomb," every schoolboy also has access to a working understanding of nuclear fission and fusion. In the uproar over nuclear weaponry, the spectacular advance of biochemical knowledge resulting from the use of radioactive tracer techniques seems lost in the shouting, and certainly, tame by comparison. To these highly important and beneficial applications of nuclear science, however, we will direct our attention after first looking at the fundamental chemistry.

In the preceding chapters we have devoted our attention to atoms and molecules and to transformations involving their *electrons*. Ionization, oxidation-reduction, and other reactions are all extranuclear chemical processes throughout which the nucleus maintains its integrity and the atom preserves its identity. Now we must concern ourselves with the nucleus. The nucleus occupies only a very small part of the volume of the individual atom, its diameter being of the order of $10^{-12}$ cm; however, most of the mass of the atom is concentrated in the nucleus. The neutrons and protons of the nucleus make the major contribution to atomic weight. The number of protons, the atomic number, is the funda-

**TABLE 8.1** NUCLEAR BUILDING BLOCKS

| Unit | Symbols | Mass (in grams) | Charge |
|------|---------|-----------------|--------|
| Proton | $p$, $^1_1H$ | $1.67239 \times 10^{-24}$ | + |
| Electron | $e^-$, $^0_{-1}\beta$ | $9.1083 \times 10^{-28}$ | − |
| Neutron | $n$, $^1_0n$ | $1.67470 \times 10^{-24}$ | 0 |
| Positron | $e^+$, $^0_{+1}\beta$ | $9.1083 \times 10^{-28}$ | + |
| Deuteron | $d$, $^2_1H$ | $3.3418 \times 10^{-24}$ | + |
| Alpha particle | $\alpha$, $^4_2He$ | $6.6430 \times 10^{-24}$ | 2+ |

mental determiner of chemical properties. Any change in the positive charge of the nucleus will change the identity of the atom, that is, transform one element into another.

When we turn our attention to the nucleus and nuclear transformations, we must also effect a second change of emphasis. Instead of dealing with matter and energy as a dichotomy, carrying out a mass balance with the left hand and an energy balance with the right hand, we must recall that the First Law of Thermodynamics, as stated in modern times, not only proclaims the indestructibility of matter and energy but also their interconvertibility. The quantification of this concept was provided by Einstein in his celebrated equation

$$E = mc^2, \tag{8.1}$$

which shows the energy-mass conversion factor to be the square of the velocity of light. Mass and energy are, thus, two different manifestations of the same entity, one gram of mass being convertible into $(1)(c^2)$ energy units, or $9 \times 10^{20}$ ergs.

## 8.1 Disintegration of Radioactive Elements: Nuclear Equations

The primary nuclear units to be considered here are presented in Table 8.1. We can write chemical equations for natural radioactive disintegration reactions, and for chemical transformations resulting from various types of bombardment experiments. A special notation is used to indicate the mass number and the atomic number of the elements or particles appearing in the stoichiometry. For example, radium decomposes to yield radon (a radioactive gas) and $\alpha$ particles (helium nuclei). We express this information by writing

$$^{226}_{88}Ra \longrightarrow \, ^{222}_{86}Rn + \, ^4_2He. \tag{8.2}$$

The number of nuclear protons (or atomic number) appears as a subscript, and the atomic weight (or mass number) appears as a superscript.

The fact that the radium and radon atoms are electrically neutral, whereas the helium nucleus ($\alpha$ particle) has a charge of 2+ when ejected, is of no concern here, because the equation is restricted to nuclear changes. The number of planetary electrons is readily adjusted: the helium nuclei rapidly pick up the necessary electrons and become neutral helium atoms, just as the radon atoms immediately attain neutral status. The disintegration of radium as shown in Eq. 8.2 was used to evaluate Avogadro's number. Rutherford and Geiger found that radium emits $3.4 \times 10^{10}$ $\alpha$ particles per gram per second, and Rutherford and Boltwood discovered that helium is produced from radium at the rate of $1.07 \times 10^{-4}$ ml per gram per day, corrected to standard conditions. If one mole of a gas occupies 22,400 ml under standard conditions,

$$\frac{22,400}{1.07 \times 10^{-4}} (3.4 \times 10^{10})(24 \times 60 \times 60) = 6.15 \times 10^{23} \text{ particles per mole.}$$

Direct counting of $\alpha$ particles combined with a measurement of the total charge carried by them established the fact that each particle carries two positive charges.

A second example of a nuclear equation is the disintegration of uranium to yield thorium and $\alpha$ particles:

$$^{238}_{92}U \rightarrow {}^{4}_{2}He + {}^{234}_{90}Th. \tag{8.3}$$

In writing these equations the student should always make sure that the sum of all superscript terms on the left is equal to the sum of superscript terms on the right and that the same type of balance holds for the subscript terms. Now let us take an examples in which the emitted radiation is composed of $\beta$ particles (electrons).

Thorium emits $\beta$ particles and is converted to protactinium. The nuclear equation is

$$^{234}_{90}Th \rightarrow {}^{0}_{-1}\beta + {}^{234}_{91}Pa. \tag{8.4}$$

When the nucleus ejects a $\beta$ particle a neutron is converted into a proton and the atomic number is increased by one unit.

The emission of $\gamma$ rays (high-energy electromagnetic radiation) by an atom does not change the atomic weight or number. Gamma radiation often accompanies the emission of $\alpha$ or $\beta$ particles, and thus enables the nucleus to release excess energy. Whereas the energies of $\beta$ particles may be sufficiently different so as to fall into a normal distribution pattern (a bell-shaped curve), the energies of $\gamma$ rays will be limited to only one or a few characteristic values. This property of $\gamma$ rays results from the fact that a $\gamma$ ray represents a discrete transition between energy levels in the nucleus. Thus, the amount of energy released is fixed by the nature of the transition.

**TABLE 8.2** THE URANIUM SERIES

| Element | Atomic weight | Atomic number | Type of emission | Half-life |
|---|---|---|---|---|
| Uranium (U$_I$) | 238 | 92 | $\alpha$ | $4.51 \times 10^9$ years |
| ↓ | | | | |
| Thorium (UX$_1$) | 234 | 90 | $\beta$ | 24.1 days |
| ↓ | | | | |
| Protactinium (UX$_2$) | 234 | 91 | $\beta$ | 1.18 minutes |
| ↓ | | | | |
| Uranium (U$_{II}$) | 234 | 92 | $\alpha$ | $2.48 \times 10^5$ years |
| ↓ | | | | |
| Thorium | 230 | 90 | $\alpha$ | $7.52 \times 10^4$ years |
| ↓ | | | | |
| Radium | 226 | 88 | $\alpha$ | 1622 years |
| ↓ | | | | |
| Radon | 222 | 86 | $\alpha$ | 3.825 days |
| ↓ | | | | |
| Polonium | 218 | 84 | $\alpha$ | 3.05 minutes |
| ↓ | | | | |
| Lead | 214 | 82 | $\beta$ | 26.8 minutes |
| ↓ | | | | |
| Bismuth | 214 | 83 | $\beta$ | 19.7 minutes |
| ↓ | | | | |
| Polonium | 214 | 84 | $\alpha$ | $1.6 \times 10^{-4}$ seconds |
| ↓ | | | | |
| Lead | 210 | 82 | $\beta$ | 22 years |
| ↓ | | | | |
| Bismuth | 210 | 83 | $\beta$ | 5.01 days |
| ↓ | | | | |
| Polonium | 210 | 84 | $\alpha$ | 138.4 days |
| ↓ | | | | |
| Lead | 206 | 82 | | stable isotope |

*Note:* Elements representing less than 1 percent of products have been omitted. For all members of the series, see Friedlander, G., J. W. Kennedy, and J. M. Miller, *Nuclear and Radiochemistry*, 2nd Ed. John Wiley and Sons, New York, 1955, p. 9.

Three major decay series of the naturally occurring radioactive elements are known: the uranium, thorium, and actinium series. A series is named for the parent element which, by a series of disintegration reactions, produces a sequence of radioactive elements terminating in one of the stable isotopes of lead. Radium and radon occur in the uranium series, which is reproduced in Table 8.2.

The nuclear equations considered thus far describe naturally occurring disintegrations. In 1932, the same year that heralded the discovery of the positron, the neutron, and the deuteron, Irene Curie and her husband, F. Joliot, discovered how to produce radioactive elements artificially. By bombarding light elements such as boron, aluminum, and magnesium with $\alpha$ particles, radioactive isotopes of naturally stable elements were produced. For example, the bombardment of magnesium with $\alpha$ particles produced radioactive silicon, which rapidly decayed to a stable isotope of aluminum. The nuclear equations for this process are

$$\,_{12}^{24}\text{Mg} + \,_{2}^{4}\text{He} \longrightarrow \,_{14}^{27}\text{Si} + \,_{0}^{1}n \tag{8.5}$$

and

$$\,_{14}^{27}\text{Si} \longrightarrow \,_{13}^{27}\text{Al} + \,_{+1}^{0}\beta. \tag{8.6}$$

The two equations tell us that in the first step, the magnesium nucleus is hit by the $\alpha$ particle, retains two protons and a neutron, and rejects a neutron. The nuclear charge is thus increased to 14+, and the new element that is produced is silicon of atomic weight 27. This light isotope of silicon is radioactive and rapidly disintegrates into a more stable element, ordinary aluminum, by emitting a positron.

The emission of electrons or positrons is characteristic of artificial radio-elements, although other types of disintegrations also occur. In some instances the nucleus may capture an electron from the $1s$ orbital lying nearest to the nucleus, sometimes referred to as the $K$ shell. This phenomenon is called $K$-capture. The addition of an electron to the nucleus reduces the atomic number by one unit, converting the element into its nearest neighbor in the decreasing periodic sequence. The vacancy in the electron orbital is filled by other planetary electrons, and X-rays are emitted during these transformations. The emission of $\gamma$ rays following particle emission, electron capture, or some other nuclear process enables the nucleus to release energy and acquire greater stability. The emission of $\alpha$ particles is most prevalent among the heavy elements, although it is known to occur among the rare earths, e.g., $\,_{58}^{142}\text{Ce}$.

Isotopes of hydrogen, carbon, and phosphorus are frequently used by life scientists. The isotope of hydrogen that contains a proton and a neutron in its nucleus is written $\,_{1}^{2}\text{H}$. It is frequently referred to by the symbol D, from its common name deuterium. Water in which all of the hydrogen is $\,_{1}^{2}\text{H}$ can be represented by the formula $D_2O$. Since deuterium is a stable isotope, it is not radioactive. Nevertheless, it is frequently used because its presence alters the physical properties of compounds. A comparison of some of the physical properties of $\,_{1}^{1}\text{H}_2\text{O}$ and $\,_{1}^{2}\text{H}_2\text{O}$ or $D_2O$ is given in Table 8.3. Substantial changes occur in certain spectral

**TABLE 8.3**  COMPARISON OF SOME PHYSICAL PROPERTIES OF $_1^1H_2O$ AND $_1^2H_2O$

| Property | $_1^1H_2O$ | $_1^2H_2O$ |
|---|---|---|
| Molecular Weight | 18.016 g/mole | 20.032 g/mole |
| Boiling Point (1 atm) | 100.00°C | 101.44°C |
| Enthalpy of Vaporization | 9.7171 kcal/mole | 9.927 kcal/mole |
| Melting Point (1 atm) | 0.00°C | 3.82°C |
| Enthalpy of Fusion | 1.4363 kcal/mole | 1.501 kcal/mole |
| Temperature of Maximum Density | 3.984°C | 11.185°C |
| Maximum Density | 0.999972 g/cm³ | 1.10600 g/cm³ |
| Visiosity at 30°C | 0.007975 poise | 0.00969 poise |

*Source:* D. Eisenberg and W. Kauzmann, *The Structure and Properties of Water,* Oxford Univ. Press, Oxford, 1969.

properties (e.g., infrared, Raman, and nuclear magnetic resonance spectra) whenever $_1^2H$ is substituted for $_1^1H$. This substitution will also affect the rate of a chemical reaction if the rate-determining step involves the breaking of the bond to the deuterium atom.

The radioactive isotopes most frequently used by life scientists are $_1^3H$, $_6^{14}C$, and $_{15}^{32}P$. The isotope $_1^3H$ is also known as tritium, and may be represented by the symbol T. Water—in which all of the hydrogen is $_1^3H$—can be represented by the formula $T_2O$. The isotopes $_6^{14}C$ and $_{15}^{32}P$ do not have common names. All three of these isotopes decay by emitting a negatively charged $\beta$ particle, but they differ markedly in the rate at which they decay and in the energy of the emitted $\beta$, as can be seen in Table 8.4. The energy of the $\beta$-particles increases by a factor of approximately ten upon going from T to $_6^{14}C$ to $_{15}^{32}P$.

Nitrogen and oxygen are abundant in molecules obtained from living cells. Radioactive isotopes of these two elements are known, but they are seldom used by life scientists because they are so short-lived. The most stable radioactive isotope of nitrogen, $_7^{13}N$, has a half-life of only 10.1 minutes, and the most stable radioactive isotope of oxygen, $_8^{15}O$, has a half-life of 124 seconds. Two stable isotopes of these elements, $_7^{15}N$ and $_8^{18}O$, have seen appreciable use by biochemists.

**TABLE 8.4**  COMPARISON OF $_1^3H$, $_6^{14}C$, AND $_{15}^{32}P$

| Isotope | Energy of $\beta^-$ (electron volts × 10⁻⁶) | Half-life |
|---|---|---|
| $_1^3H$ (or T) | 0.0186 | 12.26 years |
| $_6^{14}C$ | 0.156 | 5730 years |
| $_{15}^{32}P$ | 1.710 | 14.3 days |

*Source: Handbook of Chemistry and Physics,* 53rd Ed., Chemical Rubber Publishing Co., Cleveland, Ohio, 1972.

**TABLE 8.5** TYPES OF NUCLEAR REACTIONS

| | Incident particle | | | | | | | |
|---|---|---|---|---|---|---|---|---|
| | Intermediate nuclei (30 < A < 90) | | | | Heavy nuclei (A > 90) | | | |
| Energy of incident particle | n | p | α | d | n | p | α | d |
| Low 0–1 kev | n(el) γ (res) | no appreciable reaction | no appreciable reaction | no appreciable reaction | γ n(el) (res) | no appreciable reaction | no appreciable reaction | no appreciable reaction |
| Intermediate 1–500 kev | n(el) γ (res) | n γ α (res) | n γ p (res) | p n | n(el) γ (res) | very small reaction cross-section | very small reaction cross-section | small reaction cross-section |
| High 0.5–10 Mev | n(el) n(inel) p α (res for lower energies) | n p(inel) α (res for lower energies) | n p α(inel) (res for lower energies) | p n pn 2n | n(el) n(inel) p γ | p(inel) γ | n p γ | p n pn 2n |
| Very high 10–50 Mev | 2n n(inel) n(el) p np 2p α three or more particles | 2n n p(inel) np 2p α three or more particles | 2n n p np 2p α(inel) three or more particles | p 2n pn 3n d(inel) tritons three or more particles | 2n n(inel) n(el) p pn 2p α three or more particles | 2n n p(inel) np 2p α three or more particles | 2n n p np 2p α(inel) three or more particles | p 2n np 3n d(inel) tritons three or more particles |

*Abbreviations:* kev, 1000 electron volts; Mev, 1,000,000 electron volts; (el), elastic scattering; (inel), inelastic scattering; (res), individual resonance levels; n, neutron; p, proton; α, alpha particle; d, deuteron; A, atomic number.

*Note:* When a bombarded nucleus is not transformed in a collision, the result of the process is spoken of as elastic scattering. The probability of various reactions or scattering processes may be characterized by *cross-sections.* Thus, a small reaction cross-section means that the reaction has a low probability of occurrence. For further discussion, the reader should consult the original source.

*Source:* Reprinted from *Nuclear Structure* by Eisenbud, L., and E. P. Wigner, by permission of Princeton University Press. Copyright © 1958, by Princeton University Press.

In Table 8.5 are presented the many possible types of nuclear reactions. An individual isotope of an element is called a *nuclide*. If it is radioactive, it is called a *radionuclide*.

## 8.2  Kinetics of Nuclear Disintegrations .

The disintegration of radionuclides, such as those illustrated in Eqs. 8.3 and 8.4, is a first-order process. The rate of decay, as it is frequently called, is dependent upon the number of nuclei present:

$$\frac{-dn}{dt} = \lambda n, \tag{8.7}$$

where $n$ is the number of radioactive atoms present, $-dn/dt$ is the decrease in this number with time, and the proportionality constant, $\lambda$, is called the decay or disintegration constant. The differential equation shown as 8.7 integrates to give

$$2.303 \log \frac{n_0}{n} = \lambda t, \tag{8.8}$$

which is completely analogous to the integration of the first-order rate expression given in Chapter 6.

The radionuclides are constantly described in terms of their half-life characteristics. We established in Chapter 6 that the half-life expression for the first-order reaction is

$$t_{1/2} = \frac{0.693}{k} \quad \text{or} \quad t_{1/2} = \frac{0.693}{\lambda}. \tag{8.9}$$

From Eq. 8.9 we see that the half-life is inversely proportional to the decay constant and may be easily calculated from it.

*Example 8.1*  The decay constant for a certain radionuclide was found to be $1.65 \times 10^{-6}$ second$^{-1}$. Calculate the half-life of the element.

$$t_{1/2} = \frac{0.693}{\lambda} = \frac{0.693}{1.65 \times 10^{-6}} = 4.2 \times 10^5 \text{ sec., or 4.9 days.}$$

There are nine radionuclides that have half-lives comparable in time to the known age of the earth, *i.e.*, about $3.5 \times 10^9$ years. It must be assumed that they have been in existence since the earth was formed. Other naturally occurring nuclides with shorter half-lives have to be regenerated by some natural process, such as the disintegration of long-lived parent elements. Still other radioelements such as tritium, $^3_1$H, and radiocarbon, $^{14}_6$C, are constantly produced in small amounts by the action of cosmic rays on the earth's atmosphere.

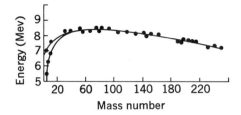

**Figure 8.1**  Binding energy of the elements as a function of mass number.

### 8.3  Nuclear Fission and Reactors

The term fission is normally applied to a reaction in which a nucleus splits into two nuclei of much lower mass. In the fission of uranium (mass number 233 to 238), one fragment will have a mass number from 82 to 100 and the other, from 128 to 150. In addition, a number of rapidly moving neutrons will be emitted. Symmetrical fission yielding two fragments of approximately equal mass will occur about one out of 1000 times.

If we inspect the periodic table we will notice that as atomic weight increases, there is a tendency for the ratio of neutrons to protons to increase in naturally occurring nuclei. In addition, it is known that any combination of neutrons and protons which results in a stable nucleus involves a great deal of binding energy. By possessing large binding or stabilization energy we mean that a great amount of energy is necessary to disrupt such nuclei, not that these nuclei are in high energy states. In fact, elements of intermediate atomic weight are in a lower energy state from the standpoint of nuclear energy than elements of either low or high atomic weight. If this were not true, neither fission nor fusion would be spontaneous processes under the conditions of their occurrence. If there is a large stabilization energy, the mass of the nuclide will be less than the expected sum of its components. The relationship between the mass difference and the stabilization energy is given by the Einstein Equation (8.1). As the atomic weight of the elements increases from about 60 upward, the binding energy decreases slightly, as may be seen from Fig. 8.1. Also from Fig. 8.1 we can see that the fission of a heavy nucleus into two medium-weight nuclei would necessitate an increase in total binding energy. We can predict that such a fragmentation would produce free neutrons, since the neutron to proton ratio is lower for the medium-weight elements. The discovery of the fission process was the result of experiments designed to synthesize the hitherto unknown elements of atomic number greater than 92. Among the heavy elements produced by neutron bombardment of uranium were medium-weight elements: isotopes of barium, technetium, krypton, and zenon. Soon after the publication of these findings by Hahn

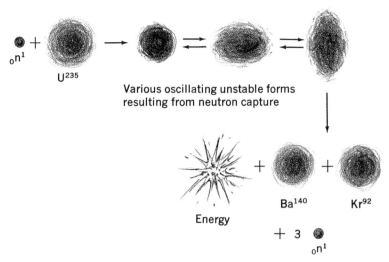

Various oscillating unstable forms
resulting from neutron capture

$U^{235}$

$_0n^1$

$Ba^{140}$  $Kr^{92}$

Energy

$+ 3$ $_0n^1$

**Figure 8.2**  A schematic representation of the fission of $^{235}U$ resulting
from the capture of a thermal neutron. Other isotopes of barium and
krypton are possible reaction products.

and Strassmann (1939), nuclear fission was visualized as a new source
of energy. World War II was in progress, and the exploitation of the fission reaction to create a nuclear weapon was undertaken in the United
States in 1941. The successful culmination of this project is now history.

The fission process is initiated by imparting an instability to the heavy
uranium nucleus. The nucleus has been likened to a drop of liquid since
this geometry will result in the lowest surface area per volume. If the
spherical nucleus can be distorted sufficiently and set into oscillation
(the energy necessary for a critical distortion will vary from one nuclide
to another), spontaneous fission will result. Some nuclei are naturally
so unstable that fission occurs without any external impetus, *e.g.*,
$^{249}$californium. The energy required to produce fission is called the
fission threshold. For one of the light isotopes of uranium, $^{235}U$, the fission threshold is quite low, and fragmentation will result from the capture of a slow neutron (thermal neutron). See Fig. 8.2. Nuclear reactors
are based on the fission of $^{235}U$.

Naturally occurring uranium is 99.3 percent $^{238}U$, which requires fast
neutron capture to attain its fission threshold. However, it can be converted into a fissionable substance — namely, plutonium — when it is
irradiated by neutrons. Nuclides which are not fissionable by thermal
neutrons, but can be converted by neutron irradiation into fissionable
nuclides, are called the *fertile material* of a reactor. The fissionable material, usually $^{235}U$, is the essential ingredient of the reactor *fuel*. As the
fissionable and fertile materials are irradiated in the course of the reactor
operation, atoms of the fissionable material are consumed and new

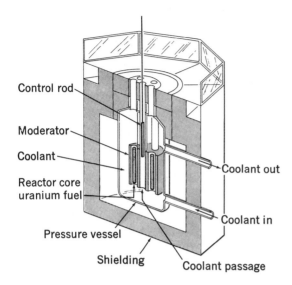

Control rod

Moderator

Coolant

Coolant out

Reactor core
uranium fuel

Coolant in

Pressure vessel

Shielding

Coolant passage

**Figure 8.3** Schematic diagram of a reactor. Coolant is circulated between the reactor components to remove heat generated during fission. [Used by permission of the Division of Technical Information of the U.S. Atomic Energy Commission.]

fissionable atoms are formed from the fertile material. The fuel is usually solid, either metallic uranium or a ceramic such as uranium oxide or uranium carbide. The neutrons which are liberated in the chain reaction are slowed down by the presence of a *moderator* such as ordinary water, heavy water, beryllium or graphite rods. The reactor is regulated by the insertion of *control rods* of material which has a high capacity for absorbing neutrons. Boron or cadmium is frequently used for this purpose. These rods are used for regulating the reactor when it is in operation and for shutting it down whenever desirable. A diagram of this type of reactor is shown in Fig. 8.3.

For a fission process to occur spontaneously in either plutonium or $^{235}$U the amount of material must exceed a certain critical mass and be arranged in a certain structure. If the mass present is insufficient, too many neutrons escape before they are captured by a fissionable atom. The mass must be arranged so that neutron leakage is at a minimum. In the atomic bomb, the critical mass is attained by quickly bringing together several pieces of fissionable material. Ordinary explosives such as TNT are used for this purpose.

The chemical sequence by which $^{238}$U is converted into plutonium is given in the following equations:

$$^{238}_{92}\text{U} + ^{1}_{0}n \longrightarrow ^{239}_{92}\text{U}; \tag{8.10}$$

$$^{239}_{92}\text{U} \longrightarrow ^{239}_{93}\text{Np} + ^{0}_{-1}\beta; \tag{8.11}$$

$$^{239}_{93}\text{Np} \underset{\text{2.3 days}}{\longrightarrow} ^{239}_{94}\text{Pu} + ^{0}_{-1}\beta. \tag{8.12}$$

## 8.4  Particle Accelerators

Neptunium, californium, berkelium, and other elements of higher atomic number are called the *transuranium elements*. They are prepared by bombarding $^{238}U$ and $^{239}Pu$ with various highly accelerated particles, such as $\alpha$ particles, carbon ions, and nitrogen ions. Charged particles can be accelerated by various methods and a number of expensive devices have been designed for this purpose. The path along which the particle moves may be straight or curved; the corresponding installations are called linear accelerators or cyclotrons. In the cyclotron, the same accelerating station is used over and over, and the particles spiral outward from the center with ever-increasing velocities. The longer the path, the higher the energy of the particles. Since the magnets must extend over the entire path area, extremely large and expensive ones are required to produce high energy particles. The limitations of the cyclotron were overcome by the design of accelerators with circular rather than spiral paths. The large central magnets were replaced by smaller individual magnets placed around the circumference of the path. The field strength of the magnets increases synchronously with the increasing velocity of the particles. These machines are called synchrotrons. Extremely high energies are obtained by focusing the particle beam, an additional important feature of the largest synchrotrons.

The first practical installation was the University of California cyclotron, completed in 1936 at a cost of $50,000. It was a positive particle accelerator and produced streams of fast-moving $\alpha$ particles and deuterons. Since 1936, many other positive particle accelerators have been constructed, most of them proton accelerators. The basic parts of a proton accelerator are diagrammed in Fig. 8.4. The two most recently constructed are the proton-synchrotron at Geneva, Switzerland, at a cost of 35 million dollars and the alternating gradient proton-synchrotron at the Brookhaven National Laboratory, Upton, New York, at a cost of 33 million dollars. There are five large electron accelerators in the United States.*

## 8.5  The Measurement of Radioactivity

As is beautifully illustrated in the operation of the Wilson Cloud Chamber, charged particles moving through matter ionize atoms in the medium through which they pass. In the Wilson Cloud Chamber, the new

---

*Details of the construction of these instruments and other information of interest is presented in a lucid and fascinating account of the development of atomic energy, appearing in Chapter 8 of the book by R. E. Lapp that is included in the listing in *Suggested Additional Reading*. Appearing in the same chapter is a discussion of the hydrogen bomb (the "fusion" reaction), which will not be considered in this presentation because it is not presently germane to useful, controlled applications of atomic energy.

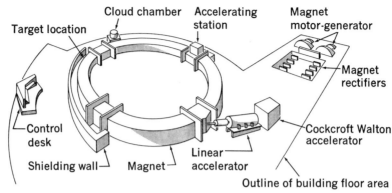

**Figure 8.4** Schematic diagram of the Bevatron, giant proton-synchrotron at the University of California Radiation Laboratory. The protons are preaccelerated to 480 kev (a kev is 1000 electron volts) in a Cockcroft-Walton accelerator and then to 19.3 Mev (a Mev is one million electron volts) in a 40-foot linear accelerator before being injected into the Bevatron itself. A proton travels about 394 feet per circuit and makes about four million revolutions in 1.8 seconds for a total distance of 300,000 miles. It finally attains a speed that is 99 percent that of light and an energy of 6.2 Bev (a Bev is a billion electron volts). The strength of the magnetic field in the Bevatron rises in step with the momentum of the particles, keeping the radius of their orbits nearly constant. [Used by permission of the Division of Technical Information of the U.S. Atomic Energy Commission.]

ions serve as nuclei for the condensation of vapor, and thin lines of fog or "tracks" may be seen. This technique, however, is not used for quantitative measurement of the radioactivity of various specimens, because more convenient methods are available. The production of ions is exploited in other ways to obtain quantitation.

One of the earliest devices (used by Becquerel and the Curies) for measuring radioactivity was the electroscope. An electroscope contains an insulated conductor to which is attached a flexible indicator unit, such as a piece of gold leaf or a gold-plated quartz fiber. The ions from the radioactive specimen place a charge on the conductor and the flexible part is repelled by the charge. The repulsion is proportional to the charge. As the indicator unit is discharged, the flexible piece moves closer to the rest of the conductor, the rate of movement being proportional to the rate of production of ions in the gas near the gold leaf or quartz fiber. An electroscope is shown in Fig. 8.5. Today, such an instrument is more of a student teaching device or laboratory curiosity than anything else.

The ionization principle has been used in the design of instruments which record the ion bursts as audible clicks or which tally them electronically. The change in voltage or in current flow in the counting chamber as pulses of ions are produced is the basis of many types of counters, including the Geiger-Müller counter. With the availability of radioiso-

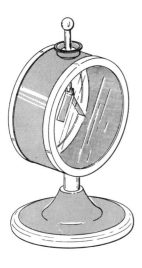

**Figure 8.5**  A leaf electroscope, one of the earliest devices for detecting and measuring radioactivity. The metal case has glass windows so that the leaves may be easily seen. The leaves, usually of aluminum or gold foil, are attached to a rod which is insulated from the metal case by a collar of amber. When a charged body is brought into contact with the knob, some of the charge will be conducted to the leaves, causing them to repel each other and swing apart.

topes for medical, chemical, biochemical, and industrial research, instrumentation in this area has become highly sophisticated. One needs only to visit an instrument display in conjunction with one of the national meetings of the large scientific societies to become aware of the variety of highly complex devices now available for radioactive tracer research.

In addition to the ionization effect, $\alpha$ particles produce flashes of light when they strike certain substances, such as zinc sulfide, anthracene, stilbene, or sodium iodide. These scintillations can be received by a photomultiplier tube which produces a measurable electric current in response to the flashes of light. The current fluctuations are then counted by a recording device. A similar technique for measuring $\beta$ and $\gamma$ rays is based on the production of light from a crystal exposed to such radiation.

One of the most useful techniques for detecting radioactive materials is based on the darkening of photographic film by nuclear particles. It will be recalled that this property of uranium ore led Becquerel to the discovery of radioactivity. The sample to be radioautographed is placed in contact with a photographic film, suitably protected from light, and left for an appropriate period of time, depending on the intensity of the radiation. Radioautography is particularly useful in detecting radioactive compounds on paper chromatograms.

The unit of radioactivity is the *curie*, which is that amount of material that gives $3.7 \times 10^{10}$ disintegrations per second. The shorter the half-life, the smaller the weight of a given element corresponding to a curie. One-thousandth of a curie is called a millicurie; one-millionth, a microcurie. The unit of radiation dosage or radiation received is the *roentgen*. The roentgen is defined as the total dosage of radiation that will produce one electrostatic unit of charge in one cubic centimeter of air at standard conditions. This is equal to the release of 85 ergs per gram of air. Since the ionization produced by various types of particles varies with the nature and the energy of the particles, there is no simple relationship between the number of particles and the exposure in roentgens. Research personnel who are exposed to radiation wear badges which contain unexposed pieces of photographic film. The darkening of the film is used as an index of total exposure to radiation over a determined time interval.

## 8.6 Biological Research Applications

**Tracer studies** The research biochemist who endeavors to follow a sequence of chemical events as they occur in living plants or animals, is all too frequently aware of the admonition that "it's mighty dark inside of a cow." To trace a series of molecular steps in the complexity of living protoplasm seems a task so formidable as to be foolish. Early efforts to find suitable tracers led to the use of unnatural isomers or halogenated compounds which, hopefully, would be metabolized to easily recognizable end products. This approach always ran the risk of detouring up an abnormal or detoxifying pathway rather than following the principal metabolic route characteristic of the natural counterpart of the tracer. A major breakthrough was achieved by Schoenheimer and colleagues at Columbia University in the 1930's when sufficient concentrations of the heavy isotopes of hydrogen and nitrogen were obtained to permit the synthesis of amino acids and fatty acids bearing the heavy isotope label. Deuterium and $^{15}N$ are not radioactive and must be detected in an instrument called the mass spectrograph, which magnetically separates and measures the amount of each isotope in a gaseous sample. The metabolic end products or tagged molecules must be degraded to water (or $H_2$) and $N_2$ prior to analysis. The relative amount of $^2_1H_2O$ in a drop of water can be determined from its density. Schoenheimer's experiments revealed an unsuspected high rate of turnover of body metabolites and the theory of dynamic metabolism was born.

Following World War II and the successful completion of the Manhattan Project, reactors and particle accelerators were available for the synthesis of a great many radioactive isotopes. For the biochemist, the most useful have been $^3H$ (tritium), $^{14}C$, $^{32}P$, and $^{35}S$. Carbon and hydrogen can be used in tracer experiments involving any phase of metabolism, phosphorus is particularly useful in the study of nucleic acids and their derivatives, and sulfur—found principally in the sulfur amino acids—is most widely used in experiments concerning proteins and amino acids.

The use of any isotope involves special planning and precautions related to the half-life of the isotope and the type of radiation emitted. The isotope $C^{14}$ has an extremely long half-life (5730 years) and, although it is only a weak $\beta$ ray emitter, its radiation would be harmful if accumulated in the human body. Experiments utilizing $^{24}Na$ must be

**TABLE 8.6**  HALF-LIVES AND EMISSION CHARACTERISTICS OF SOME OF THE USEFUL RADIOISOTOPES

| Isotope | Half-life | Type of emission or decay |
|---------|-----------|---------------------------|
| $^3$H | 12.46 years | Beta |
| $^{14}$C | $5.73 \times 10^3$ years | Beta |
| $^{24}$Na | 15.05 hours | Beta, gamma |
| $^{32}$P | 14.3 days | Beta |
| $^{35}$S | 89.0 days | Beta |
| $^{36}$Cl | $3.08 \times 10^5$ years | Beta |
| $^{42}$K | 12.46 hours | Beta, gamma |
| $^{45}$Ca | 165 days | Beta |
| $^{54}$Mn | 310 days | Gamma, electron capture |
| $^{55}$Fe | 2.94 years | X-ray, electron capture |
| $^{59}$Fe | 44.3 days | Beta, gamma |
| $^{58}$Co | 71.3 days | Positron, gamma, electron capture |
| $^{60}$Co | 5.24 years | Beta, gamma |
| $^{63}$Ni | 125 years | Beta |
| $^{64}$Cu | 12.9 hours | Beta, positron, gamma, electron capture |
| $^{65}$Zn | 246.4 days | Positron, gamma, electron capture |
| $^{76}$As | 26.8 hours | Beta, gamma |
| $^{75}$Se | 119.9 days | Gamma, electron capture |
| $^{82}$Br | 35.55 hours | Beta, gamma |
| $^{89}$Sr | 50.5 days | Beta |
| $^{99}$Mo | 67 hours | Beta, gamma |
| $^{111}$Ag | 7.5 days | Beta, gamma |
| $^{109}$Cd | 1.3 years | Gamma, electron capture |
| $^{113}$Sn | 119 days | Gamma, electron capture |
| $^{124}$Sb | 60.9 days | Beta, gamma |
| $^{131}$I | 8.05 days | Beta, gamma |
| $^{133}$Ba | 10.7 years | Gamma, electron capture |
| $^{203}$Hg | 45.4 days | Beta, gamma |
| $^{210}$Bi | 5.00 days | Beta |

*Source:* Data are quoted from the *Oak Ridge National Laboratory Catalog of Radio and Stable Isotopes,* Isotopes Development Center, Oak Ridge National Laboratory, 4th Revision, 1963.

designed for completion within a relatively short period of time since the half-life of this element is approximately 15 hours. A major advantage of the radioisotope tracer lies in the fact that the labeled compound can be detected without being destroyed, whereas compounds containing the heavy stable isotope must be degraded prior to detection. On the other hand, the stable isotope does not present a health hazard to the experimenter. A list of the more useful isotopes and their half-lives is given in Table 8.6.

**Isotope dilution**   In addition to tracer studies, radioisotopes can be used to determine the concentration of a substance in a mixture which cannot be quantitatively separated into its components. All that is required is that the compound to be measured be isolatable in pure form. The technique is called *isotope dilution*, and was first used at Columbia University with the heavy isotopes. An example will serve to explain how the method works. The *specific activity* of a radioactive compound is defined as the disintegrations per minute per unit of substance, *e.g.*, per mg, per mole, etc. When such compounds are diluted with identical nonlabeled molecules, the dilution factor can be calculated from the change in specific activity:

$$\text{Dilution Factor} = \frac{\text{Specific Activity of Original Compound}}{\text{Specific Activity of Diluted Compound}}.$$

Let us suppose that we are analyzing a protein hydrolysate for its component amino acids and wish to determine the amount of valine present. A small amount of labeled ($^{14}$C) valine of known specific activity is carefully weighed and added to the mixture. The solution is stirred until homogeneous. A small amount of pure valine is now separated from the hydrolysate by an isolation technique that has been previously perfected. The specific activity of the diluted valine is accurately measured and the dilution factor calculated. Let us say that it turns out to be 75. Suppose that the weight of labeled valine originally added to the hydrolysate would be 150.00 mg.

Before the development of the Moore-Stein amino acid analyzer, the isotope dilution method was widely used for the analysis of the more soluble amino acids in protein hydrolysates. It is still frequently used to express the precursor relation of a compound in the biosynthesis of a second compound. For example, in the sequence

$$A \rightarrow B \rightarrow C \rightarrow D \rightarrow E,$$

the dilution factor for D to E would very likely be small whereas for A to E it should be quite large. Such differences enable one to judge which of two compounds added to a tissue is the more likely precursor of a subsequent compound of interest.

**Nuclear activation analysis**   Bombardment of an element by a stream of nuclear particles gives rise to artificially radioactive products which undergo decay in a characteristic manner (the discovery of Irene Curie and F. Joliot). The type and intensity of each radiation of interest induced in a sample can be measured, although the emission may not be entirely definitive for each element. However, all elements have been determined by this method, and the samples can be solids, liquids, or gases. The method is both qualitative and quantitative, no elaborate sample preparation is needed, and in most cases, the method is nondestructive. Other advantages are: there is no reagent contamination, the procedures are generally rapid, the method can be automated, and in many cases, techniques are available for high specificity.

The preferred particle source is a deuteron accelerator. This instrument has many advantages over a reactor in that it can be turned on and off with a switch, no critical chain reaction is needed, and there is no buildup of fission products. Portable sources are available, which makes the method highly appealing to industry. A further advantage is that the use of a deuteron accelerator does not require an A.E.C. licensed operator. Large neutron sources permit analyses in the submicrogram range with 5 percent accuracy.

The principal problem in nuclear activation analysis is the lack of specificity that sometimes occurs. This situation results from either or both of two effects: the principal elements in the sample matrix may form such high levels of radioactive products that the radiation from the element of interest is completely overshadowed, and the same product radionuclide may be obtained from more than one element.

Improved specificity is obtained by chemical separation of the desired nuclides following radiation, by varying the irradiation and/or decay times, by use of multichannel analyzers, and by using the technique of specific activations. In this technique, an activation procedure, which will produce the maximum radioactivity in the element of interest and the minimum radioactivity in the matrix, is used.

The method of nuclear activation analysis involves the following steps:*

1. Selection of the most suitable nuclear reaction.
2. Selection of an appropriate irradiation facility.
3. Preparation of the sample.
4. Determination of the irradiation time.
5. Planning of a chemical separation step, if necessary.
6. Selection of an appropriate postirradiation assay method.
7. Evaluation of data.

Nuclear activation analysis has been used for the detection of minor components and trace elements in a wide variety of materials. For example, it can be used to determine nitrogen in organic compounds; oxygen in petroleum fractions; sulfur in various feedstocks; chlorine in crop extracts; manganese, vanadium, arsenic, and other trace elements in feedstocks; and selenium in organic compounds.

---

*A full discussion of principles and techniques will be found in "Nuclear Activation Analysis," by E. L. Steele, in *Proc. of the Symposium on Physics of Nondestructive Testing*, Oct. 1, 1963, San Antonio, Texas, p. 315, edited by McGonnagle, W. J.

**Other applications** The application of radiotracer methods to specific laboratory problems is, in general, limited only by the facilities available and the ingenuity of the experimenter. A few additional applications from the field of protein chemistry will serve to illustrate the point.

In studying the primary structure of proteins and peptides we must learn not only the numbers and kinds of various amino acids present, but also their sequence in the polypeptide chain. Several methods have been devised for specifically labeling the N-terminal residue, one of these being the *pipsyl chloride method*. (Pipsyl is an acronymic designation for *para*-iodophenyl sulfonyl.) This reagent reacts with the terminal amino residue of a protein to form a sulfonamide that is stable to hydrolysis. If the reagent is synthesized from $^{131}I_2$, the terminal amino acid can be so tagged and easily detected on paper chromatograms by radioautography. The chemical reactions are

$$NH_2—CHR—CONH—CHR' \ldots \cdot \xrightarrow[\text{OH}^-]{\text{I—C}_6\text{H}_4\text{—SO}_2\text{Cl}}$$

$$I—C_6H_4—SO_2—NH—CHR—CONH—CHR' \ldots \ldots \quad (8.13)$$

$$\text{HCl} \big| \text{H}_2\text{O} \Big\downarrow$$

$$I—C_6H_4—SO_2—NHCHRCOOH + NH_2—CHR'COOH + \ldots.$$

The *p*-iodophenylsulfonamide standards are prepared for the twenty-odd common amino acids, and the labeled N-terminal residue is identified by paper chromatography and/or electrophoresis versus the standards.

The sulfhydryl functional group of proteins appears to be of ever-enlarging importance in the mechanism of action of specific enzymes. Since most of these enzymes contain several −SH groups, the identification of the catalytically important one and the location of its position in the amino acid chain may be a difficult research problem. In some fortunate instances, the catalytically essential −SH group will also be the most active chemically, either because of its exposed position in the tertiary structure of the enzyme or because of its environment. The addition of a single equivalent of a radioactive tagging agent will, under these circumstances, place the label on the −SH group of interest. The more usual circumstance, however, is that the −SH group at the active site is equally or less reactive than the other −SH groups and must be labeled by a more sophisticated technique. If the essential −SH group is protected from alkylating agents by substrate, it may be possible to alkylate the noncatalytic groups in the presence of substrate and then, following the removal of substrate and excess alkylating agent, label the essential −SH group with a radioactive alkyl group. Such an approach would have to be feasible with native enzyme, since the addition of a denaturing agent would unfold the polypeptide chain, destroy the specific conformation of the active site, and render the substrate ineffectual in protecting $-$SH. One alkylating agent that has been useful in experiments of this type is $^{14}$C iodoacetamide; another is $^{14}$C N-ethylmaleimide. The chemical reactions of each of these with RSH are

$$R-SH + I-CH_2-C{\overset{O}{\underset{NH_2}{\big\langle}}} \rightarrow R-S-CH_2-\overset{\overset{O}{\|}}{C}-NH_2 + HI \quad (8.14)$$

and

$$R-SH + \begin{matrix} CH-C \\ \| \\ CH-C \end{matrix}{\overset{O}{\underset{O}{\big\langle}}}N-C_2H_5 \rightarrow \begin{matrix} R-S-CH-C \\ | \\ CH_2-C \end{matrix}{\overset{O}{\underset{O}{\big\langle}}}N-C_2H_5 \quad (8.15)$$

Tritiated reagents and metabolites are extensively used in enzyme research because they can be handled at low levels of radioactivity with a high degree of safety by laboratory personnel. Furthermore, tritiated compounds are easy to make. An example of the synthesis of a tritiated compound is the tritium gas-exchange method of labeling, devised by Wilzbach, in which a hydrogen-containing compound is placed in an atmosphere of tritium gas which has a high specific activity, and maintained for several weeks. The radioactive compound is then freed of any highly labile tritium by dissolving it in a hydroxylic solvent. Considerable destruction of the target compound can occur during Wilzbach labeling, and careful purification of the desired material is advised. Reductions involving tritium gas with an appropriate catalyst, or metal hydrides labeled with tritium, are effective means of preparing specifically labeled tritiated compounds.

Tritium-labeled water is readily available to permit measurements of incorporation of solvent hydrogens into products during enzymatic reactions. One principal drawback, however, to such use is the possibility of a discrimination effect between hydrogen of mass 1 and tritium of mass 3.

An elegant example of the role that tritium can play in investigating the details of enzymatic mechanisms is seen in the experiments that led to the conclusion that dihydroxyacetone phosphate activation by fructose diphosphate aldolase proceeds through the formation of a stereospecific enzyme-bound carbanion.* Only one atom of tritium was

---

* Rose, I. A., and S. V. Rieder, *J. Biol. Chem.*, **231**:315, 1958.

incorporated back into dihydroxyacetone phosphate when this substrate was incubated with the enzyme in tritiated water $(T_2O)$:

$$
\begin{array}{c}
\text{CH}_2\text{OH} \\
| \\
\text{C}=\text{O} \\
| \\
\text{CH}_2\text{OPO}_3^{2-}
\end{array}
\;+\;\text{Aldolase}\;\xrightarrow{-\text{H}^+}\;
\left[
\begin{array}{c}
\text{H}\bar{\text{C}}\text{OH} \\
| \\
\text{C}=\text{O} \\
| \\
\text{CH}_2\text{OPO}_3^{2-}
\end{array}
\right]
\cdot\text{Aldolase}
$$

$$+\text{T}^+$$

$$
\text{Aldolase}\;+\;
\begin{array}{c}
\text{T} \\
| \\
\text{H}\text{C}\text{OH} \\
| \\
\text{C}=\text{O} \\
| \\
\text{CH}_2\text{OPO}_3^{2-}
\end{array}
$$

## Suggested Additional Reading

### General: Nuclear Chemistry

Friedlander, G., J. W. Kennedy, and J. M. Miller, *Nuclear and Radiochemistry,* 2nd Ed. John Wiley and Sons, New York, 1955.

Lapp, R. E., editor, *Matter.* Time, Inc., New York, 1963.

Overman, R. T., *Basic Concepts of Nuclear Chemistry.* Reinhold, New York, 1963.

Seaborg, G. T., *Man-made Transuranium Elements.* Prentice-Hall, Englewood Cliffs, N. J., 1963.

### Radioisotope Techniques

Chase, G. D., and J. L. Rabinowitz, *Principles of Radioisotope Methodology,* 2nd Ed. Burgess, Minneapolis, 1962.

Moses, A. J., *Nuclear Techniques in Analytical Chemistry.* Macmillan, New York, 1964.

Overman, R. T., and H. M. Clark, *Radioisotope Techniques.* McGraw-Hill, New York, 1960.

U.S. Department of Health, Education, and Welfare, *Radiological Health Handbook.* Washington, D.C., 1957.

## Problems

**8.1** What is the number of protons and neutrons in each of the following?

(a) $^{3}_{1}\text{H}$      (d) $^{19}_{8}\text{O}$

(b) $^{9}_{3}\text{Li}$      (e) $^{68}_{30}\text{Zn}$

(c) $^{19}_{9}\text{F}$      (f) $^{64}_{30}\text{Zn}$

**8.2** The radioactivity from a sample of a transuranium nuclide is being counted on a radiation detection device. Compute the half-life in days from the following data.

| Time (days) | Activity (counts per min) |
|---|---|
| 0 | 5640 |
| 5 | 5300 |
| 10 | 4981 |
| 15 | 4681 |
| 20 | 4398 |
| 25 | 4133 |

**8.3** A certain nuclide has a first-order decay constant of 0.0276 min$^{-1}$. Calculate the half-life.

*Answer.* 25 min.

**8.4** Rutherford and Boltwood reported that one gram of radium produced 0.107 mm$^3$ of helium per day at standard conditions. The same weight of radium emits $3.7 \times 10^{10}$ *alpha* particles per sec. Calculate Avogadro's number from these data. What are sources of error in these measurements?

**8.5** A sample of uranium ore from a Colorado mine was found to contain $^{206}$Pb equivalent (in moles) to 41.6 percent of the $^{238}$U. Assuming that all the lead originated from the uranium and assuming a half-life of $4.5 \times 10^9$ years for the uranium, what is the approximate age of the sample of ore?

*Answer.* 3.5 billion years.

**8.6** A sample of radium weighs 1/8 gram. If the half-life of radium is 1,620 years, what weight of radium was present 4,860 years ago?

**8.7** Iodine ($^{131}_{53}$I) decays by beta emission and has a half-life of 8 days. Consider the following hypothetical problem. Two moles of HI (all of the iodine is $^{131}$I) are placed in an evacuated 22.4 liter container at 0°C. At the end of the sixteenth day, what is the pressure in the container if the temperature was held at 0°C?

*Answer.* 2.75 atm.

**8.8** Calculate the energy in Mev necessary to cause the nuclear reaction

$$\text{energy} + {}^4_2\text{He} \rightarrow 4^1_1\text{H}.$$

Is this reaction possible? What is the possibility of the reverse reaction? One atomic mass corresponds to 931 Mev. Use an atomic mass of 1.007825 for $^1_1$H and 4.002604 for $^4_2$He.

**8.9** The following equations illustrate different types of nuclear reactions. Supply the missing element or nuclear particle, and identify the reaction as *alpha* emission, *beta* emission, *gamma* emission, $K$-capture, etc.

(a) $^{14}_{7}N + \underline{\hspace{1.5cm}} \rightarrow {}^{17}_{8}O + {}^{1}_{1}H;$

(b) $^{9}_{4}Be + {}^{4}_{2}He \rightarrow {}^{12}_{6}C + \underline{\hspace{1.5cm}};$

(c) $^{6}_{3}Li + {}^{1}_{0}n \rightarrow {}^{4}_{2}He + \underline{\hspace{1.5cm}};$

(d) $^{39}_{19}K + \underline{\hspace{1.5cm}} \rightarrow {}^{39}_{20}Ca + {}^{1}_{0}n;$

(e) $^{55}_{25}Mn + \underline{\hspace{1.5cm}} \rightarrow {}^{55}_{26}Fe + {}^{1}_{0}n;$

(f) $^{30}_{15}P + \underline{\hspace{1.5cm}} \rightarrow {}^{30}_{14}Si;$

(g) $^{35}_{16}S \rightarrow {}^{35}_{17}Cl + \underline{\hspace{1.5cm}};$

(h) $^{2}_{1}H + {}^{3}_{1}H \rightarrow {}^{4}_{2}He + \underline{\hspace{1.5cm}};$

(i) $^{7}_{4}Be + \underline{\hspace{1.5cm}} \rightarrow {}^{7}_{3}Li.$

**8.10** The neutral aliphatic amino acids were the most difficult to determine quantitatively before the advent of ion-exchange chromatography. It was frequently necessary to use the isotope dilution technique to verify a result obtained by another method, *e.g.*, microbiological assay. Let us suppose that one micromole of a protein of molecular weight 60,000 was acid hydrolyzed overnight, and the resulting solution was filtered to remove the humin, neutralized, and diluted to 50 ml. To this neutral solution was added 0.1 mg of $^{14}C$ threonine having a specific activity of 5500 counts per min per mg. After careful fractionation, a sample of pure threonine was isolated with a specific activity of 575 counts per min per mg. How many threonine residues were there per mole of protein?

*Answer.* 8.

**8.11** Two isotopes of approximately the same atomic weight have different half-lives. Which weighs more, one microcurie of the long-lived isotope or one microcurie of the short lived isotope?

**8.12** On May 5, 1964, 1.02 millicuries of $^{45}Ca$, contained in 20.0 ml of solution, was stored in a suitably shielded container. The half-life of this isotope is 165 days. How much activity remained on March 17, 1965? What was the activity per ml?

*Answer.* 0.27 millicurie; 0.0135 mc per ml.

**8.13** Synthetic poly-$\alpha$-amino acids can be prepared in the laboratory by the initiation of the polymerization of an amino acid N-carboxy anhydride (abbreviated NCA) by a base, such as *n*-hexylamine. Two possible mechanisms are,

**Mechanism I**

$$n\text{-}C_6H_{13}NH_2 + NCA \rightarrow n\text{-}C_6H_{13}NH\overset{\displaystyle O}{\overset{\|}{-C}}\underset{\displaystyle R}{\overset{\phantom{x}}{C}}HNH_2 + CO_2$$

$$n\text{-}C_6H_{13}NH\overset{\displaystyle O}{\overset{\|}{-C}}\underset{\displaystyle R}{C}HNH_2 + NCA \rightarrow n\text{-}C_6H_{13}NH\overset{\displaystyle O}{\overset{\|}{-C}}\underset{\displaystyle R}{C}HNH\overset{\displaystyle O}{\overset{\|}{-C}}\underset{\displaystyle R}{C}HNH_2$$

etc.

**Mechanism II**

$$n\text{-}C_6H_{13}NH_2 + NCA \rightarrow n\text{-}C_6H_{13}NH_3^+ + \overset{\displaystyle O}{\underset{\displaystyle O}{C}}-N^- \cdots \overset{\displaystyle }{C}-\overset{\displaystyle }{C}-R$$

$$\overset{\displaystyle O}{\underset{\displaystyle O}{C}}-N^- + NCA \rightarrow \overset{\displaystyle O}{\underset{\displaystyle O}{C}}-N-\overset{\displaystyle O}{\overset{\|}{C}}CHNH^- + CO_2$$

$$\overset{\displaystyle O}{\underset{\displaystyle O}{C}}-N-\overset{\displaystyle O}{\overset{\|}{C}}CHNH^- + NCA \rightarrow \overset{\displaystyle O}{\underset{\displaystyle O}{C}}-N-\overset{\displaystyle O}{\overset{\|}{C}}CHNH-\overset{\displaystyle O}{\overset{\|}{C}}CHNH^-$$

etc.

Describe an experiment, making use of radioactive compounds, that would allow you to distinguish between these two mechanisms. For an actual approach to this problem, see Terbojevich *et al., J. Amer. Chem. Soc.*, **89**, 2733 (1967).

# 9 | ELECTROMAGNETIC RADIATION AND MATTER

As I looked at her I knew she wanted to tell me—she was dying. And then, as I got her message, there came a light from her eyes—powerful beams of light.

Yes . . . It was a real light, a powerful, dazzling, blinding light, a light more intense than I had ever produced by the most powerful lamps in my laboratory . . .

*From* Prodigal Genius: The Life of Nikola Tesla, *by John J. O'Neill*

The word, "light," refers to a segment of the electromagnetic spectrum that is visible to the human eye and lies roughly between 400 and 800 nanometers (1 nm = $10^{-9}$ meters) in wavelength. The familiar yellow "D line" (really a doublet) in the emission spectrum of metallic sodium is at 5893 Å (589.3 nanometers). The term, electromagnetic, describes the nature of the radiation precisely in that it is constituted by an electrical wave and a magnetic wave in phase and at right angles to each other and to the direction of propagation, which is shown schematically for one ray of plane polarized light in Fig. 9.1. The letters $E$ and $H$ are universally used to indicate the magnitudes of the electric and magnetic vectors respectively. An ordinary beam of light consists of many randomly oriented plane-polarized components, all of which are being propagated in the same direction.

A broad array of *anything*, whether ideas, colors, masses or whatever, as long as all the components are arranged in increasing or decreasing order of *magnitude* or quality, may properly be called a spectrum, *e.g.*, the expression, "the candidates represent a wide spectrum of political beliefs," is not unfamiliar to most of us. The electromagnetic spectrum is known to range from gamma rays near the short wavelength (high energy) end to the familiar radio waves near the long wavelength (low energy) end. Wavelengths are usually stated in units that are most con-

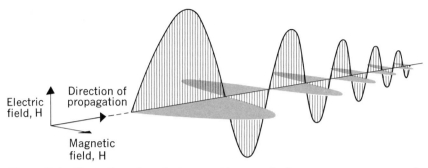

**Figure 9.1**   Illustration of an electromagnetic wave. Such waves are transverse and constituted by magnetic and electric fields at right angles to each other and to the direction of propagation.

veniently used for that particular segment of the electromagnetic spectrum. Usual units for the wavelength, $\lambda$, are,

| | |
|---|---|
| meters (m) | micrometers ($\mu$m) |
| centimeters (cm) | nanometers (nm) |
| millimeters (mm) | angstrom units (Å) |

where

$$1 \text{ Å} = 10^{-8} \text{ cm}$$

$$1 \text{ nm} = 10^{-3} \ \mu\text{m} = 10^{-6} \text{ mm} = 10^{-7} \text{ cm} = 10^{-9} \text{ m}.$$

Two alternative units which are frequently encountered are micron ($\mu$) for micrometer and millimicron (m$\mu$) for nanometer.

Table 9.1 gives the approximate ranges for the various regions of the electromagnetic spectrum in convenient units.

Many scientists prefer to talk in terms of frequency, $\nu$, or of wave number, $\bar{\nu}$. The frequency is the number of oscillations per second in a wave or may be thought of as the number of full cycles of the wave passing a point in space per unit time.* It is related to the wavelength, $\lambda$ (the distance required for a full oscillation) by the equation,

$$\nu = c/\lambda \tag{9.1}$$

where $c$ is the velocity of light, $3 \times 10^{10}$ cm/sec. The wave number is simply the number of waves or complete cycles per cm and has the units of cm$^{-1}$. It is the reciprocal of the wavelength expressed in centimeters.

---

*Modern usage is that the Hertz, symbol Hz, is the unit of frequency, equal to a cycle per second.

**TABLE 9.1** APPROXIMATE WAVELENGTH RANGES
FOR THE VARIOUS REGIONS OF THE
ELECTROMAGNETIC SPECTRA

| Name of radiation | | Wavelength range |
|---|---|---|
| Gamma rays | | .003–0.3 Å |
| X-rays | | 0.3–100 Å |
| Far ultraviolet | | 100–2000 Å |
| Ultraviolet | | 200–400 nm |
| Visible | | 400–800 nm |
| Near infrared | | 0.8–2.5 $\mu$m |
| Infrared | ENERGY | 2.5–15 $\mu$m |
| Far infrared | | 15–200 $\mu$m |
| Microwave | | 0.2–7.0 mm |
| Radar | | 7–100 mm |
| Very high frequency | | 10–1000 cm |
| Ultra high frequency | | 10–100 m |
| Radio waves | | 100–10,000 m |

Some authors use the name Kaiser to represent one reciprocal centi-meter. In general, wavelength is more frequently used for visible and ultraviolet spectra, whereas frequency and wavenumber are customarily used in the infrared. The energy of radiation of frequency, $\nu$, is given by Eq. 9.2

$$E = h\nu \tag{9.2}$$

where $h$ is Planck's constant, $6.6256 \times 10^{-27}$ erg-sec. The student should note carefully that the energies of the radiations listed in Table 9.1 are greatest for the short wavelengths.

Studies of the interaction of radiation with matter have been thus far the most important tool devised for increasing our knowledge of the universe. Sight itself represented our first detection of electromagnetic radiation from which we drew certain conclusions regarding our surroundings, shape, size, color, terrain, etc. Diffraction, reflection, refraction, scattering, and polarization of electromagnetic radiation as it interacts with matter are some observable phenomena that result from the wave nature of the radiation. Certain others, such as the photoelectric effect, seem to require that the radiation be composed of fundamental particles called photons. This realization was one of the early contributing evidences for a quantum theory that has developed into the quantum mechanics of today.

Although the treatment of quantum mechanics is well outside the scope of this book, it will be of interest to the student that the seemingly major variance between a wave theory and a particle theory of electro-magnetic radiation was resolved by de Broglie through his equation

$$\lambda = \frac{h}{mv} \qquad (9.3)$$

where $\lambda$ is the wavelength associated with a particle of momentum, $mv$. Again, the constant $h$ is Planck's constant.

Chemists are interested in many segments of the electromagnetic spectrum. In the early days of spectroscopy, almost any information that could be obtained was useful even though workers then often did not know the physical basis for their observations. A classical example was the excellent work of Moseley. Without knowing why or how x-rays originated from elements when they were bombarded with cathode rays, he was able to establish the concept of atomic number from observations of the characteristic frequencies in the x-ray spectra of elements, and thereby remove all ambiguities in Mendeleyev's Periodic Table. In spectroscopy, as in most areas of science, empirical results have often been exceedingly useful. Now, however, scientists know what structural aspect of matter responds to a particular frequency range of the electro-magnetic spectrum. Those ranges of most current interest to chemists, with the name of the technique and the kind of information obtained are shown in Table 9.2.

**X-Ray diffraction**   The technique of x-ray diffraction has yielded much information of interest to the life scientist. The amount of information that can be obtained depends upon the state of the sample. Three possi-bilities can be distinguished: (1) The sample is a single perfect crystal, with a regular repeating pattern in three dimensions. (2) The sample contains many small crystalline regions, but the crystalline regions are randomly oriented. This condition might be obtained, for example, by crushing a single large crystal into a multitude of tiny crystals. (3) The sample contains many small crystalline regions, and the crystalline re-gions have a preferred orientation in one dimension. An example would be a collection of small crystals, each of which has one long dimension and two short dimensions, in which there is a tendency for the crystals to have their long dimension oriented in the same direction.

The most detailed information can be obtained from single perfect crystals. This x-ray diffraction pattern is made up of distinct small dots. If the molecules are small enough, the position of each atom can be pre-cisely located, giving information on the shape of the molecule, the lengths of chemical bonds, and the angles between bonds. From among the many examples that could be cited are the determination of the

**TABLE 9.2** SELECTED RANGES OF ELECTROMAGNETIC RADIATION
AND THEIR APPLICATIONS IN CHEMISTRY

| Kind of radiation | Name of technique | Information obtained |
|---|---|---|
| X-rays | X-ray diffraction | Spacing in crystals Bond distances Molecular shape |
| Ultraviolet and visible | Visible and UV spectroscopy | Transitions among energy levels of valence electrons |
| Infrared | Infrared spectroscopy | Molecular vibrations and rotations |
| Microwave | Microwave spectroscopy | Molecular rotations |
| Ultra high frequency and radiowaves | Nuclear magnetic resonance spectroscopy° | Magnetic properties of nuclei and, in particular, their chemical environment within the molecule, i.e., structure of molecules |

°Nuclear magnetic resonance spectroscopy has become a major tool of the chemist and biochemist during the decade of the 1960's. The life scientist who needs to make application of this valuable technique should consult any of the several suggested additional readings listed at the end of this chapter.

structures of amino acids, peptides,° complexes of metals with peptides,† and vitamin $B_{12}$.‡

Single crystals can also be prepared from some extremely large molecules, notably globular proteins. Additional difficulties are encountered
in protein crystal studies: the crystals are approximately half solvent
by weight and crack upon drying, the crystals are frequently damaged
by the long, though necessary, exposure to x-rays, and an enormous
number of reflections must be measured and processed. Furthermore,
owing to a difficulty known as the phase problem,§ it is necessary to
study crystals of well-defined derivatives prepared from the protein and
a heavy atom in addition to studying crystals of the protein itself. The

°Freeman, H. C., *Advan. Protein Chem.*, **22**, 257 (1967).

†Marsh, R. E., and J. Donohue, *Advan. Protein Chem.*, **22**, 235 (1967).

‡Hodgkins, D. C., J. Kamiper, J. Findsey, M. MacKay, J. Pickworth, J. H. Robertson, C. B. Shoemaker, J. G. White, R. J. Posen, and K. N. Trueblood, *Proc. Roy. Soc., Part A*, **242**, 228 (1957).

§For further information, see R. D. B. Fraser and T. P. MacRae, in *Physical Principles and Techniques of Protein Chemistry, Part A*, S. J. Leach, Ed., Academic Press, New York, 1969, p. 59.

problems are such that it is not feasible to attempt to determine the distances between atoms to $\pm 0.01$ Å, as can be done with x-ray diffraction studies of perfect crystals of small molecules. The goal instead is to determine the course of the polypeptide backbone through the protein molecule and to locate the position of as many side chains as possible. A frequently used procedure is to determine the location of the $\alpha$-carbon atoms from the amino terminus to the carboxyl terminus from x-ray studies. The side chains can then be placed on the proper $\alpha$-carbon atom if the amino acid sequence has been determined by an independent chemical method.

The first globular protein for which detailed structural information was obtained was sperm whale myoglobin, studied by J. C. Kendrew and coworkers. A very readable account of the result has been presented in *Scientific American.*[*] Four recent publications can be highly recommended for descriptions of the structures found for specific proteins. Myoglobin, hemoglobin, lysozyme, ribonuclease, chymotrypsin, papain, and carboxypeptidase A are discussed in an illustrated book by Dickerson and Geis.[†] Carboxypeptidase A, chymotrypsinogen, chymotrypsin, elastase, papain, and subtilisin are presented in a recent volume of *The Enzymes.*[‡] The structures of hemoglobin and myoglobin, as well as many other properties of these proteins, have been the subject of a recent book.[§] Finally, the structure of cytochrome $c$ has recently been described.[**] Many of these articles contain figures that can be seen in three dimensions with the aid of a viewer.

Valuable information has also been obtained about the structure of biological macromolecules by the x-ray study of oriented fibers, where the sample is not a perfect crystal, but the long axis of the crystalline regions is preferentially oriented in the direction of the fiber. As a consequence of imperfect orientation, the x-ray reflections are not as well defined nor are they as numerous as those obtained from a single perfect crystal. Reflections corresponding to specific spacing in the fiber can still be measured, however, and it is possible to determine whether the spacings giving rise to a particular reflection are in the direction of the fiber or perpendicular to the direction of the fiber. This information is not sufficient to lead to an unambiguous structure for the crystalline

---

[*]Kendrew, J. C., *Scientific American* (December 1961), p. 96. (Available from W. H. Freeman and Company as *Scientific American* Offprint 121, "Three Dimensional Structure of a Protein.")

[†]Dickerson, R. E., and I. Geis, *The Structure and Action of Proteins*, Harper and Row, New York, 1969.

[‡]Boyer, P. D., Editor, *The Enzymes*, Academic Press, New York, 1971, 3rd Edition, Vol. III.

[§]Antonini, E., and M. Brunori, *Hemoglobin and Myoglobin in Their Reactions with Ligands*, American Elsevier, New York, 1971.

[**]Dickerson, R. E., *Scientific American*, April, 1972, p. 58. (Available from W. H. Freeman and Company as *Scientific American* Offprint 1245, "The Structure and History of an Ancient Protein.")

region without the use of additional data. The common procedure is to start with the covalent structure of the molecule, and together with all the pertinent conformational information obtained by other techniques attempt to discover a conformation of the molecule which would lead to the observed x-ray pattern. By far the best-known example of this procedure is the determination of the double-helical structure of DNA by Watson and Crick.[*] The technique has also been successfully applied to collagen fibers,[†] muscle fibers,[‡] and synthetic polypeptides.[§]

The x-ray diffraction study of unoriented samples yields comparatively little structural information. An x-ray reflection will appear as a circle from which the corresponding spacing can be calculated, but all directional information is lost. The technique can be applied to determine whether the material examined is at all crystalline by seeing whether any sharp circular reflections are obtained.

### 9.1  Emission Spectra

Almost without exception atoms and molecules emanate discontinuous spectra when they have somehow acquired enough energy so that they may be described as being in an excited state. Convenient equipment for supplying this energy primarily have been the electric arc (for lower excitations), the electric spark (for higher degrees of excitation) and the gas discharge tube (for gases). The gas discharge tube is similar to the original Geissler tube; the other two use any appropriate conducting solid electrodes which may be atomized by electrical discharge. Alternatively carbon electrodes may be used to hold nonconducting or powdered material in a hollowed-out "cup" in the lower electrode. Figure 9.2 illustrates schematically how a spectrograph serves to obtain line spectra of elements or discontinuous band spectra of molecules. The spectrum seen on a developed plate (or recorded on chart paper if an electronic detection system is used) will be characteristic of electronic transitions within the atoms or molecules that constitute the radiation source. Each frequency line within the spectrum represents the return of electrons from some energy level of an excited state to a lower level that may or may not be the one that it occupied in the ground or unexcited state.

---

[*]Watson, J. D., and F. H. C. Crick, *Nature*, **171**, 737, 964 (1953); F. H. C. Crick and J. D. Watson, *Proc. Roy. Soc., Part A*, **223**, 80 (1954).

[†]Ramachandran, G. N., and G. Kartha, *Nature*, **176**, 593 (1955); F. H. C. Crick and A. Rich, *Nature*, **176**, 780 (1955); P. M. Cowan, S. McGavin, and A. C. T. North, *Nature*, **176**, 1062 (1955).

[‡]Huxley, H. E., and W. Brown, *J. Mol. Biol.*, **30**, 383 (1967).

[§]Elliott, A., in *Poly-α-amino Acids*, G. D. Fasman, Ed., Marcel Dekker, New York, 1967, p. 1.

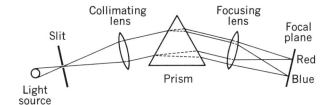

**Figure 9.2** Schematic arrangement of components of a simple spectrograph.

## 9.2 Dispersion of Radiation

The prism is the basic element of the *spectroscope* and is the dispersing device, *i.e.*, it gives different emergent angles to the different wavelengths that pass through it. Since the substance of the prism — usually quartz or glass for visible light — has a different refractive index for each wavelength, the different wavelengths are bent in varying amounts upon angular entry at the face of the prism and again upon exit into the air. Longer wavelengths are bent less and the shorter wavelengths are bent more with the resulting effect being that the emerging light is dispersed into its component wavelengths. The entire optical system from entrance slit to detector is called a spectroscope. If the detector is a photographic plate, it is called a spectrograph.

Another common dispersing device widely used in the infrared region of the spectrum is the diffraction grating. These gratings serve to disperse either transmitted or reflected light, and the principle by which they operate may be seen quantitatively in the following discussion.

If a transparent glass plate has a large number of fine parallel lines ruled on one of its surfaces and is illuminated from the back side by monochromatic radiation (light of one wavelength only), there will result an undeviated beam passing through the plate in addition to beams of the light at some other angles determined by spacing of the lines and the wavelength of the light. The described phenomenon will occur if the apertures created between rulings are uniformly spaced and their width is negligible in comparison to the distance, $d$, between them. Each aperture becomes a new light source from which the emanating light interferes with or reinforces the light coming from other apertures.

The student will recall the familiar Bragg Law for the diffraction of x-rays from crystals. The distance between particles (spacing between planes containing atoms or ions) corresponds to the distance between lines on a diffraction grating for visible or infrared spectra and the same basic equation, $n\lambda = 2d \sin \theta$, applies. The angle of diffraction or reinforcement of radiation is $\theta$, the distance between lines is $d$, the wavelength of the radiation is $\lambda$, and $n$ is an integer called the *order*.

When light of many wavelengths, *i.e.*, polychromatic radiation, is passed through or reflected from a diffraction grating there will result a series of spectra with all light of the same wavelength being diffracted through a number of differently spaced angles. Each wavelength, however, will have its principal maximum: the array of these first principal maxima of the different wavelengths is the first order spectrum, $n = 1$; the array of the second principal maxima is the second-order spectrum, $n = 2$, etc. Therefore gratings not only disperse light but also provide several orders of the resulting spectrum. The intensity of the orders decreases exponentially as $n$ increases, but at times there may be enough overlap of wavelengths from different orders to offer some lack of spectral purity. Modern instruments have ways of eliminating unwanted wavelengths from any particular order of the spectrum. Modern diffraction gratings of both the transmission and reflection types have their lines ruled as closely together as possible and over as wide an area as practicable. Ruling engines available today can rule up to 25,000 lines per mm on appropriate surfaces. A review of the theory of the diffraction grating may be found in any college level physics book.

### 9.3 Interaction of Electromagnetic Radiation with Matter

When matter in some convenient form — gas, liquid, solid or in solution — is placed in the path of a beam of electromagnetic radiation and interrupts it, the intensity of the radiation after the beam has passed through the material is always diminished to a lesser or greater degree. Students often ask, "What happened to the lost radiant energy?", "What constitutes the energy sink?" Answers to these questions are not always simple and depend primarily upon the frequency range of the radiation and the nature and state of the matter being irradiated.

The primary process always is that the radiation is absorbed by the individual atoms, ions, or molecules (in some cases even free electrons) in such a way that their *potential* energy is increased. The process may happen in several ways: for example, electronic transitions occur when electrons in the atoms or molecules of the absorbing substance are raised to higher energy levels. This is the kind of activated or excited state that results from the absorption of energy in the ultraviolet-visible range of the spectrum. If the energy is in the infrared range, the activation takes the form of changing the vibrational or rotational levels of pairs of atoms or groups of atoms within molecules. Although rotational levels occur at lower frequencies than do vibrational levels, they are so interrelated that the absorption frequencies are really characteristic of the absorbing molecule itself. It should be remembered that all these absorptions are quantized, meaning that only certain absorptions are possible for a particular substance in a given range of the spectrum and that the absorbed energy, $\Delta E$, equals $h\nu$.

We have not yet answered the question, "What is the energy sink?" Do the absorbers stay in the activated state? The answer to the second question is, "No." Re-emission of the radiation can be observed for gases at low pressures and often from molecules in solution when the solution is at liquid nitrogen temperatures or below. Certain molecules can undergo fluorescence, wherein radiation is re-emitted at longer wavelenths than the exciting radiation. Fluorescence results from the excited states not returning directly to the ground state, but rather returning to the ground state by a series of steps with each step being lower in energy than the exciting step.

The suggestion that re-emission can occur from excited molecules frozen into the "glass" of the solvent at very low temperatures is a clue to the ultimate "energy sink" for most cases of absorption of electromagnetic radiation. In liquids, collisions occur with a frequency of about $10^{12}$ per second. The average lifetime of an activated molecule is the order of one to ten nanoseconds (1 nsec $= 10^{-9}$ sec), so an activated molecule will experience many collisions during its existence. The collisions can serve to redistribute the energy of the activated molecule among many molecules and the absorbed energy is dissipated as heat.

The characteristic absorption frequencies are the same as those observed for emission spectra under circumstances described earlier and may be used as the basis for theoretical structure studies as well as for qualitative and quantitative analysis. It is this last feature which we will explore more fully in the remainder of this chapter as we investigate its value to the practicing life scientist. We should not leave our general discussion of absorption without examining one or two other processes which may occur as the result of absorption of electromagnetic radiation.

Of importance in physical chemical investigations of macromolecules is the phenomenon of light scattering, which was probably first recognized by Lord Rayleigh[*] as accounting for the blue of the sky and as accounting for the "Tyndall effect" from colloidal solutions. Debye[†] has extended light scattering as an important investigative tool to solutions of large molecules. Major workers applying this technique to biological macromolecules have been Benoit,[‡] Doty,[§] Timasheff[**] and Zimm.[††] A variation on this technique, small-angle x-ray scattering, is frequently of greater usefulness in the study of compact biological macromolecules such as globular proteins and transfer RNAs. These

[*]Lord Rayleigh, *Phil. Mag.* (4), **41**, 447 (1871).

[†]Debye, P., *Am. Physik.* (4), **46**, 809 (1915).

[‡]Benoit, H., L. Freund, and G. Spach in *Poly-α-Amino Acids*, G. D. Fasman, Ed., Marcel Dekker, New York, 1967, p. 105.

[§]Doty, P., J. H. Bradbury, and A. M. Holtzer, *J. Amer. Chem. Soc.*, **78**, 947 (1956); J. Eigner and P. Doty, *J. Mol. Biol.*, **12**, 549 (1965).

[**]Timasheff, S. N., and R. Townsend in *Physical Principles and Techniques of Protein Chemistry*, Part B, S. J. Leach, Ed., Academic Press, New York, 1970, p. 147.

[††]Harpst, J., A. Krasna, and B. Zimm, *Biopolymers*, **6**, 585 (1968).

molecules are much smaller than the wavelength ($\sim$4000 Å) of the light typically used in light scattering measurements. More detailed information on their structure can be obtained if the wavelength of the electromagnetic radiation used in the scattering experiment is reduced to 1.54 Å, *i.e.*, to the wavelength of x-rays. Kratky[*] has been active in applying this technique to biological systems.

Although the theory and practice of light scattering is presented briefly in Chapter 7, it is appropriate to recall certain important considerations at this time. The electric vector of light traversing any medium induces oscillating electric moments in the particles of that medium. If the frequency of oscillation is far removed from a characteristic absorption frequency and, preferably, the incident light is long in wavelength compared to the size of the scattering particles, the latter will act as secondary emitting sources of the same wavelength (as the incident light) and the light will be scattered in all directions with respect to the incident beam. It is the analysis of this "scattering envelope" which serves as the basis for useful results from light scattering. The method has certain empirical applications which are also of value, *e.g.*, some protein and enzyme analytical methods are based on turbidimetric methods (nephelometry) which relate the intensity of scattered light to concentrations of the scattering species.

At the core of the substantial field of photochemistry is an alternative energy sink to the heat dissipation mentioned earlier. Photochemistry is the study of chemical reactions that occur as the result of the absorption of light. The basic law of photochemistry was proposed by Einstein and in its statement utilizes a concept from the corpuscular theory of light, namely that the fundamental particle, the *photon*, possesses one quantum of energy equal to $h\nu$. Einstein suggested that the photochemical process involved the absorption of one photon (one quantum) of radiant energy by each absorbing molecule. In photochemical reactions, rather than return to the ground state, the activated molecules have become so reactive that they may experience rearrangements or dissociation or possess the activation energy necessary to undergo reaction with other molecules. An activated molecule may also pass on its excitation energy to another molecule which, in turn, may react. There are many photochemical processes of importance in biological systems, the most significant being photosynthesis[†] and vision.[‡] References to this major area of chemical concern may be found in the recommended readings at the end of the chapter.

An interesting corollary to photochemical processes is that of chemiluminescence in biological systems, sometimes called biological lu-

---

[*] Kratky, O., in *Progress in Biophysics*, **13**, 105 (1963).

[†] Bishop, N. I., *Ann. Rev. Biochem.*, **40**, 197 (1971).

[‡] Hubbard, R., and A. Kropf, *Scientific American*, June 1967, p. 64. (Available from W. H. Freeman and Company as *Scientific American* Offprint 1075, "Molecular Isomers in Vision.")

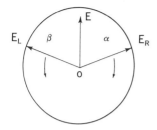

**Figure 9.3** Relationship between the left- and right-hand circularly polarized components, $OE_L$ and $OE_R$, and the resultant elective vector of plane polarized ray, OE. If the two circularly polarized components are of equal size and rotate at the same speed, but in opposite directions, the absolute magnitude of the resultant OE will vary from zero to twice the length of $OE_L$, and will always remain in the same plane.

minescence which, in a sense, is the reverse of a photochemical process. The emission of light by organisms is the result of a chemical reaction catalyzed by an enzyme. Such organisms are probably numbered in the thousands and are amazingly diverse. A most fascinating article on this subject was published by McElroy and Seliger (William D. McElroy and Howard H. Seliger, *Scientific American*, December 1962, pp. 76–89). Its reading is highly recommended. The topic also has been the subject of a recent review.[*]

**Polarized light**  When unpolarized light is passed through a properly cut and cemented crystal of Iceland spar ($CaCO_3$), known as a Nicol prism, it will be resolved into two plane polarized beams. One of these beams can be further resolved into two beams of circularly polarized rays by a device known as a Pockels cell.[†] The two circularly polarized rays are of equal intensity but of opposite sense as illustrated in Fig. 9.3. When these circularly polarized rays of light are passed through an anisotropic medium, one will be retarded more than the other because of the difference between the indices of refraction of the medium for the right-hand and left-hand ray. The observable result will be a rotation of the plane of polarization since the sum of $\alpha$ and $\beta$ will no longer be the original $OE$ plane. If the medium absorbs one component ($OE_L$ or $OE_R$) more strongly than the other, there will still result a rotation of the plane polarized light. The two components, however, will differ in intensity after having passed through the medium, and the resultant light is said to be elliptically polarized. In general if there is a difference in absorption of the left- and right-hand rays, there is a difference in index of refraction for them as well, as the two effects are not independent.

Substances that rotate the plane of polarized light are optically active, and the activity may be a property of the bulk phase as found in certain crystals, or a property of the asymmetric molecule. It is this latter feature, of course, which is of special interest to chemists. The differential ab-

[*]Hastings, J. W., *Ann. Rev. Biochem.*, **37**, 597 (1968).

[†]Velluz, L., M. LeGrand, and M. Grosjean, *Optical Circular Dichroism*, Academic Press, New York, New York, 1965, p. 40.

sorbance of the circularly polarized light resulting from a difference in absorptivity of the two rays is called *circular dichroism*, and the frequency dependence of the rotation is called *optical rotatory dispersion.*

### 9.4 Beer-Lambert Law

Quantitative measurements of the absorption of monochromatic radiation by a collection of molecules requires the utilization of the following arrangement:

The incident radiation of the appropriate wavelength has the intensity $I_0$. Appropriate units for the intensity of the radiation could be the number of photons per second which pass through a cross-section of unit area positioned perpendicular to the direction of propagation. The intensity after passing through a collection of molecules, at least some of which absorb radiation of the wavelength of the incident beam, will be reduced to the value $I$. The means used to generate the incident radiation and to measure the intensities $I_0$ and $I$ will depend upon the spectral region under investigation, and will not be discussed at this time. Our present objective is to determine the relationship between $I_0$, $I$, and the sample.

It will be assumed that all of the absorbing molecules are identical and that they are uniformly distributed in the sample. Since all of the molecules and all of the photons are respectively identical, it will also be assumed that the probability that a molecule will absorb a photon to which it is exposed will be the same for all encounters between molecules and photons. For the present, it will also be assumed that all of the reduction in the intensity of the radiation can be attributed to the ensemble of identical absorbing molecules. This means that reduction in intensity due to the interaction of the radiation with the sample container or the solvent, if a solvent is used, is assumed not to occur.

$$dA \quad dr$$

$$I \longrightarrow \quad I + dI \longrightarrow$$

Consider the result of the passage of radiation through an infinitesimally small volume element in the sample. The volume element has a thickness $dr$ in the direction of propagation of the radiation and an area $dA$ perpendicular to the direction of propagation. Passage of radiation through this volume element is accompanied by $dI$ — an infinitesimally small change in intensity. Since the intensity is reduced by the absorption of some of the photons by the molecules contained in the volume element, $dI$ will be negative. Recalling that intensity is expressed as photons/sec/unit area, the number of photons absorbed per second by the molecules in the volume element will be $-dIdA$. Letting $W$ be the probability that absorption will occur upon the encounter of a photon with a molecule, we obtain $-dIdA = W \times$ (number of encounters per second). The number of encounters per second in the volume element is proportional to the product of the number of photons/sec/unit area and the number of molecules in the volume element. Designating the proportionality constant by $\delta$, we obtain, $-dIdA = W\delta I \times$ (number of molecules). The number of molecules in the volume element is the product of the number of molecules per unit volume, designated by $n'$, and the volume of the volume element, which is $dAdr$. Making these substitutions, we obtain,

$$-dIdA = W\delta I n' \, dAdr. \tag{9.4}$$

The area of the volume element perpendicular to the direction of propagation of the radiation turns out to be inconsequential because $dA$ can be cancelled from each side of the equation.

Equation 9.4 must be integrated before it can be applied to a macroscopic sample. For this purpose, both sides of Eq. 9.4 will be divided by $IdA$ in order to cancel $dA$ and collect all terms in $I$ on the same side. The limits of the integration are from an intensity of $I_0$ when $r = 0$ to an intensity of $I$ when $r = r$. Since $W$, $\delta$,

$$-\int_{I_0}^{I} \frac{dI}{I} = W\delta n' \int_{0}^{r} dr \tag{9.5}$$

and $n'$ are constants, they can be brought outside the integral sign. Integration yields

$$-\ln \frac{I}{I_0} = W\delta n' r$$

$$-\log \frac{I}{I_0} = \frac{W\delta n' r}{2.303} \tag{9.6}$$

Frequently used units for concentration and length are molarity and cm, respectively. Conversion of $n'$ and $r$ to these units would simply require the introduction of a new constant on the right side of Eq. 9.6. A convenient and compact relationship is

$$-\log \frac{I}{I_0} = \epsilon M l = \frac{W\delta}{2.303} n' r \tag{9.7}$$

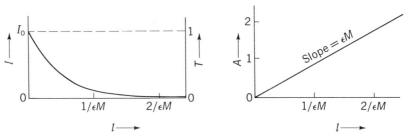

**Figure 9.4** Relationship of intensity, $I$, and transmittance, $T$, to length, $l$, when molarity, $M$, is constant (top curve) and Eq. 9.8 is obeyed.
   Relationship of absorbance, $A$, to length, $l$, when molarity, $M$, is constant (bottom curve) and Eq. 9.8 is obeyed.

where $l$ is the length of the light path in cm, $M$ is the molarity, and the single constant $\epsilon$ is used to represent all of the other constants. The units of $\epsilon$ are cm²/millimole. Since the light intensities only enter this equation as the ratio $I/I_0$, the units that are used for the intensities are inconsequential provided that the units used for the incident and emergent beam are the same. Furthermore, the intensities can be replaced by any quantity directly proportional to intensity. The ratio $I/I_0$ is called the transmittance, $T$, and the negative logarithm of the transmittance is defined as the absorbance, $A$. The term optical density is occasionally used as a synonym for absorbance. The final equation, known as the Beer-Lambert Law, is

$$- \log T = A = \epsilon M l \tag{9.8}$$

   Two particular examples are of special interest. If the light path is allowed to vary while the concentration is held constant, the results presented in Fig. 9.4 are obtained. Holding the light path constant and allowing the concentration to vary leads to the results in Fig. 9.5. The transmittance decreases rapidly at small $l$ (or $M$) and approaches, but never quite reaches, zero in the limit of large $l$ (or $M$). The absorbance, on the other hand, is a linear function of $l$ (or $M$) and increases without limit.

   It was stated earlier that no decrease in light intensity was assumed to occur due to the interaction of the radiation with the sample container or the solvent. This assumption will not generally be valid. Equation 9.8 can still be used, however, if $I_0$ is defined as the intensity sensed by the detector when only the sample container and solvent molecules are in the light path and $I$ is the intensity when sample container and solvent plus absorbing molecules are in the light path.

   Real systems frequently exhibit deviations from the Beer-Lambert Law when the absorbance is large. Alternatively stated, the Beer-Lambert Law will accurately describe the absorption of light by a

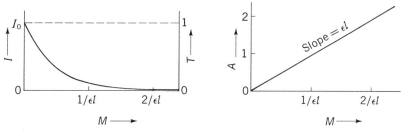

**Figure 9.5** Relationship of intensity, $I$, and transmittance, $T$, to molarity, $M$, when length, $l$, is constant (top curve) and Eq. 9.8 is obeyed.

Relationship of absorbance, $A$, to molarity, $M$, when length, $l$, is constant (bottom curve) and Eq. 9.8 is obeyed.

system as the absorbance approaches zero. The range of absorbance over which the Beer-Lambert Law is valid will depend upon the absorbing system studied and also upon the characteristics of the measuring instrument.

Let us first inquire what properties of the absorbing system might lead to deviations[*] from the Beer-Lambert Law. According to Fig. 9.5, the effect of varying concentration at constant path length should yield a linear plot of absorbance *vs.* molarity with a slope of $\epsilon l$. Since we have already specified that the light path is of constant length, the slope will be constant provided that $\epsilon$ is independent of concentration. However, the probability, $W$, that a molecule will absorb radiation of a particular wavelength is in general dependent upon the molecular environment. As the concentration of an absorbing molecule in a transparent solvent increases from an infinitely dilute solution through more concentrated solutions and finally to pure solute, the environment of an individual solute molecule changes from pure solvent to pure solute. Consequently $W$ should change continuously as the concentration is increased. Since $\epsilon$ includes $W$, $\epsilon$ will also change with concentration, leading to curvature in a plot of absorbance *vs.* molarity at constant light path length, as shown in Fig. 9.6. For this reason we expect that the absorbance will be a linear function of molarity, at constant light path length, only when the solutions are sufficiently dilute so that each solute molecule is in an immediate environment consisting of essentially pure solvent. The

---

[*]It is to be expected that the Beer-Lambert Law apply only in systems where the solute neither associates, dissociates, interacts or reacts with the solvent (to varying degrees depending on the concentration of solute). To our minds there are no *deviations* from the Beer-Lambert Law, but only *apparent* deviations which reflect our ignorance of the processes going on in the system we are attempting to understand and describe. Those observations which we have termed *deviations* from the Beer-Lambert Law have been so described because of common usage of the word, whereas in fact they emphasize how important it is that the chemist properly describe (and understand) the state of the system with which he works.

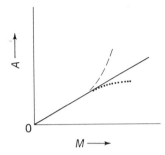

**Figure 9.6** Positive (---) and negative ($\cdot\cdot\cdot$) deviations from the Beer-Lambert Law.

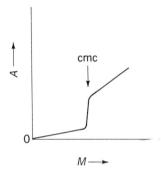

**Figure 9.7** Diagrammatic representation to the absorption of ultraviolet radiation by a system that can form micelles.

concentration range over which this condition will be satisfied may be extremely small if the solute molecules tend to associate with each other.

Dramatic deviations from the Beer-Lambert Law can arise if the solute molecules associate into complexes large enough to cause appreciable scattering of the radiation used. If some of the radiation is scattered, there will necessarily be a reduction in the intensity of the radiation reaching the detector and an increase in the absorbance. A particularly interesting manifestation of this phenomenon occurs in the formation of micelles. Over an extremely small concentration range designated the critical micelle concentration (cmc), certain kinds of solute molecules cooperatively associate into large particles, called micelles, which may contain 100 or more molecules per particle. The micelles cause appreciable scattering of radiation in the x-ray to visible region, leading to the result presented diagrammatically in Fig. 9.7. Micelles are frequently formed by molecules, one end of which has a high affinity for the solvent and another end a very low affinity for solvent. Examples are synthetic detergents,* such as sodium dodecylsulfate, and ionic lipids, such as ionized fatty acids* and phospholipids.†

---

*Winsor, P.A., *Chem. Rev.*, **68**, 1 (1968).
†Perrin, J. H., and L. Saunders, *Biochim. Biophys. Acta*, **84**, 216 (1963); D. Attwood and L. Saunders, *ibid.*, **98**, 344 (1965).

Apparent deviations from the Beer-Lambert Law can also be caused by imperfections in the instrument used for the measurements. Of particular interest in this regard is stray light. In the derivation of the Beer-Lambert Law, it was assumed that the radiation was monochromatic. Most instruments provide a light source which produces a broad spectrum of radiation. By some combination of filters, prisms, and diffraction gratings, it is endeavored to select radiation of the appropriate wavelength and to reject all other radiation for transmission through the sample. This procedure is never completely successful since the beam of radiation passing through the sample always will contain some stray light that has a wavelength different from that of the radiation desired. One of the important criteria in judging the quality of a spectrophotometer is the amount of stray light permitted through the sample.

To illustrate the deviation from the Beer-Lambert Law caused by stray light, consider a molecule that absorbs light at a wavelength designated by $\lambda_1$. Assume that the absorbance of monochromatic radiation of wavelength $\lambda_1$ is an exact linear function of molarity when the length of the light path is held constant. We wish to determine the relationship between the absorbance and the molarity when, for example, 99% of the intensity through the sample has wavelength $\lambda_1$ and 1% of the intensity is of wavelength $\lambda_2$, where $\lambda_2$ corresponds to a region of the spectrum in which the molecule does not absorb. When the concentration is zero, all of the radiation is transmitted, and the absorbance is zero. When the molarity is $1/(\epsilon l)$, only 10% of the intensity of wavelength $\lambda_1$ is transmitted. If the incident radiation were truly monochromatic, the transmittance would be 0.1 and the absorbance would be 1.000. In the example with 1% stray light, however, the total intensity of the light reaching the detector is 10% of that of the incident radiation of $\lambda_1$, plus the intensity of radiation which has $\lambda_2$. The transmittance is therefore 0.109, and the absorbance is 0.963 instead of 1.000. Deviations caused by the stray light increase as the concentration increases. When the concentration is $2/(\epsilon l)$, for example, the total intensity reaching the detector will be 1% of the incident intensity of the radiation with $\lambda_1$ plus all of the intensity of the radiation with $\lambda_2$. The transmittance will consequently be 0.02, and the absorbance is 1.699 instead of 2.000, which would have been the result if the incident radiation were truly monochromatic. The general result which would be obtained by the model described at the beginning of this paragraph is shown in Fig. 9.8 as absorbance *vs.* molarity at constant length of light path. Stray light leads to negative deviation from the Beer-Lambert Law, and the deviation becomes marked as the absorption from the incident beam increases.

In Fig. 9.9 there appears an optical diagram of the Beckman Acculab 1 Infrared Spectrophotometer. The instrument has a double-beam optical system and three scanning speeds. Samples may be gas, liquid, or solid,

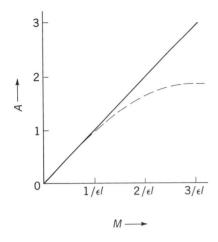

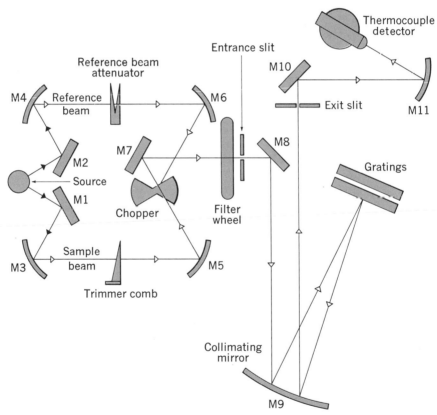

**Figure 9.8** Absorbance *vs.* molarity for monochromatic radiation (—), and absorbance *vs.* molarity when 1% of the intensity of the incident beam is stray light corresponding to a region of the spectrum where absorption does not occur (---). It is assumed that the only deviation from the Beer-Lambert Law is due to the stray light.

**Figure 9.9** Optical Diagram of the Beckman Acculab 1 Infrared Spectrophotometer. M designates mirror. [Courtesy, Beckman Instruments, Inc.; Scientific Instruments Division; 2500 Harbor Boulevard; Fullerton, California.]

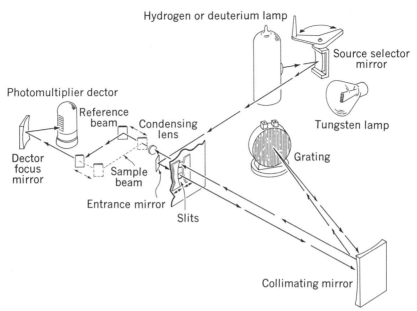

Hydrogen or deuterium lamp

Source selector mirror

Tungsten lamp

Photomultiplier dector

Reference beam

Condensing lens

Grating

Dector focus mirror

Sample beam

Entrance mirror

Slits

Collimating mirror

**Figure 9.10** Optical Diagram of the Beckman Model DB* Grating Spectrophotometer. [Courtesy, Beckman Instruments, Inc.; Scientific Instruments Division; 2500 Harbor Boulevard; Fullerton, California.]
* Registered trademark.

and the recorded output is linear transmittance *vs.* linear wavenumber or time (for reaction rate studies). For this particular model the spectral range is from 4000 to 600 cm$^{-1}$. Other excellent spectrophotometers for covering various spectral ranges are commercially available.

The optical diagram of the Beckman Model DB-Grating visible ultraviolet range Spectrophotometer is shown in Fig. 9.10. The instrument offers excellent performance through a precision filter-grating monochromator. It employs a precision replica grating as the dispersive element for high accuracy, repeatability, and resolution. The wavelength range covered by the 1403 Spectrophotometer is 190 to 700 nm with the standard equipped 1P28-A photomultiplier detector. The range can be extended to 800 nm with the Red-Sensitive Photomultiplier Accessory. A tungsten lamp source may be utilized above 320 nm and a hydrogen lamp source below 360 nm.

## 9.5  Infrared Spectroscopy

Part of the information required to specify the molecular dimensions of a molecule is knowledge of the bond lengths and the angles between adjacent bonds. Numerical values for these quantities in a large number of molecules have been obtained from x-ray diffraction studies of perfect crystals and from the analysis of the absorption of electromagnetic radiation in the microwave region by gases. For the water molecule, the appropriate values are 0.95718 Å for the $O-H$ bond length and 104.523° for the $H-O-H$ bond angle.[*] These numbers refer to a hypothetical equilibrium state in which there is no vibrational or rotational energy.

In reality molecules are continually vibrating. They experience fluctuations in their bond distance of about ±0.05 Å and in their bond angles of about ±5° at physiological temperatures. The energy associated with these vibrations cannot assume all possible values, but is instead quantized. The separation between adjacent energy levels corresponds to the energies of electromagnetic radiation in the infrared region of the spectrum. Consequently, a molecule can absorb infrared radiation of an appropriate frequency, accompanied by the promotion of the molecule to an excited vibrational state. The typical method of presenting infrared spectra is as percent transmission (transmittance × 100) *vs.* wavenumber. A representative spectrum is shown in Fig. 9.11. The precise separation for the energy levels of a particular vibration depends upon the masses of the atoms which are affected as well as the resistance of the vibrating unit to distortion. The classical result for a molecule consisting of two atoms leads to Eq. 9.9,

$$\bar{\nu} = \frac{1}{2\pi c} \sqrt{\frac{k\,(m_1 + m_2)}{m_1 m_2}} \tag{9.9}$$

where $\bar{\nu}$ is the wavenumber of infrared radiation whose absorption will promote the molecule to the nearest excited vibrational state, $c$ is the speed of light, $k$ is the force constant reflecting the resistance of the bond to distortion, and $m_1$ and $m_2$ are the masses of the two atoms in the molecule. A vibrational mode will lead to the absorption of infrared radiation if the radiation has the appropriate frequency and if the vibration gives rise to an oscillating dipole moment. The oscillating electric vector in the infrared radiation interacts with the oscillating dipole moment leading to the absorption of radiation and an excited vibrational state.

Since a non-linear molecule consisting of $n$ atoms will possess $3n - 6$ vibrational degrees of freedom, most molecules of biological interest will absorb light at many frequencies in the infrared. In particular, nucleic acids, proteins, and polysaccharides will possess such an enormous number of vibrational degrees of freedom that a rigorous explanation of their infrared spectra is unattainable. Even for molecules as complex as a protein, however, some valuable information can be obtained from infrared spectra because various functional groups, regardless of whether

---

[*]Eisenberg, D. and W. Kauzmann, *The Structure and Properties of Water*, The Clarendon Press, Oxford, 1969, p. 4.

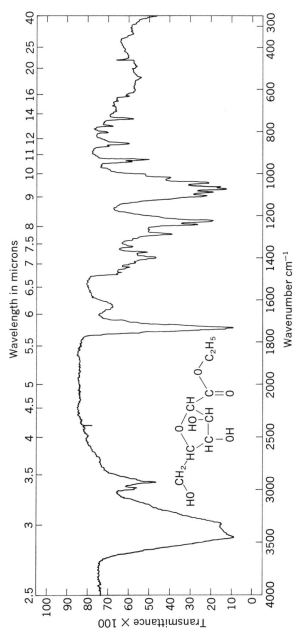

**Figure 9.11** Representative infrared spectrum. The sample is a derivative of the sugar mannose (2,5-anhydro-D-mannoate ethyl ester) in a potassium bromide pellet. The spectrum was obtained using a Beckman IR 10 Infrared Spectrometer (Courtesy of Mr. T. Koerner and Dr. E. S. Younathan).

**TABLE 9.3** CHARACTERISTIC WAVENUMBERS FOR
BOND STRETCHING

| Bond | $\bar{\nu}$ |
|---|---|
| O—H, N—H | 3100–3500 |
| C—H | 2800–3100 |
| Double bonds involving carbon | 1600–1800 |
| P—O | 1200–1300 |
| Single bonds from carbon to elements other than hydrogen | 800–1300 |

they occur in large or small molecules, absorb radiation in particular
regions of the infrared spectrum. For example, the transition known as
the amide I vibration, which primarily is caused by the stretching of the
carbonyl group in amides, leads to an intense absorption of radiation
near 1650 cm$^{-1}$ in amides as simple as N-methylacetamide and as complex as the protein $\beta$-lactoglobulin A.[*] Characteristic wavenumbers
for the vibrations of many groups can be found in a number of texts on
qualitative organic chemistry.[†] The regions of the stretching vibrations
for several bonds of particular interest to life scientists are presented
in Table 9.3.

Life scientists have applied infrared spectroscopy to four types of
problems: identification of compounds, following the kinetics of a reaction, studying the conformation of a molecule, and investigating the
interaction between molecules. In the following paragraphs each of
these will be considered in turn.

Infrared spectroscopy can be used in two ways to aid in the identification of unknown compounds. Information concerning the presence or
absence of various functional groups can be obtained by noting in which
regions of the infrared spectrum absorption occurs. Assume that an unknown substance had been shown to have the elemental formula $C_{14}H_{28}O$.
The ratio of exactly two protons per carbon atom requires the presence
of either one ring or one double bond. These possibilities could be distinguished by noting the presence or absence of an absorption at 1600–
1800 cm$^{-1}$, the region where double bonds involving a carbon atom
absorb. Another question is posed by the lone oxygen atom: Is it present
as a carbonyl group, an ether, or a hydroxyl group? If the hydroxyl
group is the answer, an absorption should be observed at 3100–3500

---

[*]Timasheff, S. N., H. Susi, R. Townend, L. Stevens, M. J. Gorbunoff, and T. F. Kumo-
sinski, in *Conformation of Biopolymers,* G. N. Ramachandran, Ed., Academic Press,
New York, New York, 1967, Vol. 1, p. 173.

[†]For example, see R. L. Shriner, R. C. Fuson, and D. Y. Curtin, *The Systematic Identi-
fication of Organic Compounds,* Wiley, New York, New York, 1964, 4th Ed., or R. M.
Silverstein and G. C. Bassler, *Spectrometric Identification of Organic Compounds,* Wiley,
New York, New York, 1967, 2nd Ed.

$cm^{-1}$, the region in which O—H stretching occurs. An example of a biological compound with the molecular formula $C_{14}H_{28}O$ is the sex pheromone from the female fall army worm, which has the structure $CH_3(CH_2)_3CH=CH(CH_2)_7CH_2OH.$* The discussion above shows that the presence of the hydroxyl group and the carbon-carbon double bond could be deduced from the molecular formula and the infrared spectrum.

A further use of infrared spectra in the identification of compounds takes advantage of the large number of vibrational degrees of freedom possessed by polyatomic molecules, since the detailed infrared spectrum serves as a finger print of the molecule. If there is reason to suspect that an unknown substance may be identical to a known sample, the hypothesis can be tested by obtaining infrared spectra of both compounds. If the two components are identical, their infrared spectra must be identical.

Since different functional groups absorb light in different areas of the infrared, the rates of reactions involving the alterations of functional groups can be followed by infrared spectroscopy. As a simple illustrative example, consider the oxidation of a secondary alcohol to a ketone. The reaction could be followed by measuring the rate of disappearance of the O—H stretching vibration of the hydroxyl group or by the rate of formation of the C=O stretching vibration of the ketone.

This technique has proven particularly useful to life scientists in measuring the rate of hydrogen-deuterium exchange reactions. Whenever a biological molecule, such as a protein, is placed in $D_2O$, the protons attached to nitrogen, oxygen, and sulfur atoms will exchange with the deuterons in the solvent. These reactions involve the substitution of an atom with

$$-N-H + D_2O \rightleftharpoons -N-D + HDO$$

$$-O-H + D_2O \rightleftharpoons -O-D + HDO$$

$$-S-H + D_2O \rightleftharpoons -S-D + HDO$$

mass number 2 for an atom with mass number 1, and, according to Eq. 9.9, the reaction should lead to a decrease in the wavenumber of the infrared radiation required to excite the vibration of the bond to the exchanged hydrogen atom. The rate of hydrogen-deuterium exchange of the peptide groups in proteins can be studied by following the amide II band, which is located at about 1550 $cm^{-1}$ for —CONH— and at about 1450 $cm^{-1}$ for —COND—.† As expected from Eq. 9.9, the substitution of deuterium for the proton causes a shift in the band to lower wavenumber. The shift is not as great as would be predicted by using the

---

*Sekul, A. A., and A. N. Sparks, *J. Econ. Entomol.*, **63**, 779 (1967).
†Hvidt, A., and S. O. Nielsen, *Adv. Protein Chem.*, **21**, 287 (1966).

masses of hydrogen and nitrogen in Eq. 9.9 because the amide II band is caused not only by the vibration of the nitrogen and hydrogen atoms, but also by the motion of other atoms in the peptide unit. The interest in the kinetics of hydrogen-deuterium exchange of proteins derives from the observation that small peptides and completely denatured proteins at room temperature and neutral pH exchange their protons by a first-order process with half-lives of about 0.1–1.0 minute, but proteins in their native state require hours or days for complete exchange. Analysis of the kinetics of the exchange reaction in proteins can lead to information concerning the kinetics and thermodynamics of their conformational transitions.[*]

The usefulness of infrared spectroscopy in the determination of the conformation of biological molecules perhaps is best illustrated by the application of infrared dichroism to fibrous structures. The type of dichroism with which we are concerned here is not to be confused with circular dichroism, mentioned earlier in this chapter. Circular dichroism refers to the difference in the absorption of right- and left-handed circularly polarized light. Optically active molecules will exhibit circular dichroism even when they are oriented at random in the sample. Existence of the type of dichroism at issue here—occasionally referred to as linear dichroism—requires that the molecules in the sample be partially or completely oriented in one direction, but does not require that they be optically active. Linear dichroism signifies that the absorption spectrum obtained with plane polarized light depends upon whether the E vector of the plane polarized radiation is parallel to or perpendicular to the direction in which the molecules in the sample are preferentially oriented.

Focusing our attention on a single vibration in one particular molecule, the absorption will be maximal when the electric vector of the infrared radiation is colinear with the oscillating dipole (or, more precisely, the transition dipole moment) of that vibration which is to be excited. On the other hand, if the electric vector forms an angle of 90° with the oscillating dipole, there will be no interaction between the radiation and the vibration, and absorption will not occur. These two examples and ones in which the angle, $\theta$, between the electric vector and the oscillating dipole is different from 0° or 90°, are described by Eq. 9.10. The absorbance is $A$

$$A = A_0 \cos^2 \theta \qquad (9.10)$$

at the angle $\theta$, and is $A_0$ when $\theta = 0°$. We wish to generalize this result for a particular vibration in one molecule to the same vibration in a large collection of identical molecules. If the molecules are oriented entirely at random, then regardless of how we position the electric

[*]Hvidt, A., and S. O. Nielsen, *Adv. Protein Chem.*, **21**, 287 (1966).

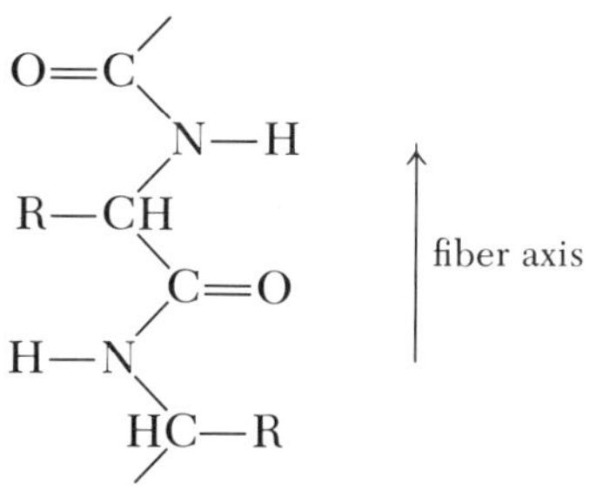

O=C
N—H
R—CH
C=O
H—N
HC—R

| fiber axis

**Figure 9.12** Preferential orientation of the polypeptide chain in silk, as determined by infrared dichroism.

vector, there will be some number of molecules for which $\theta$ is between 0–1°, and an identical number for which $\theta$ is 89–90°, and so on for all possible intervals of 1°. Therefore the absorbance $A$ will be independent of how we arrange the electric vector. If, on the other hand, there is a preferred orientation of the molecules, the number of molecules for which $\theta = 0$–1° and for which $\theta = 89$–90° will depend upon the geometric relationship between the electric vector and the sample. Furthermore, the maximum absorbance will be observed when the direction of the electric vector is the same as the preferential orientation of the oscillating dipoles of the vibration in question.

Infrared dichroism studies have been reported for a number of fibrous protein systems such as silk suture, porcupine quill, elephant hair, and rat tail tendon.[*] In each of these the infrared spectrum is different when the electric vector is parallel and perpendicular to the fiber axis, showing that there is a preferential orientation of the molecules in the fibers. With silk suture, the intensities of the absorptions at 1640 cm$^{-1}$ and 3300 cm$^{-1}$, corresponding to the peptide group C=O and N—H stretching frequencies respectively, are substantially greater when the electric vector is oriented perpendicularly to the fiber axis. It can be concluded that the N—H and C=O bonds in the peptide backbone must be preferentially oriented perpendicularly to the fiber axis, as shown schematically in Fig. 9.12, which agrees with current ideas on the structure of silk.[†]

Infrared dichroism has also been observed for oriented samples of DNA.[‡] The bands at 1600–1750 cm$^{-1}$, owing to the stretching vibrations of the aromatic rings, exhibit perpendicular dichroism, which means that their intensities are greater when the electric vector is oriented perpendicularly to the molecular axis than when the orientation is parallel.

---

[*]Ambrose, E. J., and A. Elliott, *Proc. Roy. Soc. (London), Part A,* **206**, 206 (1951).

[†]Dickerson, R. E., and I. Geis, *The Structure and Action of Proteins,* Harper and Row, New York, New York, p. 34.

[‡]Bradbury, E. M., W. C. Price, and G. R. Wilkinson, *J. Mol. Biol.,* 3, 301 (1961).

This observation requires that the bases themselves are oriented perpendicularly to the molecular axis—the result predicted by the Watson-
Crick double-helical model for DNA.

One of the best known examples of interaction between two biological molecules is the formation of hydrogen bonded pairs of purine and
pyrimidine derivatives in the double helical structure of DNA. Hydrogen bonds can also occur between different polypeptide chains, as in
the structures of collagen and silk. The infrared spectra of molecules
which form hydrogen bonds are altered by hydrogen bond formation.
For this reason infrared spectroscopy is a useful tool for studying
hydrogen bond formation.

In order to present the effects of hydrogen bond formation on the
infrared spectrum, let us consider the formation of a hydrogen bond between an N—H and an O=C group. The effect of the formation of the
hydrogen bond

$$\text{\textbackslash}N\!-\!H + O\!=\!C\diagup \rightarrow \text{\textbackslash}N\!-\!H \cdots O\!=\!C\diagup$$

is to reduce the resistance of the N—H bond, and, to a lesser extent,
the O=C bond, to stretching. In terms of Equation 9.9, the force constant k for the stretching of the N—H bond, and, to a lesser extent, for the
stretching of the O=C bond, has been reduced by hydrogen bond formation. It follows from Eq. 9.9 that the wavenumber at which these stretching vibrations are observed will also be reduced. The magnitude of the
change in the wavenumber at which the N—H stretching vibration is
observed depends upon several factors, such as the enthalpy of formation
of the hydrogen bond, the length of the N—H bond, and the distance
between the nitrogen and oxygen atoms. Whereas the largest changes
in the infrared spectrum will be observed for the stretching vibration,
smaller changes can also be seen in the wavenumbers corresponding
to the bending of these bonds. Here hydrogen bond formation increases
the resistance to bending, and accordingly the bending vibrations
shift to higher wavenumbers. These effects apply also to other types of
hydrogen bonds, such as

$$O\!-\!H + O\!=\!C \rightarrow O\!-\!H \cdots O\!=\!C\diagup$$
$$O\!-\!H + O\diagup \rightarrow O\!-\!H \cdots O\diagup$$

In addition to the effects on the wavenumber, hydrogen bond formation
also leads to alterations in band shapes and intensities.

The interaction of the aromatic bases occurring in nucleic acids has
been studied using infrared spectroscopy. A particularly noteworthy
example is the investigation of the infrared spectra of 9-ethyladenine
and 1-cyclohexyluracil in deuterated chloroform.[*] The spectrum in

---

[*]Hamlin, R. M., Jr., R. C. Lord, and A. Rich, *Science*, **148**, 1734 (1965).

the region of the N—H stretching vibrations was found to be sensitive to the molar ratio of the two compounds. The results could be successfully interpreted in terms of the formation of a 1:1 complex between the adenine and uracil derivative which involved the formation of hydrogen bonds between the bases. This result is exactly what would be expected from the Watson-Crick base pairing scheme that predicts the formation of hydrogen-bonded pairs between adenine and thymine (in DNA) and between adenine and uracil (in RNA).

Before leaving infrared and turning our attention to ultraviolet-visible spectroscopy, we should point out two disadvantages of infrared spectroscopy for the many applications which might be attempted by life scientists. The first relates to the desirability of determining the absorption spectrum of a solute when it is in a transparent solvent. The solvent of choice for many biological applications would be water, which, unfortunately, absorbs intensely throughout much of the infrared. On the other hand, water is transparent throughout the entire visible region of the spectrum and in the ultraviolet down to ~180 nm.

The second disadvantage is that quantitative spectral data in the infrared can be obtained but only with great effort. This problem arises because of the difficulty in finding a completely transparent solvent. In order to reduce the absorption by the solvent, extremely short light paths (less than 1 mm) are customarily used. To eliminate solvent completely, one technique is to disperse the sample in a thin disk of a suitable inorganic salt, usually potassium bromide. The sample is mixed with the powdered crystalline salt which is then pressed into a transparent disk approximately 1/2 mm thick and 10 mm in diameter. Finally the disk is mounted in a suitable holder and placed in the infrared beam for the absorption experiment. Though widely used in infrared absorption spectroscopy, undesirable scattering of radiation may occur if the particle size is too large, and some thermal damage to the sample may result from the compression process. Under any circumstances quantitative infrared absorption measurements should be approached with great care and excellent technique, in part because of the difficulties of establishing and maintaining a stable base line. The investigator also may be

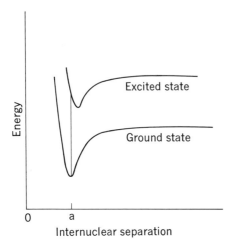

**Figure 9.13** Diagrammatic representation of the potential energy curve for the ground state and an excited state. The vertical line designates the transition from the ground state to the excited state.

faced with the task of accurately determining the length of the light path in the cell. Finally, the short light path requires the utilization of high solute concentrations, so that deviations from the Beer-Lambert Law can be anticipated. These problems are generally less severe for spectroscopy in the visible and ultraviolet regions of the spectrum.

### 9.6 Visible-Ultraviolet Spectroscopy

In the preceding section we examined the absorption of radiation in the infrared region of the spectrum. If more energetic radiation is used, a different type of transition can be observed—one that corresponds to the promotion of an electron from its ground state to an excited state. The energies required for these transitions correspond to radiation in the visible and ultraviolet regions of the spectrum. The spectra are generally recorded as absorbance (also known as optical density) *vs.* wavelength (commonly expressed in nm or Å).

The basis of electronic absorption is presented schematically for a diatomic molecule in Fig. 9.13. The curve labeled "ground state" is similar in shape to the representation in Fig. 1.2 of energy as a function of internuclear separation for a sodium ion and a chloride ion. The most stable arrangement would have the internuclear separation *a*, corresponding to the minimum in the curve for the ground state. There will also be a number of excited electronic states for this molecule. Figure 9.13 includes the curves for only one of these, the excited state of lowest energy. If the molecule is exposed to radiation corresponding to the energy represented by the solid vertical line in Fig. 9.13, absorption of

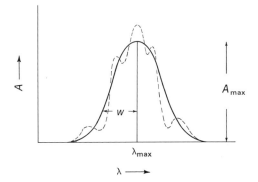

**Figure 9.14** Diagrammatic absorption band for a substance as a vapor $(\cdots)$, showing its vibrational structure, and in an aqueous solution $(—)$.

the radiation will be accompanied by the transition of an electron from the ground state to the excited state. The transition from the ground state to the excited state is drawn as a vertical line, rather than a slanted one, because absorption occurs much more rapidly than the motion of the nuclei and, therefore, occurs at an essentially constant internuclear distance.

The promotion of an electron to an excited electronic state may also lead to an excited vibrational state. If the absorbing molecules are in the vapor state or are being studied at extremely low temperatures, the electronic absorption band will frequently exhibit a number of smaller maxima and minima as illustrated in Fig. 9.14. These maxima reflect the changes in the vibrational states accompanying the electronic excitation, and are frequently called vibrational structure or fine structure. If the absorbing molecule is studied in a solution near room temperature, there is a broadening of the vibrational structure. In polar solvents, such as water, the broadening is so great that the vibrational structure frequently becomes indiscernable, and the electronic absorption band appears as a smooth curve (see Fig. 9.14). The shape of the curve can usually be approximated by a Gaussian distribution curve, expressed mathematically as

$$A_\lambda = A_{max}e^{-[(\lambda - \lambda_{max})/w]^2}$$

In this equation $A_\lambda$ is the absorbance at wavelength $\lambda$, $A_{max}$ is the absorbance at the maximum, which occurs at $\lambda_{max}$, and $w$ is half the width of the band where $A = A_{max}/e$, with e being the base of the natural logarithms. The Gaussian approximation requires the specification of three parameters, $A_{max}$, $\lambda_{max}$, and $w$, in order to quantitatively describe one absorption band.

A quantity of considerable theoretical interest in discussion of the absorption of radiation is the transition dipole moment, designated $\mu$. If the Gaussian distribution curve is a satisfactory approximation for a particular absorption band, the transition dipole moment is related to the Gaussian parameters by Eq. 9.11,

$$\mu^2 = 1.63 \times 10^{-38} \frac{\epsilon_{max} w}{\lambda_{max}} \qquad (9.11)$$

where $\epsilon_{max}$ is the extinction coefficient at $\lambda_{max}$. The transition dipole moment is a vector, and $\mu^2$ in Eq. 9.11 is the square of the length of this vector.

The transition dipole moment is defined by an integral over all space of a function containing (a) the change in charge distribution produced in the molecule as a consequence of its interaction with the radiation, and (b) the properties of the electrons in the ground and excited states, as reflected by their wave functions. Wave functions are not treated in this text, but we note briefly that the probability of finding an electron at a particular location in space is given by the square of the value of its wave function at that location. For the transition dipole moment to be nonzero, it is necessary that the interaction of the radiation with the molecule lead to some alteration in charge distribution. A change in charge distribution is not in itself a sufficient condition for a nonzero transition dipole moment, however, because of the further dependence of the transition dipole moment on the wave functions.

The probability that a given molecule will absorb light depends upon the geometrical relationship of its transition dipole moment and the electric vector of the radiation. Therefore, the absorption of plane polarized light by a partially oriented sample depends upon the geometric relationship between the direction of orientation and the electric vector of the light. This phenomenon is known as ultraviolet dichroism, and is similar in its applications to infrared dichroism. In more common applications – ones utilizing unpolarized radiation and a sample of randomly oriented absorbing molecules dispersed in a transparent solvent – dichroism is not observed.

The discussion of ultraviolet and visible spectroscopy thus far has concerned only a single absorption band. Most molecules will exhibit more than one absorption band, corresponding to more than one electronic transition. Consider, for example, the absorption spectrum of camphorsulfonic acid. In water there is a weak absorption band near 285 nm and a stronger absorption band near 200 nm. Similar absorption

bands are observed in acetone as well, making it reasonable to assume that the electrons taking part in the excitation belong to the carbonyl group. Thus the carbonyl group is the chromophore* in both acetone and camphorsulfonic acid. This chromophore gives rise to two absorption bands between 200 and 300 nm. The absorption at higher wavelength is caused by the promotion of a non-bonding electron to an anti-bonding $\pi$ orbital (abbreviated n-$\pi$* transition), and the more intense absorption at lower wavelength is caused by the promotion of a $\pi$ electron to an anti-bonding $\pi$ orbital (abbreviated $\pi$-$\pi$* transition). The diagrammatic representation of the electronic energy levels in the absorption of light by camphorsulfonic acid is shown below. The great majority of the absorption bands of interest to life scientists are of these two

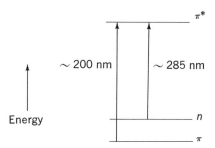

types. One important distinguishing feature of these transitions is the intensity of the absorption. The extinction coefficient for an n-$\pi$* transition is typically on the order of magnitude of 10, whereas for a $\pi$-$\pi$* transition the extinction coefficient is generally of magnitude $10^3$ to $10^4$.

We might inquire why camphorsulfonic acid does not exhibit many more absorption bands, corresponding, for example, to the promotion of the many $\sigma$-electrons in the molecule to excited states. These transitions do exist, but the energy required is substantially greater than that required in the n-$\pi$* and $\pi$-$\pi$* transitions of the carbonyl group. The excitation of the $\sigma$-electrons occurs below 180 nm, which is an extremely difficult region of the spectrum to examine because nearly everything absorbs ultraviolet light of such high energy. Nearly all solvents, including water, absorb light at these low wavelengths. In addition, the oxygen gas in the atmosphere absorbs at wavelengths below about 190 nm. Spectra can still be obtained at these low wavelengths by removing the oxygen gas and by using no solvent, *i.e.*, examining the sample as a vapor or a film. We will not concern ourselves with these studies, but will instead examine those applications that can be carried out in solution without the removal of oxygen gas from the spectrophotometer.

The contribution of a particular chromophore to the absorption spectrum of a molecule can be dramatically altered by conjugation with other chromophores. An isolated carbon-carbon double bond, such as

---

*The word "chromophore" refers to the localized part of a molecule which gives rise to an absorption band.

that found in ethylene, exhibits a $\pi$-$\pi^*$ transition at about 180 nm. No absorption occurs at higher wavelengths. If a molecule contains multiple non-conjugated carbon-carbon double bonds, such as $CH_2{=}CH{-}CH_2{-}CH_2{-}CH{=}CH_2{-}CH_2{-}CH{=}CH{-}CH_3$, each double bond makes approximately the same contribution to the absorption spectrum as does the double bond in ethylene. Conjugation of double bonds can lead to drastically different results, as illustrated by $\beta$-carotene, a precursor of vitamin A. The only double bonds in $\beta$-carotene are carbon-carbon double bonds, but eleven of these double bonds form a conjugated system. Dilute solutions of $\beta$-carotene are yellow, showing that light is being absorbed in the visible region of the spectrum. Absorption bands can be observed at 450–500 nm. What is seen here is a general effect in which conjugation of double bonds results in the appearance of absorption at longer wavelengths.

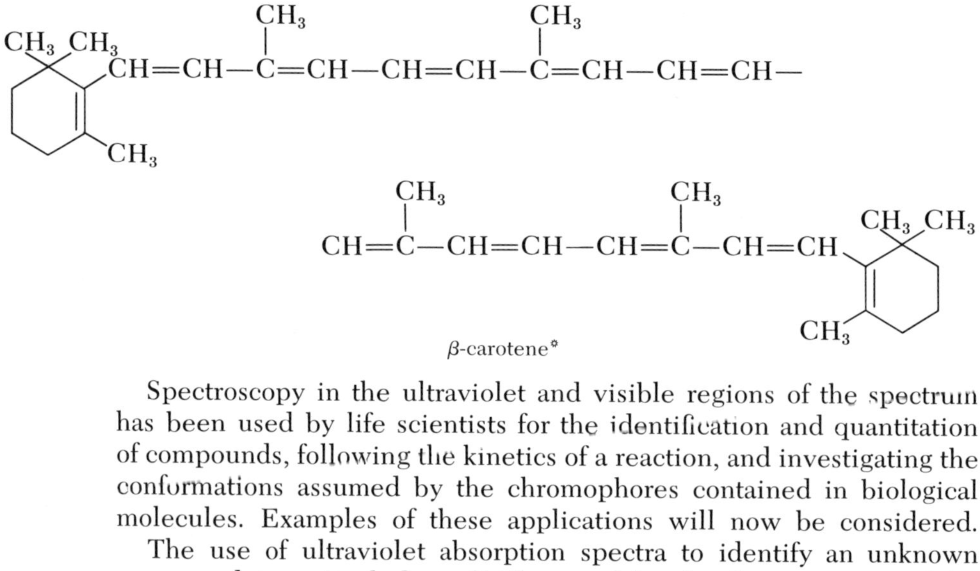

$\beta$-carotene[*]

Spectroscopy in the ultraviolet and visible regions of the spectrum has been used by life scientists for the identification and quantitation of compounds, following the kinetics of a reaction, and investigating the conformations assumed by the chromophores contained in biological molecules. Examples of these applications will now be considered.

The use of ultraviolet absorption spectra to identify an unknown compound is particularly well illustrated by the determination of the nucleotide sequence of a transfer ribonucleic acid (tRNA).[†] Individual ribonucleotides are obtained in the course of these studies, and it is necessary to identify which purine or pyrimidine derivative is contained in the product. Usually it is adenine, guanine, cytosine, or uracil, although derivatives of these bases are also found.[‡] The ultraviolet absorption spectra of the 3'-ribonucleotides of these four bases are shown in Fig. 9.15. The presence of conjugated double bonds in these

---

[*]The chain for $\beta$-carotene is, of course, continuous; it has been broken here in order to fit on the page.

[†]Holley, R. W., *Scientific American*, Feb. 1966, p. 30.

[‡]Hall, R. H., *The Modified Nucleotides in Nucleic Acids*, Columbia University Press, New York, 1971, Chapt. 4.

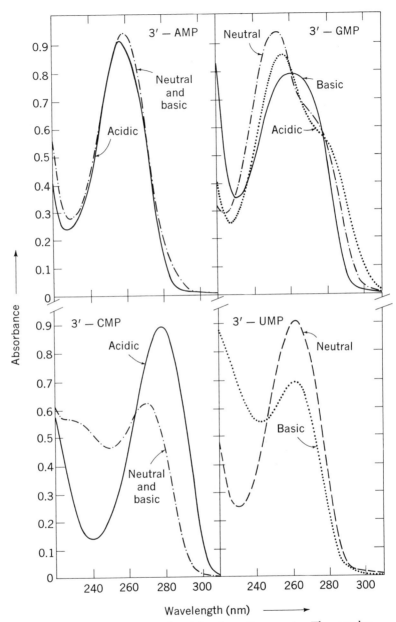

**Figure 9.15** Representative ultraviolet absorption spectra. The samples are 3'-mononucleotides obtained by the enzymatic degradation of tRNA. The identifications are based upon the location of the absorption band at longest wavelength in neutral aqueous solution and the effects of extreme pH upon the absorption spectra. The spectra were obtained using a Cary 14 Spectrophotometer (Courtesy of Dr. S. Chang).

molecules gives rise to absorption bands between 190 and 280 nm. The absorption spectra are sufficiently different to enable the base to be identified.

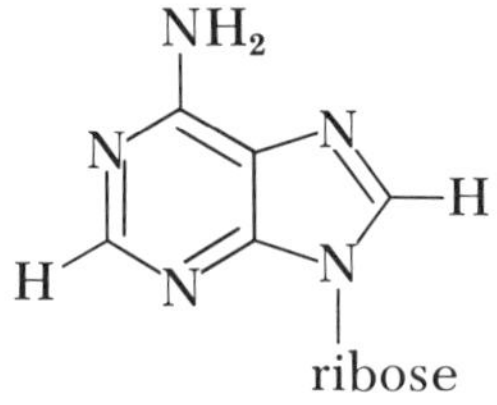

Adenosine

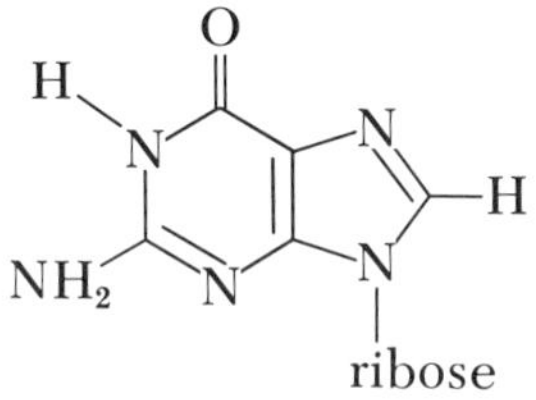

Guanosine

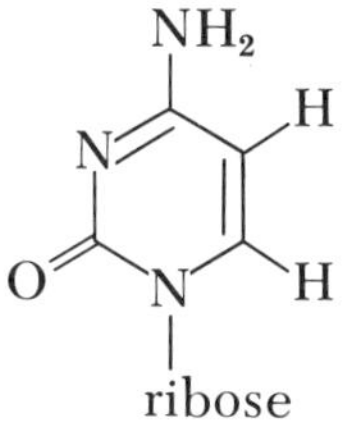

Cytidine

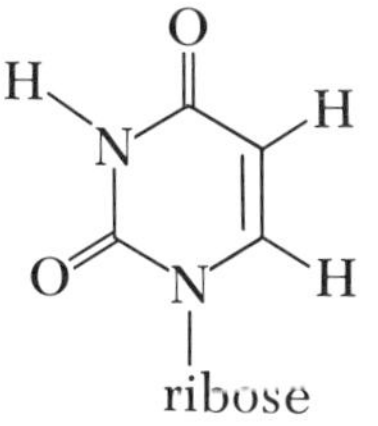

Uridine

For example, the absorption maximum at highest wavelength in an aqueous solution of neutral pH occurs at 255 nm with GMP, 271 nm with CMP, 262 nm with UMP, and 259 nm with AMP.* By simply locating the first absorption maximum, a tentative identification of GMP or CMP is possible, although it would be more difficult to distinguish between AMP and UMP. Any ambiguity can be resolved by attaining additional absorption spectra at extreme pH. The protons attached to the ring nitrogen atoms in uracil and guanine are substantially dissociated at high pH, leading to alternations in the $\pi$-electron systems in these bases. Consequently the absorption spectra of UMP and GMP, but not CMP and AMP, are markedly different at pH 7 and pH 12, as is shown in Fig. 9.15. Measurement of absorption spectra at pH 7 and pH 12 therefore provides an unambiguous distinction between AMP, CMP, GMP, and UMP. In order also to identify the modifications of adenine, cytosine, guanine, and uracil in t-RNA, the identification procedure includes an additional adsorption spectrum obtained at low pH,† and the results are usually supplemented by another technique, such as paper chromatography.‡

---

*Ts'o, P. O. P., in *Fine Structure of Proteins and Nucleic Acids*, G. D. Fasman and S. N. Timasheff, Ed., Marcel Dekker, New York, New York, 1970, p. 49.

†Hall, R. H., *The Modified Nucleotides in Nucleic Acids*, Columbia University Press, New York, New York, 1971, Chapt. 2.

‡Chang, S. H., C. Harmon, K. Munninger, and N. Miller, *FEBS Letters*, **11**, 81 (1970).

In many applications the identity of the absorbing substance is known, and it is desired to determine its concentration in a particular sample. If the absorption of radiation of a particular wavelength by the substance follows the Beer-Lambert Law, and if we may assume that no other compounds are contributing to the absorption at the wavelength used, the absorbance will be directly proportional to the concentration. The proportionality constant can be determined by measuring the absorbance of a sample of known concentration. In the event that there are deviations from the Beer-Lambert Law, it will be necessary to construct a standard curve by measuring the absorbance of several samples of known concentration, and by spanning the concentration range to be encountered in the analytical procedure.

There are numerous assay procedures based upon the absorption of light in the visible or ultraviolet region of the spectrum. For example, the reaction of ninhydrin with an amino acid yields a product that is purple for most amino acids found in proteins. Exceptions are yellow products yielded by proline and hydroxyproline. Quantitative determination of the absorbance at 570 nm (for purple products) and at 440 nm (for yellow products) is the basis of a widely utilized procedure for amino acid analysis.*

The life scientist frequently has to work with materials which can be obtained only in minute quantities. It may be inconvenient, or impossible, to weigh the total amount at hand accurately on available balances. If the substance possesses a strongly absorbing chromophore in the ultraviolet or visible region of the spectrum, the concentration of even very dilute solutions can be determined accurately without the loss or degradation of valuable material once the extinction coefficient is known. Consider, for example, the problem of determining the exact concentration of a solution consisting of roughly 30 $\mu$g of yeast tRNA

---

*Spackman, D. H., W. H. Stein, and S. Moore, *Anal. Chem.*, **30**, 1190 (1958).

dissolved in 3 ml. of 0.15 M potassium chloride plus 0.01 M potassium cacodylate buffer, pH 7.0, with small amounts of magnesium chloride and EDTA. The extinction coefficient at 258 nm for yeast tRNA in this solvent has been determined.[*] An absorbance of 0.209 at 258 nm would be obtained with a one cm light path if the concentration were exactly 10 $\mu$g/ml. An absorbance of this size can be measured quickly and accurately using one ml. of solution. Furthermore, the tRNA in the solution used for the measurement is not altered by the measuring process and could be utilized in further studies of yeast tRNA.

Another application of quantitative analysis employing ultraviolet and visible spectroscopy is the differentiation between possible courses of a reaction, some of which involve alterations in a chromophore. A simple example is the titration curve of tyrosine near pH 9-10. Both the amino and phenolic protons titrate in this pH range. Their p$K$'s are so close together that separate inflections are not observed in the titration curve. The titration curve alone does not allow discrimination between the fraction of the tyrosine which titrates in the upper and lower sequence in the following diagram:

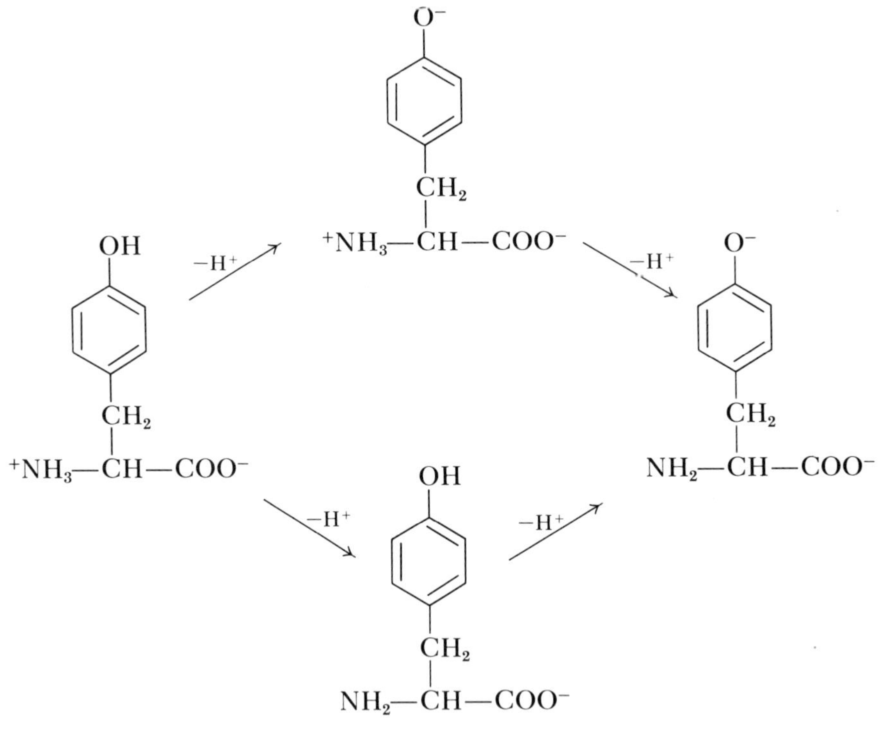

[*] Adams, A., T. Lindahl, and J. R. Fresco, *Proc. Nat'l. Acad. Sci., U.S.*, **57**, 1684 (1967).

the transition dipole moment of the chromophore. It is not possible to conduct measurements on a single molecule, but one can examine a large assembly of identical molecules. This assembly will be oriented if it can be obtained as a single perfect crystal, and the orientation of the molecules in the crystal can be established by x-ray diffraction. The orientation of a transition dipole moment in the molecule can be deduced by studying the absorption of polarized radiation of an appropriate wavelength under different geometrical relationships between the electric vector of the light and the crystal. Application of this procedure to a number of purine and pyrimidine derivatives has shown that the transition dipole moments for the $\pi$-$\pi^*$ transitions are in the plane of the aromatic rings.* The spatial orientation of transition dipole moments for some chromophores can be assigned differently, in a manner based upon theoretical considerations beyond the scope of this text.

The spectral effects expected for different relative orientations of two transition dipole moments for absorption bands with large extinction coefficients has been investigated by Tinoco.† He found that the intensity of the absorption was markedly sensitive to the orientation of the transition dipole moments when they were close together. Of particular interest to life scientists is the case where the transition dipole moments are close together and oriented identically, similar to the relationship between the index and middle finger in the extended hand. This geometry was found by Tinoco to lead to a decreased intensity of absorption for the $\pi$-$\pi^*$ transition occurring at longest wavelengths, and an increase in intensity for the absorption at lower wavelengths. The transition at longest wavelength is said to exhibit hypochromism (a decrease in the intensity of absorption) whereas a transition further in the ultraviolet is predicted to exhibit hyperchromism (an increase in the intensity of absorption).

The intensity of the absorption of light near 260 nm by DNA can be understood in terms of the concepts presented here.‡ The absorption near 260 nm is caused by the lowest energy $\pi$-$\pi^*$ transition in adenine, guanine, cytosine, and thymine. At room temperature this absorption is much less intense ($\sim$40%) than the absorption predicted by assuming that the aromatic rings have the same spectral properties in DNA as they have in mononucleotides. Hypochromism is predicted for this absorption band by Tinoco if the transition dipole moments are stacked. Stacking of the transition dipole moments will occur if the bases themselves are stacked, as in the double-helical structure of DNA. The intensity

*Ts'o, P. O. P., in *Fine Structure of Proteins and Nucleic Acids*, G. D. Fasman and S. N. Timasheff, Ed., Marcel Dekker, New York, New York, 1970, p. 49.

†Tinoco, I., Jr., *J. Chem. Phys.*, **33**, 1332 (1960); **34**, 1087 (1961); *J. Amer. Chem. Soc.*, **82**, 4785 (1961).

‡Many such studies have been made. See, for example, P. Doty, H. Boedtker, J. Fresco, R. Haselkorn, and M. Litt, *Proc. Nat'l. Acad. Sci., U.S.*, **45**, 482 (1959).

of the absorption near 260 nm is essentially unaffected by heating up to a certain critical temperature, the value of which depends upon the DNA under investigation. Upon reaching this critical temperature, the absorption increases by $\sim 40\%$ over a temperature range of only a few degrees. This increase in absorption reflects the thermal disruption of the double helix. In the resulting denatured state the aromatic rings are free to rotate. The correlation between the transition dipole moments of neighboring bases and the consequent hypochromism therefore disappears.

Many small biological molecules, such as DPNH[*] and FAD,[†] also exhibit hypochromism at room temperature, signifying a tendency toward stacking of these aromatic rings as well. Here absorbance increases gradually with increasing temperature over a broad temperature range, indicating a continuously increasing disruption of the tendency of the aromatic rings to stack with increasing temperature. This result is in contrast to the cooperative disruption of the stacking interactions in DNA.

An illustration of the advantage of considering these phenomena in terms of the transition dipole moments, rather than the actual molecular structure, is provided by turning our attention to a system in which the pertinent chromophore is not an aromatic ring. A large number of synthetic polypeptides, as well as parts of several proteins, have been found to exist in a specific conformation known as the $\alpha$-helix.[‡] The $\pi$-$\pi^*$ transition of the peptide unit, located near 200 nm, exhibits marked hypochromism when it exists in an $\alpha$-helix.[†] The $\alpha$-helical geometry has the effect of stacking the transition moments for the $\pi$-$\pi^*$ transition of the peptide unit.[§]

### 9.7 Spectroscopy Utilizing Flame Sources

Atomic vapors that exist at high flame temperatures can serve as a source of characteristic emission lines or as an absorbing medium for these lines. When emission lines are used in a photometric experiment, the excited atoms are serving as a primary light source and the technique is known as *flame photometry*. Correspondingly, if the vapors absorb light from another source and the instrument records the frequencies and extent of absorption, the technique is known as *atomic absorption spectrometry*. In both techniques the sample under investigation is

---

[*] Siegel, J. M., G. A. Montgomery and R. M. Bock, *Arch. Biochem. Biophys.*, **82**, 288 (1959).

[†] Whitby, L. G., *Biochem. J.*, **54**, 437 (1953); O. Warburg and W. Christian, *Biochem. Z.*, **296**, 294 (1938).

[‡] Dickerson, R. E., and I. Geis, *The Structure and Action of Proteins*, Harper & Row, New York, New York, 1969, p. 29.

[§] Gratzer, W. B., in *Poly-α-Amino Acids*, G. E. Fasman, Ed., Marcel Dekker, New York, New York, 1967, p. 177.

atomized in a flame at the highest feasible flame temperatures in order to make the atomization process as efficient as possible.

The former method is particularly useful for the quantitative determination of elements in the first two groups of the Periodic Table: such elements as sodium, potassium, calcium, barium, etc. The technique is of special interest to life scientists in the fields of medicine and plant sciences.* Atomic absorption spectroscopy is a particularly appropriate method for the determination of trace metals in liquids. When solid samples are used, they first must be dissolved and the solution analyzed. The types of samples that can be effectively analyzed quantitatively by the technique include blood, urine, water pollutants, animal tissue, plant materials, etc.* It must already have become apparent to the reader that quantitative application of this latter technique will be hindered by the fact that atoms of the absorbing material may also be emitting simultaneously at the same frequencies. Atomic absorption spectrophotometers are invariably equipped with a modulating system which serves to remove the emission signal from the transmitted light.

Those who are interested in further reading and application of these techniques should refer to the appropriate *Suggested Additional Readings* or to any current textbook in instrumental analysis.

## Suggested Additional Reading

### General

Alpert, N. L., W. E. Keiser, and H. A. Szymanski, *IR Theory and Practice of Infrared Spectroscopy*, Plenum Press, New York, 1970, 2nd Ed.

Dyer, John R., *Applications of Absorption Spectroscopy of Organic Compounds*, Prentice-Hall, Inc., Englewood Cliffs, New Jersey, 1965.

Fasman, G. D., Ed., *Poly-α-amino Acids*, Marcel Dekker, New York, New York, 1967.

Feinberg, Gerald, "Light"; Weisskopf, V. F., "How Light Interacts with Matter"; Connes, Pierre, "How Light is Analyzed"; Oster, Gerald, "The Chemical Effects of Light"; Hendricks, S. B., "How Light Interacts with Living Matter"; Neisser, Ulric, "The Processes of Vision," all in *Scientific American*, Vol. 219, No. 3, September, 1968. (Neisser available from W. H. Freeman and Company as Offprint No. 519.)

Leach, S. J., Ed., *Physical Principles and Techniques of Protein Chemistry*, Academic Press, New York, New York, Part A, 1969, Part B, 1970.

Shriner, R. L., R. C. Fuson, and D. Y. Curtin, *The Systematic Identification of Organic Compounds*, Wiley, New York, New York, 1964, 4th Ed.

Silverstein, R. M., and G. C. Bassler, *Spectrometric Identification of Organic Compounds*, Wiley, New York, New York, 1967, 2nd Ed.

Urry, D. W., Ed., *Spectroscopic Approaches to Biomolecular Conformation*, American Medical Association, Chicago, Illinois, 1970.

---

*Butler, L. R., *Method. Phys. Anal.*, **7**, 3 (1971); Leaton, J. R., *J. Ass. Offic. Anal. Chem.*, **53**, 237 (1970); Skinner, J. M., *Proc. Soc. Anal. Chem.*, **6**, 131 (1969); Vallee, B. L., *Clin. Chim. Acta*, **25**, 307 (1969).

### Flame Photometry and Atomic Absorption Spectroscopy

Ewing, Galen W., *Instrumental Methods of Chemical Analysis,* McGraw-Hill Book Co., New York, 1969, 3rd Ed.

Joslyn, M. A., Ed., *Methods in Food Science,* Academic Press, New York, N.Y., 1970.

Mavrodineanu, R., Ed., *Analytical Flame Spectroscopy,* Macmillan, Ltd., London, 1970.

Robinson, James W., *Atomic Absorption Spectroscopy,* Marcel Dekker, Inc., New York, 1966.

Robinson, James W., *Undergraduate Instrumental Analysis,* Marcel Dekker, Inc., New York, 1970.

### Nuclear Magnetic Resonance

Bovey, F. A., *Nuclear Magnetic Resonance Spectroscopy,* Academic Press, New York, 1969.

Bovey, F. A., *High Resolution NMR of Macromolecules,* Academic Press, New York, 1972.

### Photochemistry

McLaren, A. D., and D. Shugar, *Photochemistry of Proteins and Nucleic Acids,* Macmillan, New York, 1964.

Simons, J. P., *Photochemistry and Spectroscopy,* Wiley, New York, 1971.

### X-Ray Diffraction

Glusker, J. P., and K. N. Trueblood, *Crystal Structure Analysis: A Primer,* Oxford University Press, New York, 1972.

## Problems

**9.1** A spectrophotometer is equipped with a meter that gives a reading directly proportional to the intensity of the radiation reaching the detector. The meter reading is zero when the light source is off. When the light source is on, the meter reading is 72 with pure solvent in the light path, and 48 with a solution in the light path. What are the transmittance and absorbance of the solution?

*Answer.* 0.667, 0.176.

**9.2** An aqueous solution containing 0.94 g oxygenated horse myoglobin/100 ml gave a transmittance of 0.847 at 580 nm using a cell with 10.00 cm light path length. What is the extinction coefficient at this wavelength for oxygenated horse myoglobin? The molecular weight of horse myoglobin is 18,800 g/mole.

*Answer.* 14.4 cm²/millimole.

**9.3** The extinction coefficient for deoxygenated human hemoglobin in aqueous solution of pH 7 has been reported to be 532 cm²/millimole at 430 nm (E. Antonini and M. Brunori, *Hemoglobin and Myoglobin in Their Reactions with Ligands*, American Elsevier, New York, New York, 1971, p. 19), based on a molecular weight of 64,000 g/mole. A solution of deoxygenated hemoglobin was found to have an absorbance of 0.108 at 430 nm in a cell with a light path of 1.000 cm. What is the concentration of the deoxygenated hemoglobin in the solution?

*Answer.* 13.1 mg/ml.

**9.4** Three spectrophotometer cells, one with a light path length of 2.000 cm and two with light paths of 1.000 cm, are filled with an aqueous solution that is $10^{-4}$ molar in DPNH. What will be the relationship of the absorbances recorded at 338 nm (1) when the cell with a 2.000 cm light path is in the spectrophotometer, and (2) when both cells of 1.000 cm light path are in the spectrophotometer and arranged so that the light beam must transverse both cells before reaching the detector? Justify your answer.

**9.5** A spectrophotometer was suspected to allow significant stray light to pass to the detector when the monochromator was set at 260 nm. In order to investigate this possibility, an aqueous solution of AMP was prepared. The following absorbances were obtained, using water as a reference, for various dilutions of the stock solution of AMP.

| Dilution of stock solution | Absorbance at 260 nm |
|---|---|
| 1 to 10 | 0.186 |
| 2 to 10 | 0.371 |
| 3 to 10 | 0.553 |
| 5 to 10 | 0.903 |
| undiluted | 1.57 |

Do the results indicate the presence of stray light? Why?

**9.6** Design an assay procedure to permit the determination of the concentration of ethyl alcohol in a dilute aqueous solution by measuring an absorbance in the ultraviolet region of the spectrum.

**9.7** An aromatic molecule was observed to give a colorless solution at neutral pH and a yellow solution in 0.1 M sodium hydroxide. In order to determine the pK of the ionization involved, the absorbances at 420 nm were determined for several solutions in a

1.000 cm cell at 25°. All solutions contained the same total concentration of the aromatic molecule, but differed in pH. From the results obtained, evaluate the pK for the ionization.

| pH | A |
|-----|-------|
| 7.0 | 0.003 |
| 8.0 | 0.029 |
| 8.5 | 0.085 |
| 8.8 | 0.149 |
| 9.1 | 0.240 |
| 9.4 | 0.346 |
| 9.7 | 0.443 |
| 10.0 | 0.517 |
| 10.5 | 0.582 |
| 11.0 | 0.608 |
| 11.5 | 0.616 |

*Answer.* 9.3.

**9.8** What is the probability of observing a strong absorption band for which the ground state and excited state have permanent dipole moments whose size and location in the chromophore are identical? Why?

**9.9** Describe a spectrophotometric means for distinguishing between the following pairs of molecules:

(a) N-methylacetamide and N,N-dimethylacetamide

(b)

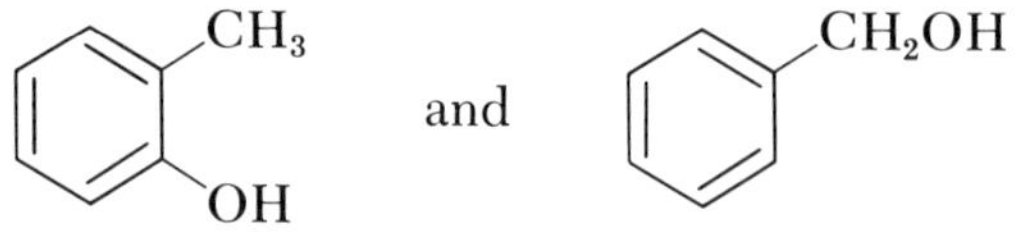

and

(c) L-lactic acid and D-lactic acid

(d) The dominant tautomer in the lactam-lactim tautomerism of uracil, *i.e.*, the dominant form in the following equation

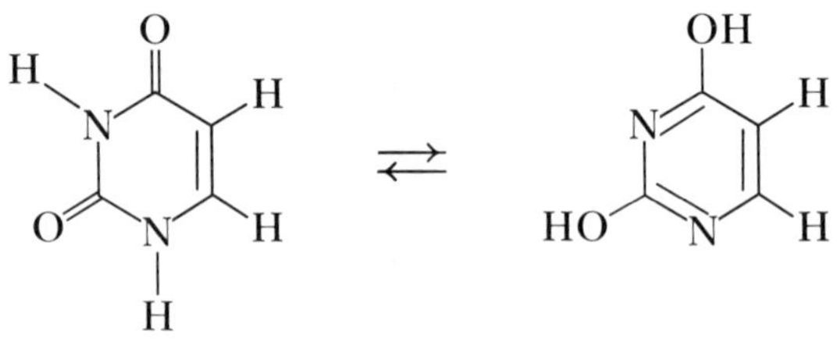

Hint. Consider the determination of the spectral properties of a suitable derivative of uracil.

**9.10** By a spectrophotometric difference titration, one can determine the number of phenolate anions and the number of —S⁻ groups of a protein molecule. The denatured protein is placed both on the reference and on the sample side of a double-beam cell holder. The pH of the reference side is 7.5 and the pH of the sample side is 11.00. At a wavelength of 295 m$\mu$, all of the differential absorbance will be due to tyrosine (phenolate form). At a wavelength of 243 m$\mu$, both titrated tyrosine and cysteine will contribute to the differential absorbance. In the tabulation below are given the molar difference extinction coefficients for tyrosine absorbance at both wavelengths and for cysteine at the lower wavelength (anionic forms absorb; protonated forms do not interfere). The protein solution is $5.5 \times 10^{-6}$ $M$. The difference absorbance at 295 m$\mu$ is 0.013. The difference absorbance at 243 m$\mu$ is 0.185. Calculate the number of tyrosine residues titrated. Assume that the cell length is 1.00 cm.

| *Substance* | *Molar Difference* 295 m$\mu$ | *Extinction Coefficients* 243 m$\mu$ |
|---|---|---|
| Tyrosine (phenolate) | 2,300 | 11,020 |
| Cysteine (sulfide) | —— | 2,700 |

*Answer.* 8 cysteines, 1 tyrosine.

**9.11** The ground state and lowest excited electronic state for a particular molecule in the vapor state are separated by 100 kcal/mole. Upon solution in water, the energy of the ground state decreases by 1 kcal/mole, and the energy of the lowest excited electronic state decreases by 3 kcal/mole. At what wavelength will this electronic transition be observed in the vapor state? in aqueous solution?

*Answer.* 286 nm, 292 nm.

**9.12** The most intense absorption in the spectral region where O—H stretching occurs is reported to be at ~3200 cm⁻¹ in ice, at 3567 and 3756 cm⁻¹ in water vapor, and at ~3490 cm⁻¹ in liquid water (D. Eisenberg and W. Kauzmann, *The Structure and Properties of Water*, The Clarendon Press, Oxford, 1969, p. 228). The location of the maximal absorbance by liquid water shifts to higher wavenumber as the temperature increases. Suggest a mechanism to account qualitatively for the change in the spectral location of the O—H stretching vibration with physical state and temperature, and explain your reasoning.

# APPENDIX A

# MATHEMATICS

## Introduction

The purpose of this appendix is to give the student an intuitive introduction to the mathematical concepts that are pertinent to the material presented in the text. A rigorous mathematical presentation is not the goal, nor is there any attempt at a complete review of the calculus. Proofs are omitted along with many additional topics, which the interested reader may find in any good calculus book. The examples are included to help the student understand the concepts that are presented, and should be studied with this aim in mind.

## Logarithms

**Definition 1** Let $a$ and $b$ be positive real numbers ($b \neq 1$). The exponent, $p$, or the power to which $b$ must be raised to obtain $a$ is called the "logarithm to the base $b$ of the number $a$."
Notation: $\log_b a = p$ means that $b^p = a$.

*Example A.1* Since $2^3 = 8$, $\log_2 8 = 3$.

*Example A.2* Since $8^{2/3} = 4$, $\log_8 4 = 2/3$.

*Example A.3* Since $10^{-2} = .01$, $\log_{10} .01 = -2$.

*Example A.4* If $\log_b 9 = 2$, find $b$.

*Solution:*

It follows from Definition 1 that $b^2 = 9$; therefore $b = 3$.

(*Note:* $b = -3$ is also a solution of the equation $b^2 = 9$, but the base must be positive by definition.)

*Example A.5*  If $\log_{10} x = -1/2$, find $x$.

*Solution:*

$$x = 10^{-1/2} = \frac{1}{\sqrt{10}}.$$

In Definition 1, the restriction that the base of a logarithm be different from 1 is necessary because any power of 1 is 1. Also, since any power of a positive number is positive, the logarithm of a negative number is undefined. The logarithm of zero is also undefined.

There are two systems of logarithms ordinarily used in scientific work — "natural logarithms" and "common logarithms." Common logarithms have the number 10 as base; natural logarithms have the number $e$ as base. (The number $e$ is an irrational number, approximately equal to 2.718, which arises in the mathematical description of natural phenomena. A definition of $e$ will be given in the section on limits.) To avoid writing the subscript to indicate the base we adopt the convention that $\ln x$ and $\log x$ denote $\log_e x$ and $\log_{10} x$, respectively.

## Properties of logarithms

1. $\log_b x + \log_b y = \log_b (xy)$.
2. $\log_b x - \log_b y = \log_b (x/y)$.
3. $k \log_b x = \log_b (x^k)$.
4. $b^{\log_b x} = x$.
5. $\log_a x = (\log_a b)\log_b x$.

Since $\ln 10 \approx 2.3026$ (read "the natural logarithm of 10 is approximately equal to 2.3026"), we have the following useful special case of (5): $\ln x \approx 2.3026 \log x$.

*Example A.6*  If $\log x = 3$, find $\log \sqrt{x}$.

*Solution:*

$$\log \sqrt{x} = \log x^{1/2} = 1/2 \log x = (1/2)(3) = 1.5.$$

*Example A.7*  If $G = \log (1 - x^2)$ and $H = \log (1 + x)$ show that

$$G - H = \log (1 - x).$$

*Solution:*

$$G - H = \log (1 - x^2) - \log (1 + x)$$

$$= \log \left(\frac{1 - x^2}{1 + x}\right) = \log (1 - x).$$

***Example A.8*** If $Z_1 = -2 \log x$, $Z_2 = 3 \log y$, and $Z_3 = 1/2 \log (x + y)$, express $Z_1 + Z_2 + Z_3$ as a single logarithm.

*Solution:*

$$Z_1 = -2 \log x = -\log x^2$$
$$Z_2 = 3 \log y = \log y^3$$
$$Z_3 = 1/2 \log (x + y) = \log (x + y)^{1/2} = \log \sqrt{x + y}$$

$$Z_1 + Z_2 + Z_3 = \log \frac{y^3 \sqrt{x + y}}{x^2}.$$

***Example A.9*** If $1 + x = \log_b (1 - y)$, express $y$ in terms of $x$.

*Solution:*

By the definition of a logarithm, $(1 + x)$ is the exponent that must be applied to $b$ to give $(1 - y)$. Therefore $b^{(1+x)} = 1 - y$. This equation yields

$$y = 1 - b^{(1+x)}.$$

***Example A.10*** Show that if $ax = e^{y/2}$, then $y = \ln (a^2 x^2)$.

*Solution:*

Taking the natural logarithm of both sides of the equation $ax = e^{y/2}$, we obtain

(1)   $\ln ax = \ln e^{y/2}$.

Since $\ln e^{y/2} = \log_e e^{y/2} = y/2$, equation (1) becomes

$$\ln ax = y/2.$$

Thus,

$$y = 2 \ln ax = \ln (ax)^2 = \ln (a^2 x^2).$$

Examples A.1–A.5 involved numbers whose logarithms could easily be obtained by direct reference to Definition 1. Generally, the numerical value of the logarithm of a number can be expressed by a finite decimal only approximately, and tables of logarithms are compilations of such decimal approximations. (The tables themselves are constructed by various numerical methods, such as expansion of the logarithm into an infinite series.) For instance, a four-place table of common logarithms would give the approximate value .3010 for log 2.

Any number expressed as a decimal may be written in "standard" form as the product of an integral power of 10 and a number between 1 and 10. For example, $.002 = 2 \times 10^{-3}$, $352 = 3.52 \times 10^2$, $4.32 = 4.32 \times 10^0$. Because of property (1) of logarithms (see page 470),

$$\log .002 = \log 2 + \log 10^{-3} = \log 2 - 3,$$
$$\log 352 = \log 3.52 + \log 10^2 = \log 3.52 + 2,$$
$$\log 4.32 = \log 4.32 + \log 10^0 = \log 4.32 + 0.$$

These remarks are illustrations of the general statement that every common logarithm can be expressed as the sum of an integer and the logarithm of a number between 1 and 10. The latter logarithm is itself a number between 0 and 1 and is thus approximated by a positive decimal fraction. This decimal fraction is called a *mantissa*, and tables of common logarithms are tables of mantissas. The exponent of 10 that occurs in the standard form of the number is called the *characteristic*. Thus, the characteristic of .002 is −3, the characteristic of 352 is 2, and that of 4.32 is 0.

In computing with logarithms it is customary to write negative characteristics in the so-called $9-10$ form. In this form, $\log .002 = 7.3010 - 10$. (Some authors use $\bar{3}.3010$, where the bar over the 3 means that the characteristic is −3 and the decimal part is +.3010.) Both notations mean that $\log .002 = .3010 - 3 = -2.6990$.

**Example A.11**  Find $\log 352$.

*Solution:*

$$352 = 3.52 \times 10^2.$$

$$\log 352 = \log 3.52 + \log 10^2.$$

From the tables, $\log 3.52 = .5464$ and, by definition, $\log 10^2 = 2$. Thus,

$$\log 352 = 2 + .5464 = 2.5465.$$

**Example A.12**  Find $\log .002$.

*Solution:*

$$\log (.002) = \log (2 \times 10^{-3}) = \log 2 + \log 10^{-3}$$

$$= .3010 + (-3) = -2.6990.$$

**Example A.13**  The pH of a solution is defined as the negative of the common logarithm of $[H^+]$, where $[H^+]$ is the hydrogen ion activity (approximately equal to the hydrogen ion concentration in gram ions per liter), *i.e.*, $pH = -\log [H^+]$. If $[H^+]$ is .000243, find pH.

*Solution:*

$$pH = -\log (.000243) = -\log (2.43 \times 10^{-4})$$

$$= -[.386 - 4] = 3.614.$$

To find a number when its logarithm is known, a similar approach may be used.

**Example A.14**  If $pH = 2.602$, find $[H^+]$.

*Solution:*

$$-\log [H^+] = 2.602.$$

$$\log [H^+] = -2.602 = .398 - 3.$$

We find from the tables that $.398 = \log 2.5$, so that

$$[H^+] = 2.5 \times 10^{-3} = .0025.$$

Before going to the tables we expressed $-2.602$ as the sum of the positive decimal fraction $.398$ and the integer $-3$, since the tables contain only positive decimal fractions.

***Example A.15*** To obtain the pH on a certain pH meter, the voltage, $\mathscr{E}$, is read and the formula $\mathrm{pH} = \dfrac{\mathscr{E} - .336}{.059}$ yields the pH value. If $\mathscr{E} = 0.525$, find $[H^+]$.

*Solution:*

$$\mathrm{pH} = \frac{.525 - .336}{.059} = 3.20.$$

$$-\log [H^+] = 3.20.$$

$$\log [H^+] = -3.20 = .80 - 4.$$

$$[H^+] = 6.3 \times 10^{-4}.$$

***Example A.16*** For the meter in the example above, what value of $\mathscr{E}$ would correspond to a hydrogen ion concentration of $2.3 \times 10^{-7}$?

*Solution:*

$$\mathrm{pH} = -\log [H^+] = -\log (2.3 \times 10^{-7})$$

$$= -(.3617 - 7)$$

$$= -(-6.6383)$$

$$= 6.638$$

Substituting into the formula for pH given in Example A.15, we get

$$\frac{\mathscr{E} - .336}{.059} = 6.638$$

$$\mathscr{E} = .059(6.638) + .336$$

$$= .728.$$

**Logarithmic graphs** Expressing the hydrogen ion concentration in terms of logarithms (pH) furnishes an example of how logarithms can be used in simplifying problems that arise in the graphical representation of relationships between physical quantities. A graph in which $[H^+]$ is plotted as one of the variables would be impractical to scale for a likely range of values such as $10^{-4}$ to $10^4$ for $[H^+]$. However, for these values of $[H^+]$, $-\log [H^+]$ the scale varies only from $-4$ to $+4$.

In representing physical quantities graphically it is frequently desirable to obtain straight line (linear) graphs. This may sometimes be accomplished by plotting the logarithm of one (or both) of the variables.

**Example A.17** Figure A.1 shows the general form of the graph of the equation $y = ab^x$, where $a$ and $b$ are positive constants.

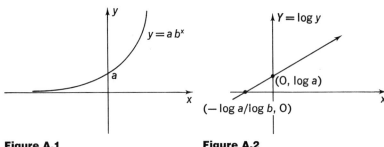

**Figure A.1**                    **Figure A.2**

In Fig. A.2 we have the graph obtained by plotting $Y$ against $x$, where $Y = \log y$. Since $\log y = \log a + x \log b$ is linear in $x$, then $Y = \log a + x \log b$ yields a straight line graph. (This is equivalent to plotting $y$ against $x$ using semilog paper.)

**Example A.18** The graph of the equation $y = kx^a$, where $a$ and $k$ are positive constants, has the general form shown in Fig. A.3.

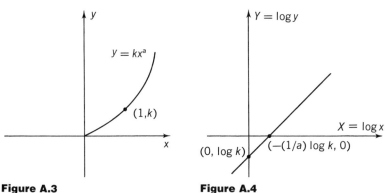

**Figure A.3**                    **Figure A.4**

Taking the logarithm of both sides of the equation, we get $\log y = \log k + a \log x$. If we set $Y = \log y$ and $X = \log x$, the resulting equation is linear in $X$ and $Y$. The graph of $Y$ against $X$, which is a straight line, is shown in Fig. A.4. (This procedure is equivalent to plotting $y$ against $x$ on log-log paper.)

**Functions, Limits, and Continuity**

Suppose that a number may be chosen at will from some given set of real numbers. It is convenient to use a letter, say $x$, as a "placeholder" to designate such a number and to call $x$ an *independent variable*. Suppose further that to each number of the given set there is made to correspond a unique real number from a second set for which we use the letter $y$ as the variable. Then we say that $y$ *is a function of* $x$, and we call $y$ a *dependent variable*. The set of allowable values of $x$ is called the *domain* and the set of corresponding values of $y$ is called the *range* of the function.

For example, if the domain is all real numbers, then equation $y = x^2$ defines $y$ as a function of $x$ with the range consisting of all nonnegative numbers. Here, the equation $y = x^2$ serves to describe the correspondence between the $x$-values and the $y$-values. More generally, such a correspondence may be given by one or more formulas, by a table (such as a table of logarithms), by a graph, or by a verbal statement.

Frequently, the domain (the set of allowable values of the independent variable), rather than being specifically described, is understood from the nature of the physical quantities involved, or is implicit in the rule of correspondence. Thus, $y = \sqrt{4 - x}$ would imply that the domain consists of all real numbers not greater than 4, unless some smaller domain is explicitly given. Similarly, if we are dealing with all circles, then the formula $A = \pi r^2$ defines $A$ as a function of $r$, the domain of this function being all positive real numbers, since the radius $r$ must obey the requirement $r > 0$.

In order to deal with functions in any general fashion, we adopt the custom of denoting the law of correspondence by the equation $y = f(x)$, which we read "$y$ equals $f$ of $x$." We then speak of the function $f$ and regard $f(x)$ as being the $y$-value corresponding to a given $x$ from the domain. We may, of course, use other letters to designate a function, providing we adhere to the preceding conventions. Thus in the above paragraphs, we might have written $f(x) = x^2$, $g(x) = \log x$, and $F(r) = \pi r^2$ to describe the respective laws of correspondence.

Since $f(x)$ stands for the value of the function $f$ corresponding to the value $x$, the notation $f(a)$ must mean the value of $f$ for $x = a$, and $f(3)$ means the value of $f$ for $x = 3$. In other words, the notation $f(*)$ means the value of $f$ for whatever meaningful expression replaces the $*$. Thus, if $f(x) = x^2$, then $f(a) = a^2$, $f(3) = 3^2 = 9$, $f(x + 1) = (x + 1)^2$, etc. Similarly, if $F(r) = \pi r^2$, then $F(4) = 16\pi$ and $F(r + x) = \pi(r + x)^2$.

***Example A.19*** If $f(x) = 2 \log x$ and $g(x) = (1/2) \log x$, find the value of $\dfrac{f(100)}{g(.01)}$.

*Solution:*

$$f(100) = 2 \log 100 = 2(2) = 4.$$

$$g(.01) = (1/2) \log (.01) = (1/2)(-2) = -1.$$

$$\frac{f(100)}{g(.01)} = \frac{4}{-1} = -4.$$

**Example A.20**   If $F(t) = t^2$, find $\dfrac{F(a+h) - F(a)}{h}$    for $h \neq 0$.

*Solution:*

$$F(a+h) = (a+h)^2$$

$$F(a) = a^2$$

$$\frac{F(a+h) - F(a)}{h} = \frac{(a+h)^2 - a^2}{h} = \frac{2ah + h^2}{h} = 2a + h, \qquad h \neq 0.$$

**Example A.21**   Graph the function $f$ where

$$f(x) = 2 \quad \text{for} \quad -2 \le x \le 0,$$

$$f(x) = x \quad \text{for} \quad x > 0.$$

*Solution:*

The required graph appears in Fig. A.5.

*Comments:* (1) This is an example of a function that is described by different formulas on two portions of its domain. (2) The domain of $f$ is $x \ge -2$; $f$ is undefined for $x < -2$. (3) The graph (Fig. A.5) has a "jump" at $x = 0$.

**Example A.22**   Graph the function defined by $f(x) = \log x$.

*Solution:*

The required graph appears in Fig. A.6.

*Comments:* Note that since $f(x)$ is defined for positive values of $x$ only, none of the graph (Fig. A.6) lies to the left of the vertical axis.

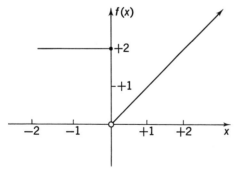

**Figure A.5**

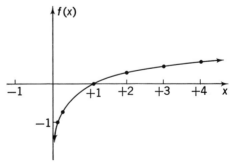

**Figure A.6**

***Example A.23*** Graph the function described by $f(x) = (1+x)^{1/x}$ for $-1 < x < 3$.

*Solution:*

Since division by zero is undefined, $x = 0$ is excluded. However, we are interested in the behavior of the function as we take values of $x$ "close to zero" on both sides of zero. The values in the following table can be verified with very accurate (10-place) logarithms.

| $x$ | 1 | 0.1 | 0.01 | 0.001 | 0.0001 | $-0.1$ | $-0.01$ | $-0.001$ | $-0.0001$ |
|---|---|---|---|---|---|---|---|---|---|
| $f(x)$ | 2 | 2.594 | 2.705 | 2.717 | 2.718 | 2.868 | 2.732 | 2.719 | 2.718 |

These values are impractical to plot but are presented to show that $f(x)$ seems to be close to 2.718 for $x$ close to zero. (Using an electronic computer, we find that $f(x) = 2.718282$, correct to six decimal places, at $x = -10^{-7}$ and at $x = 10^{-7}$.) The graph is shown in Fig. A.7.

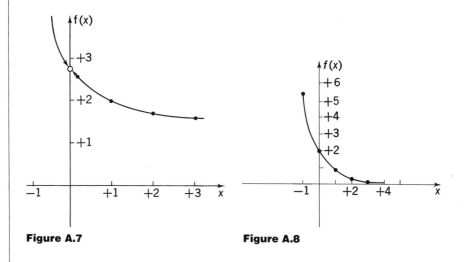

**Figure A.7**          **Figure A.8**

***Example A.24*** Graph the function defined by $f(x) = 2e^{-x}$.

*Solution:*

The graph is shown in Fig. A.8. It is convenient to read values of $e^{-x}$ from tables found in most collections of mathematical tables.

Examples A.21–A.24 are presented to introduce the idea of a limit, the basic concept of the calculus. How does a function, $f$, behave as the independent variable "approaches" a particular value, $x = a$? The answer to this question depends on the values of $f$ at values of $x$ arbitrarily close to $a$. It does *not* depend on $f(a)$; in fact, it does not even require that $f(a)$ be defined. In Example A.23, the table of values indicates that $f(x)$ approaches 2.718 as $x$ approaches zero from either the right or the left. We write this statement: $\lim_{x \to 0} f(x) = 2.718$ (read "the limit of $f$ of $x$ as $x$ approaches zero is 2.718"). More precisely, $\lim_{x \to 0}(1 + x)^{1/x} = e$, where $e$ is the base of the natural logarithms. $e \approx 2.718$.

**Definition 2**  If there is a number $L$ such that $f(x)$ is arbitrarily close to $L$ for all values of $x$ sufficiently close to $a$ (but not equal to $a$), then $f$ is said to have the limit $L$ as $x$ approaches $a$. We write this statement

$$\lim_{x \to a} f(x) = L$$

An intuitive understanding of this definition will be adequate for the purposes of this discussion, since we shall not prove the basic theorems on limits. It is to be emphasized that the limit $L$ (when it exists) is a unique real number. Thus, we may make such statements as

$$\lim_{x \to 3} x^2 = 9, \quad \lim_{x \to 2} (3x^2 + 1) = 13, \quad \lim_{x \to 0} (1 + x)^{1/x} = e.$$

Frequently the limit does not exist, but we can still convey important information about the behavior of $f(x)$ as $x \to a$. For instance (see Example A.22) as $x$ is given values closer and closer to zero, $\log x$ has larger and larger negative values. In fact, we can give $x$ values that are close enough to zero to get negative values of $\log x$ as large as we please. We indicate this fact by writing $\lim_{x \to 0} \log x = -\infty$, and say that $\log x$ decreases without bound as $x$ approaches zero. (It must be emphasized that $\infty$ is a symbol used to describe an abstract idea; it is not a number and should not be treated as a number. Arithmetic or algebraic manipulation of $\infty$ as if it were a definite number can lead to serious errors.)

Example A.24 shows another situation in which the definition of limit needs further extension. The graph indicates that as $x$ gets large, $f(x) = 2e^{-x}$ gets close to zero. Although there is no value of $x$ which will make $e^{-x}$ equal to zero, $e^{-x}$ is as close to zero as we please for large enough values of $x$. We indicate this idea by writing, $\lim_{x \to \infty} 2e^{-x} = 0$, (read "the limit of $2e^{-x}$ as $x$ increases without bound is zero").

Example A.21 illustrates a somewhat different situation. There we see that as $x$ approaches zero from the right, $f(x)$ approaches zero; but as $x$ approaches zero from the left, $f(x)$ approaches 2. We indicate approach to a number $a$ from the right by writing $x \to a^+$, and approach from the left by $x \to a^-$. Thus, in Example A.21, it is correct to write

$$\lim_{x \to 0^+} f(x) = 0 \quad \text{and} \quad \lim_{x \to 0^-} f(x) = 2.$$

The limit itself (unrestricted as to the manner of approach of the independent variable) exists if, and only if, both the one-sided limits exist and are equal. Hence, in Example A.21, $\lim_{x \to 0} f(x)$ does not exist.

Example A.21–A.24 may serve to illustrate another basic concept, that of a continuous function. Notice that in Example A.21 the graph of the function has a break or a jump at $x = 0$, whereas in Examples A.22–A.24 the curves seem to be continuous.

**Definition 3**  A function $f$ is said to be continuous at $x = a$ if $\lim_{x \to a} f(x) = f(a)$. This definition requires three things: (1) that $\lim_{x \to a} f(x)$ exist in the sense of Definition 2; (2) that $f(a)$ be defined; and (3) that the limit in (1) be the same as $f(a)$.

In Example A.21 we saw that $\lim_{x \to 0} f(x)$ does not exist because $f(x)$ does not have the same right and left hand limits as $x$ approaches zero. Hence, although $f(0)$ is defined, $f$ is discontinuous (not continuous) at $x = 0$. On the other hand, $f$ is continuous at all other values in its domain. In Example A.22, since $f(x) = \log x$ is not defined at $x = 0$, $f$ is discontinuous there. In Example A.24 the function is continuous at all values of $x$.

**Definition 4**  A function $f$ is said to be continuous over an interval if it is continuous at every point of the interval. For example, the function defined by $f(x) = \log x$ is continuous over every positive interval but not over any interval that includes zero.

### The Derivative of a Function

A convenient way of indicating an increment (or a change) in a variable $x$ is given by the delta notation $\Delta x$, read "delta $x$." Thus, $a + \Delta x$ denotes the value of $x$ that is obtained by adding the increment $\Delta x$ to the value $a$.

If $y = f(x)$, then $\Delta y$ denotes the increment in the dependent variable $y$ corresponding to the increment $\Delta x$ in the independent variable $x$. Accordingly,

$$\Delta y = f(x + \Delta x) - f(x).$$

The reader should be cautioned that $\Delta x$ may represent a positive or a negative number, so that both increases and decreases in the value of $x$ are to be regarded as increments of $x$.

**Definition 5**  Let $f$ be a function of the independent variable $x$. The derivative of $f$ at a point $a$ in the domain of $f$ is defined to be

$$\lim_{\Delta x \to 0} \frac{f(a + \Delta x) - f(a)}{\Delta x}$$

if this limit exists.

A frequently used notation for the preceding limit (when it exists) is $f'(a)$, read "$f$ prime at $a$." A function is said to be *differentiable* at $a$ if the function has a derivative at $a$. The process of finding the derivative is called *differentiation*. Many of the functions that are important in applications have derivatives at all but exceptional points of their domains. Thus, we may regard

$$f'(x) = \lim_{\Delta x \to 0} \frac{f(x + \Delta x) - f(x)}{\Delta x}$$

as defining another function, the derivative function. For each $x$ for which $f'$ is defined, its value is given by the preceding limit. Additional notations often used for the derivative are: $y'$, $D_x y$, and $dy/dx$, where $y = f(x)$. The following example illustrates the use of the definition to calculate the derivative.

**Example A.25** Find $f'(x)$ if $f(x) = 3x^2$.

*Solution:*
    Since

$$f(x) = 3x^2,$$

$$f(x + \Delta x) = 3(x + \Delta x)^2.$$

Thus,

$$\Delta f = 3(x + \Delta x)^2 - 3x^2$$

$$= 6x\Delta x + 3(\Delta x)^2.$$

Hence,

$$f'(x) = \lim_{\Delta x \to 0} \frac{6x\Delta x + 3(\Delta x)^2}{\Delta x}$$

$$= \lim_{\Delta x \to 0} (6x + 3\Delta x)$$

$$= 6x.$$

Some basic derivative formulas are listed below. These formulas are derived by applying the definition of the derivative in the same manner as in Example A.25. By this method, it is easy to show that the derivative of a constant is zero and that the derivative of a constant times a function is equal to the constant times the derivative of the function; *i.e.*, $dc/dx = 0$ and $d(cf(x))/dx = c \cdot df(x)/dx$. We shall make frequent use of the power formula, $dx^n/dx = nx^{(n-1)}$. Applying these rules to the problem in Example A.25, we get

$$\frac{d(3x^2)}{dx} = \frac{3d(x^2)}{dx} = 3(2x) = 6x.$$

Similarly,

$$\frac{d(5x^{-3})}{dx} = -15x^{-4} = \frac{-15}{x^4}.$$

In each of the following formulas where $u$ and $v$ occur, they are assumed to be differentiable functions of $x$. (Detailed derivations of these formulas may be found in any introductory calculus book.)

1. $\dfrac{dc}{dx} = 0$  ($c$ is a constant).

2. $\dfrac{dx}{dx} = 1.$

3. $\dfrac{d(cu)}{dx} = c\dfrac{du}{dx}$  ($c$ is a constant).

4. $\dfrac{d(u + v)}{dx} = \dfrac{du}{dx} + \dfrac{dv}{dx}.$

5. The chain rule

$$\frac{dy}{dx} = \frac{dy}{du} \cdot \frac{du}{dx}.$$

6. The power rule

$$\frac{d(u^n)}{dx} = nu^{n-1}\frac{du}{dx}.$$

*Example:*  $\dfrac{d[(1 - x^2)^3]}{dx} = 3(1 - x^2)^2(-2x) = -6x(1 - x^2)^2.$

7. The product rule

$$\frac{d(uv)}{dx} = u\frac{dv}{dx} + v\frac{du}{dx}.$$

*Example:*  $\dfrac{d[x^2(x^3 + 1)]}{dx} = x^2\dfrac{d(x^3 + 1)}{dx} + (x^3 + 1)\dfrac{d(x^2)}{dx}$

$$= x^2(3x^2) + (x^3 + 1)(2x)$$

$$= 5x^4 + 2x.$$

8. The quotient rule

$$\frac{d(u/v)}{dx} = \frac{v\dfrac{du}{dx} - u\dfrac{dv}{dx}}{v^2}.$$

Example:
$$\frac{d\left(\dfrac{\sqrt{2x-1}}{x}\right)}{dx} = \frac{d\left[\dfrac{(2x-1)^{\frac{1}{2}}}{x}\right]}{dx}$$

$$= \frac{\dfrac{x\,d(2x-1)^{\frac{1}{2}}}{dx} - (2x-1)^{\frac{1}{2}}\dfrac{dx}{dx}}{x^2}$$

$$= \frac{x\left(\dfrac{1}{2}\right)(2x-1)^{-\frac{1}{2}}(2) - (2x-1)^{\frac{1}{2}}(1)}{x^2}$$

$$= \frac{(2x-1)^{-\frac{1}{2}}[x-(2x-1)]}{x^2}$$

$$= \frac{1-x}{x^2(2x-1)^{\frac{1}{2}}}.$$

9. The exponential function

$$\frac{d(e^u)}{dx} = e^u \frac{du}{dx}.$$

Example:
$$\frac{d(e^{3x^2})}{dx} = (e^{3x^2})(6x) = 6xe^{3x^2}.$$

10. The natural logarithm function

$$\frac{d(\ln u)}{dx} = \frac{1}{u}\frac{du}{dx}.$$

Example:
$$\frac{d[\ln(1-4x)]}{dx} = \frac{1}{1-4x}(-4) = \frac{-4}{1-4x}.$$

Differentiation formulas are frequently utilized in the differential form, as, for example,

$$d(uv) = u\,dv + v\,du \qquad \text{and} \qquad d(\ln u) = \frac{du}{u}.$$

**A geometric interpretation of the derivative**  Let the curve $C$ in Fig. A.9 be the graph of $y = f(x)$, where $f$ is a continuous function of $x$. Let $P$ be the point $(x_1, y_1)$ and let $Q$ be the point $(x_1 + \Delta x, y_1 + \Delta y)$; ($\Delta x$ is positive in the figure but could just as well be negative, in which case $Q$ would be to the left of $P$). Clearly $\Delta y/\Delta x$ is the slope of the line $PQ$. Let us consider holding $P$ fixed and choosing a new point $Q$ by taking $\Delta x$ smaller. We are interested in the following question: "If we continue to take $\Delta x$ smaller, but not zero, does the line segment $PQ$ tend toward some limiting position?" We can attempt to answer this question by drawing $PQ$ for smaller and smaller values of $\Delta x$. An alternative would be to tie a string to a thumb tack at $P$ and, keeping the string taut, rotate the string, noting that the intersection of the curve with the string at any position is the point $Q$. This demonstration does not prove that $PQ$ approaches a limiting position, but it does suggest the following definition.

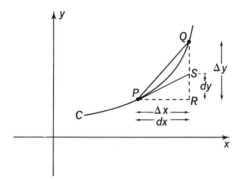

**Figure A.9**

**Definition 6** If the line through the points $P(x_1, y_1)$ and $Q(x_1 + \Delta x, y_1 + \Delta y)$ on the curve $C$ approaches a limiting position as $\Delta x$ approaches zero, then this limiting position defines the *tangent line* to $C$ at the point $P$.

Since $\Delta y/\Delta x$ is the slope of $PQ$, then $\lim_{\Delta x \to 0} \Delta y/\Delta x$ (if it exists) is the slope of the tangent line at $P$. Now this limit is the derivative of $y$ with respect to $x$ at $x = x_1$; therefore, $f'(x) =$ the slope of the tangent to $C$ at $P$. From Example A.25 we see that, if $y = 3x^2$, then $\lim_{\Delta x \to 0} \Delta y/\Delta x = 6x$. Since this limit exists at any $x$, the curve does have a tangent line at every point. For example, at $x = 1/2$, the slope of the tangent line is 3. *Note:* The slope of a curve at a point is defined to be the slope of the tangent line to the curve at that point. Thus the curve $y = 3x^2$ has a slope of 3 at $x = 1/2$.

**Differentials** In applications, the derivative, $dy/dx$, is sometimes treated as the quotient of two quantities $dy$ and $dx$, defined as follows.

**Definition 7** Let $y$ be a function of $x$: $y = f(x)$. The *differential of y,* denoted by $dy$, is defined by the formula

$$dy = f'(x)\Delta x$$

at any point where the derivative exists. The differential, $dx$, of the independent variable $x$ is defined to be the actual increment, *i.e.,* $dx = \Delta x$.

Figure A.9 shows geometrically the distinction between $\Delta y$ and $dy$. The increment in the independent variable $x$ is represented by the line segment $PR$. Since the slope of $PQ$ is $\Delta y/\Delta x$ or $QR/PR$, $\Delta y = QR$. The slope of the tangent line, $PS$, is the derivative $f'(x)$. But this slope is clearly $RS/PR$ or $RS/\Delta x$, so that $RS = f'(x)\Delta x = dy$. Accordingly, the difference $\Delta y - dy = SQ$. This difference depends on the shape of the curve (*i.e.*, on the relationship $y = f(x)$), on the choice of the point $P$, and on $\Delta x$. If $f$ is a differentiable function, then as $\Delta x$ approaches zero, this difference also approaches zero. In applications we sometimes write $dy \approx \Delta y$, if $\Delta x$ is small. These notations will not be confusing if the relationship between $\Delta y$ and $dy$ is understood.

*Example A.26*  Let $x$ be the length of a side of a square, and let $y$ denote the area of the square:

$$y = x^2.$$

If the side is increased from 10 ft to 10.1 ft, how much is the area changed?

*Solution:*
Since $y = x^2$, then at any $x$,

$$y + \Delta y = (x + \Delta x)^2 = x^2 + 2x\,\Delta x + (\Delta x)^2,$$

and

$$\Delta y = 2x\,\Delta x + (\Delta x)^2.$$

For $x = 10$ and $\Delta x = .1$, $\Delta y = 2 + .01 = 2.01$. Furthermore, $\dfrac{dy}{dx} = 2x$, so that $dy = 2x\,dx$, and for $x = 10$, $dx = .1$, we get $dy = 2$.

As the sides of the square change from 10 ft to 10.1 ft, the actual change in the area is given by $\Delta y = 2.01$ ft$^2$, while the approximate change is given by $dy = 2$ ft$^2$.

**The rate interpretation of the derivative**  In the study of chemistry the most frequently used interpretation of the derivative is perhaps that of a "rate of change." Consider, again, a square with sides $x$ inches in length. The area, $A$ sq in., is the function of $x$ given by $A = x^2$. For a particular $x$ and $\Delta x$, the quantity $\Delta A$ represents the change in the area corresponding to a change $\Delta x$ in the length of the sides and is given by

$$\Delta A = 2x\,\Delta x + (\Delta x)^2.$$

It is important to note that $\Delta A$ depends on both $x$ and $\Delta x$. For instance, if $x$ is changed from 10 in. to 12 in., $A$ increases 44 sq in., whereas, if $x$ is changed from 20 in. to 22 in., $\Delta A$ is 84 sq in. In either case, we may consider the ratio $\Delta A/\Delta x$ to be the average rate of change of the area with respect to the length of the side. In the first case above we saw that the area increases 44/2 or 22 sq in. per inch as $x$ changes from 10 in. to

12 in.; in the second case the average rate of change in $A$, $\Delta A/\Delta x$, is 42 sq in. per inch. To relate the derivative to these remarks, recall that the derivative of $A$ is the limit of the ratio, $\Delta A/\Delta x$, as $\Delta x$ approaches zero. Let us interpret this limit in terms of average rates by examining the average rate for smaller and smaller values of $\Delta x$ in the following table.

| $x$ | $\Delta x$ | $\Delta A$ | $\Delta A/\Delta x$ |
|-----|-----|-----|-----|
| 10 | 2 | 44 | 22 |
| 10 | 1 | 21 | 21 |
| 10 | .1 | 2.01 | 20.1 |
| 10 | .01 | .2001 | 20.01 |
| 10 | .001 | .020001 | 20.001 |

From the table it appears that the average rate of change of $A$ with respect to $x$ approaches 20 sq in. per inch as $\Delta x$ approaches zero. We have already shown that the derivative of $f(x) = x^2$ at any $x$ is $2x$; hence, at $x = 10$, the ratio $\Delta A/\Delta x$ does indeed approach the value 20. It then seems reasonable to define the rate of change of $A$ with respect to $x$ at any $x$, as the derivative $dA/dx$, evaluated at that $x$.

If the independent variable in a functional relationship is the time, $y = f(t)$, then $\Delta y/\Delta t$ is the average rate of change in $y$ per unit of time and we say that the derivative is the "instantaneous rate of change of $y$." For example, suppose we know that the rate of decay of a radioactive substance is proportional to the amount of the substance present at any time, $t$. This law may be expressed as $dQ/dt = -kQ$, where $t$ is the time in seconds, $Q$ is the number of grams of the substance present at time $t$, $k$ is a constant of proportionality and $dQ/dt$ is the instantaneous rate of change of $Q$ (in grams per second) at time $t$. (Since $k$ is usually regarded as a positive physical constant, the minus sign indicates that $Q$ is decreasing.)

**The chain rule**   We shall introduce this topic by referring again to the problem of the change in the area of a square as the lengths of the sides change. However, we now wish to consider that the side $x$ is changing with the time $t$, and we ask for the rate of change of $A$ with respect to $t$. Suppose that $x$ changes by $\Delta x$ inches in $\Delta t$ seconds. Then $\Delta x/\Delta t$ is the average rate of change of $x$ in inches per second during the time $\Delta t$. Also, $\Delta A/\Delta t$ is the average rate of change of the area in sq in. per sec, during the time $\Delta t$. But,

$$\frac{\Delta A}{\Delta t} = \frac{\Delta A}{\Delta x} \frac{\Delta x}{\Delta t}, \quad \Delta x \neq 0.$$

This algebraic identity makes plausible the result

$$\frac{dA}{dt} = \frac{dA}{dx}\frac{dx}{dt},$$

which is proved in many calculus books. The preceding formula, which is called the *chain rule*, holds for more general cases than that of an area. Thus, let $y$ be any differentiable function of $u$, say $y = f(u)$, and let $u$ be a differentiable function of $x$, $u = g(x)$. Then,

$$\frac{dy}{dx} = \frac{dy}{du}\frac{du}{dx}.$$

*Example A.27* Suppose a cube of ice is melting in such a way that the length of the edges decreases uniformly at the rate of 1/2 cm per sec. At what rate is the volume changing at the instant when the edge is 10 in.?

*Solution:* Let $V$ and $e$, respectively, denote the volume (cc) and edge (cm) at any time $t$ (sec). The rate of change of $e$ with respect to $t$ is given: $de/dt = -0.5$ cm/sec. Using the chain rule, we get $dV/dt = (dV/de)(de/dt)$. Since the volume of a cube is given by $V = e^3$, $dV/de = 3e^2$. Thus, at the instant when $e = 10$, $dV/dt = (300 \text{ cm}^2)(-.5 \text{ cm/sec}) = -150 \text{ cc/sec}$.

**Higher derivatives**   When the derivative of $f(x)$ is a differentiable function of $x$, one may consider the derivative of the derivative. The new function is called the *second derivative*. Notations for the second derivative of $y$ with respect to $x$, when $y = f(x)$, include $y''$, $f''(x)$, $d^2y/dx^2$, and $D_x^2 y$. Obviously, this idea can be extended to obtain higher order derivatives.

*Example A.28*   If $y = xe^{2x}$, find $\dfrac{d^2y}{dx^2}$.

*Solution:*
     Using the product rule, we get

$$\frac{dy}{dx} = 2xe^{2x} + e^{2x}.$$

Differentiating again, we obtain

$$\frac{d^2y}{dx^2} = 4xe^{2x} + 2e^{2x} + 2e^{2x} = 4(x+1)e^{2x}.$$

*Example A.29*   Find $f^{(3)}(x)$ if $f(x) = \ln x$.

*Solution:*
     The symbol $f^{(3)}(x)$ is another notation for $f'''(x)$, the third derivative.

Therefore, differentiating successively, we obtain

$$f(x) = \ln x.$$

$$f'(x) = \frac{1}{x}.$$

$$f''(x) = -\frac{1}{x^2}.$$

$$f^{(3)}(x) = \frac{2}{x^3}.$$

**Antiderivatives or indefinite integrals** In applications we frequently need to find a function whose derivative is known. For example, if $f'(x) = 4x^3$ what is $f(x)$? By trial and error we can show that $f(x) = x^4$ is such a function. But is this the only function whose derivative is $4x^3$? How about $(x^4 + 10)$? Or, $(x^4 - 5)$? Clearly, the derivative of $(x^4 + c)$ is $4x^3$, where $c$ is any constant. In this example the function, $x^4 + c$, is called the indefinite integral (or, the antiderivative) of the function, $4x^3$. The term "indefinite" is used because every continuous function has many antiderivatives, any two of which differ only by an additive constant. The terminology "definite integral" is used for an entirely different, but closely related, concept.

**Definition 8** If $dF(x)/dx = f(x)$ then $\int f(x)dx = F(x) + c$, where the notation, $\int f(x)dx$, is read "the indefinite integral of $f(x)$." The constant $c$ is called the *constant of integration*, the symbol, $\int$, is an *integral sign*, $f(x)$ is said to be the *integrand*, and $dx$ indicates that $x$ is the variable of integration.

Restated in differential form the definition becomes: if $dF(x) = f(x)dx$, then $F(x) + c$ is the indefinite integral of $f(x)$. The differential form is often encountered in connection with elementary differential equations. The theorem used in finding a function with a given derivative states that if the differentials of two functions are equal then the difference of the functions is a constant. Thus, if $dg = df$, then $g = f + c$ regardless of the independent variables for $f$ and $g$. For example, suppose that $y\ dy = dx/x$. We know that $y\ dy = d(\frac{1}{2}y^2)$ and $dx/x = d(\ln x)$. Therefore, $d(\frac{1}{2}y^2) = d(\ln x)$, so that $\frac{1}{2}y^2 = \ln x + c$. Note that the last equation is equivalent to the integral form

$$\int y\ dy = \int \frac{1}{x}\ dx + c.$$

In order to find the indefinite integrals, we seek functions whose differentials are $y\ dy$ and $dx/x$.

## Integration formulas

1. $\int c\, du = c \int du$ ($c$ is a constant).

2. $\int (u + v)dx = \int u\, dx + \int v\, dx.$

   *Example:* $\int (x^2 + 4)dx = \int x^2\, dx + \int 4dx = \dfrac{x^3}{3} + 4x + c.$

3. If $n \neq -1 \int u^n\, du = \dfrac{u^{n+1}}{n+1} + c.$

   *Example:* $\int \dfrac{dx}{x^3} = \int x^{-3}\, dx = \dfrac{x^{-2}}{-2} + c = -\dfrac{1}{2x^2} + c.$

4. $\int u^{-1}\, du = \int \dfrac{du}{u} = \ln |u| + c, u \neq 0.$

   *Example:* $\int \dfrac{dp}{p - 4} = \ln |p - 4| + c.$

   *Note:* The symbol $|u|$ is read "The absolute value of $u$." The vertical bars mean that we are to use $u$ if $u \geq 0$ and $-u$ if $u < 0$. If we are concerned with a set of values of $p$ greater than 4, then $\ln |p - 4| = \ln (p - 4)$, but if the values of $p$ are all less than 4, then $\ln |p - 4| = \ln (4 - p)$.

5. $\int e^u\, du = e^u + c.$

   *Example:* $\int e^{3x}\, dx = \dfrac{1}{3} \int e^{3x}(3dx) = \dfrac{1}{3} e^{3x} + c.$

### The Definite Integral

**Definition 9** Let $f$ be a real single-valued function defined over the interval $a \leq x \leq b$. Suppose the interval is subdivided into $N$ subintervals. Let $x_i$ denote any arbitrary value of $x$ in the $i$th subinterval and let $\Delta x_i$ denote the length of the $i$th subinterval. Form the sum,

$$f(x_1)\Delta x_1 + f(x_2)\Delta x_2 + \cdots + f(x_N)\Delta x_N,$$

which is customarily abbreviated as

$$\sum_{i=1}^{N} f(x_i)\Delta x_i.$$

Consider subdividing the interval further in such a manner that: (1) $N \to \infty$ and (2) the lengths of all the subintervals approach zero. If the sum approaches the same limiting value for every mode of subdivision for which requirements (1) and (2) obtain and for arbitrary choice of the points $x_i$ within their respective subintervals, then this limit is called the *definite integral of $f(x)$* over the interval $a \le x \le b$, and is denoted by the symbol $\int_a^b f(x)\, dx$; i.e.,

$$\int_a^b f(x)dx = \lim_{N \to \infty} \sum_{i=1}^{N} f(x_i)\Delta x_i.$$

The numbers $a$ and $b$ in the symbol $\int_a^b$ are called the *lower* and *upper limits* of integration, respectively. It is proved in the more rigorous calculus books that if a function $f$ is continuous over the interval $a \le x \le b$, then $\int_a^b f(x)dx$ exists in the sense of Definition 9.

**Example A.30** Assume that $\int_b^a x\, dx$ exists and evaluate it by the use of Definition 9.

*Solution:* If the integral exists, then the limit of the sum is the same for all allowable modes of subdivision and choice of $x_i$'s. Accordingly, we simplify the problem by dividing the interval $a \le x \le b$ into $N$ equal subintervals and

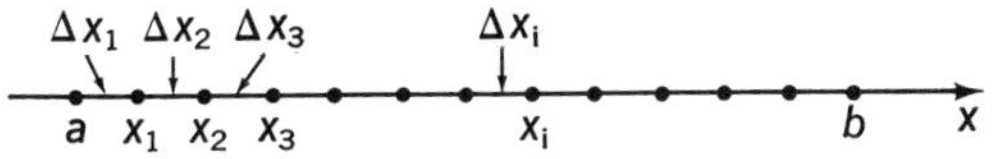

**Figure A.10**

by choosing the $x_i$'s as the right-hand end points of the respective subintervals. From Figure A.10 it appears that

$$x_1 = a + \Delta x, \; x_2 = a + 2\Delta x, \; x_3 = a + 3\Delta x, \; \cdots, \; x_N = a + N\,\Delta x.$$

Furthermore, $f(x) = x$, so that

$$f(x_1) = a + \Delta x, \; f(x_2) = a + 2\Delta x, \; \cdots, \; f(x_N) = a + N\,\Delta x.$$

The sum $\sum_{i=1}^{N} f(x_i)\Delta x_i$ is thus

$$(a + \Delta x)\Delta x + (a + 2\Delta x)\Delta x + (a + 3\Delta x)\Delta x + \cdots + (a + N\,\Delta x)\Delta x$$

$$= \Delta x[Na + (\Delta x + 2\Delta x + 3\Delta x + \cdots + N\,\Delta x)]$$

$$= \Delta x[Na + \Delta x(1 + 2 + 3 + \cdots + N)]$$

$$= \Delta x\left[Na + \Delta x\left(\frac{N(N+1)}{2}\right)\right].$$

Substituting $\dfrac{b-a}{N}$ for $\Delta x$ and taking the limit as $N \to \infty$, we get

$$\lim_{N \to \infty} \sum_{i=1}^{N} f(x_i)\Delta x_i = \lim_{N \to \infty} \left[\frac{Na(b-a)}{N} + \left(\frac{b-a}{N}\right)^2 \frac{N(N+1)}{2}\right]$$

$$= \lim_{N \to \infty} \left[a(b-a) + \frac{(b-a)^2}{2} \frac{N^2 + N}{N^2}\right].$$

Now,

$$\lim_{N \to \infty} \frac{N^2 + N}{N^2} = \lim_{N \to \infty} \left(1 + \frac{1}{N}\right) = 1.$$

Therefore,

$$\lim_{N \to \infty} \sum_{i=1}^{N} f(x_i)\Delta x = a(b - a) + \frac{(b - a)^2}{2}$$

$$= \frac{b^2 - a^2}{2}.$$

The preceding example illustrates that the evaluation of a definite integral by means of the definition is not practical. However, the following basic theorem frequently resolves this difficulty.

**Basic theorem**  If $f(x)$ is continuous and has an antiderivative $F(x)$ over the interval $a \le x \le b$, then

$$\int_a^b f(x)\,dx = F(b) - F(a).$$

*Example A.31*  Evaluate $\int_a^b x\,dx$ using the preceding theorem.

*Solution:*

Since $\int x\,dx = \frac{x^2}{2} + C$, the theorem gives

$$\int_a^b x\,dx = \left(\frac{b^2}{2} + C\right) - \left(\frac{a^2}{2} + C\right)$$

$$= \frac{b^2 - a^2}{2}.$$

*Note:* Compare this result with that of Example A.30.

The preceding basic theorem allows us to evaluate a definite integral by finding an antiderivative $F(x)$ and then taking the difference between $F(x)$ evaluated at the upper and lower limits of integration. We should not, however, lose sight of the definition of the definite integral as being the limit of a sum. In physical or geometric applications, the terms in the sum have a meaning which may be overlooked in evaluating a definite integral by means of the antiderivative.

*Example A.32*  Find the area $A$ bounded by the curve $y = x^{1/2}$ and the lines $x = 1$, $x = 4$, $y = 0$.

*Solution:* The required area is shown in Fig. A.11.

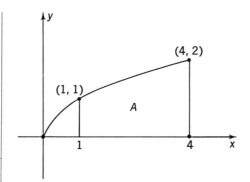

**Figure A.11**

We can approximate the desired area by finding the sum of the areas of the $N$ rectangles in Fig. A.12 or of Fig. A.13. In each case we divide the interval $1 \le x \le 4$ into $N$ subintervals and construct a rectangle on each subinterval. The area of the rectangle on the $i$th subinterval is $f(x_i)\Delta x_i$, where $f(x_i)$ is the height of the curve at $x_i$ and $\Delta x_i$ is the length of the subinterval. In Fig. A.12, the $x_i$'s were chosen as the right ends of the subintervals whereas

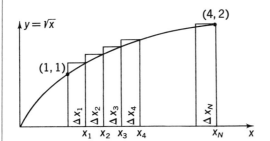

**Figure A.12**

in Fig. A.13 the $x_i$'s are the left ends. For this example, the sum of the areas of the rectangles in one case exceeds the area $A$ and in the other case is less than $A$. (We could consider other cases by taking the $x_i$'s as intermediate points of the subintervals.) It is intuitively clear that all the sums will approach the area $A$ as $N \to \infty$. (From a rigorous mathematical point of view,

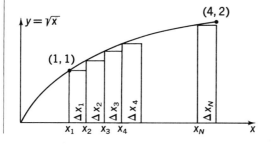

**Figure A.13**

the area must be defined as the limit of the sum.) Since $A = \lim_{N \to \infty} \sum_{i=1}^{N} f(x_i) \Delta x_i$, and $f(x) = \sqrt{x}$ is continuous over the interval, then

$$A = \int_{1}^{4} x^{1/2}\, dx = \left[\frac{2}{3} x^{3/2}\right]_{1}^{4} = \frac{2}{3}(8-1) = \frac{14}{3}.$$

*Note:* The notation $[F(x)]_a^b$ means $F(b) - F(a)$.

**Example A.33** Evaluate $\displaystyle\int_{0}^{1} \frac{1}{x+1}\, dx$.

*Solution:*

Since $\displaystyle\int \frac{1}{x+1}\, dx = \ln(x+1) + C$, we obtain

$$\int_{0}^{1} \frac{1}{x+1}\, dx = [\ln(x+1)]_0^1 = \ln 2 - \ln 1 = \ln 2.$$

The next example shows how the use of an integral between limits can render unnecessary the evaluation of the constant of integration.

**Example A.34** On page 348, the law of decomposition of a radioactive substance was expressed as $dQ/dt = -kQ$, where $Q$ is the amount (in grams) of the undecomposed substance present at any time $t$ (secs). Suppose at $t = 0$, $Q = Q_0$, and at $t = 10$, $Q = .6Q_0$. Find $t$ when $Q = .5Q_0$. (This value of $t$ is the so-called half-life of $Q$.)

*Solution:*

From $dQ/dt = -kQ$ we obtain an equality of two differentials, $dQ/Q = -k\, dt$. Integrating both sides between corresponding limits, we have

$$\int_{Q_0}^{.6Q_0} \frac{dQ}{Q} = \int_{0}^{10} -k\, dt.$$

Thus,

$$\left[\ln Q\right]_{Q_0}^{.6Q_0} = \left[-kt\right]_{0}^{10},$$

and

$$\ln .6Q_0 - \ln Q_0 = -10k + 0,$$

or

$$k = -\frac{1}{10} \ln .6 \approx .05.$$

Similarly,

$$\int_{Q_0}^{.5Q_0} \frac{dQ}{Q} = \int_{0}^{t_2} -k\, dt.$$

Integrating, substituting $k = .05$, and solving for $t_2$, we find a half-life of 13.8 secs. To obtain an expression for $Q$ at any time $t_1$, we write

$$\int_{Q_0}^{Q_1} \frac{dQ}{Q} = \int_0^{t_1} - .05 dt.$$

$$\left[\ln Q\right]_{Q_0}^{Q_1} = -.05\left[t\right]_0^{t_1}.$$

$$\ln Q_1 - \ln Q_0 = -.05 t_1.$$

$$\ln \frac{Q_1}{Q_0} = -.05 t_1.$$

$$\frac{Q_1}{Q_0} = e^{-.05 t_1}.$$

$$Q_1 = Q_0 e^{-.05 t_1}.$$

**Example A.35**  Solve the preceding example without using definite integrals.

*Solution:*

By integrating

$$\frac{dQ}{Q} = -k \, dt,$$

we get

$$\ln Q = -kt + C.$$

At $t = 0$, $Q = Q_o$, so that $C = \ln Q_o$, and we have $\ln Q = -kt + \ln Q_0$ or,

$$\ln Q - \ln Q_0 = -kt.$$

Therefore,

$$Q = Q_0 e^{-kt}.$$

Using the given condition that $Q = .6 Q_0$ at $t = 10$ sec, we find $k \approx .05$. Substituting this value for $k$, we see that $Q = Q_0 e^{-.05 t}$.

The last two examples illustrate that integration between limits and evaluating a constant of integration are ⟨ quivalent. In applications both methods are used.

*Remark*  The letter used for the variable of integration in a definite integral is arbitrary inasmuch as the value of the integral depends on the limits of integration and the functional form of the integrand. To clarify this remark consider $\int_0^2 2t \, dt = 4$. The variable of integration is $t$. However, it is clear that $\int_0^2 2x \, dx = 4$ and that $\int_0^2 2u \, du = 4$. The letter used for the variable of integration in a definite integral is sometimes called a "dummy." In order to avoid confusion it is best to change the variable of integration to some other symbol when the limits on the definite integral involve the letter that designates the variable of integration.

Thus, $\int_0^t f(t)\,dt$ should be written as $\int_0^t f(u)\,du$ or $\int_0^t f(x)\,dx$. In Example A.34, we could avoid using the subscript on $t_1$ by changing $\int_0^{t_1} - k\,dt$ to $\int_0^t - k\,dz$.

**Remark**   In applications, integrals of the form $\int_a^b y\,dx$ are frequently encountered. It should be emphasized that such an integral cannot be evaluated by antidifferentiation unless a relationship between $y$ and $x$ is known. For example, in thermodynamic studies of gases $\int_{V_1}^{V_2} P\,dV$ occurs, where $P$ and $V$ represent pressure and volume, respectively. If $P$ depends on $V$ then the evaluation of this integral requires a knowledge of the functional relationship. For example, for an ideal gas at constant temperature, $P = \dfrac{k}{V}$, where $k$ is constant; thus,

$$\int_{V_1}^{V_2} P\,dV = \int_{V_1}^{V_2} \frac{k}{V}\,dV = k\,\ln\frac{V_2}{V_1}.$$

(The integration could be done using $P$ as the variable of integration since $V = \dfrac{k}{P}$ gives $dV = -\dfrac{k}{P^2}\,dP$. The integral then becomes

$$\int_{V_1}^{V_2} P\,dV = \int_{P_1}^{P_2} P\left(-\frac{k}{P^2}\right)\,dP = \int_{P_1}^{P_2} -\frac{k\,dP}{P},$$

where $P_1$ and $P_2$ correspond to $V_1$ and $V_2$, respectively.   It is not difficult to show that a result equivalent to the preceding one is obtained.)

### Functions of More Than One Variable

In applications, we find that a given physical quantity usually depends on more than one other quantity. For example, the volume of a gas depends on the temperature as well as the pressure. This leads to the need for mathematical techniques that treat a variable that is regarded as a function of two or more other variables. The concepts of function, limit, continuity, and derivatives are analogous extensions of these notions in the one variable case.

We say that a variable $z$ (the *dependent* variable) is a single-valued function of the variables $x$ and $y$ (the *independent* variables) if to each pair $(x, y)$ in some set of points (the *domain* of definition) of the $x$-$y$ plane there corresponds a unique value of $z$. We denote this relationship by $z = f(x, y)$. As before, $f(x_1, y_1)$ means the value of the function at $x = x_1$, $y = y_1$. It is implied that $x$ may be assigned values independent of the values assigned to $y$ and vice versa.

**Example A.36** If $f(x, y) = 4 - x^2 - y^2$, find (a) $f(0, 1)$, (b) $f(a, b)$, (c) $f(a+1, b)$.

*Solution:*

(a) $f(0, 1) = 4 - 0 - 1 = 3$.

(b) $f(a, b) = 4 - a^2 - b^2$.

(c) $f(a + 1, b) = 4 - (a + 1)^2 - b^2$.

**Definition 10** The partial derivative of $f(x, y)$ with respect to $x$ is

$$\frac{\partial f}{\partial x} = \lim_{\Delta x \to 0} \frac{f(x + \Delta x, y) - f(x, y)}{\Delta x}$$

if this limit exists. Other common notations for this limit are $f_x$, $(\partial f/\partial x)_y$, and $f_1$, where the subscript 1 denotes the 1st independent variable. Similarly,

$$\frac{\partial f}{\partial y} = \lim_{\Delta y \to 0} \frac{f(x, y + \Delta y) - f(x, y)}{\Delta y}.$$

We see from the definition that the partial derivative of a function of two or more variables with respect to any one of the variables is obtained by differentiation with respect to this variable with all other variables held constant. Thus, to find $\partial f/\partial y$, we treat $x$ as a constant and differentiate with respect to $y$.

**Example A.37** If $f(x, y) = x^2 + 3xy$ find $f_x$ and $f_y$.

*Solution:*

$$f_x = 2x + 3y; \; f_y = 3x.$$

**Example A.38** If $g(p, v, t) = p^2 + v^2 - t^3 pv$, find $\dfrac{\partial g}{\partial v}$.

*Solution:*

$$\frac{\partial g}{\partial v} = 2v - t^3 p.$$

Partial derivatives may be given the same physical interpretation as the ordinary derivative except that all but the one independent variable are held fixed, so that the change in the dependent variable is attributed to a change in only one of the independent variables. Suppose, for example, the temperature $T$ at a point in a rectangular plate varies with the time, $t$, as well as with the location of the point; *i.e.*, $T = f(x, y, t)$. Then $\partial T/\partial t$ is the rate of change of $T$ with respect to the time (degrees/sec) with the point $(x, y)$ fixed. But $\partial T/\partial x$ is the rate of change of temperature per unit distance $x$ along the $x$-axis (degrees/cm) with $t$ and $y$ fixed.

For a function of two variables a simple geometric interpretation is possible. Let $z = f(x, y)$, where $x$, $y$, and $z$ are rectangular coordinates in ordinary three-space, represent a surface. For a point $(x_1, y_1)$ in the x-y plane $f(x_1, y_1)$ is the height of the surface at that point (Fig. A.14). The plane $y = y_1$ intersects the surface in a curve $C$, and $\partial x/\partial x$ evaluated at $(x_1, y_1)$ is the slope of the line tangent to this curve at the point $(x_1, y_1, z_1)$. Likewise, $\partial z/\partial y$ may be interpreted as the slope of the tangent to the curve of intersection of the surface with the plane $x = x_1$.

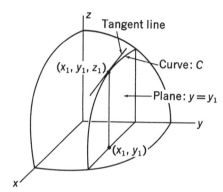

**Figure A.14**

It frequently happens that $z = f(x, y)$ and that both $x$ and $y$ are functions of some other variable $u$. Then if $u$ is specified, $z$ is determined since $x$ and $y$ are determined for any allowable value of $u$. Thus $z$ is indirectly a function of $u$, and one could validly ask for the rate of change of $z$ with respect to $u$. This question leads to a generalization of the chain rule discussed on page 485.

**Total derivative: Definition 11**  If $f(x, y)$ is a real single-valued function of $x$ and $y$ having continuous partial derivatives and if $x$ and $y$ are differentiable functions of $u$, then

$$\frac{df}{du} = \frac{\partial f}{\partial x}\frac{dx}{dy} + \frac{\partial f}{\partial y}\frac{dy}{du}.$$

**Differential form**  A differential form, analogous to the differential of a function of one variable, occurs frequently in physical applications. We shall define it for the three-variable case but generalization to any number of variables should be obvious.

**Definition 12**  If $f(P, V, N)$ is a function of $P$, $V$, and $N$, then the differential of $f$ is

$$df = \frac{\partial f}{\partial P}dP + \frac{\partial f}{\partial V}dV + \frac{\partial f}{\partial N}dN$$

whether or not $P$, $V$, and $N$ are independent. The differential of an independent variable is defined to be its increment.

Returning to the two-variable case, we define $\Delta f$ as the change in $f(x, y)$ as $x$ changes from $x_1$ to $(x_1 + \Delta x)$ and $y$ changes from $y_1$ to $(y_1 + \Delta y)$; i.e., $\Delta f = f(x_1 + \Delta x, y_1 + \Delta y) - f(x_1, y_1)$. For "small" $\Delta x$ and $\Delta y$, $\Delta f$ is approximately equal to $df$ evaluated at $(x_1, y_1)$. As we noted in the one-variable case, the closeness of the approximation depends on the size of the increments in the independent variables.

**Example A.39** If the radius, $r$, of a right circular cylinder is increasing 1/4 in./sec and the height, $h$, is decreasing 1/2 in./sec, at what rate is the volume, $V$, changing at the instant when the radius is 6 in. and the height is 20 in.?

*Solution:*

$$V = f(r, h) = \pi r^2 h.$$

$$\frac{dV}{dt} = (2\pi r h)\frac{dr}{dt} + (\pi r^2)\frac{dh}{dt}.$$

Evaluating for $r = 6$, $h = 20$, $\dfrac{dr}{dt} = 0.25$, and $\dfrac{dh}{dt} = -.5$, we get

$$\frac{dV}{dt} = 42\pi \text{ cu in./sec.}$$

The term total derivative is applied to the chain rule above since $f$ is actually a function of only one variable, $u$. In fact, if $x$ and $y$ in the functional relation for $f(x, y)$ are replaced by $x = p(u)$ and $y = q(u)$, so that $f(x, y)$ explicitly becomes some function $F(u)$ then the total derivative at a point is seen to be the same as the ordinary derivative defined on page 479. This idea is illustrated in the next example.

**Example A.40** If $z = f(x, y) = x^2 - 2y$ and $x = 4u$, $y = u^2 + 1$, find $dz/du$ at $u = 2$ by the chain rule and also by first finding $z$ as a function of $u$.

*Solution:*
By the chain rule,

$$\frac{dz}{du} = \frac{\partial z}{\partial x}\frac{dx}{du} + \frac{\partial z}{\partial y}\frac{dy}{du}$$

$$= (2x)(4) + (-2)(2u).$$

At $u = 2$, we find $x = 8$, and $y = 5$, so $dz/du = 56$. By substituting $x = 4u$ and $y = u^2 + 1$, we get

$$z = x^2 - 2y$$

$$= 16u^2 - 2u^2 - 2$$

$$= 14u^2 - 2.$$

Therefore, $dz/du = 28u$, which is equal to 56 at $u = 2$.

**Generalized chain rule**   If $x = p(u, v)$ and $y = q(u, v)$ then $z = f(x, y)$ is indirectly a function of $u$ and $v$. The derivative $dz/du$ is now meaningless since this notation implies that $z$ is a function of only one variable, $u$. However, it is possible to derive an analogy to the chain rule that will yield the partial derivatives of $z$, $\partial z/\partial u$ and $\partial z/\partial v$. This generalized chain rule gives

$$\frac{\partial z}{\partial u} = \frac{\partial z}{\partial x}\frac{\partial x}{\partial u} + \frac{\partial z}{\partial y}\frac{\partial y}{\partial u},$$

and

$$\frac{\partial z}{\partial v} = \frac{\partial z}{\partial x}\frac{\partial x}{\partial v} + \frac{\partial z}{\partial y}\frac{\partial y}{\partial v}.$$

*Example A.41*   Suppose $G(r, s, t) = r^2 \ln (s - t)$, $r = x^2 + y^2$, $s = x^2 - t^2$, and $t = xy$. Find $G_x$ and $G_y$.

*Solution:*

$$G_x = \frac{\partial G}{\partial x} = \frac{\partial G}{\partial r}\frac{\partial r}{\partial x} + \frac{\partial G}{\partial s}\frac{\partial s}{\partial x} + \frac{\partial G}{\partial t}\frac{\partial t}{\partial x}$$

$$= [2r \ln (s - t)](2x) + \frac{r^2}{s - t}(2x) + \frac{-r^2}{s - t}(y).$$

$$G_y = \frac{\partial G}{\partial y} = \frac{\partial G}{\partial r}\frac{\partial r}{\partial y} + \frac{\partial G}{\partial s}\frac{\partial s}{\partial y} + \frac{\partial G}{\partial t}\frac{\partial t}{\partial y}$$

$$= [2r \ln (s - t)](2y) + \frac{r^2}{s - t}(-2y) + \frac{-r^2}{s - t}(x).$$

**Notation**   The notation used in the chain rule above is ambiguous in that the symbol $\partial z/\partial v$ specifies that $z$ is considered a function of $v$ and one or more other variables but does not make it clear what the other variables are. One way to avoid this ambiguity is through the use of functional notation. Thus, we would write

$$\frac{\partial z(u, v)}{\partial v} = \frac{\partial z(x, y)}{\partial x}\frac{\partial x(u, v)}{\partial v} + \frac{\partial z(x, y)}{\partial y}\frac{\partial y(u, v)}{\partial v}.$$

Another notation, particularly common in chemistry and engineering, is

$$\left(\frac{\partial z}{\partial v}\right)_u = \left(\frac{\partial z}{\partial x}\right)_y\left(\frac{\partial x}{\partial v}\right)_u + \left(\frac{\partial z}{\partial y}\right)_x\left(\frac{\partial y}{\partial v}\right)_u.$$

In this notation the subscript on the parenthesis indicates the independent variable that is held fixed. Thus, $(\partial z/\partial v)_u$ means that $z$ is considered a function of $u$ and $v$ and that the differentiation is with respect to $v$ with $u$ held constant; whereas $(\partial z/\partial y)_x$ means that $z$ is considered a function of $x$ and $y$, and $x$ is constant in the differentiation.

***Example A.42*** Let $H = f(u, v, w)$, where $u$, $v$, and $w$ are each functions of $x$, $y$, and $z$. Write the chain rule expression for $(\partial H/\partial y)_{x,z}$, using subscript notation.

*Solution:*

$$\left(\frac{\partial H}{\partial y}\right)_{x,z} = \left(\frac{\partial H}{\partial u}\right)_{v,w}\left(\frac{\partial u}{\partial y}\right)_{x,z} + \left(\frac{\partial H}{\partial v}\right)_{u,w}\left(\frac{\partial v}{\partial y}\right)_{x,z} + \left(\frac{\partial H}{\partial w}\right)_{u,v}\left(\frac{\partial w}{\partial y}\right)_{x,z}.$$

In general the chain rule involves a function of two or more variables each of which is a function of other variables. Relationships among these variables may make it difficult to determine which variables are independent. A general solution to this problem requires implicit function theory from advanced calculus.

Suppose $z = u^2 + v^3$ and $u = 3x + 2v$. Clearly, by substitution, we can write $z$ as a function of $x$ and $v$, so that $\partial z/\partial x$ and $\partial z/\partial v$ could be obtained directly. Thus,

$$z = (3x + 2v)^2 + v^3,$$

and

$$\frac{\partial z}{\partial x} = 2(3x + 2v)(3) = 6u.$$

We could obtain this result by way of the chain rule but must keep in mind that $x$ and $v$ have been chosen as independent variables, $\partial v/\partial x = 0 = \partial x/\partial v$. Thus,

$$\left(\frac{\partial z}{\partial x}\right)_v = \left(\frac{\partial z}{\partial u}\right)_v\left(\frac{\partial u}{\partial x}\right)_v + \left(\frac{\partial z}{\partial v}\right)_u\left(\frac{\partial v}{\partial x}\right)_v$$

$$= (2u)(3) + (3v^2)(0) = 6u.$$

Similarly,

$$\left(\frac{\partial z}{\partial v}\right)_x = \left(\frac{\partial z}{\partial u}\right)_v\left(\frac{\partial u}{\partial v}\right)_x + \left(\frac{\partial z}{\partial v}\right)_u\left(\frac{\partial v}{\partial v}\right)_x$$

In this example we could ask for $(\partial z/\partial u)_x$ which implies that we consider $u$ and $x$ as independent variables rather than $v$ and $x$.

***Example A.43*** Suppose $V = f(P, T)$ and $T = g(P, E)$. What is meant by $(\partial V/\partial P)_T$? $(\partial V/\partial P)_E$?

*Solution:*

If the functions $f$ and $g$ were given explicitly we could obtain $(\partial V/\partial P)_T$ simply by differentiating $f$ with respect to $P$, holding $T$ constant. The $g$ function would not be needed. However, $(\partial V/\partial P)_E$ implies that we regard $V$

as a function of $P$ and $E$. We must either (1) eliminate $T$ by substituting $T = g(P, E)$ into $V = f(P, T)$ obtaining $V$ explicitly in terms of $P$ and $E$, or (2) use the chain rule. In certain theoretical derivations that are encountered in thermodynamics and other applications, formulas for $f$ and $g$ are not given so that recourse to the chain rule is necessary. Thus

$$\left(\frac{\partial V}{\partial P}\right)_E = \left(\frac{\partial V}{\partial P}\right)_T \left(\frac{\partial P}{\partial P}\right)_E + \left(\frac{\partial V}{\partial T}\right)_P \left(\frac{\partial T}{\partial P}\right)_E$$

$$= \left(\frac{\partial V}{\partial P}\right)_T + \left(\frac{\partial V}{\partial T}\right)_P \left(\frac{\partial T}{\partial P}\right)_E$$

since $\partial P / \partial P = 1$.

**Exact differentials**   In the discussion of the indefinite integral for a function of one variable we pointed out that finding $\int f(x)\, dx$ is equivalent to finding a function $F(x)$ such that the differential of $F(x)$ is $f(x)\, dx$; i.e., given $f(x)$ we find $F(x)$ such that $dF(x) = f(x)\, dx$. We also emphasized that if the differentials of two functions are equal then the functions can differ only by a constant; e.g., if $d(\ln x) = dz$, then $z = \ln x + c$.

The analogous situation holds for functions of two or more variables. We defined the differential of $f(x, y)$ as $df = (\partial f/\partial x)\, dx + (\partial f/\partial y)\, dy$. Now in general $\partial f/\partial x$ and $\partial f/\partial y$ will be functions of $x$ and $y$ so that $df$ has the form, $df = M(x, y)\, dx + N(x, y)\, dy$ where $M(x, y) = f_x$ and $N(x, y) = f_y$.

*Example A.44*   If $f(x, y) = x^2 e^{3y}$, express $df$ in the form $M(x, y)\, dx + N(x, y)\, dy$.

*Solution:*

$$df(x, y) = f_x\, dx + f_y\, dy$$

$$= 2xe^{3y}\, dx + 3x^2 e^{3y}\, dy,$$

so

$$M(x, y) = 2xe^{3y} \quad \text{and} \quad N(x, y) = 3x^2 e^{3y}.$$

In the one variable case we considered the question, "Given $f(x)\, dx$, is there a function $g(x)$ such that $f(x)\, dx = dg$?" The analogous problem is, "Given $M(x, y)\, dx + N(x, y)\, dy$, is there a function $f(x, y)$ such that $M(x, y)\, dx + N(x, y)\, dy = df$?"

*Example A.45*   If possible, find a function whose differential is $2xy^3\, dx + 3x^2 y^2\, dy$.

*Solution:*

We must find a function $f(x, y)$, if there is one, such that $\partial f/\partial x = 2xy^3$ and $\partial f/\partial y = 3x^2 y^2$. Integrating $2xy^3$, with $y$ held constant, we obtain $x^2 y^3$ and we can verify that the partial derivatives of this function are the given $M(x, y)$ and $N(x, y)$. Therefore, $f(x, y) = x^2 y^3$ is a function whose differential is the expression given.

**Definition 13** The expression $M(x, y) \, dx + N(x, y) \, dy$ is said to be an *exact differential* if there exists a function $f(x, y)$ such that $df(x, y) = M(x, y) \, dx + N(x, y) \, dy$. From the definition of $df$ this is equivalent to the two conditions: $f_x = M(x, y)$ and $f_y = N(x, y)$.

> **Example A.46** Is $y \, dx + x^3 y \, dy$ an exact differential?
>
> Solution:
>
> We must find a function $f$, if there is one, such that $f_x = y$ and $f_y = x^3 y$. If we integrate $f_x$ holding $y$ constant we obtain $xy$ but if we take $f = xy$ then $f_y = x$ which is not the desired result. Can we conclude that the given expression is not an exact differential? The answer to this question is that just because we do not find the proper function by a particular technique we cannot conclude that one does not exist.

This example illustrates the need for a means of deciding if a differential expression is an exact differential. The following theorem furnishes such a test.

**Theorem 1: The Euler Reciprocity Relationship** A necessary and sufficient condition that $M(x, y) \, dx + N(x, y) \, dy$ be an exact differential is that

$$\frac{\partial M(x, y)}{\partial y} = \frac{\partial N(x, y)}{\partial x}.$$

We note that in Example A.45, $M_y = N_x = 6xy^2$, whereas in Example A.46, $M_y = 1$ and $N_x = 3x^2 y$. We also note that the above theorem enables one to determine whether a differential form is exact or not. In case the form is exact, the theorem does not give a means for finding a function having the given differential. (Proof of the theorem and a technique for finding $f(x, y)$ may be found in advanced calculus texts.) Our primary interest here is in recognizing an exact differential.

> **Example A.47** Is the differential form, $\ln y \, dx + (x/y) \, dy$, an exact differential?
>
> Solution:
>
> $$M(x, y) = \ln y \quad \text{and} \quad N(x, y) = \frac{x}{y} \quad \text{so,} \quad M_y = \frac{1}{y} = N_x.$$
>
> Therefore, by the Euler Reciprocity Relationship, the expression is an exact differential.

> **Example A.48** Is $xe^y \, dx + ye^x \, dy$ an exact differential?
>
> Solution:
>   Since
>
> $$\frac{\partial (xe^y)}{\partial y} \neq \frac{\partial (ye^x)}{\partial x},$$
>
> the given expression is not an exact differential.

**Line integrals** In our discussion of functions of a single variable we saw that the problem of finding a function when its differential is known (indefinite integral) was related to the definite integral by way of the fundamental theorem of integral calculus (page 490). A somewhat analogous situation exists for differential forms of functions of more than one variable. We shall need to generalize the concept of the definite integral.

**Definition 14** Let $f(x, y)$ be a real, single-valued function defined over some region, $R$, of the $x$-$y$ plane. Let $C$ be a curve in $R$, with points $A$ and $B$ on $C$ (Fig. A.15). Subdivide the arc of the curve from $A$ to $B$ into $N$ subarcs. Let $\Delta s_i$ denote the length of the $i$th subarc and let $(x_i, y_i)$ be an arbitrary point on the $i$th subarc. Form the sum

$$\sum_{i=1}^{N} f(x_i, y_i) \Delta s_i.$$

Let the number of subdivisions increase without limit in such a manner that all the lengths, $\Delta s_i$, approach zero. If the sum approaches the same limit for every mode of subdivision and for arbitrary choice of the points $(x_i, y_i)$ in their respective subarcs, then this limit is called a *line integral* of $f(x, y)$ along $C$ from $A$ to $B$. The notations generally used are $\int_A^B f(x, y)\ ds$ or $\int_C f(x, y)\ ds$.

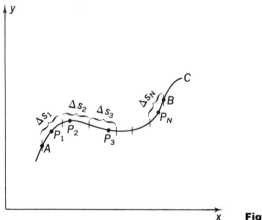

**Figure A.15**

**Other forms of the line integral** In the preceding definition, if $\Delta s_i$ is replaced by $\Delta x_i$ one gets an integral of the form $\int_A^B f(x, y)\ dx$; similarly, one could define $\int_A^B f(x, y)\ dy$. Whichever form is used it is most important to note that the value of the integral generally depends on the curve $C$ as well as on the points $A$ and $B$. A combination of the last two forms occurs most often in applied work, $\int_A^B M(x, y) \, dx + \int_A^B N(x, y) \, dy$ along $C$. Usually, this is written as the single integral $\int_A^B [M(x, y) \, dx + N(x, y) \, dy]$. It is shown in advanced calculus that if the equation of the curve $C$ is given as $y = f(x)$, then we may substitute for $y$ and for $dy$ to obtain the integrand as a function of one variable and thus reduce the line integral to an ordinary integral. A similar technique may be applied if the curve is defined by $x = g(y)$ or in parametric form by $x = f(t)$, $y = g(t)$.

***Example A.49*** Evaluate $\int_{(0,0)}^{(2,4)} (xy \, dx + x^2 \, dy)$ along the curve $C$,
(a)   where $C$ is the parabola $y = x^2$;
(b)   where $C$ is the straight line $y = 2x$.

*Solution:*
(a)   At any point on the curve $y = x^2$, $dy = 2x \, dx$. Hence

$$\int_{(0,0)}^{(2,4)} (xy \, dx + x^2 \, dy) = \int_0^2 (x \cdot x^2 \, dx + x^2 \cdot 2x \, dx)$$

$$= \int_0^2 3x^3 \, dx$$

$$= 12.$$

The line integral along the curve $y = x^2$ between the points $(0, 0)$ and $(2, 4)$ was evaluated as a definite integral between $x = 0$ and $x = 2$. We could just as well have evaluated the line integral as a definite integral between $y = 0$ and $y = 4$. Substituting $x = \sqrt{y}$ and $dx = dy/2\sqrt{y}$, we have

$$\int_{(0,0)}^{(2,4)} (xy \, dx + x^2 \, dy) = \int_0^4 \left( \sqrt{y} \, y \frac{dy}{2\sqrt{y}} + y \, dy \right)$$

$$= \int_0^4 \frac{3}{2} y \, dy$$

$$= 12.$$

(b)   Along the line $y = 2x$, $dy = 2dx$;

$$\int_{(0,0)}^{(2,4)} (xy \, dx + x^2 \, dy) = \int_0^2 (x \cdot 2x \, dx + x^2 \cdot 2dx)$$

$$= \int_0^2 4x^2 \, dx$$

$$= \frac{32}{3}.$$

***Example A.50*** Evaluate $\int (y\ dx + x\ dy)$ between the points $(0,0)$ and $(2,4)$ using the two paths in Example A.49.

*Solution:*

(a) For $y = x^2$, $\int (y\ dx + x\ dy) = \int_0^2 (x^2\ dx + 2x^2\ dx) = 8.$

(b) For $y = 2x$, $\int (y\ dx + x\ dy) = \int_0^2 (2x\ dx + 2x\ dx) = 8.$

These two examples illustrate that the value of a line integral between two points may be different for different curves or may be the same. Note that in Example A.49 the differential form, $xy\ dx + x^2\ dy$, is not exact, since

$$\frac{\partial(xy)}{\partial y} \neq \frac{\partial(x^2)}{\partial x};$$

whereas, in Example A.50 the differential form is exact, since

$$\frac{\partial M(x,\ y)}{\partial y} = \frac{\partial N(x,\ y)}{\partial x} = 1.$$

In fact, the line integral in Example A.50 has the value 8 for any path between the two given points.

In order to state a useful theorem in this connection, we need an auxiliary definition. A region $R$ in the $xy$-plane is said to be *simply connected* if every closed curve in the region can be continuously shrunk to a point without passing out of $R$. For instance, the interior of a circle or a square is simply connected, but the region between two circles, one inside the other, is not simply connected. Generally a region with "holes" in it is not simply connected because a closed path around a hole could not be shrunk to a point without going outside the region.

**Theorem 2**   If in a simply connected region $R$, $M(x,y)$, $N(x,y)$, $M_y(x,y)$, $N_x(x,y)$ are all continuous and $\partial M(x,y)/\partial y = \partial N(x,y)/\partial x$, then $\int (M\ dx + N\ dy)$ is independent of the path of integration in $R$.

If there is a function $f(x, y)$ whose differential is $M\ dx + N\ dy$, then the value of $\int (M\ dx + N\ dy)$ between $(a, b)$ and $(c, d) - f(a, b)$, regardless of the path of integration in $R$, provided that the continuity conditions of Theorem 2 are satisfied. Returning to Example A.50 we can show that $y\ dx + x\ dy$ is the differential of $f(x, y) = xy$ and the continuity conditions are satisfied everywhere. Therefore, $\int (y\ dx + x\ dy)$ between $(0,0)$ and $(2, 4)$ is $f(2, 4) - f(0,0) = 8$ for all paths in the $xy$-plane.

**Integral around a closed path**  In many applications, such as the Carnot Cyle, one is interested in the value of a line integral along a loop or closed curve. Such an integral is usually denoted by an integral sign with a circle $\oint$. It is easy to show that if the integrand is an exact differential, then the integral around a closed path is zero. To see this, consider $\int_A^B (M\,dx + N\,dy)$ along a curve $C_1$ and the $\int_B^A (M\,dx + N\,dy)$ along a different curve $C_2$. The "path" $A$ to $B$ along $C_1$ and $B$ to $A$ along $C_2$ forms a closed loop. But for an exact differential $\int_A^B df = f(B) - f(A)$ and $\int_B^A df = f(A) - f(B)$; so $\oint df = f(B) - f(A) + f(A) - f(B) = 0$. It is assumed here that the loop or closed curve does not pass through or enclose any point where $M$, $N$, $M_y$, $N_x$ are not all continuous. This condition is usually, but not always, satisfied in the application of line integrals.

*Example A.51*  Evaluate the line integral $\oint (y^2\,dx + x^2\,dy)$ along the following triangular path: from $(0, 0)$ to $(1, 1)$ along the line $y = x$, from $(1, 1)$ to $(0, 1)$ along $y = 1$, and back to $(0, 0)$ along the $y$-axis.

*Solution:*

Denote the "legs" of the path by (I), (II), and (III). Then, for (I), $y = x$ and $dy = dx$. Thus,

$$\int_I (y^2\,dx + x^2\,dy) = \int_0^1 2x^2\,dx = \frac{2}{3}.$$

For (II), $y = 1$, $dy = 0$. So,

$$\int_{II} (y^2\,dx + x^2\,dy) = \int_1^0 (dx + 0) = -1.$$

For (III), $x = 0$, $dx = 0$, and

$$\int_{III} (y^2\,dx + x^2\,dy) = \int_1^0 (y^2 \cdot 0 + 0 \cdot dy) = 0.$$

$$\oint (y\,dx + x\,dy) = \int_I + \int_{II} + \int_{III} = -\frac{1}{3}.$$

*Example A.52*  Evaluate $\oint (xy^2\,dx + yx^2\,dy)$ along the closed path in Example A.51.

*Solution:*

Since $\partial(xy^2)/\partial y = 2xy = \partial(yx^2)/\partial x$, the integral is independent of the path and around a closed path is zero. We shall, however, verify this by computing the value along the three legs as in the last example.

$$\int\limits_{\mathrm{I}} (xy^2\, dx + yx^2\, dy) = \int\limits_0^1 (x^3\, dx + x^3\, dx) = \frac{1}{2}.$$

$$\int\limits_{\mathrm{II}} (xy^2\, dx + yx^2\, dy) = \int\limits_1^0 (x\, dx + x^2 \cdot 0) = -\frac{1}{2}.$$

$$\int\limits_{\mathrm{III}} (xy^2\, dx + yx^2\, dy) = \int\limits_1^0 (0 + 0 \cdot dy) = 0.$$

Thus,

$$\oint (xy^2\, dx + yx^2\, dy) = 0.$$

**Higher order partial derivatives**   The partial derivatives of a function $f(x, y)$ are also functions of $x$ and $y$ so that one could consider derivatives of these derivatives. However, in contrast to the one-variable case, a function of two variables has four second partial derivatives. The second partial derivative with respect to $x$ is denoted by $\partial^2 f/\partial x^2$ or by $f_{xx}$; similarly, $\partial^2 f/\partial y^2$ means the second partial derivative with respect to $y$. The second partial derivative first with respect to $x$ and then with respect to $y$ is written $\partial^2 f/\partial y\, \partial x$ or $f_{xy}$; of course, $\partial^2 f/\partial x\, \partial y$ denotes the second partial derivative in the reverse order from that of $\partial^2 f/\partial y\, \partial x$.

**Theorem 3**   If $f_{xy}$ and $f_{yx}$ are continuous then $f_{xy} = f_{yx}$.

In other words, if the mixed higher order partial derivatives are continuous the order of differentiation is immaterial. In most elementary applications the required continuity conditions are met so that the same result is obtained whether one differentiates with respect to $x$ first and then $y$, or $y$ first and then $x$.

*Example A.53*   Verify that $\partial^2 f/\partial x\, \partial y = \partial^2 f/\partial y\, \partial x$ if $f(x, y) = x^2 y^3 + 2x - 3y$.

*Solution:*

$$f_x = \frac{\partial f}{\partial x} = 2xy^3 + 2.$$

$$f_{xy} = \frac{\partial^2 f}{\partial y\, \partial x} = 6xy^2.$$

$$f_y = \frac{\partial f}{\partial y} = 3x^2 y^2 - 3.$$

$$f_{yx} = \frac{\partial^2 f}{\partial x\, \partial y} = 6xy^2 = f_{xy}.$$

## Power Series Expansions

In applications, the power series representation of a function is some-times needed. The two types of series most commonly used in elemen-tary work are Taylor's expansion and Maclaurin's expansion. Taylor's expansion is used to obtain a series which represents the function in the neighborhood of $x = a$; Maclaurin's expansion is the special case of Taylor's expansion where $a = 0$.

The range of values of $x$ for which a power series represents a function is called the region of validity of the series. Applications of series repre-sentation should not be made without investigation of the region of validity. A detailed discussion of infinite series may be found in standard calculus texts.

### Taylor's expansion of a function

$$f(x) = f(a) + f'(a)(x - a) + f''(a)\frac{(x - a)^2}{2!} + f^{(3)}(a)\frac{(x - a)^3}{3!} + \cdots$$
$$+ f^{(k)}(a)\frac{(x - a)^k}{k!} + \cdots$$

### Maclaurin's expansion of $f(x)$

$$f(x) = f(0) + f'(0)x + f''(0)\frac{x^2}{2!} + f^{(3)}(0)\frac{x^3}{3!} + \cdots + f^{(k)}(0)\frac{x^k}{k!} + \cdots$$

*Example A.54*  Find the Maclaurin series for $f(x) = \ln(1 + x)$.

*Solution:*

$$f(x) = \ln(1 + x), \quad f(0) = \ln 1 = 0.$$

$$f'(x) = \frac{1}{1 + x}, \quad f'(0) = 1.$$

$$f''(x) = \frac{-1}{(1 + x)^2}, \quad f''(0) = -1.$$

$$f^{(3)}(x) = \frac{2}{(1 + x)^3}, \quad f'''(0) = 2.$$

$$f^{(4)}(x) = \frac{-(2)(3)}{(1 + x)^4}, \quad f^{(4)}(0) = -(2)(3).$$

$$f^{(5)}(x) = \frac{(2)(3)(4)}{(1 + x)^5}, \quad f^{(5)}(0) = (2)(3)(4).$$

It can be shown that for $k > 1$

$$f^{(k)}(x) = (-1)^{k-1}\frac{(1)(2)(3)(4)\;\cdots\;(k-1)}{(x+1)^k}, f^{(k)}(0) = (-1)^{k-1}(k-1)!$$

Thus Maclaurin's series becomes

$$\ln(1+x) = x - \frac{x^2}{2} + \frac{x^3}{3} - \frac{x^4}{4} + \cdots \qquad \text{for } -1 < x \le 1.$$

## Improper Integrals

Integrals such as $\int_{-\infty}^{\infty} f(x)\,dx$ and $\int_a^{\infty} f(x)\,dx$, are meaningless in terms of our definition of a definite integral. These symbols are called improper integrals and are given meaning by the following definitions:

$$\int_a^{\infty} f(x)\,dx = \lim_{t \to \infty} \int_a^t f(x)\,dx, \qquad \text{if the limit exists.}$$

$$\int_{-\infty}^{\infty} f(x)\,dx = \lim_{\substack{t \to \infty \\ z \to -\infty}} \int_z^t f(x)\,dx, \qquad \text{if the limit exists.}$$

*Example A.55*  Find the "area" bounded by the coordinate axes and the curve $y = 2e^{-x}$.

*Solution:*
     This curve is given in Figure A.8. The "area" required is the limit of the area under the curve between $x = 0$ and $x = t$ as $t \to \infty$. Thus,

$$A = \int_0^{\infty} 2e^{-x}\,dx$$

$$= \lim_{t \to \infty} \int_0^t 2e^{-x}\,dx$$

$$= \lim_{t \to \infty} [-2e^{-x}]_0^t$$

$$= \lim_{t \to \infty} [-2e^{-t} + 2e^0]$$

$$= 2.$$

This result shows that the area from $x = 0$ to $x = t$ approaches 2 as $t$ increases without bound, or in other words, one can go far enough out on the $x$-axis to make the area as close to 2 as one pleases.

Another type of improper integral occurs if the integrand is not continuous at a value of $x$ in the interval of integration. If, for example, $f(x)$ is discontinuous at $x = b$,

$$\int_a^b f(x)\ dx = \lim_{t \to b^-} \int_a^t f(x)\ dx, \qquad \text{if this limit exists.}$$

Consider Example A.21 in which the function is discontinuous at $x = 0$. It seems intuitively clear that the area under the curve exists. We can obtain the area by considering the value of the integral as we approach the point of discontinuity:

$$\int_{-2}^3 f(x)\ dx = \int_{-2}^0 f(x)\ dx + \int_0^3 f(x)\ dx,$$

$$\int_{-2}^0 f(x)\ dx = \lim_{z \to 0^-} \int_{-2}^z f(x)\ dx$$

$$\int_0^3 f(x)\ dx = \lim_{t \to 0^+} \int_t^3 f(x)\ dx$$

In this example, $f(x) = 2$ for $-2 \le x \le 0$; thus

$$\int_{-2}^0 f(x)\ dx = \lim_{z \to 0^-} \int_{-2}^z 2dx = \lim_{z \to 0^-} [2x]_{-2}^z = \lim_{z \to 0^-} [2z + 4] = 4.$$

For $x$

$$\int_0^3 f(x)\ dx = \lim_{t \to 0^+} \int_t^3 x\ dx = \lim_{t \to 0^+} \left[\frac{x^2}{2}\right]_t^3 = \lim_{t \to 0^+} \left[\frac{9}{2} - \frac{t^2}{2}\right] = \frac{9}{2}.$$

Therefore, $A = 4 + \frac{9}{2} = \frac{17}{2}$.

**Example A.56**  Evaluate $\int_0^1 (1/x)\ dx$, if the integral exists.
*Solution:*
 The function is not continuous at $x = 0$, so we consider

$$\int_0^1 \frac{1}{x}\ dx = \lim_{z \to 0^+} \int_z^1 \frac{1}{x}\ dx = \lim_{z \to 0^+} [\ln x]_z^1 = \lim_{z \to 0^+} (-\ln z).$$

We have seen (see Example A.22) that as $z \to 0^+$, $\ln z \to -\infty$. Therefore, the improper integral does not exist, or the value of the integral is not defined.

### L'Hôpital's Rule for Indeterminate Forms

The problem of finding $\lim_{x \to a} f(x)$ is simplified in certain special situations by the use of results obtained by L'Hôpital.

**The indeterminate form 0/0**   If $f(x) = p(x)/q(x)$ where $p$ and $q$ are continuous and have continuous derivatives at $x = a$, and if $p(a) = q(a) = 0$, then

$$\lim_{x \to a} \frac{p(x)}{q(x)} = \lim_{x \to a} \frac{p'(x)}{q'(x)}, \qquad \text{if this limit exists.}$$

**The indeterminate form $\infty/\infty$**   If $\lim_{x \to a} p(x) = \infty$ and $\lim_{x \to a} q(x) = \infty$, and $p$ and $q$ are continuous and have continuous derivatives for all sufficiently large values of $x$, then

$$\lim_{x \to a} \frac{p(x)}{q(x)} = \lim_{x \to a} \frac{p'(x)}{q'(x)}, \qquad \text{if this limit exists.}$$

*Example A.57*   Find $\lim\limits_{x \to 1} \dfrac{x^2 - 2x + 1}{x - 1}$.

*Solution:*
At $x = 1$, $x^2 - 2x + 1 = 0$ and $x - 1 = 0$.

$$\lim_{x \to 1} \frac{x^2 - 2x + 1}{x - 1} = \lim_{x \to 1} \frac{2x - 2}{1} = 0.$$

*Example A.58*   Find $\lim\limits_{x \to \infty} (xe^{-x})$.

*Solution:*

$$\lim_{x \to \infty} xe^{-x} = \lim_{x \to \infty} \frac{x}{e^x} = \lim_{x \to \infty} \frac{1}{e^x} = 0.$$

# APPENDIX B

# SYMBOLS AND NOTATION

## I Non-Greek Characters

| | |
|---|---|
| $a$ | van der Waals constant, activity, molar excess of solute, number of active centers |
| A | absorbance |
| $A$ | area, work content, Helmholtz free energy |
| $b$ | van der Waals constant, Langmuir constant $(\beta/\alpha)$ |
| $c$ | speed of light, number of components |
| $C$ | heat capacity, concentration |
| $d$ | density, thickness of double layer, deuteron |
| $D$ | dielectric constant, diameter |
| $\mathcal{D}$ | diffusion coefficient |
| $e$ | charge on the electron, electron, positron |
| E | electric vector |
| $E$ | energy, enzyme |
| $\mathcal{E}$ | electromotive force or potential |
| $f$ | frictional coefficient, degrees of freedom, fraction |
| $F$ | force |
| $\mathcal{F}$ | Faraday |
| $g$ | acceleration of gravity, gram, gaseous state |
| $G$ | Gibbs free energy, flow gradient |
| $h$ | height, Planck constant, BET constant |
| $H$ | enthalpy, light scattering proportionality constant |
| H | magnetic vector |
| $i$ | intensity of light |
| $I$ | intensity of light, inhibitor, ionic strength |
| $J$ | flux |
| $k$ | Boltzmann constant, Henry's Law constant, Warburg flask constant, velocity constant |

| | |
|---|---|
| k | force constant |
| K | Kelvin, equilibrium constant |
| *l* | length, liquid state |
| *L* | length |
| *m* | mass, molality |
| *M* | molarity |
| M | molecular weight |
| *n* | number (of moles, molecules, atoms, etc.), index of refraction, neutron |
| *N* | normality, number |
| 𝔑 | Avogadro's number |
| *p* | number of phases, pressure, proton |
| *P* | pressure, reaction product |
| *q* | heat |
| *Q* | charge in electrostatic units |
| *r* | radius, distance, rate |
| *R* | gas constant |
| *s* | solid state, sedimentation coefficient or Svedberg coefficient |
| *S* | entropy, solubility, substrate |
| 𝕊 | Svedberg unit |
| *t* | time, temperature Celsius |
| *T* | temperature absolute |
| T | transmittance |
| *u* | velocity of a gas molecule, the electrophoretic mobility |
| U | electrochemical potential |
| *v* | velocity of a solute particle, vapor state, reaction velocity |
| V | volume, maximum velocity |
| *w* | work |
| W | probability, weight |
| *x* | distance, special term in Freundlich isotherm |
| X | mole fraction, electrical field strength |
| Z | charge in valence units |

## II Greek Characters

| | | |
|---|---|---|
| $\alpha$ | (alpha) | solubility coefficient, degree of dissociation, fraction of acid titrated, optical rotation, polarizability, alpha particle |
| $\beta$ | (beta) | beta particle, Langmuir proportionality constant |
| $\gamma$ | (gamma) | $C_P/C_V$, activity coefficient, surface tension, gamma particle |
| $\Delta$ | (delta) | change in a function, stabilization energy |
| $\epsilon$ | (epsilon) | extinction coefficient |
| $\zeta$ | (zeta) | zeta potential, rotary frictional coefficient |

| | | |
|---|---|---|
| $\eta$ | (eta) | viscosity |
| $\theta$ | (theta) | angle |
| $\Theta$ | (theta) | rotational diffusion coefficient |
| $\kappa$ | (kappa) | Debye-Hückel parameter |
| $\lambda$ | (lambda) | wavelength, disintegration constant |
| $\mu$ | (mu) | chemical potential, dipole moment, transition dipole moment |
| $\nu$ | (nu) | frequency |
| $\bar{\nu}$ | (nu-bar) | wavenumber |
| $\pi$ | (pi) | 3.14 . . . |
| $\Pi$ | (pi) | osmotic pressure |
| $\rho$ | (rho) | density of solvent |
| $\sigma$ | (sigma) | density of solute |
| $\Sigma$ | (sigma) | sum |
| $\tau$ | (tau) | turbidity |
| $\phi$ | (phi) | volume fraction, angle of rotation |
| $\chi$ | (chi) | extinction angle |
| $\psi$ | (psi) | the electrical potential |
| $\omega$ | (omega) | angular velocity |

# APPENDIX C

# FUNDAMENTAL CONSTANTS

| Constant | Symbol | Value |
|---|---|---|
| Acceleration of gravity | $g$ | 980.665 cm sec$^{-2}$ |
| Atmosphere, pressure | atm | 1,013,250 dyne cm$^{-2}$ |
| Absolute temperature of the "ice point" of water under air at 1 atm* | $T_{0°C}$ | 273.15°K |
| Thermochemical calorie | cal | 4.184 joules |
| Velocity of light *in vacuo* | $c$ | 2.997925 × 10$^{10}$ cm sec$^{-1}$ |
| Avogadro's number | $\mathfrak{N}$ | 6.02252 × 10$^{23}$ molecules mol$^{-1}$ |
| Faraday constant | $\mathscr{F}$ | 96,487.0 coulombs equiv$^{-1}$ |
| | | 23,060.9 cal volt$^{-1}$ equiv$^{-1}$ |
| Gas constant | $R$ | $\begin{cases} 8.31433 \times 10^7 \text{ ergs deg}^{-1} \text{ mol}^{-1} \\ 1.98717 \text{ cal deg}^{-1} \text{ mol}^{-1} \\ 0.082055 \text{ liter-atm deg}^{-1} \text{ mol}^{-1} \end{cases}$ |
| Planck constant | $h$ | 6.6256 × 10$^{-27}$ erg-sec |
| Charge on the electron | $e = \mathscr{F}/\mathfrak{N}$ | 4.80298 × 10$^{-10}$ cm$^{3/2}$ g$^{1/2}$ sec$^{-1}$ esu |
| Boltzmann constant | $k = R/\mathfrak{N}$ | 1.38054 × 10$^{-16}$ erg deg$^{-1}$ molecule$^{-1}$ |

*The "ice point" is the temperature at which ice and liquid water are in equilibrium.
Source: Rossini, F. D., *Pure and Applied Chemistry*, 9:453, 1964.

# APPENDIX D

# INTERNATIONAL ATOMIC WEIGHTS

Based on C¹² = 12 exactly

| Element | Symbol | Atomic number | Atomic weight | Element | Symbol | Atomic number | Atomic weight |
|---------|--------|---------------|---------------|---------|--------|---------------|---------------|
| Actinium | Ac | 89 | [227]° | Einsteinium | Es | 99 | [254]° |
| Aluminum | Al | 13 | 26.9815 | Erbium | Er | 68 | 167.26 |
| Americium | Am | 95 | [243]° | Europium | Eu | 63 | 151.96 |
| Antimony | Sb | 51 | 121.75 | Fermium | Fm | 100 | [253]° |
| Argon | Ar | 18 | 39.948 | Fluorine | F | 9 | 18.9984 |
| Arsenic | As | 33 | 74.9216 | Francium | Fr | 87 | [223]° |
| Astatine | At | 85 | [210]° | Gadolinium | Gd | 64 | 157.25 |
| Barium | Ba | 56 | 137.34 | Gallium | Ga | 31 | 69.72 |
| Berkelium | Bk | 97 | [247]° | Germanium | Ge | 32 | 72.59 |
| Beryllium | Be | 4 | 9.0122 | Gold | Au | 79 | 196.967 |
| Bismuth | Bi | 83 | 208.980 | Hafnium | Hf | 72 | 178.49 |
| Boron | B | 5 | $10.811^a$ | Helium | He | 2 | 4.0026 |
| Bromine | Br | 35 | $79.909^b$ | Holmium | Ho | 67 | 164.930 |
| Cadmium | Cd | 48 | 112.40 | Hydrogen | H | 1 | $1.00797^a$ |
| Calcium | Ca | 20 | 40.08 | Indium | In | 49 | 114.82 |
| Californium | Cf | 98 | [247]° | Iodine | I | 53 | 126.9044 |
| Carbon | C | 6 | $12.01115^a$ | Iridium | Ir | 77 | 192.2 |
| Cerium | Ce | 58 | 140.12 | Iron | Fe | 26 | $55.847^b$ |
| Cesium | Cs | 55 | 132.905 | Krypton | Kr | 36 | 83.80 |
| Chlorine | Cl | 17 | $35.453^b$ | Lanthanum | La | 57 | 138.91 |
| Chromium | Cr | 24 | $51.996^b$ | Lawrencium | Lw | 103 | [257]° |
| Cobalt | Co | 27 | 58.9332 | Lead | Pb | 82 | 207.19 |
| Copper | Cu | 29 | 63.54 | Lithium | Li | 3 | 6.939 |
| Curium | Cm | 96 | [247]° | Lutetium | Lu | 71 | 174.97 |
| Dysprosium | Dy | 66 | 162.50 | Magnesium | Mg | 12 | 24.312 |

[a] The atomic weight varies because of natural variations in the isotopic composition of the element. The observed ranges are: boron, ±0.003; carbon, ±0.00005; hydrogen, ±0.00001; oxygen, ±0.0001; silicon, ±0.001, sulfur, ±0.003.

[b] The atomic weight is believed to have an experimental uncertainty of the following magnitude: bromine, ±0.002; chlorine, ±0.001; chromium, ±0.001; iron, ±0.003; silver, ±0.003. For other elements the last digit given is believed to be reliable to ±0.5.

° A number in brackets designates the mass number of a selected isotope of the element, usually the one of longest known half-life. Atomic weights: Courtesy the International Union of Pure and Applied Chemistry.

Bracketed numbers: Courtesy the National Bureau of Standards.

| | | | | | | | | |
|---|---|---|---|---|---|---|---|---|
| Manganese | Mn | 25 | 54.9380 | Ruthenium | Ru | 44 | 101.07 |
| Mendelevium | Md | 101 | [256]° | Samarium | Sm | 62 | 150.35 |
| Mercury | Hg | 80 | 200.59 | Scandium | Sc | 21 | 44.956 |
| Molybdenum | Mo | 42 | 95.94 | Selenium | Se | 34 | 78.96 |
| Neodymium | Nd | 60 | 144.24 | Silicon | Si | 14 | 28.086$^a$ |
| Neon | Ne | 10 | 20.183 | Silver | Ag | 47 | 107.870$^b$ |
| Neptunium | Np | 93 | [237]° | Sodium | Na | 11 | 22.9898 |
| Nickel | Ni | 28 | 58.71 | Strontium | Sr | 38 | 87.62 |
| Niobium | Nb | 41 | 92.906 | Sulfur | S | 16 | 32.064$^a$ |
| Nitrogen | N | 7 | 14.0067 | Tantalum | Ta | 73 | 180.948 |
| Nobelium | No | 102 | [254]° | Technetium | Tc | 43 | [97]° |
| Osmium | Os | 76 | 190.2 | Tellurium | Te | 52 | 127.60 |
| Oxygen | O | 8 | 15.9994$^a$ | Terbium | Tb | 65 | 158.924 |
| Palladium | Pd | 46 | 106.4 | Thallium | Tl | 81 | 204.37 |
| Phosphorus | P | 15 | 30.9738 | Thorium | Th | 90 | 232.038 |
| Platinum | Pt | 78 | 195.09 | Thulium | Tm | 69 | 168.934 |
| Plutonium | Pu | 94 | [242]° | Tin | Sn | 50 | 118.69 |
| Polonium | Po | 84 | [210] | Titanium | Ti | 22 | 47.90 |
| Potassium | K | 19 | 39.102 | Tungsten | W | 74 | 183.85 |
| Praseodymium | Pr | 59 | 140.907 | Uranium | U | 92 | 238.03 |
| Promethium | Pm | 61 | [147]° | Vanadium | V | 23 | 50.942 |
| Protoactinium | Pa | 91 | [231]° | Xenon | Xe | 54 | 131.30 |
| Radium | Ra | 88 | [226]° | Ytterbium | Yb | 70 | 173.04 |
| Radon | Rn | 86 | [222]° | Yttrium | Y | 39 | 88.905 |
| Rhenium | Re | 75 | 186.2 | Zinc | Zn | 30 | 65.37 |
| Rhodium | Rh | 45 | 102.905 | Zirconium | Zr | 40 | 91.22 |
| Rubidium | Rb | 37 | 85.47 | | | | |

# INDEX OF NAMES

Generally names of authors whose work is footnoted are not included in this Index. Laws, theories, equations, methods, rules, etc., that are identified with names of persons will be found in the Index of Subjects.

# INDEX OF SUBJECTS

# Four-place Common Logarithms of Numbers

| N | 0 | 1 | 2 | 3 | 4 | 5 | 6 | 7 | 8 | 9 |
|---|---|---|---|---|---|---|---|---|---|---|
| 10 | 0000 | 0043 | 0086 | 0128 | 0170 | 0212 | 0253 | 0294 | 0334 | 0374 |
| 11 | 0414 | 0453 | 0492 | 0531 | 0569 | 0607 | 0645 | 0682 | 0719 | 0755 |
| 12 | 0792 | 0828 | 0864 | 0899 | 0934 | 0969 | 1004 | 1038 | 1072 | 1106 |
| 13 | 1139 | 1173 | 1206 | 1239 | 1271 | 1303 | 1335 | 1367 | 1399 | 1430 |
| 14 | 1461 | 1492 | 1523 | 1553 | 1584 | 1614 | 1644 | 1673 | 1703 | 1732 |
| 15 | 1761 | 1790 | 1818 | 1847 | 1875 | 1903 | 1931 | 1959 | 1987 | 2014 |
| 16 | 2041 | 2068 | 2095 | 2122 | 2148 | 2175 | 2201 | 2227 | 2253 | 2279 |
| 17 | 2304 | 2330 | 2355 | 2380 | 2405 | 2430 | 2455 | 2480 | 2504 | 2529 |
| 18 | 2553 | 2577 | 2601 | 2625 | 2648 | 2672 | 2695 | 2718 | 2742 | 2765 |
| 19 | 2788 | 2810 | 2833 | 2856 | 2878 | 2900 | 2923 | 2945 | 2967 | 2989 |
| 20 | 3010 | 3032 | 3054 | 3075 | 3096 | 3118 | 3139 | 3160 | 3181 | 3201 |
| 21 | 3222 | 3243 | 3263 | 3284 | 3304 | 3324 | 3345 | 3365 | 3385 | 3404 |
| 22 | 3424 | 3444 | 3464 | 3483 | 3502 | 3522 | 3541 | 3560 | 3579 | 3598 |
| 23 | 3617 | 3636 | 3655 | 3674 | 3692 | 3711 | 3729 | 3747 | 3766 | 3784 |
| 24 | 3802 | 3820 | 3838 | 3856 | 3874 | 3892 | 3909 | 3927 | 3945 | 3962 |
| 25 | 3979 | 3997 | 4014 | 4031 | 4048 | 4065 | 4082 | 4099 | 4116 | 4133 |
| 26 | 4150 | 4166 | 4183 | 4200 | 4216 | 4232 | 4249 | 4265 | 4281 | 4298 |
| 27 | 4314 | 4330 | 4346 | 4362 | 4378 | 4393 | 4409 | 4425 | 4440 | 4456 |
| 28 | 4472 | 4487 | 4502 | 4518 | 4533 | 4548 | 4564 | 4579 | 4594 | 4609 |
| 29 | 4624 | 4639 | 4654 | 4669 | 4683 | 4698 | 4713 | 4728 | 4742 | 4757 |
| 30 | 4771 | 4786 | 4800 | 4814 | 4829 | 4843 | 4857 | 4871 | 4886 | 4900 |
| 31 | 4914 | 4928 | 4942 | 4955 | 4969 | 4983 | 4997 | 5011 | 5024 | 5038 |
| 32 | 5051 | 5065 | 5079 | 5092 | 5105 | 5119 | 5132 | 5145 | 5159 | 5172 |
| 33 | 5185 | 5198 | 5211 | 5224 | 5237 | 5250 | 5263 | 5276 | 5289 | 5302 |
| 34 | 5315 | 5328 | 5340 | 5353 | 5366 | 5378 | 5391 | 5403 | 5416 | 5428 |
| 35 | 5441 | 5453 | 5465 | 5478 | 5490 | 5502 | 5514 | 5527 | 5539 | 5551 |
| 36 | 5563 | 5575 | 5587 | 5599 | 5611 | 5623 | 5635 | 5647 | 5658 | 5670 |
| 37 | 5682 | 5694 | 5705 | 5717 | 5729 | 5740 | 5752 | 5763 | 5775 | 5786 |
| 38 | 5798 | 5809 | 5821 | 5832 | 5843 | 5855 | 5866 | 5877 | 5888 | 5899 |
| 39 | 5911 | 5922 | 5933 | 5944 | 5955 | 5966 | 5977 | 5988 | 5999 | 6010 |
| 40 | 6021 | 6031 | 6042 | 6053 | 6064 | 6075 | 6085 | 6096 | 6107 | 6117 |
| 41 | 6128 | 6138 | 6149 | 6160 | 6170 | 6180 | 6191 | 6201 | 6212 | 6222 |
| 42 | 6232 | 6243 | 6253 | 6263 | 6274 | 6284 | 6294 | 6304 | 6314 | 6325 |
| 43 | 6335 | 6345 | 6355 | 6365 | 6375 | 6385 | 6395 | 6405 | 6415 | 6425 |
| 44 | 6435 | 6444 | 6454 | 6464 | 6474 | 6484 | 6493 | 6503 | 6513 | 6522 |
| 45 | 6532 | 6542 | 6551 | 6561 | 6571 | 6580 | 6590 | 6599 | 6609 | 6618 |
| 46 | 6628 | 6637 | 6646 | 6656 | 6665 | 6675 | 6684 | 6693 | 6702 | 6712 |
| 47 | 6721 | 6730 | 6739 | 6749 | 6758 | 6767 | 6776 | 6785 | 6794 | 6803 |
| 48 | 6812 | 6821 | 6830 | 6839 | 6848 | 6857 | 6866 | 6875 | 6884 | 6893 |
| 49 | 6902 | 6911 | 6920 | 6928 | 6937 | 6946 | 6955 | 6964 | 6972 | 6981 |
| 50 | 6990 | 6998 | 7007 | 7016 | 7024 | 7033 | 7042 | 7050 | 7059 | 7067 |
| 51 | 7076 | 7084 | 7093 | 7101 | 7110 | 7118 | 7126 | 7135 | 7143 | 7152 |
| 52 | 7160 | 7168 | 7177 | 7185 | 7193 | 7202 | 7210 | 7218 | 7226 | 7235 |
| 53 | 7243 | 7251 | 7259 | 7267 | 7275 | 7284 | 7292 | 7300 | 7308 | 7316 |
| 54 | 7324 | 7332 | 7340 | 7348 | 7356 | 7364 | 7372 | 7380 | 7388 | 7396 |
| N | 0 | 1 | 2 | 3 | 4 | 5 | 6 | 7 | 8 | 9 |